Undergraduate Texts in Mathematics

Editors

S. Axler
F.W. Gehring
K.A. Ribet

Springer

New York
Berlin
Heidelberg
Barcelona
Hong Kong
London
Milan
Paris
Singapore
Tokyo

Undergraduate Texts in Mathematics

Abbott: Understanding Analysis.

Anglin: Mathematics: A Concise History and Philosophy.
Readings in Mathematics.

Anglin/Lambek: The Heritage of Thales.
Readings in Mathematics.

Apostol: Introduction to Analytic Number Theory. Second edition.

Armstrong: Basic Topology.

Armstrong: Groups and Symmetry.

Axler: Linear Algebra Done Right. Second edition.

Beardon: Limits: A New Approach to Real Analysis.

Bak/Newman: Complex Analysis. Second edition.

Banchoff/Wermer: Linear Algebra Through Geometry. Second edition.

Berberian: A First Course in Real Analysis.

Bix: Conics and Cubics: A Concrete Introduction to Algebraic Curves.

Brémaud: An Introduction to Probabilistic Modeling.

Bressoud: Factorization and Primality Testing.

Bressoud: Second Year Calculus.
Readings in Mathematics.

Brickman: Mathematical Introduction to Linear Programming and Game Theory.

Browder: Mathematical Analysis: An Introduction.

Buchmann: Introduction to Cryptography.

Buskes/van Rooij: Topological Spaces: From Distance to Neighborhood.

Callahan: The Geometry of Spacetime: An Introduction to Special and General Relavitity.

Carter/van Brunt: The Lebesgue–Stieltjes Integral: A Practical Introduction.

Cederberg: A Course in Modern Geometries. Second edition.

Childs: A Concrete Introduction to Higher Algebra. Second edition.

Chung: Elementary Probability Theory with Stochastic Processes. Third edition.

Cox/Little/O'Shea: Ideals, Varieties, and Algorithms. Second edition.

Croom: Basic Concepts of Algebraic Topology.

Curtis: Linear Algebra: An Introductory Approach. Fourth edition.

Devlin: The Joy of Sets: Fundamentals of Contemporary Set Theory. Second edition.

Dixmier: General Topology.

Driver: Why Math?

Ebbinghaus/Flum/Thomas: Mathematical Logic. Second edition.

Edgar: Measure, Topology, and Fractal Geometry.

Elaydi: An Introduction to Difference Equations. Second edition.

Exner: An Accompaniment to Higher Mathematics.

Exner: Inside Calculus.

Fine/Rosenberger: The Fundamental Theory of Algebra.

Fischer: Intermediate Real Analysis.

Flanigan/Kazdan: Calculus Two: Linear and Nonlinear Functions. Second edition.

Fleming: Functions of Several Variables. Second edition.

Foulds: Combinatorial Optimization for Undergraduates.

Foulds: Optimization Techniques: An Introduction.

Franklin: Methods of Mathematical Economics.

Frazier: An Introduction to Wavelets Through Linear Algebra.

Gamelin: Complex Analysis.

Gordon: Discrete Probability.

Hairer/Wanner: Analysis by Its History.
Readings in Mathematics.

Halmos: Finite-Dimensional Vector Spaces. Second edition.

Halmos: Naive Set Theory.

Hämmerlin/Hoffmann: Numerical Mathematics.
Readings in Mathematics.

(continued after index)

Michael W. Frazier

An Introduction to Wavelets Through Linear Algebra

With 46 Illustrations

 Springer

Michael W. Frazier
Michigan State University
Department of Mathematics
East Langsing, MI 48824
USA

Mathematics Subject Classification (2000): 42-01, 46CXX, 65F50

Library of Congress Cataloging-in-Publication Data
Frazier, Michael, 1956-
    An introduction to wavelets through linear algebra / Michael W. Frazier
      p.   cm. – (Undergraduate texts in mathematics)
    Includes bibliographical references and index.
    ISBN 0-387-98639-1 (hardcover)
    1. Wavelets (Mathematics)   2. Algebras, Linear.   I. Title.
    II. Series.
    QA403.3.F73   1999
    515′.2433—dc21   98-43866

Printed on acid-free paper.

Production managed by Jenny Wolkowicki; manufacturing supervised by Joe Quatela.
Typeset by The Bartlett Press, Inc. from the author's TeX files.
Printed and bound by R.R. Donnelly and Sons, Harrisonburg, VA.
Printed in the United States of America.

9 8 7 6 5 4 3 2 (Corrected second printing, 2001)

ISBN 0-387-98639-1   SPIN 10841115

Springer-Verlag   New York Berlin Heidelberg
A member of BertelsmannSpringer Science+Business Media GmbH

Preface

Mathematics majors at Michigan State University take a "Capstone" course near the end of their undergraduate careers. The content of this course varies with each offering. Its purpose is to bring together different topics from the undergraduate curriculum and introduce students to a developing area in mathematics. This text was originally written for a Capstone course.

Basic wavelet theory is a natural topic for such a course. By name, wavelets date back only to the 1980s. On the boundary between mathematics and engineering, wavelet theory shows students that mathematics research is still thriving, with important applications in areas such as image compression and the numerical solution of differential equations. The author believes that the essentials of wavelet theory are sufficiently elementary to be taught successfully to advanced undergraduates.

This text is intended for undergraduates, so only a basic background in linear algebra and analysis is assumed. We do not require familiarity with complex numbers and the roots of unity. These are introduced in the first two sections of chapter 1. In the remainder of chapter 1 we review linear algebra. Students should be familiar with the basic definitions in sections 1.3 and 1.4. From our viewpoint, linear transformations are the primary object of study;

a matrix arises as a realization of a linear transformation. Many students may have been exposed to the material on change of basis in section 1.4, but may benefit from seeing it again. In section 1.5, we ask how to pick a basis to simplify the matrix representation of a given linear transformation as much as possible. We then focus on the simplest case, when the linear transformation is diagonalizable. In section 1.6, we discuss inner products and orthonormal bases. We end with a statement of the spectral theorem for matrices, whose proof is outlined in the exercises. This is beyond the experience of most undergraduates.

Chapter 1 is intended as reference material. Depending on background, many readers and instructors will be able to skip or quickly review much of this material. The treatment in chapter 1 is relatively thorough, however, to make the text as self-contained as possible, provide a logically ordered context for the subject matter, and motivate later developments.

The author believes that students should be introduced to Fourier analysis in the finite dimensional context, where everything can be explained in terms of linear algebra. The key ideas can be exhibited in this setting without the distraction of technicalities relating to convergence. We start by introducing the Discrete Fourier Transform (DFT) in section 2.1. The DFT of a vector consists of its components with respect to a certain orthogonal basis of complex exponentials. The key point, that all translation-invariant linear transformations are diagonalized by this basis, is proved in section 2.2. We turn to computational issues in section 2.3, where we see that the DFT can be computed rapidly via the Fast Fourier Transform (FFT).

It is not so well known that the basics of wavelet theory can also be introduced in the finite dimensional context. This is done in chapter 3. The material here is not entirely standard; it is an adaptation of wavelet theory to the finite dimensional setting. It has the advantage that it requires only linear algebra as background. In section 3.1, we search for orthonormal bases with both space and frequency localization, which can be computed rapidly. We are led to consider the even integer translates of two vectors, the mother and father wavelets in this context. The filter bank arrangement for the computation of wavelets arises naturally here. By iterating this filter bank structure, we arrive in section 3.2 at a multilevel wavelet basis.

Examples and applications are discussed in section 3.3. Daubechies's wavelets are presented in this context, and elementary compression examples are considered. A student familiar with MatLab, Maple, or Mathematica should be able to carry out similar examples if desired.

In section 4.1 we change to the infinite dimensional but discrete setting $\ell^2(\mathbb{Z})$, the square summable sequences on the integers. General properties of complete orthonormal sets in inner product spaces are discussed in section 4.2. This is first point where analysis enters our picture in a serious way. Square integrable functions on the interval $[-\pi, \pi)$ and their Fourier series are developed in section 4.3. Here we have to cheat a little bit: we note that we are using the Lebesgue integral but we don't define it, and we ask students to accept certain of its properties. We arrive again at the key principle that the Fourier system diagonalizes translation-invariant linear operators. The relevant version of the Fourier transform in this setting is the map taking a sequence in $\ell^2(\mathbb{Z})$ to a function in $L^2([-\pi, \pi))$ whose Fourier coefficients make up the original sequence. Its properties are presented in section 4.4. Given this preparation, the construction of first stage wavelets on the integers (section 4.5) and the iteration step yielding a multilevel basis (section 4.6) are carried out in close analogy to the methods in chapter 3. The computation of wavelets in the context of $\ell^2(\mathbb{Z})$ is discussed in section 4.7, which includes the construction of Daubechies's wavelets on $\mathbb{Z}$. The generators u and v of a wavelet system for $\ell^2(\mathbb{Z})$ reappear in chapter 5 as the scaling sequence and its companion.

The usual version of wavelet theory on the real line is presented in chapter 5. The preliminaries regarding square integrable functions and the Fourier transform are discussed in sections 5.1 and 5.2. The facts regarding Fourier inversion in $L^2(\mathbb{R})$ are proved in detail, although many instructors may prefer to assume these results. The Fourier inversion formula is analogous to an orthonormal basis representation, using an integral rather than a sum. Again we see that the Fourier system diagonalizes translation-invariant operators. Mallat's theorem that a multiresolution analysis yields an orthonormal wavelet basis is proved in section 5.3. The aforementioned relation between the scaling sequence and wavelets on $\ell^2(\mathbb{Z})$ allows us to make direct use of the results of chapter 4. The conditions under which wavelets on $\ell^2(\mathbb{Z})$ can be used to generate a multiresolution analysis, and hence wavelets on $\mathbb{R}$, are considered in section 5.4. In section 5.5, we construct Daubechies's wavelets of compact support, and show how the wavelet transform is implemented using filter banks.

We briefly consider the application of these results to numerical differential equations in chapter 6. We begin in section 6.1 with a discussion of the condition number of a matrix. In section 6.2, we present a simple example of the numerical solution of a constant coefficient ordinary differential equation on $[0, 1]$ using finite differences. We see that although the resulting matrix is sparse, which is convenient, it has a condition number that grows quadratically with the size of the matrix. By comparison, in section 6.3, we see that for a wavelet-Galerkin discretization of a uniformly elliptic, possibly variable-coefficient, differential equation, the matrix of the associated linear system can be preconditioned to be sparse and to have bounded condition number. The boundedness of the condition number comes from a norm equivalence property of wavelets that we state without proof. The sparseness of the associated matrix comes from the localization of the wavelet system. A large proportion of the time, the orthogonality of wavelet basis members comes from their supports not overlapping (using wavelets of compact support, say). This is a much more robust property, for example with respect to multiplying by a variable coefficient function, than the delicate cancellation underlying the orthogonality of the Fourier system. Thus, although the wavelet system may not exactly diagonalize any natural operator, it nearly diagonalizes (in the sense of the matrix being sparse) a much larger class of operators than the Fourier basis.

Basic wavelet theory includes aspects of linear algebra, real and complex analysis, numerical analysis, and engineering. In this respect it mimics modern mathematics, which is becoming increasingly interdisciplinary.

This text is relatively elementary at the start, but the level of difficulty increases steadily. It can be used in different ways depending on the preparation level of the class. If a long time is required for chapter 1, then the more difficult proofs in the later chapters may have to be only briefly outlined. For a more advanced group, most or all of chapter 1 could be skipped, which would leave time for a relatively thorough treatment of the remainder. A shorter course for a more sophisticated audience could start in chapter 4 because the main material in chapters 4 and 5 is technically, although not conceptually, independent of the content of chapters 2 and 3. An individual with a solid background in Fourier analysis could learn the basics of wavelet theory from sections 4.5, 4.7, 5.3, 5.4, 5.5, and 6.3 with only occasional references to the remainder of the text.

This volume is intended as an introduction to wavelet theory that is as elementary as possible. It is not designed to be a thorough reference. We refer the reader interested in additional information to the Bibliography at the end of the text.

Michigan State University   M. Frazier

April 1999

Acknowledgments

This text owes a great deal to a number of my colleagues and students. The discrete presentation in Chapters 3 and 4 was developed in joint work (Frazier and Kumar, 1993) with Arun Kumar, in our early attempt to understand wavelets. This was further clarified in consulting work done with Jay Epperson at Daniel H. Wagner Associates in California. Many of the graphs in this text are similar to examples done by Douglas McCulloch during this consulting project. Additional insight was gained in subsequent work with Rodolfo Torres.

My colleagues at Michigan State University provided assistance with this text in various ways. Patti Lamm read a preliminary version in its entirety and made more than a hundred useful suggestions, including some that led to a complete overhaul of section 6.2. She also provided computer assistance with the figures in the Prologue. Sheldon Axler supplied technical assistance and made suggestions that improved the style and presentation throughout the manuscript. T.-Y. Li made a number of helpful suggestions, including providing me with Exercise 1.6.20. Byron Drachman helped with the index.

I have had the opportunity to test preliminary versions of this text in the classroom on several occasions. It was used at Michigan State University in a course for undergraduates in spring

1996 and in a beginning graduate course in summer 1996. The administration of the Mathematics Department, especially Jon Hall, Bill Sledd, and Wei-Eihn Kuan, went out of their way to provide these opportunities. The students in these classes made many suggestions and corrections, which have improved the text. Gihan Mandour, Jian-Yu Lin, Rudolf Blazek, and Richard Andrusiak made large numbers of corrections.

This text was also the basis for three short courses on wavelets. One of these was presented at the University of Puerto Rico at Mayagüez in the spring of 1997. I thank Nayda Santiago for helping arrange the visit, and Shawn Hunt, Domingo Rodríguez, and Ramón Vásquez for inviting me and for their warm hospitality. Another short course was given at the University of Missouri at Columbia in fall 1997. I thank Elias Saab and Nakhlé Asmar for making this possible. The third short course took place at the Instituto de Matemáticas de la UNAM in Cuernavaca, Mexico in summer 1998. I thank Professors Salvador Pérez-Esteva and Carlos Villegas Blas for their efforts in arranging this trip, and for their congeniality throughout. The text in preliminary form has also been used in courses given by Cristina Pereyra at the University of New Mexico and by Suzanne Tourville at Carnegie-Mellon University. Cristina, Suzanne, and their students provided valuable feedback and a number of corrections, as did Kees Onneweer.

My doctoral students Kunchuan Wang and Mike Nixon made many helpful suggestions and found a number of corrections in the manuscript. My other doctoral student, Shangqian Zhang, taught me the mathematics in Section 6.3. I also thank him and his son Simon Zhang for Figure 35.

The fingerprint examples in Figures 1–3 in the Prologue were provided by Chris Brislawn of the Los Alamos National Laboratories. I thank him for permission to reproduce these images. Figures 36e and f were prepared using a program (Summus 4U2C 3.0) provided to me by Björn Jawerth and Summus Technologies, Inc, for which I am grateful. Figures 36b, c, and d were created using the commercially available software WinJPEG v.2.84. The manuscript and some of the figures were prepared using LaTeX. The other figures were done using MatLab. Steve Plemmons, the computer manager in the mathematics department at Michigan State University, aided in many ways, particularly with regard to the images in Figure 36. I thank Ina Lindemann, my editor at Springer-Verlag, for her assistance, encouragement, and especially her patience.

I take this opportunity to thank the mathematicians whose aid was critical in helping me reach the point where it became possible for me to write this text. The patience and encouragement of my thesis advisor John Garnett was essential at the start. My early collaboration with Björn Jawerth played a decisive role in my career. My postdoctoral advisor Guido Weiss encouraged and helped me in many important ways over the years.

This text was revised and corrected during a sabbatical leave provided by Michigan State University. This leave was spent at the University of Missouri at Columbia. I thank the University of Missouri for their hospitality and for providing me with valuable resources and technical assistance.

At a time when academic tenure is under attack, it is worth commenting that this text and many others like it would not have been written without the tenure system.

Contents

Prologue: Compression of the FBI Fingerprint Files

When your local police arrest somebody on a minor charge, they would like to check whether that person has an outstanding warrant, possibly in another state, for a more serious crime. To check, they can send his or her fingerprints to the FBI fingerprint archive in Washington, D.C. Unfortunately, the FBI cannot compare the received fingerprints with their records rapidly enough to make an identification before the suspect must be released. A criminal wanted on a serious charge will most likely have vacated the area by the time the FBI has provided the necessary identification.

Why does it take so long? The FBI fingerprint files are stored on fingerprint cards in filing cabinets in a warehouse that occupies about an acre of floor space. The logistics of the search procedure make it impossible to proceed sufficiently rapidly.

The solution to this seems obvious—the FBI fingerprint data should be computerized and searched electronically. After all, this is the computer age. Why hasn't this been done long ago?

Data representing a fingerprint image can be stored on a computer in such a way that the image can be reconstructed with sufficient accuracy to allow positive identification. To do this, the fingerprint image is scanned and digitized. Each square inch of the fingerprint image is broken into a 500 by 500 grid of small boxes,

FIGURE 1   Original fingerprint image (Courtesy of Chris Brislawn, Los Alamos National Laboratory)

called *pixels*. Each pixel is given a gray-scale value corresponding to its darkness, on a scale from 0 to 255. Because the integers from 0 to 255 can be represented in base 2 using eight places (that is, each integer between 0 and 255 corresponds to an 8-digit sequence of zeros and ones), it takes eight binary data bits to specify the darkness of one pixel. (One digit in base 2 represents a single data bit, which electronically corresponds to the difference between a switch being on or off.) A portion of a fingerprint scanned in this way is exhibited in Figure 1.

Consider the amount of data required for a single fingerprint card. Each rolled fingerprint is about 1.5 inches by 1.6 inches, with $500^2 = 250,000$ pixels per square inch, each requiring eight data bits (one data *byte*). So each fingerprint requires about 600,000 data bytes. A card includes all 10 rolled fingerprints, plus 2 unrolled thumb impressions and 2 impressions of all 5 fingers on a hand. The result is that each card requires about 10 megabytes of data (a megabyte is one million bytes). This is still manageable for modern computers, which frequently have several gigabytes of memory (a gigabyte is a billion, or 10^9, bytes). Electronic transmission of the data on a card is feasible, although slow. So it is possible for the police to send the necessary data electronically to the FBI while the suspect is still in custody.

However, the FBI has about 200 million fingerprint cards in its archive. (Many are for deceased individuals, and there are some duplications—apparently the FBI is not good at throwing things away.) Hence digitizing the entire archive would require roughly 2 $\times$ 10^{15} data bytes, or about 2,000 terabytes (a terabyte is 10^{12} bytes) of memory. This represents more data than current computers can store. Even if we restrict to cards corresponding to current criminal suspects, we are dealing with about 29 million cards (with some duplications due to aliases), or roughly 2 $\times$ 10^{14} data bytes. Thus it would require about 60,000 3-gigabyte hard drives to store. This is too much, even for the FBI. Even if this large of a data base could be stored, it could not be rapidly searched. Yet it is not astronomically too large. If the amount of data could be cut by a factor of about 20, it could be stored on roughly 3,000 3-gigabyte hard drives. This is still a lot, but not an unimaginable amount for a government agency. Thus what is needed is a method to compress the data, that is, to represent the information using less data while retaining enough accuracy to allow positive identification.

Data compression is a major field in signal analysis, with a long history. The current industry standard for image compression was written by the Joint Photographic Experts Group, known as JPEG. Many, perhaps most, of the image files that are downloaded on the Internet are compressed with this standard, which is why they end in the suffix "jpg." The FBI solicitated proposals for compressing their fingerprint files a few years ago. Different groups proposing different methods responded to the FBI solicitation. The contract was awarded to a group at the Los Alamos National Laboratory, headed by Jonathan Bradley and Christopher Brislawn; the project leader was Tom Hopper from the FBI. They proposed compression using the recently developed theory of wavelets. An account of this project can be found in Brislawn (1995).

To see the reason the wavelet proposal was accepted instead of proposals based on JPEG, consider the images in Figures 2 and 3. Both contain compressions by a factor of about 13 of the fingerprint image in Figure 1. Figure 2 shows the compression using JPEG, and Figure 3 exhibits the wavelet compression. One feature of JPEG is that it first divides a large image into smaller boxes, and then compresses in these smaller boxes independently. This provides some advantages due to local homogeneities in the image, but the disadvantage is that the subimages may not align well at the edges of the smaller boxes. This causes the regular pattern of horizontal and vertical lines seen in Figure 2. These are called *block artifacts*, or *block lines* for short. These are not just a visual annoyance, they also are an impediment to machine recognition of fingerprints. Wavelet compression methods do not require dividing the image into smaller blocks because the desired localization properties are naturally built into the wavelet system. Hence the wavelet compression in Figure 3 does not show block lines. This is one of the main reasons that the FBI fingerprint compression contract was awarded to the wavelet group. We introduce both Fourier compression and wavelet compression in section 3.3 of this text.

The examples of fingerprint file compression in Figures 2 and 3 show that mathematics that has been developed recently (within the last 10 or 12 years) has important practical applications.

FIGURE 2 JPEG compression (Courtesy of Chris Brislawn, Los Alamos National Laboratory)

FIGURE 3 Wavelet compression (Courtesy of Chris Brislawn, Los Alamos National Laboratory)

1 CHAPTER

Background: Complex Numbers and Linear Algebra

1.1 Real Numbers and Complex Numbers

We start by setting some notation. The natural numbers $\{1, 2, 3, 4, \ldots\}$ will be denoted by $\mathbb{N}$, and the integers $\{\ldots, -3, -2, -1, 0, 1, 2, 3, \ldots\}$ by $\mathbb{Z}$. Complex numbers will be introduced later. We assume familiarity with the real numbers $\mathbb{R}$ and their properties, which we briefly summarize here. The basic algebraic properties of $\mathbb{R}$ follow from the fact that $\mathbb{R}$ is a field.

Definition 1.1 *A field $(\mathbb{F}, +, \cdot)$ is a set $\mathbb{F}$ with operations $+$ (called* addition) *and $\cdot$ (called* multiplication) *satisfying the following properties:*

A1. *(Closure for addition) For all $x, y \in \mathbb{F}, x + y$ is defined and is an element of $\mathbb{F}$.*

A2. *(Commutativity for addition) $x + y = y + x$, for all $x, y \in \mathbb{F}$.*

A3. *(Associativity for addition) $x + (y + z) = (x + y) + z$, for all $x, y, z \in \mathbb{F}$.*

A4. *(Existence of additive identity) There exists an element in $\mathbb{F}$, denoted 0, such that $x + 0 = x$ for all $x \in \mathbb{F}$.*

7

A5. *(Existence of additive inverse) For each $x \in \mathbb{F}$, there exists an element in $\mathbb{F}$, denoted $-x$, such that $x + (-x) = 0$.*

M1. *(Closure for multiplication) For all $x, y \in \mathbb{F}$, $x \cdot y$ is defined and is an element of $\mathbb{F}$.*

M2. *(Commutativity for multiplication) $x \cdot y = y \cdot x$, for all $x, y \in \mathbb{F}$.*

M3. *(Associativity for multiplication) $x \cdot (y \cdot z) = (x \cdot y) \cdot z$, for all $x, y, z \in \mathbb{F}$.*

M4. *(Existence of multiplicative identity) There exists an element in $\mathbb{F}$, denoted 1, such that $1 \neq 0$ and $x \cdot 1 = x$ for all $x \in \mathbb{F}$.*

M5. *(Existence of multiplicative inverse) For each $x \in \mathbb{F}$ such that $x \neq 0$, there exists an element in $\mathbb{F}$, denoted x^{-1} (or $1/x$), such that $x \cdot (x^{-1}) = 1$.*

D. *(Distributive law) $x \cdot (y + z) = (x \cdot y) + (x \cdot z)$, for all $x, y, z \in \mathbb{F}$.*

We emphasize that in principle the operations $+$ and $\cdot$ in Definition 1.1 could be any operations satisfying the required properties. However, in our main examples $\mathbb{R}$ and $\mathbb{C}$, these are the usual addition and multiplication. In particular, with the usual meanings of $+$ and $\cdot$, $(\mathbb{R}, +, \cdot)$ forms a field. We usually omit $\cdot$ and write xy in place of $x \cdot y$. All of the usual basic algebraic properties (such as $-(-x) = x$) of $\mathbb{R}$ follow from the field properties. This is shown in most introductory analysis texts. We assume all these familiar properties in this text.

An *ordered field* is a field $\mathbb{F}$ with a relation $<$ satisfying properties O1–O4. The first two properties state that $\mathbb{F}$ is an ordered set.

O1. *(Comparison principle) If $x, y \in \mathbb{F}$, then one and only one of the following holds:*

$$x < y, \quad y < x, \quad y = x.$$

O2. *(Transitivity) If $x, y, z \in \mathbb{F}$, with $x < y$ and $y < z$, then $x < z$.*

The remaining two properties state that the operations $+$ and $\cdot$ defined on $\mathbb{F}$ are consistent with the ordering $<$:

O3. *(Consistency of $+$ with $<$) If $x, y, z \in \mathbb{F}$ and $y < z$, then $x + y < x + z$.*

O4. *(Consistency of $\cdot$ with $<$) If $x, y \in \mathbb{F}$, with $0 < x$ and $0 < y$, then $0 < xy$.*

We assume the fact that $\mathbb{R}$ with the usual relation $<$ forms an ordered field. All of the standard order properties of $\mathbb{R}$ (such as, if $0 < x$ then $-x < 0$) follow from O1–O4. We assume such basic facts as needed. We use the standard notations $x > y$, with the same meaning as $y < x$, and $x \leq y$ (equivalently $y \geq x$), meaning that either $x < y$ or $x = y$.

For $x \in \mathbb{R}$, we denote the absolute value, or magnitude, of x by $|x|$, where $|x| = x$ if $x \geq 0$ and $|x| = -x$ if $x < 0$. Then $|x| \geq 0$ for all $x \in \mathbb{R}$, and $|x| = 0$ if and only if $x = 0$.

Lemma 1.2   (Triangle inequality in $\mathbb{R}$) *If $x, y \in \mathbb{R}$, then*

$$|x + y| \leq |x| + |y|.$$

Proof
Exercise 1.1.1   ∎

We interpret $|x - y|$ as the distance between the points x and y in $\mathbb{R}$. (See Exercise 1.1.2). This leads to the notion of the convergence of a sequence.

Definition 1.3   *Let $M \in \mathbb{Z}$ and $x \in \mathbb{R}$. A sequence $\{x_n\}_{n=M}^{\infty}$ of real numbers* converges to x *if, for all $\epsilon > 0$, there exists $N \in \mathbb{N}$ such that $|x_n - x| < \epsilon$ for all $n > N$. A sequence $\{x_n\}_{n=M}^{\infty}$* converges *if it converges to some $x \in \mathbb{R}$.*

Definition 1.4   *A sequence $\{x_n\}_{n=M}^{\infty}$ of real numbers is a* Cauchy sequence *if, for all $\epsilon > 0$, there exists $N \in \mathbb{N}$ such that $|x_n - x_m| < \epsilon$ for all $n, m > N$.*

The rational numbers $\mathbb{Q}$ form an ordered field. The property that distinguishes $\mathbb{R}$ is its completeness. In many texts, this is formulated as the least upper bound property, namely that every nonempty set of real numbers that is bounded above has a least upper bound. The least upper bound property implies the following result, which we assume.

Theorem 1.5   (Cauchy criterion for convergence of a sequence) *Every Cauchy sequence of real numbers converges.*

The Cauchy criterion allows us to prove that a sequence converges without knowing the value of the limit. This is especially useful when we consider series. The converse of the Cauchy criterion (i.e., that any convergent sequence is Cauchy) is true also (Exercise 1.1.3).

We have seen that $\mathbb{R}$ (with the usual addition and multiplication) forms a complete ordered field. This characterizes $\mathbb{R}$: any other complete ordered field is essentially the same as $\mathbb{R}$ except for the choices of names or notation given to the elements and operations (more precisely, any other complete ordered field is "isomorphic" to $\mathbb{R}$). We will not prove this.

Most of the work in this text is done over the complex numbers $\mathbb{C}$. The complex numbers also form a complete field (but not an ordered field; see Exercise 1.1.4). One (somewhat mysterious) way to define $\mathbb{C}$ is to assume the existence of some sort of generalized number (not a real number) i that satisfies $i^2 = -1$. Then $\mathbb{C}$ is defined as the set of all numbers of the form $z = x + iy$ where $x, y \in \mathbb{R}$. We then give $\mathbb{C}$ the usual addition and multiplication operations: for $x_1, x_2, y_1, y_2 \in \mathbb{R}$,

$$(x_1 + iy_1) + (x_2 + iy_2) = (x_1 + x_2) + i(y_1 + y_2) \tag{1.1}$$

and

$$(x_1 + iy_1) \cdot (x_2 + iy_2) = (x_1 x_2 - y_1 y_2) + i(x_1 y_2 + x_2 y_1), \tag{1.2}$$

which is what you get if you formally multiply things out and use the relation $i^2 = -1$. (To be precise we should emphasize that we are defining the operations $+$ and $\cdot$ on $\mathbb{C}$ in the left side of equations (1.1) and (1.2), using the usual $+$, $-$, and $\cdot$ defined on $\mathbb{R}$ on the right side.) As before, we usually write zw instead of $z \cdot w$. The only problem is that none of this makes sense if the hypothesized number i does not exist.

The simplest way around this is to let

$$\tilde{\mathbb{C}} = \mathbb{R} \times \mathbb{R} = \{(x, y) : x, y \in \mathbb{R}\}$$

with operations $+$ and $\cdot$ defined by

$$(x_1, y_1) + (x_2, y_2) = (x_1 + x_2, \; y_1 + y_2) \tag{1.3}$$

and

$$(x_1, y_1) \cdot (x_2, y_2) = (x_1 x_2 - y_1 y_2, \; x_1 y_2 + x_2 y_1), \tag{1.4}$$

where the $+$, $-$, and $\cdot$ on the right side of equations (1.3) and (1.4) are the standard operations on $\mathbb{R}$. There is no question that these definitions make sense. Note that equation (1.3) is essentially (1.1) and equation (1.4) is essentially (1.2).

Observe that

$$\mathbb{R} \times \{0\} = \{(x, 0) : x \in \mathbb{R}\}$$

is a copy of $\mathbb{R}$; that is, the map $(x, 0) \rightarrow x$ is a one-to-one correspondence that identifies $\mathbb{R} \times \{0\}$ with $\mathbb{R}$. The equations

$$(x_1, 0) + (x_2, 0) = (x_1 + x_2, 0)$$

and

$$(x_1, 0) \cdot (x_2, 0) = (x_1 x_2 - 0 \cdot 0, \ x_1 \cdot 0 + x_2 \cdot 0) = (x_1 x_2, 0)$$

show that the restrictions of the operations (1.3) and (1.4) to $\mathbb{R} \times \{0\}$ are consistent with the usual operations on $\mathbb{R}$. Thus $\mathbb{R}$ is imbedded in $\tilde{\mathbb{C}}$ in a way that preserves the addition and multiplication operations.

By equation (1.4),

$$(0, 1) \cdot (0, 1) = (0 \cdot 0 - 1 \cdot 1, \ 0 \cdot 1 + 0 \cdot 1) = (-1, 0).$$

Hence the equation $z^2 = (-1, 0)$ has a solution $(0, 1)$ in $\tilde{\mathbb{C}}$ (actually two solutions, the other being $(0, -1)$), even though it has no solution in $\mathbb{R} \times \{0\}$. Since we identify $(-1, 0) \in \tilde{\mathbb{C}}$ with $-1 \in \mathbb{R}$, this says that the equation $z^2 = -1$ has a solution in the larger set $\tilde{\mathbb{C}}$ even though it has none in $\mathbb{R}$. There is nothing inconsistent or even very surprising about this.

Now we *define*

$$i = (0, 1).$$

For $x \in \mathbb{R}$, we write x in place of $(x, 0)$. Note that for $y \in \mathbb{R}$,

$$iy = yi = (y, 0) \cdot (0, 1) = (0, y),$$

by equation (1.4). Thus we obtain

$$(x, y) = (x, 0) + (0, y) = x + iy.$$

In this notation, equations (1.3) and (1.4) give us (1.1) and (1.2), and we are back where we started, but without fear of inconsistency.

It is important to go through this exercise in notation once. However, in practice nobody uses the notation (x, y) for complex numbers, preferring to keep that for the vector space $\mathbb{R}^2$ (see section 1.3). We follow the standard terminology here: we call the set of complex numbers $\mathbb{C}$, forgetting about $\tilde{\mathbb{C}}$ forever, and we denote the elements of $\mathbb{C}$ in the usual way, namely

$$z = x + iy, \quad \text{where } x, y \in \mathbb{R}.$$

We call x the *real part* of z, and y the *imaginary part* (a particularly poor name, undoubtedly coming from a failure on somebody's part to understand the construction we have just considered). We sometimes write

$$\operatorname{Re} z \quad \text{and} \quad \operatorname{Im} z$$

to denote the real and imaginary parts of z, respectively.

We regard points $z = x + iy$ as points in the plane, where one axis (the *real axis*) contains the points $x \in \mathbb{R}$, and the perpendicular axis (the *imaginary axis*) contains the points iy, for $y \in \mathbb{R}$. In this plane (the *complex plane*), the point $x + iy$ occupies the same position that (x, y) holds in $\mathbb{R}^2$.

Definition 1.6 *Let $z = x + iy \in \mathbb{C}$. Define the* complex conjugate $\overline{z}$ *of z by*

$$\overline{z} = x - iy,$$

the modulus squared *of z by*

$$|z|^2 = z\overline{z} = (x + iy)(x - iy) = x^2 + y^2,$$

and the modulus *or* magnitude $|z|$ *of z by*

$$|z| = \sqrt{|z|^2} = \sqrt{x^2 + y^2}.$$

These definitions yield the following properties.

Lemma 1.7 *Suppose $z, w \in \mathbb{C}$. Then*

$$\overline{\overline{z}} = z,$$

$$\operatorname{Re} z = \tfrac{z + \overline{z}}{2}, \quad \operatorname{Im} z = \tfrac{z - \overline{z}}{2i},$$

$$\overline{z + w} = \overline{z} + \overline{w}, \quad \overline{z \cdot w} = \overline{z} \cdot \overline{w},$$

$$|\overline{z}| = |z|, \quad |zw| = |z||w|,$$

$$|\operatorname{Re} z| \le |z|, \quad and \quad |\operatorname{Im} z| \le |z|.$$

Proof
Exercise 1.1.5. ∎

When proving statements about complex numbers, students often first reformulate these statements in terms of real numbers. This temptation should usually be avoided because the proof in complex notation is often simpler.

Lemma 1.8 (Triangle inequality for $\mathbb{C}$) *Suppose $z, w \in \mathbb{C}$. Then*

$$|z + w| \le |z| + |w|.$$

Proof
Exercise 1.1.6. ∎

Similarly to the case for $\mathbb{R}$ above, we think of $|z-w|$ as the distance in the complex plane between the points z and w (see Exercise 1.1.7). Note that if $z_1 = x_1 + iy_1$ and $z_2 = x_2 + iy_2$, then

$$|z_1 - z_2| = |x_1 - x_2 + i(y_1 - y_2)| = \sqrt{(x_1 - x_2)^2 + (y_1 - y_2)^2},$$

which is the same as the usual distance in $\mathbb{R}^2$ between the points (x_1, y_1) and (x_2, y_2).

We can now check that $(\mathbb{C}, +, \cdot)$ is a field (Exercise 1.1.8). The additive identity is $0 = 0 + i0$, and the multiplicative identity is $1 = 1 + i0$. The additive inverse of $z = x + iy$ is $-z = -x - iy$. To find the multiplicative inverse of a nonzero $z = x + iy$, we guess

$$\frac{1}{z} = \frac{1}{z} \cdot \frac{\bar{z}}{\bar{z}} = \frac{\bar{z}}{|z|^2} = \frac{x - iy}{x^2 + y^2} = \frac{x}{x^2 + y^2} + i\frac{-y}{x^2 + y^2}.$$

This does not make sense yet because we have not defined the division of two complex numbers, but we can check that the complex number

$$\frac{x}{x^2 + y^2} + i\frac{-y}{x^2 + y^2}$$

is in fact the multiplicative inverse of $x + iy$ (assuming $x + iy \ne 0$). This determines z^{-1} for nonzero $z \in \mathbb{C}$, and we define

$$\frac{z}{w} = z \cdot w^{-1} \quad \text{for } z, w \in \mathbb{C} \text{ with } w \ne 0.$$

Lemma 1.9   *Suppose $z, w \in \mathbb{C}$ with $w \neq 0$. Then*

$$\overline{\left(\frac{z}{w}\right)} = \frac{\bar{z}}{\bar{w}},$$

and

$$\left|\frac{z}{w}\right| = \frac{|z|}{|w|}.$$

Proof
Exercise 1.1.9.   ∎

The definitions relating to the convergence of a sequence of complex numbers are formally the same as Definitions 1.3 and 1.4.

Definition 1.10   *Let $M \in \mathbb{N}$ and $z \in \mathbb{C}$. A sequence $\{z_n\}_{n=M}^{\infty}$ of complex numbers* converges *to z if, for all $\epsilon > 0$, there exists $N \in \mathbb{N}$ such that $|z_n - z| < \epsilon$ for all $n > N$. We say $\{z_n\}_{n=M}^{\infty}$* converges *if it converges to some $z \in \mathbb{C}$.*

The notations $\lim_{n \to +\infty} z_n = z$ and $z_n \to z$ are used to indicate that $\{z_n\}_{n=M}^{\infty}$ converges to z.

Definition 1.11   *A sequence $\{z_n\}_{n=M}^{\infty}$ of complex numbers is a* Cauchy *sequence if, for all $\epsilon > 0$, there exists $N \in \mathbb{N}$ such that $|z_n - z_m| < \epsilon$ for all $n, m > N$.*

This leads to the Cauchy criterion for the convergence of a sequence of complex numbers.

Theorem 1.12   (Completeness of $\mathbb{C}$) *A sequence of complex numbers converges if and only if it is a Cauchy sequence.*

Proof
Exercise 1.1.10.   ∎

A sequence $\{x_n\}_{n=M}^{+\infty}$ of real numbers can be regarded as a sequence of complex numbers. However, it is easy to see that the sequence converges in the real sense if and only if it converges in the complex sense, with the same limit (compare with Exercise 1.1.10). Hence there is no ambiguity in the definitions, and we write $\lim_{n \to \infty} x_n$ without specifying the field in which convergence takes place.

Exercises

1.1.1. Prove Lemma 1.2.

1.1.2. Let $\mathbb{X}$ be a set. A *metric*, or *distance function*, on $\mathbb{X}$ is a map
$d : \mathbb{X} \times \mathbb{X} \to \{t \in \mathbb{R} : t \geq 0\}$ satisfying the properties:
Me1. *(Symmetry)* $d(x, y) = d(y, x)$ *for all* $x, y \in \mathbb{X}$;
Me2. *(Nondegeneracy)* $d(x, y) = 0$ *if and only if* $x = y$;
Me3. *(Metric triangle inequality)* $d(x, z) \leq d(x, y) + d(y, z)$ *for all* $x, y, z \in \mathbb{X}$.
A *metric space* $(\mathbb{X}, d)$ is a set $\mathbb{X}$ with a metric d.
For $x, y \in \mathbb{R}$, define $d(x, y) = |x - y|$. Prove that d is a metric on $\mathbb{R}$.

1.1.3. Prove that a convergent sequence $\{x_n\}_{n=M}^{\infty}$ of real numbers is a Cauchy sequence.

1.1.4. Let $\mathbb{F}$ with the relation $<$ be an ordered field.
  i. Suppose $x \in \mathbb{F}$ and $x \neq 0$. Prove that $x^2 > 0$.
  ii. Prove that there is no ordering $<$ on the field $\mathbb{C}$ that makes $\mathbb{C}$ an ordered field. Hint: Suppose by contradiction that $<$ is such an ordering. Use part i to obtain $0 < -1$. Argue that this is a contradiction, keeping in mind that $<$ is not necessarily the usual ordering when restricted to $\mathbb{R}$.

1.1.5. Prove Lemma 1.7.

1.1.6. Prove Lemma 1.8. Suggestion: Do not write it out in terms of the real and imaginary parts. Instead, prove that

$$|z + w|^2 = (z + w)(\overline{z + w}) = |z|^2 + 2\operatorname{Re}(\overline{z}w) + |w|^2$$

and use Lemma 1.7.

1.1.7. For $z, w \in \mathbb{C}$, define $d(z, w) = |z - w|$. Prove that $(\mathbb{C}, d)$ is a metric space (see Exercise 1.1.2 for the definition). Draw a picture in the complex plane to show why condition Me3 in Exercise 1.1.2 is called the triangle inequality.

1.1.8. Verify that $\mathbb{C}$ with the operations (1.1) and (1.2) is a field by checking properties A1–A5, M1–M5, and D.

1.1.9. Prove Lemma 1.9.

1.1.10. Let $\{z_n\}_{n=M}^{\infty}$ be a sequence of complex numbers. For each n, let $z_n = x_n + iy_n$, where $x_n, y_n \in \mathbb{R}$.

    i. Prove that $\{z_n\}_{n=M}^{\infty}$ is a Cauchy sequence of complex numbers (Definition 1.11) if and only if $\{x_n\}_{n=M}^{\infty}$ and $\{y_n\}_{n=M}^{\infty}$ are Cauchy sequences of real numbers (Definition 1.4).

    ii. Prove that $\{z_n\}_{n=M}^{\infty}$ converges to some $z = x + iy \in \mathbb{C}$ (Definition 1.10), where $x, y \in \mathbb{R}$, if and only if $\{x_n\}_{n=M}^{\infty}$ converges to x and $\{y_n\}_{n=M}^{\infty}$ converges to y (Definition 1.3).

    iii. Assuming Theorem 1.5, prove Theorem 1.12.

1.2 Complex Series, Euler's Formula, and the Roots of Unity

We begin with series of complex numbers. Particular cases of interest are geometric series and the power series for $\sin z$, $\cos z$, and e^z. Using these we establish Euler's formula $e^{i\theta} = \cos\theta + i\sin\theta$. This will lead to the polar representation of complex numbers and allow us to calculate N^{th} roots of complex numbers, especially the Nth "roots of unity," the roots of the number 1. In chapter 2 we write Fourier expansions of vectors using the complex exponentials introduced here.

    We begin with the definition of convergence of a series of complex numbers, which is formally the same as for a series of real numbers.

Definition 1.13 *A series of complex numbers, or complex series, is an expression of the form*

$$\sum_{n=M}^{\infty} z_n,$$

where each z_n is a complex number and $M \in \mathbb{Z}$. For $k \geq M$, let

$$s_k = \sum_{n=M}^{k} z_n$$

be the kth partial sum *of the series. If the complex sequence* $\{s_k\}_{k=M}^{\infty}$ *converges to some* $s \in \mathbb{C}$ (Definition 1.10), *we say the series* $\sum_{n=M}^{\infty} z_n$ *converges to* s *or* $\sum_{n=M}^{\infty} z_n = s$. *If the sequence* $\{s_k\}_{k=M}^{\infty}$ *does not converge, we say the series* diverges.

This definition together with the Cauchy criterion for convergence of a complex sequence (Theorem 1.12) imply that a series converges if and only if its partial sums form a Cauchy sequence.

Lemma 1.14 (Cauchy criterion for convergence of a series) *A series of complex numbers* $\sum_{n=M}^{\infty} z_n$ *converges if and only if for every* $\epsilon > 0$, *there exists an integer* N *such that* $\left| \sum_{n=k}^{m} z_n \right| < \epsilon$ *for all* $m \geq k > N$.

Proof
Exercise 1.2.1. ∎

Corollary 1.15 (n^{th} term test) *If a complex series* $\sum_{n=M}^{\infty} z_n$ *converges, then* $\lim_{n \to +\infty} z_n = 0$.

Proof
Exercise 1.2.2. ∎

Corollary 1.16 (Comparison test) *Let* $\sum_{n=M}^{\infty} z_n$ *be a complex series and* $\sum_{n=M}^{\infty} a_n$ *a series of nonnegative real numbers. Suppose that there exists an integer* N *such that* $|z_n| \leq a_n$ *for all* $n \geq N$, *and that* $\sum_{n=M}^{\infty} a_n$ *converges. Then* $\sum_{n=M}^{\infty} z_n$ *converges.*

Proof
Exercise 1.2.3. ∎

If the elements of the series are real numbers, they can be regarded as complex and the definitions of convergence for real and complex series are consistent. So from now on we use the term "series" without specifying whether the terms are real or complex.

Definition 1.17 *A series* $\sum_{n=M}^{\infty} z_n$ *converges absolutely if* $\sum_{n=M}^{\infty} |z_n|$ *converges.*

The comparison test shows that an absolutely convergent series is convergent. If a series is convergent but not absolutely convergent, reindexing the terms can yield a series converging to a different value (Exercise 1.2.4). This cannot happen with an absolutely convergent series.

The Cauchy criterion and the comparison test enable us to determine that a series converges without determining its value. It is rare that a series can be exactly evaluated. Geometric series are one of the exceptions.

Definition 1.18 A geometric series *is a series of the form*

$$\sum_{n=0}^{\infty} z^n = 1 + z + z^2 + z^3 + \cdots,$$

for some $z \in \mathbb{C}$.

Note that the partial sum of the geometric series is

$$s_k = 1 + z + z^2 + z^3 + z^4 + \cdots + z^{k-1} + z^k.$$

This is one of the few cases in which the partial sum can be evaluated in closed form. To do this, observe that

$$(1 - z)s_k = 1 + z + z^2 + \cdots + z^k - (z + z^2 + \cdots + z^k + z^{k+1}).$$

All terms on the right cancel out except the first and the last (this is called a *telescoping sum*), so

$$(1 - z)s_k = 1 - z^{k+1}.$$

We can divide by $1 - z$ (as long as it is not 0) to obtain

$$s_k = \sum_{n=0}^{k} z^n = 1 + z + z^2 + \cdots + z^k = \frac{1 - z^{k+1}}{1 - z} \quad \text{for } z \neq 1. \quad (1.5)$$

When $z = 1$, the definition yields $s_k = k + 1$. Using relation (1.5), we obtain the following result.

Theorem 1.19 (Geometric series test) *Let* $z \in \mathbb{C}$. *The geometric series* $\sum_{n=0}^{\infty} z^n$ *converges to* $1/(1 - z)$ *if* $|z| < 1$, *and diverges if* $|z| \geq 1$. *The convergence for* $|z| < 1$ *is absolute.*

Proof
Exercise 1.2.5. ■

We remark that relation (1.5) is a useful formula that we apply for other purposes in chapter 2. We now consider power series.

Definition 1.20 *Fix a point $z_0 \in \mathbb{C}$. A power series about z_0 is a series of the form*

$$\sum_{n=0}^{\infty} a_n(z - z_0)^n,$$

where $a_n \in \mathbb{C}$ for each integer $n \geq 0$.

A power series has a *radius of convergence*, which is determined by the coefficients $\{a_n\}$ by the formula in Exercise 1.2.7. A function f defined on an open set $O \subseteq \mathbb{C}$ (a set having the property that any point in it has a ball of positive radius around it that is contained in the set) is said to be *analytic* if at every point $z \in O$, f is represented by a power series about z with a positive radius of convergence. We barely touch the rich subject of *complex analysis*, the study of analytic functions.

From calculus we recall that

$$\sin x = \sum_{n=0}^{\infty} (-1)^n \frac{x^{2n+1}}{(2n+1)!}, \quad \cos x = \sum_{n=0}^{\infty} (-1)^n \frac{x^{2n}}{(2n)!}, \quad \text{and } e^x = \sum_{n=0}^{\infty} \frac{x^n}{n!},$$
$$(1.6)$$

in the sense that these series converge absolutely to the stated function values at every $x \in \mathbb{R}$. Because these series converge absolutely, the series

$$\sum_{n=0}^{\infty} \frac{r^{2n+1}}{(2n+1)!}, \quad \sum_{n=0}^{\infty} \frac{r^{2n}}{(2n)!}, \quad \text{and } \sum_{n=0}^{\infty} \frac{r^n}{n!}$$

converge for every real number r. By replacing r with $|z|$, we see that the complex series

$$\sum_{n=0}^{\infty} (-1)^n \frac{z^{2n+1}}{(2n+1)!}, \quad \sum_{n=0}^{\infty} (-1)^n \frac{z^{2n}}{(2n)!}, \quad \text{and } \sum_{n=0}^{\infty} \frac{z^n}{n!}$$

converge absolutely for all complex numbers z. This can also be seen by the ratio test (Exercise 1.2.6). In any case, the following definition makes sense.

Definition 1.21 *For all $z \in \mathbb{C}$, let*

$$\sin z = \sum_{n=0}^{\infty} (-1)^n \frac{z^{2n+1}}{(2n+1)!}, \quad \cos z = \sum_{n=0}^{\infty} (-1)^n \frac{z^{2n}}{(2n)!}, \quad \text{and } e^z = \sum_{n=0}^{\infty} \frac{z^n}{n!}.$$
$$(1.7)$$

When z is real, relations (1.6) and (1.7) agree. So Definition 1.21 extends the usual sine, cosine, and exponential functions to all of $\mathbb{C}$. Many of the key properties of the real-valued sine and cosine functions in relation (1.6) continue to hold in the complex case. For example, relation (1.7) implies that

$$\cos(-z) = \cos z \quad \text{and} \quad \sin(-z) = -\sin z, \tag{1.8}$$

for all $z \in \mathbb{C}$. Equation (1.7) leads to Euler's formula, a very elegant and useful identity.

Theorem 1.22 (Euler's formula) *For every $z \in \mathbb{C}$,*

$$e^{iz} = \cos z + i \sin z. \tag{1.9}$$

Proof
We apply relation (1.7) with iz in place of z and collect the even and odd powers of z:

$$e^{iz} = 1 + iz + \frac{(iz)^2}{2} + \frac{(iz)^3}{3!} + \frac{(iz)^4}{4!} + \frac{(iz)^5}{5!} + \frac{(iz)^6}{6!} + \cdots$$
$$= (1 + \frac{i^2 z^2}{2} + \frac{i^4 z^4}{4!} + \frac{i^6 z^6}{6!} + \cdots) + (iz + \frac{i^3 z^3}{3!} + \frac{i^5 z^5}{5!} + \cdots)$$
$$= (1 - \frac{z^2}{2} + \frac{z^4}{4!} - \frac{z^6}{6!} + \cdots) + i(z - \frac{z^3}{3!} + \frac{z^5}{5!} - \cdots)$$
$$= \cos z + i \sin z. \qquad \blacksquare$$

This remarkable formula includes curious facts such as $-e^{i\pi} = 1$.

Applying Euler's formula with $-z$ in place of z and using equation (1.8) gives the alternate formula

$$e^{-iz} = \cos z - i \sin z. \tag{1.10}$$

Adding and subtracting equations (1.9) and (1.10) gives

$$\cos z = \frac{e^{iz} + e^{-iz}}{2} \quad \text{and} \quad \sin z = \frac{e^{iz} - e^{-iz}}{2i}. \tag{1.11}$$

Although these formulae hold for all complex numbers z, our main interest in them is in the case when z is real. For $\theta \in \mathbb{R}$, we have

$$e^{i\theta} = \cos\theta + i \sin\theta, \ e^{-i\theta} = \cos\theta - i \sin\theta, \tag{1.12}$$

$$\cos\theta = \frac{e^{i\theta} + e^{-i\theta}}{2}, \text{ and } \sin\theta = \frac{e^{i\theta} - e^{-i\theta}}{2i}. \qquad (1.13)$$

For $\theta \in \mathbb{R}$, equation (1.12) implies that

$$\overline{e^{i\theta}} = e^{-i\theta} \qquad (1.14)$$

and

$$|e^{i\theta}| = 1. \qquad (1.15)$$

Our main interest in the next result is in the special case where z and w are purely imaginary, which yields

$$e^{i\theta}e^{i\varphi} = e^{i(\theta+\varphi)} \quad \text{for all} \quad \theta, \varphi \in \mathbb{R}. \qquad (1.16)$$

This case can be proved using equation (1.12) and elementary trigonometry (Exercise 1.2.9(i)).

Lemma 1.23 *Suppose $z, w \in \mathbb{C}$. Then*

$$e^{z+w} = e^z e^w.$$

Proof
Exercise 1.2.9(ii). ∎

As an application of equation (1.16), note that for $\theta \subset \mathbb{R}$ and $n \in \mathbb{N}$,

$$(e^{i\theta})^n = e^{in\theta}. \qquad (1.17)$$

This equation can be used to obtain elementary trigonometric identities easily. For example, it is clear that by iterating the addition formulae for sine and cosine, we can write $\sin n\theta$ and $\cos n\theta$ in terms of $\sin\theta$ and $\cos\theta$ (in fact as polynomials in $\sin\theta$ and $\cos\theta$). But equation (1.17) gives a faster way.

Example 1.24
Express $\sin 4\theta$ and $\cos 4\theta$ in terms of $\sin\theta$ and $\cos\theta$.

Solution
First,

$$e^{4i\theta} = \cos 4\theta + i\sin 4\theta.$$

But we also have

$$e^{4i\theta} = (e^{i\theta})^4 = (\cos\theta + i\sin\theta)^4.$$

By expanding this expression, using $i^2 = -1$, we have

$$(\cos\theta + i\sin\theta)^4 = \cos^4\theta + 4i\cos^3\theta\sin\theta - 6\cos^2\theta\sin^2\theta$$
$$- 4i\cos\theta\sin^3\theta + \sin^4\theta.$$

By equating the real and imaginary parts of this last expression with $\cos 4\theta$ and $\sin 4\theta$, we have

$$\cos 4\theta = \cos^4\theta - 6\cos^2\theta\sin^2\theta + \sin^4\theta,$$

and

$$\sin 4\theta = 4\cos^3\theta\sin\theta - 4\cos\theta\sin^3\theta.$$

∎

Further, equation (1.12) leads to the polar representation of complex numbers. Suppose $z = x + iy \in \mathbb{C}$, with $x, y \in \mathbb{R}$ and $z \neq 0$. The point $(x/\sqrt{x^2+y^2}, y/\sqrt{x^2+y^2})$ has distance 1 from the origin in $\mathbb{R}^2$ and so lies on the unit circle. Hence there exists an angle θ (in fact infinitely many of them) such that $\cos\theta = x/\sqrt{x^2+y^2}$ and $\sin\theta = y/\sqrt{x^2+y^2}$. Using equation (1.12),

$$z = |z|\frac{x+iy}{|z|} = |z|\left(\frac{x}{\sqrt{x^2+y^2}} + i\frac{y}{\sqrt{x^2+y^2}}\right)$$
$$= |z|(\cos\theta + i\sin\theta) = |z|e^{i\theta}.$$

By letting $r = |z|$, we have

$$z = re^{i\theta} = r\cos\theta + ir\sin\theta.$$

So $re^{i\theta}$ occupies the same point in $\mathbb{C}$ that the point with polar coordinates (r, θ) occupies in $\mathbb{R}^2$. We call $re^{i\theta}$ the *polar representation* of z. We call θ the *argument* of z, written $\theta = \arg z$. Thus r is the distance in $\mathbb{C}$ from z to the origin, and θ is the angle, in radians, between the positive x-axis and the ray from 0 to z. For $z = 0$, we define the polar representation to be $r = 0$ and we do not define $\arg 0$.

As in the case of polar coordinates in $\mathbb{R}^2$, the polar representation of a nonzero $z \in \mathbb{C}$ is not unique. By equation (1.12) and the fact that the sine and cosine functions have period 2π, we have

$$re^{i\theta} = re^{i(\theta+2k\pi)} \tag{1.18}$$

for any integer k. So $\arg z$ is determined only up to adding $2k\pi$ for some $k \in \mathbb{Z}$. If we select $\arg z$ so that $-\pi < \arg z \le \pi$, we call this the *principal value of the argument*.

The polar representation gives geometric interpretation to the multiplication of complex numbers. Suppose $z_1 = r_1 e^{i\theta_1}$ and $z_2 = r_2 e^{i\theta_2}$. Then

$$z_1 z_2 = r_1 e^{i\theta_1} r_2 e^{i\theta_2} = r_1 r_2 e^{i(\theta_1 + \theta_2)},$$

by equation (1.16). Thus the modulus of the product is the product of the moduli (which we already knew by Lemma 1.7) and the argument of the product is the sum of the arguments. In other words, the effect of multiplying a complex number z by $re^{i\theta}$ is to multiply the distance from z to the origin by r and to rotate z by an angle of θ radians in the counterclockwise direction.

The polar representation makes the computation of positive integer powers of complex numbers easier, since, for $n \in \mathbb{N}$,

$$(re^{i\theta})^n = r^n (e^{i\theta})^n = r^n e^{in\theta},$$

by equation (1.17).

Example 1.25
Find $(1 + i)^{43}$.

Solution
One polar representation of $1 + i$ is $\sqrt{2} e^{i\pi/4}$, since $|1 + i| = \sqrt{2}$ and $(\cos \pi/4, \sin \pi/4) = (1/\sqrt{2}, 1/\sqrt{2})$. So

$$(1 + i)^{43} = (\sqrt{2})^{43} e^{i43\pi/4} = 2^{21} \sqrt{2} e^{i(3\pi/4 + 10\pi)}$$

$$= 2^{21} \sqrt{2} e^{i3\pi/4} = 2^{21} \sqrt{2} \left(\cos \frac{3\pi}{4} + i \sin \frac{3\pi}{4} \right) = -2^{21} + i2^{21}.$$
∎

For comparison, imagine trying to do this problem directly.

The polar representation also allows us to find roots of complex numbers. First consider an example, which we will solve below after some discussion.

Example 1.26
Find all 5th roots of $2 + 2\sqrt{3}i$; that is, find all complex numbers $a + ib$ such that $(a + ib)^5 = 2 + 2\sqrt{3}i$.

If we multiply out the left side of the equation $(a + ib)^5 = 2 + 2\sqrt{3}i$ and set the real and imaginary parts equal, we obtain two 5th degree polynomial equations in two variables a and b to solve simultaneously. Even one 5th degree polynomial equation in one variable cannot be solved by a general algebraic formula (by a deep theorem in Galois theory, a topic in abstract algebra), so another approach must be found.

Consider a nonzero complex number z. We guess its Nth roots (for $N \in \mathbb{N}$) as follows. Let $r = |z|$ and let θ be a value of $\arg z$ such that $0 \le \theta < 2\pi$. We use equation (1.18) to write $z = re^{i\theta}$ in the following N ways:

$$re^{i\theta} = re^{i(\theta+2\pi)} = re^{i(\theta+4\pi)} = \cdots = re^{i(\theta+2(N-1)\pi)}. \tag{1.19}$$

Our guesses for the Nth roots of z are

$$\{r^{1/N}e^{i\theta/N}, r^{1/N}e^{i(\theta+2\pi)/N}, r^{1/N}e^{i(\theta+4\pi)/N}, \ldots, r^{1/N}e^{i(\theta+2(N-1)\pi)/N}\} \tag{1.20}$$

or, more concisely,

$$\{r^{1/N}e^{i(\theta+2k\pi)/N}\}_{k=0}^{N-1}. \tag{1.21}$$

It is easy to check that each of these values is an Nth root of z: by equations (1.17) and (1.18),

$$(r^{1/N}e^{i(\theta+2k\pi)/N})^N = re^{i(\theta+2k\pi)} = re^{i\theta} = z.$$

Note that although the values in equation (1.19) are the same, the arguments in the set (1.20) are distinct and lie between 0 and 2π (including perhaps 0 but not 2π). Hence the N values in the set (1.20) are distinct. (If we extend our list in equation (1.19) by adding the next possibility $re^{i(\theta+2N\pi)}$, this would give one more term in the set (1.20), namely $r^{1/N}e^{i(\theta+2N\pi)/N}$, which would be the same as the first term $r^{1/N}e^{i\theta/N}$, by equation (1.18).) In fact, a general result (see Exercise 1.2.16) shows that the equation $z^N = w$ (or any polynomial equation of degree N) can have at most N distinct solutions z. So the N distinct N^{th} roots of $re^{i\theta}$ in the set (1.20) or (1.21) are a complete list.

One might think that because the N expressions in equation (1.19) are equal, taking their N^{th} roots should imply that the expressions in the set (1.20) are equal, which they are not. The point is that there is no such thing as *the* N^{th} root of a complex number z (unless $z = 0$); there are instead N different complex numbers whose N^{th} power is z.

Solution to Example 1.26
The polar representation of $2 + 2\sqrt{3}i$ is $4e^{i\pi/3}$. Thus we can write $2 + 2\sqrt{3}i$ five ways as

$$4e^{i\pi/3} = 4e^{i7\pi/3} = 4e^{i13\pi/3} = 4e^{i19\pi/3} = 4e^{i25\pi/3}.$$

Hence the 5th roots of $2 + 2\sqrt{3}i$ are:

$$\sqrt[5]{4}e^{i\pi/15}, \sqrt[5]{4}e^{i7\pi/15}, \sqrt[5]{4}e^{i13\pi/15}, \sqrt[5]{4}e^{i19\pi/15}, \text{ and } \sqrt[5]{4}e^{i25\pi/15}.$$

Using equation (1.12), we can write these in the form $a + ib$ (e.g., $\sqrt[5]{4}e^{i\pi/15} = \sqrt[5]{4}\cos(\pi/15) + i\sqrt[5]{4}\sin(\pi/15) \approx 1.2906735 + .2743411i$, determined by using a calculator). ∎

The theory of roots in $\mathbb{C}$ is much simpler than the theory in $\mathbb{R}$. In $\mathbb{R}$, a positive number has 2 real N^{th} roots if N is even and one if N is odd, whereas a negative number has no real roots if N is even and one if N is odd. In $\mathbb{C}$, every nonzero complex number has N different N^{th} roots. This is a typical phenomenon in mathematics: from the right perspective, the situation simplifies.

For our purposes, the most important roots are the Nth *roots of unity*, that is, the Nth roots of the number 1. By the procedure we have just considered, these are

$$\{e^{2\pi i k/N}\}_{k=0}^{N-1} = \{1, e^{2\pi i/N}, e^{4\pi i/N}, \ldots, e^{2(N-1)\pi i/N}\}. \tag{1.22}$$

We conclude this section with a brief discussion of the fundamental theorem of algebra.

Definition 1.27 *A polynomial of degree $N \geq 0$ is a function P of the form*

$$P(z) = a_N z^N + a_{N-1} z^{N-1} + \cdots + a_2 z^2 + a_1 z + a_0,$$

where $a_i \in \mathbb{C}$, for $i = 0, 1, 2, \ldots, N$, and $a_N \neq 0$. We call a_N the leading coefficient *of P. A* root *of P is a complex number a such that $P(a) = 0$.*

We have seen that for a nonzero $w \in \mathbb{C}$, the polynomial $z^N - w$ has a root, in fact N of them. This is a special case of Theorem 1.28. We assume this result because its proof is outside the scope of this introductory chapter (it follows easily from a result in complex analysis called Liouville's theorem, for example).

Theorem 1.28   (Fundamental theorem of algebra) *Let P be a polynomial of degree 1 or more. Then P has a root in $\mathbb{C}$.*

This theorem was proved by Gauss in his doctoral dissertation. It implies that any polynomial factors completely over $\mathbb{C}$.

Corollary 1.29   *Let P be a polynomial of degree $N \geq 1$, with leading coefficient a_N. Then there exist complex numbers $z_1, z_2, \ldots z_N$ (not necessarily distinct), such that*

$$P(z) = a_N(z - z_1)(z - z_2) \cdots (z - z_N).$$

Proof
Exercise 1.2.16. ∎

Exercises

1.2.1. Prove Lemma 1.14.

1.2.2. Prove Corollary 1.15.

1.2.3. Prove Corollary 1.16.

1.2.4. A series of complex numbers is *conditionally convergent* if it is convergent but not absolutely convergent. Let $\sum_{k=1}^{\infty} a_k$ be a conditionally convergent series of real numbers. Let α be an arbitrary real number. Show that there is a reindexing $\{a_{\pi(k)}\}_{k=1}^{\infty}$ of the sequence $\{a_k\}_{k=1}^{\infty}$ (this means that π is a permutation of the index set $\mathbb{N}$, i.e., a 1–1, onto map from $\mathbb{N}$ to $\mathbb{N}$) such that the reindexed series $\sum_{k=1}^{\infty} a_{\pi(k)}$ converges to α. Hint: Since $\sum_{k=1}^{\infty} a_k$ converges, $\lim_{k \to +\infty} a_k = 0$. Since $\sum_{k=1}^{\infty} a_k$ does not converge absolutely, the series of positive terms and the series of negative terms in $\{a_k\}$ must each diverge. List the positive terms in decreasing order (call these $\{b_k\}$) and the negative terms in order of decreasing magnitude (say $\{c_k\}$). Form the rearranged series by taking enough b_ks, in order starting at b_1, until their sum is above α. Then add enough c_ks, starting at c_1 and proceeding in order, until the sum is below α. Then add more b_ks, starting from where you left off before, until the sum is above α. Keep going back and forth, stopping each time as soon as the sum crosses the target value α. The remaining b_ks and c_ks always have an infinite sum, so one can always add enough to cross over the value α. Show that since the sequences $\{b_k\}$ and $\{c_k\}$ converge to zero, the series formed this way converges to α.

1.2.5. Prove Theorem 1.19. Suggestion: for divergence, apply Corollary 1.15.

1.2.6. (Ratio test) Let $\sum_{n=M}^{\infty} z_n$ be a series and suppose that

$$\lim_{n \to \infty} \left| \frac{z_{n+1}}{z_n} \right| = \rho \text{ exists.}$$

i. Prove that if $\rho < 1$, then $\sum_{n=M}^{\infty} z_n$ converges absolutely, whereas if $\rho > 1$, $\sum_{n=M}^{\infty} z_n$ diverges. Hint: Make a comparison with a geometric series and apply Corollary 1.16. Assume the fact that the sequence $\{r^n\}_{n=0}^{\infty}$ diverges if $r > 1$.

ii. Give an example of a convergent series for which $\rho = 1$ and a divergent series for which $\rho = 1$. Thus the ratio test gives no conclusion when $\rho = 1$. You can assume the fact that $\sum 1/n^p$ converges if and only if $p > 1$.

1.2.7. (Radius of convergence) Consider a power series

$$\sum_{n=0}^{\infty} a_n(z - z_0)^n.$$

Let $R = 1/\limsup_{n \to \infty}(|a_n|^{1/n})$, where if the denominator is 0, R is interpreted as ∞, whereas if the denominator is ∞, R is interpreted as 0. Prove that $\sum_{n=0}^{\infty} a_n(z - z_0)^n$ converges absolutely for $|z - z_0| < R$ and diverges for $|z - z_0| > R$. This R is the radius of convergence of the series. The hint is the same as for Exercise 1.2.6 although the solution is different.

1.2.8. In the proof of Theorem 1.22, we implicitly used the fact that if $\sum_{n=0}^{\infty} z_n$ and $\sum_{n=0}^{\infty} w_n$ converge, then $\sum_{n=0}^{\infty}(z_n + w_n)$ converges and

$$\sum_{n=0}^{\infty}(z_n + w_n) = \sum_{n=0}^{\infty} z_n + \sum_{n=0}^{\infty} w_n.$$

Prove this and explain how it was used in the proof of Euler's formula (i.e., what were z_n and w_n?). In addition, we used the fact that if $\sum_{n=0}^{\infty} z_n$ converges and $\alpha \in \mathbb{C}$, then $\sum_{n=0}^{\infty} \alpha z_n$ converges and

$$\sum_{n=0}^{\infty} \alpha z_n = \alpha \sum_{n=0}^{\infty} z_n.$$

Prove this also.

1.2.9. i. Prove equation (1.16). Hint: recall the addition formulae: $\cos(\theta + \varphi) = \cos\theta \cos\varphi - \sin\theta \sin\varphi$ and $\sin(\theta + \varphi) = \sin\theta \cos\varphi + \cos\theta \sin\varphi$.

ii. Prove Lemma 1.23. Hint: $e^{z+w} = \sum_{n=0}^{\infty}(z + w)^n/n!$ by definition. Expand $(z+w)^n$ using the binomial theorem. Then interchange the order of summation.

1.2.10. Suppose $\theta \in \mathbb{R}$. Express $\sin 5\theta$ and $\cos 5\theta$ in terms of $\sin\theta$ and $\cos\theta$.

1.2.11. Write each of the following complex numbers in the form $a + ib$, with $a, b \in \mathbb{R}$, where your answer is stated without using the trigonometric functions:

i. $e^{3i\pi/2}$

ii. $e^{17\pi i}$

iii. $e^{2\pi i/3}$

iv. $3e^{7\pi i/4}$

v. $5e^{-5i\pi/6}$

vi. $2e^{-32\pi i/3}$.

1.2.12. Express the following complex numbers in polar form $re^{i\theta}$, with $r > 0$ and $0 \le \theta < 2\pi$:

i. $4 + 4i$

ii. $2\sqrt{3} - 2i$

iii. $-3 + 3\sqrt{3}i$

iv. $-\sqrt{15} - \sqrt{5}i$.

1.2.13. Find $(\sqrt{3} + i)^{101}$. Write your answer in the form $a + ib$, with $a, b \in \mathbb{R}$, without using the trigonometric functions.

1.2.14. Find all cube roots of $-2 + 2i$. Write your answers in the form $a + ib$ with $a, b \in \mathbb{R}$. Pick one of your answers (your choice) and check it by cubing it directly (i.e., check without using the polar representation). Remark: you can use your calculator to get the answer to high accuracy, but in this case you can get the exact answers using the addition formulae (see Exercise 1.2.9) and writing integer multiples of $1/12$ as sums of integer multiples of $1/3$ and $1/4$, for example, $1/12 = 1/3 - 1/4$.

1.2.15. Using a calculator, find all 4th roots of $3+4i$ to several decimal places accuracy.

1.2.16.   i. (Factor theorem) Let P be a polynomial and let $a \in \mathbb{C}$. We say $z - a$ *divides* P, or $z - a$ is a *factor* of P (written $(z-a)|P$), if there exists a polynomial Q such that $P(z) = (z - a)Q(z)$. Prove that $(z - a)|P$ if and only if $P(a) = 0$; that is, $z - a$ is a factor of P if and only if a is a root of P. Hint: The "only if" direction is immediate. For the "if" direction, the result follows easily from the division algorithm for polynomials, if you are familiar with that. If not, a more elementary proof can be obtained by first showing that $(z-a)|(z^k - a^k)$ for any integer $k \geq 1$. (This can be proved by a slight generalization of the argument leading up to equation (1.5), which is the special case $a = 1$.) Then writing out $P(z) - P(a)$ leads to the general result.

  ii. Prove Corollary 1.29.

  iii. Prove that a polynomial of degree N can have at most N distinct roots.

1.3   Vector Spaces and Bases

We work with the vector space $\mathbb{C}^n$ in chapters 2 and 3. It is similar to the more familiar space $\mathbb{R}^n$ except that its vectors have complex components. Later we are concerned with certain infinite dimensional vector spaces. To give a unified treatment, we start here with the general definition of a vector space. Then we discuss the span of a set of vectors, linear independence, and bases for vector spaces.

Roughly, a vector space V over a field $\mathbb{F}$ is a set of objects (called vectors) for which two operations are defined, namely addition of vectors and multiplication of vectors by field elements (called scalars), such that these operations satisfy certain natural properties. The general definition of a field was given in Definition 1.1, but the only examples we use here are $\mathbb{R}$ and $\mathbb{C}$.

Definition 1.30 *Let $\mathbb{F}$ be a field. A vector space V over $\mathbb{F}$ is a set with operations of vector addition $+$ and scalar multiplication $\cdot$ satisfying the following properties:*

A1. *(Closure for addition) For all $u, v \in V$, $u + v$ is defined and is an element of V.*

A2. *(Commutativity for addition) $u + v = v + u$, for all $u, v \in V$.*

A3. *(Associativity for addition) $u + (v + w) = (u + v) + w$, for all $u, v, w \in V$.*

A4. *(Existence of additive identity) There exists an element in V, denoted 0, such that $u + 0 = u$ for all $u \in V$.*

A5. *(Existence of additive inverse) For each $u \in V$, there exists an element in V, denoted $-u$, such that $u + (-u) = 0$.*

M1. *(Closure for scalar multiplication) For all $\alpha \in \mathbb{F}$ and $u \in V$, $\alpha \cdot u$ is defined and is an element of V.*

M2. *(Behavior of the scalar multiplicative identity) $1 \cdot u = u$, for all $u \in V$, where 1 is the multiplicative identity in $\mathbb{F}$.*

M3. *(Associativity for scalar multiplication) $\alpha \cdot (\beta \cdot u) = (\alpha\beta) \cdot u$, for all $\alpha, \beta \in \mathbb{F}$ and $u \in V$.*

D1. *(First distributive property) $\alpha \cdot (u + v) = (\alpha \cdot u) + (\alpha \cdot v)$, for all $\alpha \in \mathbb{F}$ and $u, v \in V$.*

D2. *(Second distributive property) $(\alpha + \beta) \cdot u = (\alpha \cdot u) + (\beta \cdot u)$, for all $\alpha, \beta \in \mathbb{F}$ and $u \in V$.*

Properties A1–A5 guarantee that addition behaves reasonably. The field properties tell us that the scalars themselves behave reasonably. Properties M1–M3 and D1–D2 state that scalar multiplication is compatible with vector addition. We usually omit the symbol $\cdot$ for scalar multiplication, writing αu instead of $\alpha \cdot u$.

Example 1.31

$\mathbb{R}^n$. The field $\mathbb{F}$ in this example is $\mathbb{R}$. Let $\mathbb{R}^n$ be the set of all n-tuples of real numbers, that is, all

$$x = \begin{bmatrix} x_1 \\ x_2 \\ \vdots \\ x_n \end{bmatrix},$$

where $x_1, x_2, x_3, \ldots, x_n \in \mathbb{R}$. Vector addition and scalar multiplication are defined in the usual way: for $\alpha \in \mathbb{R}$ and $x_i, y_i \in \mathbb{R}, i = 1, 2, \ldots n,$

$$\alpha \begin{bmatrix} x_1 \\ x_2 \\ \vdots \\ x_n \end{bmatrix} = \begin{bmatrix} \alpha x_1 \\ \alpha x_2 \\ \vdots \\ \alpha x_n \end{bmatrix} \quad \text{and} \quad \begin{bmatrix} x_1 \\ x_2 \\ \vdots \\ x_n \end{bmatrix} + \begin{bmatrix} y_1 \\ y_2 \\ \vdots \\ y_n \end{bmatrix} = \begin{bmatrix} x_1 + y_1 \\ x_2 + y_2 \\ \vdots \\ x_n + y_n \end{bmatrix}.$$

It is straightforward to check that with these operations, $\mathbb{R}^n$ is a vector space over $\mathbb{R}$.

Example 1.32

$\mathbb{C}^n$. Here the field $\mathbb{F}$ is $\mathbb{C}$. Let $\mathbb{C}^n$ be the set of all n-tuples of complex numbers, that is, all

$$z = \begin{bmatrix} z_1 \\ z_2 \\ \vdots \\ z_n \end{bmatrix},$$

where $z_1, z_2, z_3, \ldots, z_n \in \mathbb{C}$. We call $z_1, z_2, \ldots, z_n$ the components of z. For $\alpha \in \mathbb{C}$ and $z_i, w_i \in \mathbb{C}$, for $i = 1, 2, \ldots n,$ let

$$\alpha \begin{bmatrix} z_1 \\ z_2 \\ \vdots \\ z_n \end{bmatrix} = \begin{bmatrix} \alpha z_1 \\ \alpha z_2 \\ \vdots \\ \alpha z_n \end{bmatrix} \quad \text{and} \quad \begin{bmatrix} z_1 \\ z_2 \\ \vdots \\ z_n \end{bmatrix} + \begin{bmatrix} w_1 \\ w_2 \\ \vdots \\ w_n \end{bmatrix} = \begin{bmatrix} z_1 + w_1 \\ z_2 + w_2 \\ \vdots \\ z_n + w_n \end{bmatrix}.$$

It is easy to check that $\mathbb{C}^n$, with these operations, is a vector space over $\mathbb{C}$.

In these examples, the vector space elements, or vectors, are of the type encountered in calculus. In general this may not be the case; see Exercise 1.3.1(i) and the following.

Example 1.33
Let $[0, 1] = \{x \in \mathbb{R} : 0 \leq x \leq 1\}$. The field is $\mathbb{C}$ in this example. Let V be the set of all complex-valued functions on $[0, 1]$, that is,

$$V = \{f : [0, 1] \rightarrow \mathbb{C}\}.$$

We define addition in the usual way: for $f, g \in V$, $f + g$ is the function on $[0, 1]$ defined by

$$(f + g)(x) = f(x) + g(x).$$

Scalar multiplication is naturally defined also: for $\alpha \in \mathbb{C}$ and $f \in V$, αf is the function on $[0, 1]$ such that

$$(\alpha f)(x) = \alpha f(x).$$

Then V is a vector space over $\mathbb{C}$ (Exercise 1.3.1(ii)).

Definition 1.34 *Let V be a vector space over a field $\mathbb{F}$, let $n \in \mathbb{N}$, and let $v_1, v_2, \ldots, v_n \in V$. A* linear combination *of $v_1, v_2, \ldots, v_n$ is any vector of the form*

$$\sum_{j=1}^{n} \alpha_j v_j = \alpha_1 v_1 + \alpha_2 v_2 + \cdots + \alpha_n v_n,$$

where $\alpha_1, \alpha_2, \ldots, \alpha_n \in \mathbb{F}$.

Note that our definition requires a linear combination to be a finite sum.

Definition 1.35 *Let V be a vector space over a field $\mathbb{F}$, and suppose $U \subseteq V$. The* span *of U (denoted* span U*) is the set of all linear combinations of elements of U. In particular, if U is a finite set, say*

$$U = \{u_1, u_2, \ldots u_n\},$$

then

$$\text{span}\, U = \left\{ \sum_{j=1}^{n} \alpha_j u_j : \alpha_j \in \mathbb{F} \text{ for all } j = 1, 2, \ldots, n \right\}.$$

To visualize the span, note that the span of a single nonzero vector u in $\mathbb{R}^n$ consists of all vectors lying on the line through the origin that contains u. If u and v are two noncollinear vectors in $\mathbb{R}^3$, then span $\{u, v\}$ is the plane through the origin containing u and v.

Definition 1.36 *Let V be a vector space over a field $\mathbb{F}$ and let $v_1, v_2, \ldots v_n$ be elements of V. We say that $v_1, v_2, \ldots v_n$ are* linearly dependent *(or $\{v_1, v_2, \ldots v_n\}$ is a* linearly dependent set*) if there exist $\alpha_1, \alpha_2, \ldots, \alpha_n \in \mathbb{F}$ that are not all zero, such that*

$$\alpha_1 v_1 + \alpha_2 v_2 + \cdots + \alpha_n v_n = 0.$$

We say that $v_1, v_2, \ldots v_n$ are linearly independent *(or $\{v_1, v_2, \ldots v_n\}$ is a* linearly independent set*) if*

$$\alpha_1 v_1 + \alpha_2 v_2 + \cdots + \alpha_n v_n = 0$$

holds only when $\alpha_j = 0$ for every $j = 1, 2, 3, \ldots, n$.

For a possibly infinite subset U of V, we say U is linearly independent *if every finite subset of U is linearly independent, and we say U is* linearly dependent *if U has a finite subset that is linearly dependent.*

In other words, $v_1, v_2, \ldots v_n$ are linearly independent if the only linear combination $\alpha_1 v_1 + \alpha_2 v_2 + \cdots + \alpha_n v_n$ that is 0 is the trivial one ($\alpha_j = 0$ for all $j = 1, 2, \ldots, n$). If there is a nontrivial linear combination of $v_1, v_2, \ldots v_n$ that is 0, then $v_1, v_2, \ldots v_n$ are linearly dependent.

The last part of Definition 1.36 is consistent with the first by Exercise 1.3.7. If a set is linearly dependent, then one element can be written as a linear combination of the others and that element can be removed from the set without changing the span (Exercise 1.3.8).

Definition 1.37 *Let V be a vector space over a field $\mathbb{F}$. A subset U of V is a* basis *for V if U is a linearly independent set such that span $U = V$.*

Bases are also characterized in the following way.

Lemma 1.38 *Let V be a vector space over a field $\mathbb{F}$, and let U be a nonempty subset of V.*

i. Suppose U is finite, say $U = \{u_1, u_2, \ldots, u_n\}$, for some $n \in \mathbb{N}$, with $u_j \neq u_k$ for $j \neq k$. Then U is a basis for V if and only if *for each $v \in V$, there exist unique $\alpha_1, \alpha_2, \ldots, \alpha_n \in \mathbb{F}$ such that $v = \sum_{j=1}^{n} \alpha_j u_j$.*

ii. *If U is infinite, then U is a basis for V if and only if for each non-zero $v \in V$, there exist unique $m \in \mathbb{N}, u_1, u_2, \ldots, u_m \in U$, and nonzero $\alpha_1, \alpha_2, \ldots, \alpha_m \in \mathbb{F}$ such that $v = \sum_{j=1}^{m} \alpha_j u_j$.*

Proof

Exercise 1.3.11. ∎

The easiest example of a basis is the following.

Definition 1.39 *Define $E = \{e_1, e_2, \ldots, e_n\}$ by*

$$
e_1 = \begin{bmatrix} 1 \\ 0 \\ 0 \\ \vdots \\ 0 \end{bmatrix}, \quad e_2 = \begin{bmatrix} 0 \\ 1 \\ 0 \\ \vdots \\ 0 \end{bmatrix}, \quad e_3 = \begin{bmatrix} 0 \\ 0 \\ 1 \\ \vdots \\ 0 \end{bmatrix}, \quad \cdots, \quad e_n = \begin{bmatrix} 0 \\ 0 \\ 0 \\ \vdots \\ 1 \end{bmatrix}.
$$

These vectors can be regarded as elements of $\mathbb{R}^n$ or $\mathbb{C}^n$. We call E the standard, or Euclidean basis for $\mathbb{R}^n$ or $\mathbb{C}^n$.

This terminology is justified because E is a basis for $\mathbb{R}^n$ or $\mathbb{C}^n$. To see this, either verify the definition, or use Lemma 1.38 and note that a vector with components $\alpha_1, \alpha_2, \ldots \alpha_n$, can be written uniquely as $\sum_{j=1}^{n} \alpha_j e_j$. It may seem strange that the same vectors span both $\mathbb{R}^n$ and the apparently larger space $\mathbb{C}^n$, but this happens because the scalar field for $\mathbb{C}^n$ is $\mathbb{C}$, instead of $\mathbb{R}$ as for $\mathbb{R}^n$.

If a vector space V has a basis consisting of finitely many elements, we say that V is *finite dimensional*. In this case, any two bases for V have the same number of elements.

Theorem 1.40 *Let V be a vector space over a field $\mathbb{F}$. Suppose that V has a basis consisting of n elements. Then any other basis of V also has n elements.*

Proof

Exercise 1.3.13. ∎

This allows us to make Definition 1.41.

Definition 1.41 *Suppose V is a finite dimensional vector space. The number of elements in a basis for V is called the* dimension *of V, written dim V. If dim V = n, we say V is n-dimensional.*

Theorem 1.42 yields the useful fact that a collection of n vectors in an n-dimensional vector space is a basis if one of the two conditions (that they are linearly independent, or that they span the space) in Definition 1.37 holds.

Theorem 1.42 *Suppose V is an n-dimensional vector space, and $v_1, \ldots, v_n$ are distinct vectors in V. Then $v_1, v_2, \ldots, v_n$ are linearly independent if and only if*

$$\mathrm{span}\{v_1, v_2, \ldots, v_n\} = V.$$

Proof
Since V is n-dimensional, V has a basis consisting of n vectors, say $w_1, w_2, \ldots, w_n$.

First suppose $v_1, v_2, \ldots, v_n$ are linearly independent. Let $u \in V$ be arbitrary. Then $u, v_1, v_2, \ldots, v_n$ are $n + 1$ vectors belonging to $\mathrm{span}\{w_1, w_2, \ldots, w_n\}$. By Exercise 1.3.12, $u, v_1, v_2, \ldots, v_n$ are linearly dependent. By Exercise 1.3.9(ii), this implies $u \in \mathrm{span}\{v_1, v_2, \ldots, v_n\}$. Since u is arbitrary, this proves that $\mathrm{span}\{v_1, v_2, \ldots, v_n\} = V$.

Now suppose that $\mathrm{span}\{v_1, v_2, \ldots, v_n\} = V$. If $v_1, v_2, \ldots, v_n$ are linearly dependent, then by Exercise 1.3.8(ii), we can find a subset of $n - 1$ vectors that still spans V. Then $w_1, w_2, \ldots, w_n$ are n linearly independent vectors belonging to the span of $n - 1$ vectors, which is impossible by Exercise 1.3.12. This contradiction shows that $v_1, v_2, \ldots, v_n$ are linearly independent. ∎

Definition 1.43 *Suppose V is a vector space over a field $\mathbb{F}$ and $S = \{v_1, v_2, \ldots, v_n\}$ is a basis for V. For any vector $v \in V$, there exist unique $\alpha_1, \alpha_2, \ldots, \alpha_n \in \mathbb{F}$ such that $v = \sum_{j=1}^{n} \alpha_j v_j$, by Lemma 1.38. We denote by $[v]_S$ the vector in $\mathbb{F}^n$ with components $\alpha_1, \alpha_2, \ldots, \alpha_n$, that is,*

$$[v]_S = \begin{bmatrix} \alpha_1 \\ \alpha_2 \\ \vdots \\ \alpha_n \end{bmatrix}. \tag{1.23}$$

We call α_j the jth component of v with respect to S.

In other words, for a basis $S = \{v_1, v_2, \ldots, v_n\}$, statement (1.23) means that $v = \sum_{j=1}^{n} \alpha_j v_j$.

One should not confuse $[v]_S$ with the vector v itself. First, they may be different types of objects. For example, if the vector space is P_n, the polynomials of degree n (see Exercise 1.3.1(i)), then v is a polynomial whereas $[v]_S$ is an n-tuple of numbers. Second, even if the vector space V is $\mathbb{R}^n$ or $\mathbb{C}^n$, so that v is an n-tuple of numbers (which is a well-defined object, without reference to any basis), v and $[v]_S$ will usually be different n-tuples of numbers (unless S is the Euclidean basis, as we note below). So we do not regard $[v]_S$ as identified with v (since two n-tuples are the same only if they have the same components). Instead $[v]_S$ is the vector whose components are the components of v with respect to the basis S.

However, the components of a vector v with respect to the Euclidean basis E (Definition 1.39) are the usual components of v. For example, if

$$
z = \begin{bmatrix} z_1 \\ z_2 \\ \vdots \\ z_n \end{bmatrix}
$$

is a vector in $\mathbb{C}^n$ (similarly for $\mathbb{R}^n$) then $z = \sum_{j=1}^{n} z_j e_j$, so

$$
z = [z]_E. \tag{1.24}
$$

Suppose that V is an n-dimensional vector space over $\mathbb{C}$ (similar remarks hold for $\mathbb{R}$). We can represent any element v in V with respect to a given basis by a vector in $\mathbb{C}^n$. So in an appropriate sense (namely in the sense of a vector space isomorphism, which we do not define), any n-dimensional vector space over $\mathbb{C}$ is equivalent to $\mathbb{C}^n$. If we are concerned only about finite dimensional vector spaces over $\mathbb{R}$ and $\mathbb{C}$, then up to isomorphism, all we need to consider are $\mathbb{R}^n$ and $\mathbb{C}^n$. So the general theory of vector spaces is needed only if we want to consider more general fields (which we do not) or infinite dimensional vector spaces (which we do, later).

In the abstract sense, one basis is pretty much as good as another. However, in particular cases (e.g., once we have a particular linear operator to study), the choice of basis can be of paramount importance. A great deal of elementary linear algebra comes down to selecting the basis that simplifies a problem as much as possible. This is also the underlying theme of Fourier analysis, wavelet theory, and this text.

Exercises

1.3.1. i. For $n \in \mathbb{N}$, let P_n be the collection of all polynomials over $\mathbb{C}$ of degree $\leq n$. Define addition and scalar multiplication by complex numbers in the sense of functions; that is, for $p, q \in P_n$ and $\alpha \in \mathbb{C}$, let $(p+q)(x) = p(x) + q(x)$ and $(\alpha p)(x) = \alpha p(x)$. Prove that P_n is a vector space over $\mathbb{C}$.

    ii. Prove that the set V in Example 1.33 with the operations defined there forms a vector space over $\mathbb{C}$.

1.3.2. Let V be a vector space over a field $\mathbb{F}$, and let U be a subset of V. We say that U is a *subspace* of V if the set U itself with the operations inherited from V forms a vector space over $\mathbb{F}$. Most of the properties of a vector space hold automatically in U just because they hold in the bigger set V. Prove that in order to verify that U is a subspace of V, one needs only to check that

$$U \neq \emptyset \quad \text{(where } \emptyset \text{ is the empty set);}$$
$$\text{if} \quad u, v \in U, \quad \text{then} \quad u + v \in U;$$

and

$$\text{if} \quad u \in U \quad \text{and} \quad \alpha \in \mathbb{F}, \quad \text{then} \quad \alpha u \in U.$$

In other words, if U is nonempty, one needs only to check that U itself is closed under addition and scalar multiplication.

1.3.3. Let V be a vector space over a field $\mathbb{F}$, and let $u_1, u_2, \ldots, u_n \in V$. Prove that span $\{u_1, u_2, \ldots, u_n\}$ is a subspace of V. (Assume the result of Exercise 1.3.2.)

1.3.4. Prove that the vectors

$$\begin{bmatrix} 1 \\ -2 \\ 3 \end{bmatrix}, \quad \begin{bmatrix} -2 \\ 3 \\ 1 \end{bmatrix}, \quad \text{and} \quad \begin{bmatrix} 5 \\ 1 \\ 2 \end{bmatrix}$$

are linearly independent in $\mathbb{R}^3$.

1.3.5. Prove that the vectors

$$\begin{bmatrix} i \\ 2+i \\ 3 \end{bmatrix}, \quad \begin{bmatrix} 2 \\ -i \\ 4-i \end{bmatrix}, \quad \text{and} \quad \begin{bmatrix} 3 \\ -1 \\ 2 \end{bmatrix}$$

are linearly independent in $\mathbb{C}^3$.

1.3.6. Prove that any collection of vectors that includes the 0 vector is linearly dependent.

1.3.7. Prove: if U is a finite collection of vectors that has a linearly dependent subset, then U is linearly dependent.

1.3.8. Suppose $v_1, v_2, \ldots, v_n$ are linearly dependent vectors in some vector space.

   i. Prove that there is some $j \in \{1, 2, 3, \ldots, n\}$ such that

$$v_j \in \text{span}\{v_1, v_2, \ldots, v_{j-1}, v_{j+1}, \ldots, v_n\}.$$

   ii. For j as in part i, prove that

$$\text{span}\{v_1, v_2, \ldots, v_j, \ldots, v_n\}$$
$$= \text{span}\{v_1, v_2, \ldots, v_{j-1}, v_{j+1}, \ldots, v_n\}.$$

1.3.9. Suppose $u, v_1, v_2, \ldots, v_n$ are vectors in some vector space.

   i. If $u \in \text{span}\{v_1, v_2, \ldots, v_n\}$, prove that $u, v_1, v_2, \ldots, v_n$ are linearly dependent.

   ii. Suppose $v_1, v_2, \ldots, v_n$ are linearly independent. If $u, v_1, v_2, \ldots, v_n$ are linearly dependent, prove that $u \in \text{span}\{v_1, v_2, \ldots, v_n\}$.

1.3.10. Suppose V is an n-dimensional vector space, $k \leq n$, and $\{v_1, v_2, \ldots, v_k\}$ is a linearly independent set in V. Show that there exist $v_{k+1}, \ldots, v_n$ in V so that $\{v_1, v_2, \ldots, v_n\}$ is a basis for V. (In other words, any linearly independent set can be extended to a basis.) Hint: If $k \neq n$, $\text{span}\{v_1, v_2, \ldots, v_k\} \neq V$. So there exists $v_{k+1} \in V \setminus \text{span}\{v_1, v_2, \ldots, v_k\}$. By Exercise

1.3.9(ii), the set $\{v_1, \ldots, v_k, v_{k+1}\}$ is linearly independent. Continue in this fashion.

1.3.11. Prove Lemma 1.38.

1.3.12. Suppose $n + 1$ vectors belong to the span of n vectors, say $w_j \in \text{span}\{v_1, v_2, \ldots, v_n\}$ for $j = 1, 2, \ldots, n + 1$. Prove that $w_1, w_2, \ldots, w_n, w_{n+1}$ are linearly dependent. Hint: Prove this by induction on n. Let p_n be the statement given. The case $n = 1$ is not difficult. Suppose p_{n-1} holds. To prove p_n, write

$$w_1 = \alpha_{1,1} v_1 + \alpha_{1,2} v_2 + \cdots + \alpha_{1,n} v_n,$$
$$w_2 = \alpha_{2,1} v_1 + \alpha_{2,2} v_2 + \cdots + \alpha_{2,n} v_n,$$
$$\vdots$$
$$w_{n+1} = \alpha_{n+1,1} v_1 + \alpha_{n+1,2} v_2 + \cdots + \alpha_{n+1,n} v_n.$$

If $w_{n+1} = 0$, we are done (Exercise 1.3.6). Otherwise, at least one of its coefficients is nonzero. By reindexing, we can assume $\alpha_{n+1,n} \neq 0$. Consider the vectors

$$u_k = w_k - \left(\frac{\alpha_{k,n}}{\alpha_{n+1,n}} \right) w_{n+1},$$

for $k = 1, 2, \ldots, n$. Writing u_k in terms of $v_1, \ldots, v_n$, show that the v_n terms cancel out. Hence $u_k \in \text{span}\{v_1, \ldots, v_{n-1}\}$. By p_{n-1}, the induction assumption, $u_1, u_2, \ldots, u_n$ are linearly dependent. Show that this implies that $w_1, \ldots, w_{n+1}$ are linearly dependent.

1.3.13. Prove Theorem 1.40. Hint: Use Exercise 1.3.12.

1.3.14. Let R be a basis for a finite dimensional vector space V over a field $\mathbb{F}$. Prove that:

i. $[u + v]_R = [u]_R + [v]_R$ for all $u, v \in V$;

ii. $[\alpha v]_R = \alpha [v]_R$ for all $\alpha \in \mathbb{F}$ and $v \in V$.

1.3.15. Suppose V is a finite dimensional vector space over $\mathbb{C}$ with basis R. Let $v_1, v_2, \ldots, v_m$ be elements of V. Prove that the set $\{v_1, v_2, \ldots, v_m\}$ is linearly independent in V if and only if the set $\{[v_1]_R, [v_2]_R, \ldots, [v_m]_R\}$ is linearly independent in $\mathbb{C}^n$.

1.4 Linear Transformations, Matrices, and Change of Basis

In mathematics we often consider the class of all objects having a certain structure, for example, groups, fields, metric spaces, or topological spaces. In each case we consider maps between these objects that are consistent with or preserve this structure, such as homomorphisms or isomorphisms of groups, isometries of metric spaces, and homeomorphisms of topological spaces. Vector spaces have the linear structure stated in Definition 1.30. Maps that respect this structure are called linear transformations.

Definition 1.44 *Let U and V be vector spaces over the same field $\mathbb{F}$. A linear transformation T is a function $T : U \to V$ having the following properties:*
  L1. (Additivity) $T(u_1 + u_2) = T(u_1) + T(u_2)$, for all $u_1, u_2 \in U$.
  L2. (Scalar homogeneity) $T(\alpha u) = \alpha T(u)$, for all $\alpha \in \mathbb{F}$ and $u \in U$.

Suppose $T : U \to V$ is a linear transformation, and U is finite dimensional with basis $\{u_1, u_2, \ldots, u_n\}$. Then T is determined by its action on the basis $\{u_1, u_2, \ldots, u_n\}$, in the following sense. Suppose $u \in U$. Then there exist unique scalars $\alpha_1, \alpha_2, \ldots, \alpha_n$ such that $u = \sum_{j=1}^{n} \alpha_j u_j$, by Lemma 1.38. By properties L1 and L2,

$$
T(u) = T\left(\sum_{j=1}^{n} \alpha_j u_j \right) = \sum_{j=1}^{n} \alpha_j T(u_j).
$$

In particular, suppose $L : U \to V$ is a linear transformation, and $L(u_j) = T(u_j)$ for all $j = 1, 2, \ldots, n$. Then $L = T$, that is, $L(u) = T(u)$ for all $u \in U$, because the steps above also show that $L(u) = \sum_{j=1}^{n} \alpha_j L(u_j)$.

Recall (Definition 1.43) that for a given basis $S = \{u_1, u_2, \ldots, u_n\}$ of a vector space U, $[u]_S$ is the vector in $\mathbb{F}^n$ (e.g., $\mathbb{R}^n$ or $\mathbb{C}^n$) whose components $\alpha_1, \alpha_2, \ldots, \alpha_n$ are the coefficients in the expansion $u = \sum_{j=1}^{n} \alpha_j u_j$. Any linear transformation $T : U \to V$ can be represented in bases for U and V by matrix multiplication, in a sense described in Lemma 1.49. First we define matrices and their natural operations.

Definition 1.45 *For $m, n \in \mathbb{N}$, an $m \times n$ matrix A over a field $\mathbb{F}$ is a rectangular array of the form*

$$
A = \begin{bmatrix}
a_{11} & a_{12} & \cdot & \cdot & \cdot & a_{1n} \\
a_{21} & a_{22} & \cdot & \cdot & \cdot & a_{2n} \\
\cdot & \cdot & \cdot & \cdot & \cdot & \cdot \\
\cdot & \cdot & \cdot & \cdot & \cdot & \cdot \\
\cdot & \cdot & \cdot & \cdot & \cdot & \cdot \\
a_{m1} & a_{m2} & \cdot & \cdot & \cdot & a_{mn}
\end{bmatrix},
$$

where $a_{ij} \in \mathbb{F}$ for all $i = 1, 2, \ldots, m$ and $j = 1, 2, \ldots, n$. We call a_{ij} the $(i, j)^{\text{th}}$ entry of A. We also denote A by $[a_{ij}]_{1 \le i \le m, 1 \le j \le n}$ or, when m and n are understood, by $[a_{i,j}]$.

Note that an $n \times 1$ matrix is a vector with n components, that is, an element of $\mathbb{R}^n$ or $\mathbb{C}^n$.

Addition of matrices is defined by the obvious addition of corresponding entries. Scalar multiplication is also defined in the natural way.

Definition 1.46 *Suppose $A = [a_{i,j}]$ and $B = [b_{i,j}]$ are two $m \times n$ matrices over the same field $\mathbb{F}$. Then $A + B$ is the $m \times n$ matrix $C = [c_{i,j}]$ with $c_{ij} = a_{ij} + b_{ij}$ for all $i = 1, 2, \ldots, m$ and $j = 1, 2, \ldots, n$. For $\alpha \in \mathbb{F}$, αA is the $m \times n$ matrix $D = [d_{ij}]$ with $d_{ij} = \alpha a_{ij}$ for all i, j.*

Matrix multiplication is a more subtle process.

Definition 1.47 *Suppose $A = [a_{i,j}]$ is an $m \times \ell$ matrix over a field $\mathbb{F}$ and $B = [b_{i,j}]$ is an $\ell \times n$ matrix over the same field $\mathbb{F}$. Then AB is the $m \times n$ matrix $C = [c_{i,j}]$ with*

$$
c_{ij} = \sum_{k=1}^{\ell} a_{ik} b_{kj}
$$

for $i = 1, 2, \ldots, m$ and $j = 1, 2, \ldots, n$.

For real matrices, c_{ij} is the dot product of the ith row of A with the jth column of B.

Notice that we define the multiplication of an $m \times \ell$ matrix only with an $\ell \times n$ matrix, which gives an $m \times n$ matrix (the ℓ "cancels"). Matrix multiplication is not commutative: first, when AB is defined, it may be that BA is not even defined; second, even if both are defined, we may have $AB \neq BA$. However, matrix multiplication is associative: for $A = [a_{ij}]_{1 \leq i \leq m, 1 \leq j \leq \ell}$, $B = [b_{ij}]_{1 \leq i \leq \ell, 1 \leq j \leq k}$, $C = [c_{ij}]_{1 \leq i \leq k, 1 \leq j \leq n}$, the $(i, j)^{\text{th}}$ entry of the $m \times n$ matrices $(AB)C$ and $A(BC)$ are both

$$\sum_{p=1}^{\ell} \sum_{q=1}^{k} a_{ip} b_{pq} c_{qj}.$$

A special case of particular interest is the product of an $m \times n$ matrix A and an $n \times 1$ matrix (i.e., a vector) x. The result $y = Ax$ is an $m \times 1$ matrix, that is, a vector with m components. The map T taking x to Ax, that is, $T : \mathbb{F}^n \to \mathbb{F}^m$ (e.g., $\mathbb{F} = \mathbb{R}$ or $\mathbb{C}$) defined by $T(x) = Ax$, is a linear transformation (Exercise 1.4.2). By Definition 1.47, the ith component of Ax is

$$(Ax)_i = \sum_{k=1}^{n} a_{ik} x_k.$$

For more general finite dimensional vector spaces, a matrix leads to a linear transformation by acting on the vector of components with respect to a given basis as follows.

Lemma 1.48   (Linear transformation associated with a matrix) *Let U and V be finite dimensional vector spaces over a field $\mathbb{F}$. Suppose $R = \{u_1, u_2, \ldots, u_n\}$ is a basis for U and $S = \{v_1, v_2, \ldots, v_m\}$ is a basis for V. Let A be an $m \times n$ matrix over $\mathbb{F}$. We define a mapping $T_A : U \to V$ as follows: for $u \in U$, let $T_A(u)$ be the element of V whose vector of components $[T_A(u)]_S$ with respect to S is $A[u]_R$, that is,*

$$[T_A(u)]_S = A[u]_R.$$

Then T_A is a linear transformation.

Proof
Exercise 1.4.3. ∎

Thus every matrix gives a linear transformation. The converse is also true: every linear transformation between finite dimensional vector spaces can be represented by a matrix.

Lemma 1.49   (Matrix representing a linear transformation) *Let U and V be finite dimensional vector spaces over a field $\mathbb{F}$. Suppose $R = \{u_1, u_2, \ldots, u_n\}$ is a basis for U, $S = \{v_1, v_2, \ldots, v_m\}$ is a basis for*

V, and $T : U \rightarrow V$ is a linear transformation. Since $T(u_j) \in V$ for each j, there are unique scalars $a_{ij}, i = 1, 2, \ldots, m,$ and $j = 1, 2, \ldots, n,$ such that

$$T(u_1) = a_{11}v_1 + a_{21}v_2 + \cdots + a_{m1}v_m,$$
$$T(u_2) = a_{12}v_1 + a_{22}v_2 + \cdots + a_{m2}v_m,$$

$$\vdots$$

$$T(u_n) = a_{1n}v_1 + a_{2n}v_2 + \cdots + a_{mn}v_m.$$

Let A be the $m \times n$ matrix whose k^{th} column consists of the scalars $\alpha_{1k}, \alpha_{2k}, \ldots, \alpha_{mk}$ in the expansion of $T(u_k)$:

$$A = \begin{bmatrix} a_{11} & a_{12} & \cdots & a_{1n} \\ a_{21} & a_{22} & \cdots & a_{2n} \\ \vdots & \vdots & \ddots & \vdots \\ a_{m1} & a_{m2} & \cdots & a_{mn} \end{bmatrix}.$$

Then

$$[T(u)]_S = A[u]_R \quad \text{for all } u \in U. \tag{1.25}$$

Moreover, A is the unique matrix satisfying relationship (1.25).

Proof
Let $u \in U$ be arbitrary. Let $c_1, c_2, \ldots, c_n$ be the components of u with respect to R, that is,

$$[u]_R = \begin{bmatrix} c_1 \\ \vdots \\ c_n \end{bmatrix},$$

or, equivalently, $u = \sum_{j=1}^{n} c_j u_j$. By definition of A, $T(u_j) = \sum_{i=1}^{m} a_{ij}v_i$. By linearity,

$$T(u) = T \left(\sum_{j=1}^{n} c_j u_j \right) = \sum_{j=1}^{n} c_j T(u_j)$$

$$= \sum_{j=1}^{n} c_j \sum_{i=1}^{m} a_{ij}v_i = \sum_{i=1}^{m} \left(\sum_{j=1}^{n} a_{ij}c_j \right) v_i.$$

Hence the ith component of $T(u)$ with respect to the basis S is $\sum_{j=1}^{n} a_{ij}c_j$, which is, by definition, the ith component of the product of the matrix A with the vector $[u]_R$. So (1.25) holds. The uniqueness of A is left to the reader (Exercise 1.4.6). ∎

In the previous lemma, we call A the *matrix representing T with respect to R and S*. We sometimes write it as A_T.

Example 1.50
Define a basis R for $\mathbb{R}^2$ by

$$R = \left\{ \begin{bmatrix} 2 \\ 1 \end{bmatrix}, \begin{bmatrix} 0 \\ 2 \end{bmatrix} \right\},$$

and a basis S for $\mathbb{R}^3$ by

$$S = \left\{ \begin{bmatrix} 1 \\ 0 \\ 2 \end{bmatrix}, \begin{bmatrix} 0 \\ 1 \\ 1 \end{bmatrix}, \begin{bmatrix} 1 \\ 0 \\ 3 \end{bmatrix} \right\}.$$

Define $T : \mathbb{R}^2 \to \mathbb{R}^3$ by

$$T\left(\begin{bmatrix} x \\ y \end{bmatrix} \right) = \begin{bmatrix} 2x - y \\ x + y \\ x - 3y \end{bmatrix}.$$

Find the matrix A that represents T with respect to R and S.

Solution
By definition,

$$T\left(\begin{bmatrix} 2 \\ 1 \end{bmatrix} \right) = \begin{bmatrix} 3 \\ 3 \\ -1 \end{bmatrix}, \quad \text{and} \quad T\left(\begin{bmatrix} 0 \\ 2 \end{bmatrix} \right) = \begin{bmatrix} -2 \\ 2 \\ -6 \end{bmatrix}.$$

According to Lemma 1.49, we should find $\{a_{ij}\}$ such that

$$\begin{bmatrix} 3 \\ 3 \\ -1 \end{bmatrix} = a_{11} \begin{bmatrix} 1 \\ 0 \\ 2 \end{bmatrix} + a_{21} \begin{bmatrix} 0 \\ 1 \\ 1 \end{bmatrix} + a_{31} \begin{bmatrix} 1 \\ 0 \\ 3 \end{bmatrix}, \qquad (1.26)$$

and

$$\begin{bmatrix} -2 \\ 2 \\ -6 \end{bmatrix} = a_{12} \begin{bmatrix} 1 \\ 0 \\ 2 \end{bmatrix} + a_{22} \begin{bmatrix} 0 \\ 1 \\ 1 \end{bmatrix} + a_{32} \begin{bmatrix} 1 \\ 0 \\ 3 \end{bmatrix}. \qquad (1.27)$$

These are linear systems; for example, equation (1.26) is the system

$$1a_{11} + 0a_{21} + 1a_{31} = 3,$$
$$0a_{11} + 1a_{21} + 0a_{31} = 3,$$
$$2a_{11} + 1a_{21} + 3a_{31} = -1,$$

which can be solved, for example, by Gaussian elimination (we omit the calculation), to give

$$a_{11} = 13, \quad a_{21} = 3, \quad \text{and} \quad a_{31} = -10.$$

Similarly, equation (1.27) can be solved to give

$$a_{12} = 2, \quad a_{22} = 2, \quad \text{and} \quad a_{32} = -4.$$

Hence

$$A = \begin{bmatrix} 13 & 2 \\ 3 & 2 \\ -10 & -4 \end{bmatrix}. \qquad \blacksquare$$

Lemmas 1.48 and 1.49 show that there is a complete correspondence between matrices and linear transformations between finite dimensional vector spaces. To understand this correspondence better, we consider certain properties that a linear transformation can have and the corresponding properties of the associated matrix.

The next definition is stated in a general form because it makes sense for any function.

Definition 1.51 *Let U and V be sets, and $T : U \rightarrow V$ a function. T is* one-to-one *(written $1 - 1$) or* injective *if $T(u_1) = T(u_2)$ implies $u_1 = u_2$. T is* onto *or* surjective *if, for every $v \in V$, there exists $u \in U$ such that $T(u) = v$. T is* invertible *or* bijective *if T is both $1 - 1$ and onto.*

In other words, T is $1 - 1$ if $u_1 \neq u_2$ implies $T(u_1) \neq T(u_2)$; that is, T cannot take two different values to the same value, and T is onto if T attains every element in V.

If T is $1 - 1$ and onto, we define $T^{-1} : V \rightarrow U$ as follows: for each $v \in V$, there exists (since T is onto) a unique (since T is $1 - 1$) $u \in U$ such that $T(u) = v$. Let $T^{-1}(v) = u$. Then it is easy to see that T^{-1} is the inverse mapping of T in the sense that $T^{-1}(T(u)) = u$ for all $u \in U$ and $T(T^{-1}(v)) = v$ for all $v \in V$. This is the reason for the term *invertible* in Definition 1.51.

Related to the terms $1 - 1$ and onto are the kernel and range of a linear transformation.

Definition 1.52   *Let U and V be vector spaces, and T : U → V a linear transformation.*

The kernel *of T is the set of all vectors that get mapped to the 0 vector by T; that is,*

$$\ker T = \{u \in U : T(u) = 0\}.$$

The range *of T is the set of all vectors in V that are the image of some vector in U under T:*

$$\text{range } T = \{T(u) : u \in U\}.$$

By definition, T is onto if and only if range $(T) = V$. Also, by the linearity of T, T is $1 - 1$ if and only if $\ker T = \{0\}$ (Exercise 1.4.7).

Corresponding to the notion of an invertible linear transformation is the notion of an invertible matrix. For this, we first need to define the identity matrix.

Definition 1.53   *The n × n identity matrix I over $\mathbb{R}$ or $\mathbb{C}$ is the matrix $[a_{ij}]_{1 \le i,j \le n}$ such that $a_{ii} = 1$ for $i = 1, 2, \ldots, n$ and $a_{ij} = 0$ if $i \ne j$.*

In other words, the identity I is the matrix with entry 1 at each position on the main diagonal and 0 everywhere else. A simple computation shows that $Ix = x$ for any vector $x \in \mathbb{R}^n$ or $\mathbb{C}^n$, which is the reason I is called the identity matrix.

Definition 1.54   *Let A be an n × n matrix over $\mathbb{R}$ or $\mathbb{C}$. We say A is* invertible *if there exists an n × n matrix, denoted A^{-1}, such that*

$$A^{-1}A = I \quad and \quad AA^{-1} = I.$$

We call A^{-1}, when it exists, the inverse *of A.*

For 2×2 matrices, it is worth remembering that if $ad - bc \ne 0$,

$$\begin{bmatrix} a & b \\ c & d \end{bmatrix}^{-1} = \frac{1}{ad - bc} \begin{bmatrix} d & -b \\ -c & a \end{bmatrix}.$$

Invertible linear transformations correspond to invertible matrices.

Lemma 1.55 *Let U and V be n-dimensional vector spaces over $\mathbb{C}$ (similarly if both are over $\mathbb{R}$). Suppose $T : U \to V$ is a linear transformation. Let R be a basis for U and S a basis for V. Let A_T be the matrix that represents T with respect to R and S, as in Lemma 1.49. Then T is an invertible linear transformation if and only if A_T is an invertible matrix.*

Proof

First suppose T is invertible. Let z be an arbitrary element of $\mathbb{C}^n$, with components $z_1, z_2, \ldots, z_n$. Suppose $R = \{u_1, u_2, \ldots, u_n\}$. Let $u = \sum_{j=1}^n z_j u_j$; in other words, $[u]_R = z$. Let $A_{T^{-1}}$ be the matrix that represents T^{-1} with respect to S and R. Then

$$z = [u]_R = [T^{-1}(T(u))]_R = A_{T^{-1}}[T(u)]_S$$
$$= A_{T^{-1}}A_T[u]_R = A_{T^{-1}}A_T z.$$

Since z is arbitrary, we deduce that $A_{T^{-1}}A_T = I$, by Exercise 1.4.6(i). A symmetric argument shows that also $A_T A_{T^{-1}} = I$. Thus A_T is invertible with inverse $A_{T^{-1}}$.

For the other direction, suppose A_T is an invertible matrix. We claim that $\ker T = \{0\}$. To see this, suppose $u \in U$ and $T(u) = 0$. Then $[T(u)]_S$ is the 0 vector in $\mathbb{C}^n$. So $A_T[u]_R = [T(u)]_S = 0$. Multiplying this equation on the left by $(A_T)^{-1}$ gives $[u]_R = (A_T)^{-1}0 = 0$. But $[u]_R = 0$ implies $u = 0$. This proves the claim that $\ker T = \{0\}$. By Exercise 1.4.7, T is $1 - 1$. Since U and V are n-dimensional, T is onto by Exercise 1.4.8(v). Hence T is invertible. ∎

Another useful notion is the rank of a matrix.

Definition 1.56 *Let A be an $m \times n$ matrix over $\mathbb{C}$ (or $\mathbb{R}$). Let T be the linear transformation associated to A in the standard basis, that is, define $T : \mathbb{C}^n \to \mathbb{C}^m$ (or $T : \mathbb{R}^n \to \mathbb{R}^m$) by $T(x) = Ax$. By Exercise 1.4.9(ii), range T is a subspace of $\mathbb{C}^m$ (or $\mathbb{R}^m$), and hence is a finite dimensional vector space. The* rank *of A is the dimension of range T.*

In the case of a square matrix, this leads to another interesting criterion for invertibility.

Lemma 1.57 *Suppose A is an $n \times n$ matrix over $\mathbb{C}$ (or $\mathbb{R}$). Then A is invertible if and only if rank $A = n$.*

Proof

Let $T : \mathbb{C}^n \to \mathbb{C}^n$ be defined by $T(z) = Az$, as in Definition 1.56.

If A is invertible, then T is invertible by Lemma 1.55. Hence range $T = \mathbb{C}^n$. So range T has dimension n, that is, rank $A = n$.

Conversely, if rank $A = n$, then range T has dimension n. This implies range $T = \mathbb{C}^n$. (Proof: Let $z \in \mathbb{C}^n$ be arbitrary. Then the basis elements $u_1, u_2, \ldots, u_n$ of range T together with z make up $n + 1$ elements in $\mathbb{C}^n$, which must be linearly dependent by Exercise 1.3.12. By Exercise 1.3.9(ii), $z \in \text{span}\{u_1, u_2, \ldots, u_n\} = \text{range } T$.) So T is onto. By Exercise 1.4.8(v), T is $1 - 1$, hence invertible. By Lemma 1.55, A is invertible.

For the proof in the case where A is a matrix over $\mathbb{R}$, replace $\mathbb{C}$ everywhere by $\mathbb{R}$. ∎

Having learned a few prerequisites about matrices, we are ready to return to our study of representing vectors in different bases. If we have two bases for the same finite dimensional vector space, how can we obtain the components of a vector with respect to one of these bases if we know its components with respect to the other?

Lemma 1.58 *Suppose V is an n-dimensional vector space over $\mathbb{R}$ or $\mathbb{C}$. Suppose $R = \{u_1, u_2, \ldots, u_n\}$ and $S = \{v_1, v_2, \ldots, v_n\}$ are two bases for V. Since S is a basis for V, there are unique scalars a_{ij}, $i = 1, 2, \ldots, n$, and $j = 1, 2, \ldots, n$ such that*

$$u_1 = a_{11}v_1 + a_{21}v_2 + \cdots + a_{n1}v_n,$$
$$u_2 = a_{12}v_1 + a_{22}v_2 + \cdots + a_{n2}v_n,$$
$$\vdots$$
$$u_n = a_{1n}v_1 + a_{2n}v_2 + \cdots + a_{nn}v_n.$$

Let A be the $n \times n$ matrix whose k^{th} column consists of the coefficients $\alpha_{1k}, \alpha_{2k}, \ldots, \alpha_{nk}$ in the representation of u_k:

$$A = \begin{bmatrix} a_{11} & a_{12} & \cdots & a_{1n} \\ a_{21} & a_{22} & \cdots & a_{2n} \\ \vdots & \vdots & \ddots & \vdots \\ a_{n1} & a_{n2} & \cdots & a_{nn} \end{bmatrix}.$$

Then

$$[x]_S = A[x]_R \quad \text{for all } x \in V,$$

and A is the unique matrix with this property.

Proof

This follows from Lemma 1.49, by letting T be the identity transformation I defined by $I(x) = x$ for all x. Clearly I is linear. Since $u_j = I(u_j)$ for each j, the matrix A is the one required in Lemma 1.49. Hence

$$[x]_S = [I(x)]_S = A[x]_R.$$

The uniqueness of A also follows from Lemma 1.49. ∎

We call the matrix A obtained in Lemma 1.58 the *R to S change of basis matrix*.

Example 1.59

Define bases R and S for $\mathbb{R}^2$ by

$$R = \left\{ \begin{bmatrix} 1 \\ -2 \end{bmatrix}, \begin{bmatrix} -1 \\ 7 \end{bmatrix} \right\},$$

and

$$S = \left\{ \begin{bmatrix} 2 \\ 1 \end{bmatrix}, \begin{bmatrix} 3 \\ 4 \end{bmatrix} \right\}.$$

Find the R to S change of basis matrix A.

Solution

By Lemma 1.58, we need to find $\{a_{ij}\}$ such that

$$\begin{bmatrix} 1 \\ -2 \end{bmatrix} = a_{11} \begin{bmatrix} 2 \\ 1 \end{bmatrix} + a_{21} \begin{bmatrix} 3 \\ 4 \end{bmatrix},$$

and

$$\begin{bmatrix} -1 \\ 7 \end{bmatrix} = a_{12} \begin{bmatrix} 2 \\ 1 \end{bmatrix} + a_{22} \begin{bmatrix} 3 \\ 4 \end{bmatrix}.$$

Solving these linear systems gives

$$a_{11} = 2, \quad a_{21} = -1,$$
$$a_{12} = -5, \quad \text{and} \quad a_{22} = 3.$$

Hence

$$A = \begin{bmatrix} 2 & -5 \\ -1 & 3 \end{bmatrix}.$$

∎

As one might suspect, the R to S change of basis matrix is always invertible and its inverse is the S to R change of basis matrix (Exercise 1.4.14).

Now we restrict our attention to a linear transformation from a finite dimensional vector space to itself. Fix a basis and consider representing the linear transformation in that basis. The following is a special case of what was considered in Lemma 1.49.

Definition 1.60 *Let V be a finite dimensional vector space over $\mathbb{R}$ or $\mathbb{C}$, with basis R. Let $A_{T,R}$ be the matrix that represents T with respect to R and R (as defined in Lemma 1.49), that is, $A_{T,R}$ is the matrix such that*

$$[T(x)]_R = A_{T,R}[x]_R,$$

for any $x \in V$. We call $A_{T,R}$ the matrix representing T with respect to R.

For the simplest example of this, suppose we start with an $n \times n$ matrix A over $\mathbb{C}$ and define the associated linear transformation $T_A : \mathbb{C}^n \to \mathbb{C}^n$ by $T_A(z) = Az$. Then in the Euclidean basis $E = \{e_1, \dots, e_n\}$ (Definition 1.39),

$$[T_A(z)]_E = T_A(z) = Az = A[z]_E,$$

by equation (1.24). In other words, A represents T_A with respect to E, or

$$A_{T_A,E} = A.$$

If we are working with a linear transformation T from V to V, we can fix a basis for V and represent T with respect to this basis by a matrix. This allows us to use matrix algebra to do computations regarding T. However, there are many choices of bases for V. What effect does the choice of basis have on the matrix that represents the transformation?

Lemma 1.61 *Suppose V is a finite dimensional vector space with bases R and S. Suppose $T : V \to V$ is a linear transformation. Let*

$A_{T,R}$ be the matrix representing T with respect to R and $A_{T,S}$ the matrix representing T with respect to S. Let P be the R to S change of basis matrix. Then

$$A_{T,R} = P^{-1}A_{T,S}P.$$

Proof

Let $x \in V$ be arbitrary. By Exercise 1.4.14, P^{-1} is the S to R change of basis matrix. Hence

$$[T(x)]_R = P^{-1}[T(x)]_S.$$

However, $A_{T,S}$ represents T in the basis S, so

$$[T(x)]_R = P^{-1}A_{T,S}[x]_S.$$

But P is the R to S change of basis matrix, so $[x]_S = P[x]_R$. Substituting this in the last equation gives

$$[T(x)]_R = P^{-1}A_{T,S}P[x]_R.$$

The result now follows from the uniqueness assertion in Lemma 1.49. ∎

This result has a natural interpretation: to obtain the action of T in the basis R, one can first change to the basis S (i.e., multiply by P), then apply T as represented in the basis S (i.e., multiply by $A_{T,S}$), and finally convert back to the basis R (multiply by P^{-1}).

Definition 1.62 *Suppose A and B are $n \times n$ matrices over $\mathbb{R}$ or $\mathbb{C}$. We say A and B are* similar, *or that A is similar to B, if there exists a matrix P such that*

$$B = P^{-1}AP.$$

It is easy to see (Exercise 1.4.16) that similarity is an equivalence relation. In particular, the roles of A and B can be interchanged in Definition 1.62.

Lemma 1.61 tells us that two matrices representing the same linear transformation with respect to different bases must be similar. A major concern of ours is to make the best choice of basis, so that the linear operator with which we are working will have a representing matrix in the chosen basis that is as simple as possible. In section

1.5, we discuss operators whose representing matrix can be made diagonal.

Exercises

1.4.1. Let U be the set of all differentiable functions on the interval $(0, 1) = \{x \in \mathbb{R} : 0 < x < 1\}$, that is,

$$U = \{f : (0, 1) \to \mathbb{R} \text{ such that } f'(x) \text{ exists for all } x \in (0, 1)\}.$$

Let V be the collection of all functions $f : (0, 1) \to \mathbb{R}$. Define addition of functions and multiplication of functions by real numbers in the usual way (as in Example 1.33).

  i. Prove that U and V are vector spaces. (Assume standard facts from calculus.)
  ii. Define $T : U \to V$ by $T(f) = f'$. Prove that T is a linear transformation.

  This example demonstrates that important linear transformations arise in the context of infinite dimensional vector spaces.

1.4.2. Let A be an $m \times n$ matrix over $\mathbb{C}$. Define $T : \mathbb{C}^n \to \mathbb{C}^m$ by $T(z) = Az$ (matrix multiplication). Prove that T is a linear transformation.

1.4.3. Prove Lemma 1.48. Suggestion: Use Exercises 1.4.2 and 1.3.14.

1.4.4. Fix an angle $\theta \in \mathbb{R}$. Let T_θ be the map from $\mathbb{R}^2$ to itself, which acts on a vector by rotating it by an angle θ in the counterclockwise direction (i.e., in polar coordinates, T_θ leaves the magnitude unchanged but increases the angle by θ). Draw pictures indicating why T_θ is a linear transformation. Confirm this by determining the matrix that represents T_θ in the standard basis for $\mathbb{R}^2$.

1.4.5. Define a basis R for $\mathbb{R}^3$ by

$$R = \left\{ \begin{bmatrix} 1 \\ 1 \\ 0 \end{bmatrix}, \begin{bmatrix} 2 \\ 0 \\ 1 \end{bmatrix}, \begin{bmatrix} 3 \\ 2 \\ 0 \end{bmatrix} \right\},$$

and a basis S for $\mathbb{R}^2$ by

$$S = \left\{ \begin{bmatrix} 1 \\ 2 \end{bmatrix}, \begin{bmatrix} 2 \\ 5 \end{bmatrix} \right\}.$$

Define $T : \mathbb{R}^3 \to \mathbb{R}^2$ by

$$T\left(\begin{bmatrix} x \\ y \\ z \end{bmatrix} \right) = \begin{bmatrix} x + y - z \\ 2x - y + 3z \end{bmatrix}.$$

  i. Find the matrix A that represents T with respect to R and S.

  ii. Suppose $u \in \mathbb{R}^3$ and

$$[u]_R = \begin{bmatrix} 1 \\ 0 \\ -1 \end{bmatrix}.$$

    Find $[T(u)]_S$.

  iii. For u as in part ii, find the Euclidean coordinates of u.

  iv. Find the Euclidean coordinates of $T(u)$ two ways: using part iii and the definition of T, and using part ii and the definition of S.

1.4.6. Let A be an $n \times n$ matrix over $\mathbb{R}$ or $\mathbb{C}$.

  i. Suppose that $Ax = x$ for all vectors $x \in \mathbb{R}^n$. Prove that A is the identity matrix I. (Hint: Consider the standard basis vectors $e_1, e_2, \ldots, e_n$ in Definition 1.39.)

  ii. Suppose B is an $n \times n$ matrix over the same field as A and $Bx = Ax$ for all $x \in \mathbb{R}^n$. Prove that $A = B$, that is, $a_{ij} = b_{ij}$ for all i, j.

  iii. Prove the uniqueness statement in Lemma 1.49.

1.4.7. Let U and V be vector spaces and $T : U \to V$ a linear transformation. Prove that T is $1-1$ if and only if $\ker T = \{0\}$.

1.4.8. Let U and V be vector spaces and $T : U \to V$ a linear transformation. Suppose U is n-dimensional, with basis $\{u_1, u_2, \ldots, u_n\}$.

  i. Prove that T is $1-1$ if and only if $\{T(u_1), T(u_2), \ldots, T(u_n)\}$ is a linearly independent set in V.

  ii. Prove that $\text{span}\{T(u_1), T(u_2), \ldots, T(u_n)\} = \text{range } T$.

iii. Prove that T is invertible if and only if $\{T(u_1), T(u_2), \ldots, T(u_n)\}$ is a basis for V.

iv. Suppose that $T : U \to V$ is an invertible linear transformation. Prove that V is also finite dimensional with dimension n.

v. Suppose that V is finite dimensional with dimension n. Prove that T is $1-1$ if and only if T is onto. (Hint: Use Theorem 1.42)

1.4.9. Let U and V be vector spaces and $T : U \to V$ a linear transformation. Recall the definition of a subspace from Exercise 1.3.2.

i. Prove that ker T is a subspace of U.

ii. Prove that range T is a subspace of V.

1.4.10. *(Rank theorem)* Suppose U and V are vector spaces, dim $U = n$, and $T : U \to V$ is a linear transformation. Prove that

$$\dim \ker T + \dim \operatorname{range} T = n.$$

Hint: Suppose $k = \dim \ker T$. Let $\{u_1, u_2, \ldots, u_k\}$ be a basis for ker T. By Exercise 1.3.10, we can find $u_{k+1}, \ldots, u_n$ such that $\{u_1, u_2, \ldots, u_n\}$ is a basis for U. Prove that $T(u_{k+1}), \ldots, T(u_n)$ are linearly independent and apply Exercise 1.4.8(ii).

1.4.11. Let A be an $m \times k$ matrix and B a $k \times n$ matrix (both over the same field, either $\mathbb{R}$ or $\mathbb{C}$). Prove that rank $(AB) \leq$ rank A and rank $(AB) \leq$ rank B.

1.4.12. Let A and B be $n \times n$ matrices over the same field, either $\mathbb{R}$ or $\mathbb{C}$.

i. Prove that AB is invertible if and only if both A and B are invertible, in which case, $(AB)^{-1} = B^{-1}A^{-1}$. (Hint: Use Lemma 1.57 and Exercise 1.4.11 for the "only if" direction.)

ii. Suppose $AB = I$. Prove that A and B are invertible, $B = A^{-1}$, and $A = B^{-1}$.

Hence in order to check that $B = A^{-1}$, it is enough to verify that $AB = I$ or that $BA = I$; it is not necessary to check both.

1.4.13. Define bases R and S for $\mathbb{R}^2$ by

$$R = \left\{ \begin{bmatrix} 1 \\ 1 \end{bmatrix}, \begin{bmatrix} 3 \\ 7 \end{bmatrix} \right\},$$

and

$$S = \left\{ \begin{bmatrix} 1 \\ 3 \end{bmatrix}, \begin{bmatrix} 2 \\ 5 \end{bmatrix} \right\}.$$

    i. Find the R to S change of basis matrix A.

    ii. Suppose $x \in \mathbb{R}^2$ and

$$[x]_R = \begin{bmatrix} 1 \\ 2 \end{bmatrix}.$$

    Find $[x]_S$.

    iii. Find the Euclidean coordinates of x two ways: using $[x]_R$ and the definition of R, and using $[x]_S$ and the definition of S.

    iv. Find the S to R change of basis matrix directly, that is, by the method of Lemma 1.58. Then check that it is the inverse of the matrix found in part i.

1.4.14. Let V be a finite dimensional vector space over $\mathbb{R}$ or $\mathbb{C}$, with bases R and S. Let A be the R to S change of basis matrix. Prove that A is invertible and its inverse is the S to R change of basis matrix. (Hint: Use Exercise 1.4.6 and, to make it easier, Exercise 1.4.12(ii).)

1.4.15. Let V be either $\mathbb{R}^n$ or $\mathbb{C}^n$. Let $E = \{e_1, \ldots, e_n\}$ be the Euclidean basis (Definition 1.39) for V and let $R = \{u_1, \ldots, u_n\}$ be another basis for V. Suppose that for each $j = 1, 2, \ldots, n$

$$u_j = \begin{bmatrix} a_{1j} \\ a_{2j} \\ \cdot \\ \cdot \\ \cdot \\ a_{nj} \end{bmatrix}.$$

Let $A = [a_{ij}]_{1 \le i,j \le n}$.

    i. Prove that A is the R to E change of basis matrix.

ii. Suppose $S = \{v_1, \ldots, v_n\}$ is also a basis for V, with

$$v_j = \begin{bmatrix} b_{1j} \\ b_{2j} \\ \cdot \\ \cdot \\ b_{nj} \end{bmatrix},$$

for each $j = 1, 2, \ldots, n$. Let $B = [b_{ij}]_{1 \leq i,j \leq n}$. Prove that $B^{-1}A$ is the R to S change of basis matrix.

iii. Check the result in part ii by solving Exercise 1.4.13(i) this way. Remark: This is usually the simplest way to do these problems.

1.4.16. For two $n \times n$ matrices A and B, define $A \sim B$ if A and B are similar (Definition 1.62). Prove that $\sim$ is an equivalence relation, that is,

i. $A \sim A$ for all $n \times n$ matrices A.

ii. If $A \sim B$, then $B \sim A$.

iii. If $A \sim B$ and $B \sim C$, then $A \sim C$. (Hint: Use Exercise 1.4.12(i).)

1.5 Diagonalization of Linear Transformations and Matrices

Consider a linear transformation $T : V \to V$, where V is a finite dimensional vector space. V has many bases; choosing a basis amounts to picking a coordinate system. Once we select a basis R for V, we can represent T with respect to R by a matrix, which we call $A_{T,R}$ (Definition 1.60). If calculations involving T are to be done with this matrix, we would like to pick the basis R in such a way that $A_{T,R}$ is as simple as possible. How much choice do we have? By Lemma 1.61, we know that for any other basis S, the matrix $A_{T,S}$ representing T with respect to S must be similar to $A_{T,R}$. This is the only restriction: any matrix similar to $A_{T,R}$ is the matrix representing T with respect to some basis for V. This fact is a corollary to Lemma 1.63, which has independent interest.

Lemma 1.63 *Suppose R is a basis for an n-dimensional vector space V over $\mathbb{R}$ or $\mathbb{C}$, and P is an $n \times n$ invertible matrix (over the same field). Then there exists a basis S for V such that P is the R-to-S change of basis matrix.*

Proof
Suppose $R = \{x_1, x_2, \ldots, x_n\}$ and $[q_{ij}] = Q = P^{-1}$. Define $S = \{v_1, v_2, \ldots, v_n\}$, where

$$v_i = \sum_{k=1}^{n} q_{ki} x_k, \tag{1.28}$$

for $i = 1, 2, \ldots, n$.

For $j = 1, 2, \ldots, n$, relation (1.28) yields

$$\sum_{i=1}^{n} p_{ij} v_i = \sum_{i=1}^{n} p_{ij} \sum_{k=1}^{n} q_{ki} x_k = \sum_{k=1}^{n} \left(\sum_{i=1}^{n} q_{ki} p_{ij} \right) x_k = x_j, \tag{1.29}$$

since $\sum_{i=1}^{n} q_{ki} p_{ij}$ is the $(k, j)^{\text{th}}$ entry of $QP = I$, and hence is 1 when $k = j$ and 0 otherwise.

Let $u \in V$ be arbitrary. Suppose

$$[u]_R = a = \begin{bmatrix} a_1 \\ a_2 \\ \cdot \\ \cdot \\ \cdot \\ a_n \end{bmatrix},$$

or, equivalently, $u = \sum_{j=1}^{n} a_j x_j$. By equation (1.29),

$$u = \sum_{j=1}^{n} a_j \sum_{i=1}^{n} p_{ij} v_i = \sum_{i=1}^{n} \left(\sum_{j=1}^{n} p_{ij} a_j \right) v_i = \sum_{i=1}^{n} (Pa)_i v_i, \tag{1.30}$$

where $(Pa)_i$ is the ith component of Pa. This shows that S spans V, and hence is a basis, by Theorem 1.42. It also shows that

$$[u]_S = Pa = P[u]_R.$$

This implies that P is the R-to-S change of basis matrix. ∎

Corollary 1.64 *Suppose V is a finite dimensional vector space with basis R, and $T : V \to V$ is a linear transformation. Let $A_{T,R}$ be the matrix representing T with respect to R. Suppose B is a matrix that is* *similar to* $A_{T,R}$. *Then there exists a basis S for V such that B represents T with respect to S (i.e., such that* $B = A_{T,S}$).

Proof

By assumption, there exists an invertible matrix P such that $B = P^{-1}A_{T,R}P$. Then P^{-1} is also invertible; applying Lemma 1.63 to P^{-1}, there exists a basis S for V such that P^{-1} is the R-to-S change of basis matrix. By Exercise 1.4.14, P is the S-to-R change of basis matrix. Applying Lemma 1.61 (with R and S interchanged in its statement),

$$A_{T,S} = P^{-1}A_{T,R}P = B.$$ ∎

Hence finding the best basis to represent a linear transformation comes down to finding the simplest matrix similar to a given one. Recall (Exercise 1.4.16) that similarity of matrices is an equivalence relation. So we are looking for the simplest representative of the equivalence class consisting of all matrices similar to a given one.

To start, we consider some features of the linear transformation T that must be shared by all matrices representing T. The eigenvalues are one such feature. We define eigenvalues for both linear transformations and matrices.

Definition 1.65   *Let V be a vector space over a field* $\mathbb{F}$, *and* $T : V \to V$ *a linear transformation. A scalar* $\lambda \in \mathbb{F}$ *is an* eigenvalue *of T if there exists a nonzero* $v \in V$ *such that*

$$T(v) = \lambda v. \tag{1.31}$$

Any vector $v \in V$ *satisfying relation (1.31) (including the 0 vector) is called an* eigenvector *of T corresponding to* λ. *The set of all such, namely*

$$E_\lambda = E_\lambda(T) = \{v \in V : T(v) = \lambda v\},$$

is called the eigenspace *of T corresponding to* λ.

Any vector $v \in V$ *satisfying relation (1.31) for some* $\lambda \in \mathbb{F}$ *is called an* eigenvector *of T.*

An eigenvector is a direction in which T acts simply by scalar multiplication. The definitions in the case of a matrix are similar.

Definition 1.66   *Let* $\mathbb{F}$ *be either* $\mathbb{R}$ *or* $\mathbb{C}$. *Let A be an* $n \times n$ *matrix over* $\mathbb{F}$. *If* $\mathbb{F} = \mathbb{R}$, *let* $V = \mathbb{R}^n$, *and if* $\mathbb{F} = \mathbb{C}$, *let* $V = \mathbb{C}^n$. *A scalar* $\lambda \in \mathbb{F}$

is an eigenvalue of A *if there exists a nonzero* $v \in V$ *such that*

$$Av = \lambda v. \tag{1.32}$$

Any vector $v \in V$ *satisfying equation (1.32) (including the 0 vector) is called an* eigenvector of A *corresponding to* λ. *The set of all such is*

$$E_\lambda = E_\lambda(A) = \{v \in V : Av = \lambda v\},$$

called the eigenspace *of A corresponding to* λ. *Any vector* $v \in V$ *satisfying equation (1.32) for some* $\lambda \in \mathbb{F}$ *is called an* eigenvector of A.

Let $T : V \to V$ be a linear transformation and λ a scalar. By definition, $v \in E_\lambda(T)$ means that $T(v) = \lambda v$, which is equivalent to $(\lambda I - T)(v) = 0$, or $v \in \ker(\lambda I - T)$, where I is the identity operator (defined by $I(v) = v$ for all v). Hence $E_\lambda(T) = \ker(\lambda I - T)$. By Exercise 1.4.9(i), it follows that $E_\lambda(T)$ is a subspace of V. We define the *geometric multiplicity of* λ (with respect to T) to be the dimension of $E_\lambda(T)$. When λ is not an eigenvalue of T, then $E_\lambda(T) = \{0\}$, and the geometric multiplicity of λ is 0.

We make a similar definition for a matrix A. By linearity,

$$E_\lambda(A) = \{v \in V : (\lambda I - A)v = 0\}$$

is a subspace of V, where here I is the identity matrix. The *geometric multiplicity of* λ (with respect to A) is the dimension of $E_\lambda(A)$. If λ is not an eigenvalue of A, the geometric multiplicity of λ is 0.

The next lemma shows that when a matrix corresponds to a linear transformation, the eigenvalues and their geometric multiplicities for the matrix and for the transformation are the same.

Lemma 1.67 *Let V be a finite dimensional vector space with basis R, let* $T : V \to V$ *be a linear transformation, and let A be the matrix representing T with respect to R (i.e.,* $A = A_{T,R}$). *Then A and T have the same eigenvalues, and*

$$v \in E_\lambda(T) \quad \text{if and only if} \quad [v]_R \in E_\lambda(A), \tag{1.33}$$

for each eigenvalue λ. *The geometric multiplicities of* λ *for T and for A are the same:*

$$\dim E_\lambda(T) = \dim E_\lambda(A).$$

Proof

First, suppose λ is an eigenvalue of T. Then there exists a nonzero v such that $T(v) = \lambda v$, that is, $v \in E_\lambda(T)$. Then $[v]_R \neq 0$ and, since A represents T in the basis R,

$$A[v]_R = [T(v)]_R = [\lambda v]_R = \lambda[v]_R,$$

using Exercise 1.3.14(ii). Hence λ is an eigenvalue of A and $[v]_R \in E_\lambda(A)$.

Conversely, suppose λ is an eigenvalue of A and y is a nonzero vector satisfying $Ay = \lambda y$ (so $y \in E_\lambda(A)$). Let the components of y be $y_1, y_2, \ldots, y_n$ and let $R = \{u_1, u_2, \ldots, u_n\}$. Define $v = \sum_{j=1}^n y_j u_j$. Then $y = [v]_R$, so

$$[T(v)]_R = A[v]_R = \lambda[v]_R = [\lambda v]_R,$$

again using Exercise 1.3.14(ii). This implies $T(v) = \lambda v$ (since $T(v)$ and λv have the same components in the basis R), that is, λ is an eigenvalue of T and $v \in E_\lambda(T)$.

The only case left to prove in relation (1.33) is when $v = 0$ (equivalently $[v]_R = 0$), which is trivial because then $v \in E_\lambda(T)$ and $[v]_R \in E_\lambda(A)$.

We now prove the statement about the geometric multiplicities. Note that λ is not an eigenvalue of T if and only if λ is not an eigenvalue of A (by what we just proved), in which case both geometric multiplicities are 0. So the result holds in this case. Now suppose $E_\lambda(T)$ has basis $\{t_1, t_2, \ldots, t_m\}$ and $E_\lambda(A)$ has basis $\{s_1, s_2, \ldots, s_k\}$, which, by what we have recently noted, can be written as $\{[v_1]_R, [v_2]_R, \ldots, [v_k]_R\}$. By relation (1.33), $[t_1]_R, [t_2]_R, \ldots, [t_m]_R$ belong to $E_\lambda(A)$ and, by Exercise 1.3.15, are linearly independent. This implies that $m \leq k$ (recall Exercise 1.3.12). Also by what we just proved, $v_1, v_2, \ldots, v_k$ belong to $E_\lambda(T)$ and are linearly independent (also by Exercise 1.3.15), so $k \leq m$. Therefore $k = m$, as desired. ∎

Lemma 1.67 yields the following result about similar matrices.

Corollary 1.68 *Suppose A and B are similar matrices. Then A and B have the same eigenvalues with the same geometric multiplicities.*

Proof

Let T be the linear transformation (on $\mathbb{R}^n$ or $\mathbb{C}^n$, depending whether the scalar field is $\mathbb{R}$ or $\mathbb{C}$) defined by $T(x) = Ax$. Then A represents

T with respect to the standard basis. Since A and B are similar, there is a basis S such that B represents T with respect to S, by Corollary 1.64. By Lemma 1.67, both A and B have the same eigenvalues, with the same geometric multiplicities, as the linear transformation T, hence the same as each other. ∎

This last proof may seem somewhat roundabout, and one can prove Corollary 1.68 by a straightforward argument involving the definition of similar matrices (Exercise 1.5.1). However this computational proof does not give much insight, and does not suggest how the result might be anticipated. The proof above emphasizes that the eigenvalues and their geometric multiplicities are properties of the transformation itself, independent of how it is realized by a matrix in some choice of basis. Similar matrices are different realizations of the same transformation, so any quantity that depends only on the underlying transformation must be the same for similar matrices. This gives us an abstract way of understanding what features similar matrices must have in common. Any quantity determined by a matrix that must be the same for any two similar matrices is called a *similarity invariant*. Corollary 1.68 states that eigenvalues and their geometric multiplicities are similarity invariants.

An important fact about eigenvectors of a linear transformation is that eigenvectors corresponding to different eigenvalues are linearly independent. A more general statement is as follows.

Lemma 1.69 *Let V be an n-dimensional vector space over $\mathbb{R}$ or $\mathbb{C}$ and $T : V \to V$ a linear transformation with distinct eigenvalues $\lambda_1, \lambda_2, \ldots, \lambda_k$. For $i = 1, 2, \ldots, k$, suppose E_{λ_i} has dimension m_i and a basis $\{v_{i,1}, v_{i,2}, \ldots, v_{i,m_i}\}$. Let*

$$A = \{v_{1,1}, v_{1,2}, \ldots, v_{1,m_1}, v_{2,1}, v_{2,2}, \ldots, v_{2,m_2}, \ldots, v_{k,1}, v_{k,2}, \ldots, v_{k,m_k}\},$$

that is, let A be the union of the bases for the eigenspaces E_{λ_i}, $i = 1, 2, \ldots, n$. Then A is linearly independent.

The sum of the geometric multiplicities of the eigenvalues is at most n, that is,

$$\sum_{i=1}^{k} m_i \leq n. \qquad (1.34)$$

In particular, T cannot have more than n distinct eigenvalues.

Proof

Suppose there is a linear combination of vectors in A that is 0:

$$\sum_{i=1}^{k}\sum_{j=1}^{m_i} a_{ij}v_{i,j} = 0. \tag{1.35}$$

For $i = 1, 2, \ldots, k$, define $u_i = \sum_{j=1}^{m_i} a_{ij}v_{i,j}$. Then equation (1.35) becomes

$$\sum_{i=1}^{k} u_i = 0. \tag{1.36}$$

Note that $u_i \in E_{\lambda_i}$, since u_i is a linear combination of the vectors $v_{i,j}, j = 1, 2, \ldots, m_i$, which all belong to the subspace E_{λ_i}. So $T(u_i) = \lambda_i u_i$.

We claim that every u_i is 0. By reindexing (specifically, by interchanging the indices i and k), it is enough to prove that u_k is 0. Apply the linear transformation $\lambda_1 I - T$ to both sides of equation (1.36). Note that $(\lambda_1 I - T)u_i = (\lambda_1 - \lambda_i)u_i$, which is 0 only for $i = 1$. Since the $i = 1$ term drops out, we obtain

$$\sum_{i=2}^{k} (\lambda_1 - \lambda_i)u_i = 0.$$

Now we apply $\lambda_2 I - T$ to both sides of this equation, which causes the $i = 2$ term to drop out. We continue in this way until only one term remains, which gives

$$(\lambda_1 - \lambda_k)(\lambda_2 - \lambda_k) \cdots (\lambda_{k-1} - \lambda_k)u_k = 0.$$

Since the coefficient is not 0, we must have $u_k = 0$.

We have shown that $u_i = \sum_{j=1}^{m_i} a_{ij}v_{i,j} = 0$ for each i. However, by assumption, each set $\{v_{i,1}, v_{i,2}, \ldots, v_{i,m_i}\}$ is linearly independent. So each a_{ij} must be 0. This implies that A is linearly independent.

The number of vectors $v_{i,j}$ is $\sum_{i=1}^{k} m_i$. Since these vectors are linearly independent and V is n-dimensional, the total number is at most n (Exercise 1.3.12), so relation (1.34) holds. In particular, since each eigenspace is at least one-dimensional, the number of eigenvalues is at most n. ∎

The number $\sum_{i=1}^{k} m_i$ in relation (1.34) is the maximum possible number of linearly independent eigenvectors of T, since no eigenspace E_{λ_i} can contribute more than $m_i = \dim E_{\lambda_i}$ eigenvectors to any linearly independent collection.

Statements corresponding to those in Lemma 1.69 hold for matrices also.

Corollary 1.70   *Let A be an $n \times n$ matrix over $\mathbb{R}$ or $\mathbb{C}$. Suppose $\lambda_1, \lambda_2, \ldots, \lambda_k$ are the distinct eigenvalues of A. For $i = 1, \ldots, k$, suppose E_{λ_i}, the eigenspace of λ_i, has dimension m_i and basis $\{v_{i,1}, v_{i,2}, \ldots, v_{i,m_i}\}$. Then the set $\{v_{i,j}\}_{1 \leq i \leq k, 1 \leq j \leq m_i}$ is linearly independent, $\sum_{i=1}^{k} m_i \leq n$, and k, the number of eigenvalues of A, is at most n.*

Proof
Associate with A the linear transformation T defined by $T(x) = Ax$. Then the eigenvalues and eigenvectors of the transformation T are the same as for the matrix A, so all of the assertions follow from Lemma 1.69. ∎

The easiest linear transformations to work with are those for which the maximal linearly independent set of eigenvectors in Lemma 1.69 is a basis, that is, for which there are enough eigenvectors to span the vector space.

Definition 1.71   *Let V be a finite dimensional vector space and $T : V \to V$ a linear transformation. If V has a basis consisting of eigenvectors of T, we say T is* diagonalizable.

A diagonalizable linear transformation T is simpler than an arbitrary linear transformation because the action of T can be broken up into the eigenvector directions, in which T acts by scalar multiplication.

The corresponding notion for matrices is the following.

Definition 1.72   *An $n \times n$ matrix $D = [d_{ij}]$ is* diagonal *if $d_{ij} = 0$ whenever $i \neq j$, that is, if all the entries of D off the main diagonal are 0.*

An $n \times n$ matrix A is diagonalizable *if A is similar to some diagonal matrix.*

In other words, a matrix A is diagonalizable if there is a diagonal matrix D and an invertible matrix P such that $P^{-1}AP = D$.

The relation between the two notions of diagonalizability is natural.

Lemma 1.73 *Suppose V is a finite dimensional vector space and $T : V \rightarrow V$ is a linear transformation.*

  i. *T is diagonalizable if and only if there exists a basis R for V such that the matrix $A_{T,R}$ representing T with respect to R is diagonal.*

  ii. *Let S be any basis for V. Let $A_{T,S}$ be the matrix representing T with respect to S. Then T is a diagonalizable linear transformation if and only if $A_{T,S}$ is a diagonalizable matrix.*

Proof

  i. First suppose T is diagonalizable. By definition, V has a basis $R = \{v_1, v_2, \ldots, v_n\}$ of eigenvectors of T, say $T(v_i) = \lambda_i v_i$, $i = 1, 2, \ldots, n$. Let $D = [d_{i,j}]_{1 \leq i,j \leq n}$ be the diagonal matrix with diagonal entries $d_{ii} = \lambda_i$. Let $v \in V$ be arbitrary. Since R is a basis for V, there exist scalars $\alpha_1, \ldots, \alpha_n$ such that $v = \sum_{i=1}^{n} \alpha_i v_i$, or

$$[v]_R = \begin{bmatrix} \alpha_1 \\ \vdots \\ \alpha_n \end{bmatrix}.$$

By linearity,

$$T\left(\sum_{i=1}^{n} \alpha_i v_i\right) = \sum_{i=1}^{n} \alpha_i T(v_i) = \sum_{i=1}^{n} \alpha_i \lambda_i v_i,$$

or, equivalently,

$$[T(v)]_R = \begin{bmatrix} \alpha_1 \lambda_1 \\ \cdot \\ \vdots \\ \alpha_n \lambda_n \end{bmatrix} = \begin{bmatrix} \lambda_1 & 0 & \cdot & \cdot & 0 \\ 0 & \lambda_2 & 0 & \cdot & 0 \\ \cdot & & \cdot & & \cdot \\ \cdot & & & \cdot & \cdot \\ 0 & \cdot & \cdot & \cdot & \lambda_n \end{bmatrix} \begin{bmatrix} \alpha_1 \\ \alpha_2 \\ \cdot \\ \cdot \\ \alpha_n \end{bmatrix} = D[v]_R.$$

Hence $A_{T,R} = D$, which is diagonal.

Conversely, suppose that $R = \{v_1, v_2, \ldots, v_n\}$ is a basis for V such that $A_{T,R}$ is a diagonal matrix $D = [d_{i,j}]_{1 \leq i,j \leq n}$. Then $[T(v)]_R = D[v]_R$ for all $v \in V$. Note that $[v_i]_R = e_i$, the ith standard basis vector (Definition 1.39), since the expansion of v_i in terms of R has a coefficient of 1 in front of v_i and 0 elsewhere. By assumption,

$$[T(v_i)]_R = D[v_i]_R = De_i = d_{ii}e_i = d_{ii}[v_i]_R = [d_{ii}v_i]_R,$$

using the diagonality of D and Exercise 1.3.14(ii). This implies that $T(v_i) = d_{ii}v_i$ for each i. Thus each v_i is an eigenvector of T, and V has a basis (namely R) of eigenvectors of T.

ii. First suppose T is diagonalizable. By part i, there is a basis R so that $A_{T,R}$ is diagonal. By Lemma 1.61, $A_{T,S}$ is similar to $A_{T,R}$, so $A_{T,S}$ is diagonalizable.

Conversely, suppose $A_{T,S}$ is diagonalizable, say similar to a diagonal matrix D. By Corollary 1.64, there is a basis R so that D represents T with respect to R, so by part i, T is diagonalizable. ■

It is much easier to do computations with a diagonal matrix than with a general one. For example, multiplying a vector by a general $n \times n$ matrix requires a total of n^2 multiplications. If the matrix is diagonal, however, it requires only n multiplications. More dramatically, computing a large power of a diagonal matrix is easy, whereas for a general matrix it requires a huge number of multiplications to do directly (compare with Example 1.82). So for a diagonalizable linear transformation T, Lemma 1.73, part i, answers our basic question of how to choose a basis to simplify computations with T: we select a basis that diagonalizes T.

It is not easy to determine whether a given linear transformation $T : V \to V$ is diagonalizable. By definition, T is diagonalizable if and only if V has a basis of eigenvectors of T. If V is n-dimensional, this means T is diagonalizable if and only if T has n linearly independent eigenvectors. By Lemma 1.69, the maximum number of linearly independent eigenvectors of T is $\sum_{i=1}^{k} m_i$, the sum of the geometric multiplicities of the eigenvalues. Thus T is diagonalizable if and only if $\sum_{i=1}^{k} m_i = n$.

The same criterion holds for a matrix A. If we consider the linear transformation T defined by $T(x) = Ax$ (so A represents T in the Euclidean basis), then by Lemma 1.67, A and T have the same eigenvalues with the same geometric multiplicities. By Lemma 1.73,

T is diagonalizable if and only if A is diagonalizable. So an $n \times n$ matrix A is diagonalizable if and only if the sum of the geometric multiplicities of the eigenvectors of A is n, that is, if and only if A has n linearly independent eigenvectors.

In the special case in which A or T has n distinct eigenvalues, each eigenspace must have dimension one (at least one, by definition, and at most one by relation (1.34)). So the sum of the geometric multiplicities is n. Hence A or T is automatically diagonalizable.

To see if A or T is diagonalizable when there are less than n distinct eigenvalues, we have to consider the eigenspaces and determine whether the sum of their dimensions is n.

Now we turn to practical matters of computation. Lemma 1.74 tells us how to carry out the diagonalization of a diagonalizable matrix, assuming we know the eigenvalues and eigenvectors. A little thought shows that a diagonal matrix D has eigenvalues equal to its diagonal entries (with eigenvectors equal to the standard basis vectors). But we have already noted that similar matrices have the same eigenvalues (Corollary 1.68). So if A is similar to D, the only possible entries for the diagonal matrix D are the eigenvalues of A. This explains part of the following result and shows why eigenvalues play a central role in diagonalization of matrices.

Lemma 1.74   *Let A be an $n \times n$ diagonalizable matrix.*

i. *Let $v_1, v_2, \ldots, v_n$ be n linearly independent eigenvectors of A (which exist, as noted above). Let $\lambda_1, \lambda_2, \ldots, \lambda_n$ be the corresponding eigenvalues. Let P be the matrix whose j^{th} column is the vector v_j. Let $D = [d_{ij}]$ be the diagonal matrix whose j^{th} diagonal entry d_{jj} is λ_j. Then $P^{-1}AP = D$.*

ii. *Conversely, if $P^{-1}AP = D$, where D is a diagonal matrix, then the columns of P are linearly independent eigenvectors of A, with corresponding eigenvalues equal to the diagonal entries of D.*

Proof

i. Note that the desired equation $P^{-1}AP = D$ is equivalent to $AP = PD$. We calculate

$$AP = A[v_1 \; v_2 \cdots v_n] = [Av_1 \; Av_2 \cdots Av_n] = [\lambda_1 v_1 \; \lambda_2 v_2 \cdots \lambda_n v_n].$$

On the other hand,

$$PD = [v_1 \ v_2 \cdots v_n] \begin{bmatrix} \lambda_1 & 0 & \cdot & \cdot & 0 \\ 0 & \lambda_2 & 0 & \cdot & 0 \\ & & \cdot & & \\ & & & \cdot & \\ 0 & \cdot & \cdot & 0 & \lambda_n \end{bmatrix} = [\lambda_1 v_1 \ \lambda_2 v_2 \cdots \lambda_n v_n].$$

This proves part i. Part ii is proved by reversing these steps; we leave this as Exercise 1.5.3. ∎

The suspicious reader will note that we have not yet proved that a matrix necessarily has any eigenvalues. In fact, a matrix over $\mathbb{R}$ may not have any real eigenvalues (Exercise 1.5.4(i)), although a matrix over $\mathbb{C}$ must have complex eigenvalues, as we will see. Also, we have not yet learned how to find the eigenvalues and eigenvectors of a matrix when they exist. For this, we assume some basic facts about determinants.

Definition 1.75 *The* determinant *of an $n \times n$ matrix A, denoted det A or det(A), is defined as follows. If A is a 1×1 matrix, say $A = [a]$, let det $A = a$. For a 2×2 matrix, define*

$$\det \begin{bmatrix} a & b \\ c & d \end{bmatrix} = ad - bc.$$

Proceeding inductively, suppose the determinant of an $(n-1) \times (n-1)$ matrix has been defined. Now let $A = [a_{ij}]$ be an $n \times n$ matrix. For $1 \le i, j \le n$, the $(i,j)^{\text{th}}$ minor M_{ij} of A is the $(n-1) \times (n-1)$ matrix obtained from A by deleting the ith row and j^{th} column of A. Define

$$\det A = \sum_{j=1}^{n} (-1)^{1+j} a_{1j} \det M_{1j}. \tag{1.37}$$

As an example, for a 3×3 matrix,

$$\det \begin{bmatrix} a_{11} & a_{12} & a_{13} \\ a_{21} & a_{22} & a_{23} \\ a_{31} & a_{32} & a_{33} \end{bmatrix} = a_{11} \det \begin{bmatrix} a_{22} & a_{23} \\ a_{32} & a_{33} \end{bmatrix}$$

$$-a_{12} \det \begin{bmatrix} a_{21} & a_{23} \\ a_{31} & a_{33} \end{bmatrix} + a_{13} \det \begin{bmatrix} a_{21} & a_{22} \\ a_{31} & a_{32} \end{bmatrix}$$

$$= a_{11} (a_{22}a_{33} - a_{23}a_{32})$$

$$-a_{12} (a_{21}a_{33} - a_{23}a_{31}) + a_{13} (a_{21}a_{32} - a_{22}a_{31}).$$

Actually we can expand $\det A$ along any row or column by an expression similar to equation (1.37), but proving this takes some work and this definition will be sufficient for our purposes. In practice, applying Definition 1.75 is a very slow way to compute the determinant of a large matrix. Instead we apply elementary row operations to the matrix and keep track of their effects (which are simple) on the determinant, until we reach an upper triangular matrix, whose determinant is the product of its diagonal elements. We do not go into this further here, as our purpose is only to compute a few simple examples by hand to illustrate the ideas we have been discussing.

We assume the following two facts about determinants, whose proof can be found in most linear algebra texts.

Theorem 1.76 *Let A be an $n \times n$ matrix. Then*
   i. A is invertible if and only if $\det A \neq 0$.
  ii. If B is another $n \times n$ matrix, then $\det(AB) = \det A \det B$.

Assuming Theorem 1.76, we can compute the eigenvalues and eigenvectors of a matrix. First we need a definition.

Definition 1.77 *Let A be an $n \times n$ matrix. The* characteristic polynomial *of A is*

$$\det(\lambda I - A),$$

regarded as a polynomial in the variable λ.

A few examples should convince you that the characteristic polynomial of an $n \times n$ matrix is a polynomial of degree n whose highest order term is λ^n. The characteristic polynomial plays a key role throughout linear algebra, but for us its main use comes from the following observation.

Lemma 1.78 *Let A be an $n \times n$ matrix. Then the eigenvalues of A are the roots of the characteristic polynomial of A.*

Proof

By definition, λ is an eigenvalue of A if and only if there is a nonzero vector v such that $(\lambda I - A)v = 0$. Equivalently, the linear transformation T defined by $T(x) = (\lambda I - A)x$ is not $1 - 1$ (Exercise 1.4.7). By Exercise 1.4.8(v), this happens if and only if

T is not invertible, hence (by Lemma 1.55) if and only if $\lambda I - A$ is not invertible. By Theorem 1.76, part i, this is equivalent to $\det(\lambda I - A) = 0$, which means that λ is a root of the characteristic polynomial of A. ∎

The characteristic polynomial of a matrix A may not have any real-valued roots (Exercise 1.5.4(i)). However, by the fundamental theorem of algebra (Theorem 1.28), any nonconstant polynomial has a complex root. This is one of the reasons we prefer to work with the field $\mathbb{C}$. So if we regard our matrices as being complex (which we can, even if all the entries are real), then every matrix has an eigenvalue. In fact, by Corollary 1.29, every polynomial splits completely into a product of linear factors over $\mathbb{C}$. Hence if A is an $n \times n$ matrix over $\mathbb{C}$, we can write

$$\det(\lambda I - A) = (\lambda - \lambda_1)(\lambda - \lambda_2) \cdots (\lambda - \lambda_n), \qquad (1.38)$$

where some λ_i may be repeated. (There is no constant in front because the coefficient of λ^n is 1.) These $\lambda_1, \lambda_2, \ldots, \lambda_n$ are the only roots of the characteristic polynomial, that is, the eigenvalues of A. To deal with the possibility that some λ_i are repeated in equation (1.38), we make the following definition.

Definition 1.79 *Let A be an $n \times n$ matrix over $\mathbb{C}$. Let $\lambda_1, \lambda_2, \ldots, \lambda_k$ be the distinct eigenvalues of A. Then the characteristic polynomial of A can be written*

$$\det(\lambda I - A) = (\lambda - \lambda_1)^{m_1}(\lambda - \lambda_2)^{m_2} \cdots (\lambda - \lambda_k)^{m_k}, \qquad (1.39)$$

where each m_j is a positive integer, called the algebraic multiplicity *of the eigenvalue λ_j.*

If A is an $n \times n$ matrix, then the characteristic polynomial of A has degree n, so the sum of the algebraic multiplicities of the eigenvalues of A is n. By Corollary 1.70, the sum of the geometric multiplicities is at most n. In fact, more is true: for each eigenvalue, its geometric multiplicity is less than or equal to its algebraic multiplicity. We do not need this fact, so we will not prove it. However, we do note that it gives another criterion for diagonalizability: a matrix is diagonalizable if and only if the geometric multiplicity of each eigenvalue is equal to its algebraic multiplicity (since that is the only way the sum of the geometric multiplicities can be n). Exercise 1.5.8 shows that this may fail.

Notice that the fact that a matrix must have a (complex) eigenvalue tells us that a linear transformation $T : V \to V$ on a finite dimensional vector space V over $\mathbb{C}$ must have an eigenvalue, since its representation by a matrix with respect to any basis must have one.

Next we see that the characteristic polynomial and the algebraic multiplicities of the eigenvalues of a matrix are similarity invariants.

Lemma 1.80 *Suppose A and B are similar matrices. Then*

$$\det(\lambda I - A) = \det(\lambda I - B).$$

In particular, the algebraic multiplicities of the eigenvalues of A and B are the same.

Proof
Let P be such that $B = P^{-1}AP$. Then

$$\lambda I - B = \lambda I - P^{-1}AP = P^{-1}\lambda IP - P^{-1}AP = P^{-1}(\lambda I - A)P,$$

since P commutes with I and with multiplication by the scalar λ, and hence cancels with P^{-1}. By Theorem 1.76, part ii,

$$\det P^{-1} \det P = \det(P^{-1}P) = \det I = 1.$$

So

$$
\begin{aligned}
\det(\lambda I - B) &= \det(P^{-1}(\lambda I - A)P) \\
&= \det(P^{-1}) \det(\lambda I - A) \det P = \det(\lambda I - A).
\end{aligned}
$$

The remark about algebraic multiplicities is now obvious because they are determined by the characteristic polynomial. ∎

We are now ready for an example and some applications.

Example 1.81
Let

$$A = \begin{bmatrix} -2 & 0 & 6 \\ -1 & 1 & 2 \\ -2 & 0 & 5 \end{bmatrix}.$$

Determine whether A is diagonalizable, and if it is, find an invertible matrix P and a diagonal matrix D such that $D = P^{-1}AP$.

Solution

We begin by computing the characteristic polynomial of A:

$$\det(\lambda I - A) = \det \begin{bmatrix} \lambda + 2 & 0 & -6 \\ 1 & \lambda - 1 & -2 \\ 2 & 0 & \lambda - 5 \end{bmatrix}$$

$$= (\lambda + 2)(\lambda - 1)(\lambda - 5) + (-6)[0 - 2(\lambda - 1)]$$

$$= (\lambda - 1)(\lambda^2 - 3\lambda + 2) = (\lambda - 1)^2(\lambda - 2).$$

So 1 is an eigenvalue of algebraic multiplicity 2 and 2 is an eigenvalue of algebraic multiplicity 1. We start by computing the eigenspace E_1 of 1, which consists of all vectors v such that $(1I - A)v = 0$, that is, all solutions of

$$\begin{bmatrix} 3 & 0 & -6 \\ 1 & 0 & -2 \\ 2 & 0 & -4 \end{bmatrix} \begin{bmatrix} a \\ b \\ c \end{bmatrix} = \begin{bmatrix} 0 \\ 0 \\ 0 \end{bmatrix}.$$

These equations all give the same constraint, namely $a = 2c$. So we can arbitrarily pick b and c, if we then set $a = 2c$. Thus E_1 consists of all vectors of the form

$$\begin{bmatrix} 2c \\ b \\ c \end{bmatrix} = b \begin{bmatrix} 0 \\ 1 \\ 0 \end{bmatrix} + c \begin{bmatrix} 2 \\ 0 \\ 1 \end{bmatrix},$$

for some scalars b and c. This shows that E_1 is two dimensional, spanned by the linearly independent vectors

$$\begin{bmatrix} 0 \\ 1 \\ 0 \end{bmatrix} \quad \text{and} \quad \begin{bmatrix} 2 \\ 0 \\ 1 \end{bmatrix}.$$

We see now that A is diagonalizable. Similar analysis shows that the eigenspace corresponding to the eigenvector 2 is spanned by the vector

$$\begin{bmatrix} 3 \\ 1 \\ 2 \end{bmatrix}.$$

Hence by Lemma 1.74, we take

$$P = \begin{bmatrix} 0 & 2 & 3 \\ 1 & 0 & 1 \\ 0 & 1 & 2 \end{bmatrix} \quad \text{and} \quad D = \begin{bmatrix} 1 & 0 & 0 \\ 0 & 1 & 0 \\ 0 & 0 & 2 \end{bmatrix}.$$

A computation shows that

$$P^{-1} = \begin{bmatrix} 1 & 1 & -2 \\ 2 & 0 & -3 \\ -1 & 0 & 2 \end{bmatrix}.$$

We can check this: direct computation shows that $P^{-1}AP = D$. ∎

It is easier to compute with a diagonalizable matrix A. For example, consider computing large powers of A. If $P^{-1}AP = D$, then $A = PDP^{-1}$. For any positive integer k,

$$A^k = PDP^{-1}PDP^{-1}PDP^{-1}\cdots PDP^{-1} = PD^kP^{-1},$$

since the middle terms $P^{-1}P$ cancel out. But D^k is the diagonal matrix whose diagonal entries are the kth power of the corresponding diagonal entries of D. In some cases we can find roots of matrices also.

Example 1.82
Let A be the matrix in Example 1.81.

    i. Compute A^{20}.

   ii. Find a matrix B such that $B^2 = A$.

Solution
i. As noted above, $A^{20} = PD^{20}P^{-1}$

$$= \begin{bmatrix} 0 & 2 & 3 \\ 1 & 0 & 1 \\ 0 & 1 & 2 \end{bmatrix} \begin{bmatrix} 1 & 0 & 0 \\ 0 & 1 & 0 \\ 0 & 0 & 2^{20} \end{bmatrix} \begin{bmatrix} 1 & 1 & -2 \\ 2 & 0 & -3 \\ -1 & 0 & 2 \end{bmatrix}$$

$$= \begin{bmatrix} 0 & 2 & 3 \\ 1 & 0 & 1 \\ 0 & 1 & 2 \end{bmatrix} \begin{bmatrix} 1 & 1 & -2 \\ 2 & 0 & -3 \\ -2^{20} & 0 & 2^{21} \end{bmatrix}$$

$$= \begin{bmatrix} 4 - 3 \cdot 2^{20} & 0 & -6 + 3 \cdot 2^{21} \\ 1 - 2^{20} & 1 & -2 + 2^{21} \\ 2 - 2^{21} & 0 & -3 + 2^{22} \end{bmatrix}.$$

ii. Since the diagonal entries of D are nonnegative, it is easy to find a matrix C such that $C^2 = D$, namely

$$C = \begin{bmatrix} 1 & 0 & 0 \\ 0 & 1 & 0 \\ 0 & 0 & \sqrt{2} \end{bmatrix}.$$

Then $B = PCP^{-1}$ satisfies

$$B^2 = PCP^{-1}PCP^{-1} = PC^2P^{-1} = PDP^{-1} = A.$$

Computation gives

$$B = \begin{bmatrix} 4 - 3\sqrt{2} & 0 & -6 + 6\sqrt{2} \\ 1 - \sqrt{2} & 1 & -2 + 2\sqrt{2} \\ 2 - 2\sqrt{2} & 0 & -3 + 4\sqrt{2} \end{bmatrix},$$

which would be difficult to guess. ∎

The next two examples give a stronger sense of the power of diagonalization. The first involves solving a *difference equation*.

Example 1.83
Define a sequence $\{x_n\}_{n=0}^{\infty}$ inductively by setting $x_0 = 1, x_1 = 1$, and, for $n \geq 0$,

$$x_{n+2} = 2x_{n+1} + 3x_n. \tag{1.40}$$

Find a closed form expression for x_n.

Solution
For each $n \geq 0$, define the vector

$$u_n = \begin{bmatrix} x_{n+1} \\ x_n \end{bmatrix}.$$

From equation (1.40), we obtain

$$u_{n+1} = \begin{bmatrix} x_{n+2} \\ x_{n+1} \end{bmatrix} = \begin{bmatrix} 2x_{n+1} + 3x_n \\ x_{n+1} \end{bmatrix} = \begin{bmatrix} 2 & 3 \\ 1 & 0 \end{bmatrix} \begin{bmatrix} x_{n+1} \\ x_n \end{bmatrix} = Au_n,$$

where

$$A = \begin{bmatrix} 2 & 3 \\ 1 & 0 \end{bmatrix}.$$

Hence

$$u_n = Au_{n-1} = A^2 u_{n-2} = \cdots = A^n u_0.$$

Since the conditions $x_0 = x_1 = 1$ give us u_0, we need to compute only A^n. A computation shows that the eigenvalues of A are $\lambda = 3$ with eigenvector $[3\ \ 1]$ and $\lambda = -1$ with eigenvector $[-1\ \ 1]$. Hence

$$A = PDP^{-1} = \begin{bmatrix} 3 & -1 \\ 1 & 1 \end{bmatrix} \begin{bmatrix} 3 & 0 \\ 0 & -1 \end{bmatrix} \frac{1}{4} \begin{bmatrix} 1 & 1 \\ -1 & 3 \end{bmatrix}.$$

Consequently,

$$
\begin{aligned}
A^n &= PD^n P^{-1} \\
&= \begin{bmatrix} 3 & -1 \\ 1 & 1 \end{bmatrix} \begin{bmatrix} 3^n & 0 \\ 0 & (-1)^n \end{bmatrix} \frac{1}{4} \begin{bmatrix} 1 & 1 \\ -1 & 3 \end{bmatrix} \\
&= \frac{1}{4} \begin{bmatrix} 3^{n+1} - (-1)^{n+1} & 3^{n+1} + 3(-1)^{n+1} \\ 3^n - (-1)^n & 3^n + 3(-1)^n \end{bmatrix}.
\end{aligned}
$$

Therefore

$$
\begin{aligned}
\begin{bmatrix} x_{n+1} \\ x_n \end{bmatrix} &= u_n = A^n u_0 \\
&= \frac{1}{4} \begin{bmatrix} 3^{n+1} - (-1)^{n+1} & 3^{n+1} + 3(-1)^{n+1} \\ 3^n - (-1)^n & 3^n + 3(-1)^n \end{bmatrix} \begin{bmatrix} 1 \\ 1 \end{bmatrix} \\
&= \frac{1}{4} \begin{bmatrix} 2 \cdot 3^{n+1} + 2(-1)^{n+1} \\ 2 \cdot 3^n + 2(-1)^n \end{bmatrix}.
\end{aligned}
$$

In particular

$$x_n = \frac{(3^n + (-1)^n)}{2}.$$

(Note that we did not need to compute the first entry of $A^n u_0$.) One can check directly that x_n satisfies the initial conditions $x_0 = x_1 = 1$ and that the recurrence relation (1.40) holds. ∎

One can use this method to determine a formula for the nth Fibonacci number (Exercise 1.5.6).

The next example shows how diagonalization can be used in solving systems of differential equations.

Example 1.84
Find the general solution to the system of differential equations

$$y_1' = -2y_1 + 6y_3$$

$$y_2' = -y_1 + y_2 + 2y_3 \qquad (1.41)$$

$$y_3' = -2y_1 + 5y_3.$$

Solution
Let y denote the vector with components y_1, y_2, y_3. Let A be the matrix in Examples 1.81 and 1.82. For P and D obtained there, we can write the given equations as

$$y' = Ay = PDP^{-1}y.$$

This is equivalent to

$$P^{-1}y' = DP^{-1}y.$$

Let $z = P^{-1}y$, and let the components of z be z_1, z_2, z_3. Then (using the linearity of the derivative) our equation just becomes $z' = Dz$, or

$$z_1' = z_1, \quad z_2' = z_2, \quad \text{and,} \quad z_3' = 2z_3.$$

By basic calculus we know that the general solutions to this are $z_1(t) = C_1e^t$, $z_2(t) = C_2e^t$, and $z_3(t) = C_3e^{2t}$ (the general solution to $f' = kf$ is $f(t) = Ce^{kt}$, where C is an undetermined constant). But $y = Pz$. Thus the solution is

$$
\begin{bmatrix} y_1(t) \\ y_2(t) \\ y_3(t) \end{bmatrix} = \begin{bmatrix} 0 & 2 & 3 \\ 1 & 0 & 1 \\ 0 & 1 & 2 \end{bmatrix} \begin{bmatrix} C_1e^t \\ C_2e^t \\ C_3e^{2t} \end{bmatrix} = \begin{bmatrix} 2C_2e^t + 3C_3e^{2t} \\ C_1e^t + C_3e^{2t} \\ C_2e^t + 2C_3e^{2t} \end{bmatrix}.
$$

One can check that $y_1(t), y_2(t)$, and $y_3(t)$ satisfy system (1.41). ∎

In this example, the diagonalizability of A allows us to change basis (to the z-coordinates) in such a way that the equations decouple into one-dimensional equations. This is the basic idea behind diagonalization: the problem is broken up into simpler, independent problems.

Not every matrix or linear transformation is diagonalizable (Exercise 1.5.8). One might ask how close one can get in general; that is, for an arbitrary matrix A, if we look among all similar matrices, how close to diagonal is the "best" one? One answer is that there is always a matrix that has the eigenvalues of A on the main diagonal (repeated in blocks according to algebraic multiplicity), either 0 or 1 at every entry on the superdiagonal (the shorter diagonal just above the main diagonal), depending on the geometric multiplicity of the eigenvalues, and 0 everywhere else in the matrix. This matrix is called the *Jordan canonical form* of A; it is a complete similarity invariant of A in the sense that each matrix is similar to only one matrix in Jordan canonical form (up to permutation of the eigenvalues), and two matrices are similar if they have the same Jordan form. This is something of the theoretical culmination of elementary linear algebra, but we do not need such a powerful and general result here. We are mainly concerned with the simpler case of diagonalizable matrices.

Exercises

1.5.1. Suppose A and B are $n \times n$ matrices over a field $\mathbb{F}$, where either $\mathbb{F} = \mathbb{R}$ or $\mathbb{F} = \mathbb{C}$. Suppose $A \sim B$, say $B = P^{-1}AP$. Let λ be a scalar, and v a vector (in $\mathbb{R}^n$ if $\mathbb{F} = \mathbb{R}$, in $\mathbb{C}^n$ if $\mathbb{F} = \mathbb{C}$).

   i. Prove that $Av = \lambda v$ if and only if $B(P^{-1}v) = \lambda P^{-1}v$.

   ii. Give a direct proof of Corollary 1.68, that is, a proof without using Lemma 1.67 or Corollary 1.64. Hint: Similarly to Exercise 1.4.8(i), prove that multiplication by an invertible matrix preserves linear independence.

1.5.2. Let A be an $n \times n$ matrix over $\mathbb{R}$ or $\mathbb{C}$. Prove that A is invertible if and only if the columns of A, regarded as vectors, are linearly independent. Hint: Let T be the linear transformation defined by $T(x) = Ax$. Let v_j be the jth column of A, regarded as a vector. Prove that span$\{v_1, \ldots, v_n\}$ = range T (e.g., by applying Exercise 1.4.8(ii) with $u_i = e_i$), and recall Lemma 1.57 and Theorem 1.42.

1.5.3. Prove Lemma 1.74, part ii: If P is an invertible matrix with columns $v_1, v_2, \ldots, v_n$, $D = [d_{ij}]$ is a diagonal matrix, and $P^{-1}AP = D$, then the columns of P are linearly independent (this follows just from the invertibility of P and Exercise 1.5.2) and $Av_i = d_{ii}v_i$ for each $i = 1, 2, \ldots, n$ (i.e., v_i is an eigenvector of A with eigenvalue equal to the ith diagonal entry of D).

1.5.4. Let

$$A = \begin{bmatrix} 0 & -1 \\ 1 & 0 \end{bmatrix}$$

i. Regarding A as a matrix over $\mathbb{R}$, show that A has no real eigenvalues.

ii. Regarding A as a matrix over $\mathbb{C}$, find the complex eigenvalues of A and a corresponding eigenvector for each eigenvalue.

iii. Find an invertible matrix P over $\mathbb{C}$ and a diagonal matrix D over $\mathbb{C}$ such that $P^{-1}AP = D$.

iv. Use part iii to calculate A^{99}.

v. Check that $A^2 = -I$. Use this to compute A^{99}, and compare with the answer in part iv.

1.5.5. Define $x_0 = 1$ and $x_1 = -1$. For $n \geq 0$, inductively define $x_{n+2} = x_{n+1} + 6x_n$. Find a formula for x_n.

1.5.6. The *Fibonacci sequence* is defined inductively by $x_0 = 0$, $x_1 = 1$, and for $n \geq 0$, $x_{n+2} = x_{n+1} + x_n$. Find a formula for the nth *Fibonacci number* x_n. (Answer: $x_n = 5^{-1/2}2^{-n}((1 + \sqrt{5})^n - (1 - \sqrt{5})^n)$.)

1.5.7. Let A be an $n \times n$ matrix over $\mathbb{C}$.

i. Prove that $\det(-A) = (-1)^n \det A$. Hint: Use induction.

ii. Suppose that $\lambda_1, \ldots, \lambda_k$ are the eigenvalues of A and the algebraic multiplicity of λ_i is m_i, for each i. Prove that the determinant of A is the product of its eigenvalues, counted according to multiplicity, that is,

$$\det(A) = \lambda_1{}^{m_1} \lambda_2{}^{m_2} \cdots \lambda_k{}^{m_k}.$$

Many books use this as a more elegant way to define the determinant, but this approach requires first proving the existence of eigenvalues. Hint: Make a good choice of λ in equation (1.39).

Remark: This implies that the determinant of a matrix is a similarity invariant (since the eigenvalues and their algebraic multiplicities are). This can be seen more directly using Theorem 1.76, part ii.

1.5.8. Let

$$A = \begin{bmatrix} 3 & 1 \\ 0 & 3 \end{bmatrix}$$

Show that 3 is a eigenvalue of A which has algebraic multiplicity 2 but geometric multiplicity 1. Conclude that A does not have a basis of eigenvectors and hence is not diagonalizable.

1.5.9. Let

$$A = \begin{bmatrix} 2 & 0 & 0 \\ 3 & 2 & 1 \\ 3 & 0 & 3 \end{bmatrix}.$$

   i. Find a diagonal matrix D and an invertible matrix P such that $P^{-1}AP = D$.

   ii. Find a matrix B such that $B^2 = A$.

1.5.10. Let A be an $n \times n$ matrix over $\mathbb{C}$. If A is diagonalizable and k is a positive integer, prove that A^k is diagonalizable. How are the eigenvalues and eigenvectors of A^k related to those of A?

1.5.11. Find the general solution to the system of differential equations

$$y_1' = 8y_1 - 15y_2$$
$$y_2' = 2y_1 - 3y_2.$$

1.5.12. We say matrices A and B are *simultaneously diagonalizable* if they are diagonalized by the same matrix, that is, if there exists an invertible matrix P such that $P^{-1}AP = D_1$ and $P^{-1}BP = D_2$, for some diagonal matrices D_1 and D_2. If A and B are simultaneously diagonalizable, prove that they commute: $AB = BA$. Hint: Note that any two diagonal matrices commute.

1.5.13. Under the assumption that the $n \times n$ matrix A has n distinct eigenvalues, prove the converse to Exercise 1.5.12: If B is an $n \times n$ matrix that commutes with A, then A and B are simultaneously diagonalizable (see Exercise 1.5.12 for the definition). Hint: If v_j is an eigenvector of A with eigenvalue λ_j, use the commutativity of A and B to show that Bv_j is an eigenvector of A with eigenvalue λ_j, hence is a multiple of v_j. Remark: This result is true without the assumption that A has n distinct eigenvalues: If A and B are diagonalizable and commute, then A and B are simultaneously diagonalizable. This result is important in quantum mechanics.

1.6 Inner Products, Orthonormal Bases, and Unitary Matrices

The notions we have considered so far, such as linear independence and bases, make sense for any vector space. However, our main examples $\mathbb{R}^n$ and $\mathbb{C}^n$, as well as certain infinite dimensional vector spaces, possess the additional structure of having an "inner product." An inner product is a generalization of the dot product for vectors in $\mathbb{R}^n$. It gives a generalized notion of perpendicularity, called orthogonality. This leads to orthonormal bases and unitary matrices, which are particularly simple to use.

For x and y in $\mathbb{R}^n$, with components $x_1, x_2, \ldots, x_n$ and $y_1, y_2, \ldots, y_n$, respectively, the *dot product* of x and y is the real number

$$x \cdot y = \sum_{j=1}^{n} x_j y_j.$$

The analog in the complex case is the following.

Definition 1.85   *Suppose $z, w \in \mathbb{C}^n$, say*

$$z = \begin{bmatrix} z_1 \\ z_2 \\ \cdot \\ \cdot \\ \cdot \\ z_n \end{bmatrix} \quad and \quad w = \begin{bmatrix} w_1 \\ w_2 \\ \cdot \\ \cdot \\ \cdot \\ w_n \end{bmatrix}.$$

The (complex) dot product of z and w is

$$z \cdot w = \sum_{j=1}^{n} z_j \overline{w_j},$$

where $\overline{w_j}$ is the complex conjugate of w_j.

We will see the reason for the conjugate in Definition 1.85 soon. This complex dot product is a special case of a (complex) inner product.

Definition 1.86   *Let V be a vector space over $\mathbb{C}$. A (complex) inner product is a map $\langle \cdot, \cdot \rangle : V \times V \to \mathbb{C}$ with the following properties:*

I1. *(Additivity) $\langle u + v, w \rangle = \langle u, w \rangle + \langle v, w \rangle$ for all $u, v, w \in V$.*

I2. *(Scalar homogeneity) $\langle \alpha u, v \rangle = \alpha \langle u, v \rangle$ for all $\alpha \in \mathbb{C}$ and all $u, v \in V$.*

I3. *(Conjugate symmetry) $\langle u, v \rangle = \overline{\langle v, u \rangle}$ for all $u, v \in V$.*

I4. *(Positive definiteness) $\langle u, u \rangle \geq 0$ for all $u \in V$, and $\langle u, u \rangle = 0$ if and only if $u = 0$.*

A vector space V with a complex inner product is called a (complex) inner product space.

Additivity (I1) in the first variable together with conjugate symmetry (I3) imply (Exercise 1.6.1) additivity in the second variable:

$$\langle u, v + w \rangle = \langle u, v \rangle + \langle u, w \rangle \text{ for all } u, v, w \in V. \tag{1.42}$$

However, scalar homogeneity (I2) in the first variable and conjugate symmetry (I3) yield (Exercise 1.6.1) conjugate linearity in the second variable:

$$\langle u, \alpha v \rangle = \overline{\alpha} \langle u, v \rangle, \text{ for all } \alpha \in \mathbb{C} \text{ and all } u, v \in V. \tag{1.43}$$

Example 1.87

For $z, w \in \mathbb{C}^n$, define $\langle z, w \rangle = z \cdot w$. Then $\langle \cdot, \cdot \rangle$ is a complex inner product on $\mathbb{C}^n$ (Exercise 1.6.2(i)). In checking this fact, one can see that the conjugate in the definition of the dot product for vectors in $\mathbb{C}^n$ is needed to obtain the positive definiteness property I4. This explains the need for the conjugate symmetry I3.

For a vector space V over $\mathbb{R}$, we define a real inner product similarly, except that $\langle u, v \rangle$ is always a real number, and we consider only $\alpha \in \mathbb{R}$ in I2. In this case, I3 becomes just $\langle u, v \rangle = \langle v, u \rangle$. As an example, define $\langle x, y \rangle = x \cdot y$, for $x, y \in \mathbb{R}^n$. Then (Exercise 1.6.2(ii)) $\langle \cdot, \cdot \rangle$ is a real inner product on $\mathbb{R}^n$.

In this text, we are primarily concerned with complex inner product spaces. Nearly all of the results we discuss also hold for real inner product spaces, when formulated appropriately. To avoid notational confusion, we discuss only the complex case because that is what we need later.

Example 1.88

Let $\ell^2(\mathbb{N})$ be the set of all square-summable complex sequences:

$$\ell^2(\mathbb{N}) = \left\{ \{z_j\}_{j=1}^{\infty} : z_j \in \mathbb{C} \text{ for all } j, \text{ and } \sum_{j=1}^{\infty} |z_j|^2 < \infty. \right\}$$

With the obvious componentwise addition and scalar multiplication, $\ell^2(\mathbb{N})$ is a vector space over $\mathbb{C}$ (Exercise 1.6.3(ii)). For $z = \{z_j\}_{j=1}^{\infty} \in \ell^2(\mathbb{N})$ and $w = \{w_j\}_{j=1}^{\infty} \in \ell^2(\mathbb{N})$, let

$$\langle z, w \rangle = \sum_{j=1}^{\infty} z_j \overline{w_j}.$$

(By Exercise 1.6.3(iii), this series converges absolutely.) Then $\langle \cdot, \cdot \rangle$ is a complex inner product on $\ell^2(\mathbb{N})$ (Exercise 1.6.3(iv)).

Example 1.89

Let $C([0, 1])$ be the set of continuous, complex-valued functions on the closed interval $[0, 1]$:

$$C([0, 1]) = \{f : [0, 1] \to \mathbb{C} \text{ such that } f \text{ is continuous on } [0, 1]\}.$$

(A complex valued function f on $[0, 1]$ is continuous if its real part $u(x) = \operatorname{Re} f(x)$ and its imaginary part $v(x) = \operatorname{Im} f(x)$ are continuous.)

With pointwise addition and scalar multiplication (as in Example 1.33), $C([0,1])$ is a complex vector space. For $f, g \in C([0,1])$, define

$$\langle f, g \rangle = \int_0^1 f(x)\overline{g(x)}\, dx.$$

(The integral over $[0,1]$ of a complex-valued function $f(x) = u(x) + iv(x)$, where u and v are real-valued, is defined to be $\int_0^1 u(x)\, dx + i \int_0^1 v(x)\, dx$.) Then $\langle \cdot, \cdot \rangle$ is a complex inner product on $C([0,1])$ (Exercise 1.6.4).

An inner product always yields a norm (Exercise 1.6.5) in the following way.

Definition 1.90   *Let V be a vector space over $\mathbb{C}$ with a complex inner product $\langle \cdot, \cdot \rangle$. For $v \in V$, define*

$$\|v\| = \sqrt{\langle v, v \rangle}.$$

(The square root is defined as a nonnegative real number because of property I4 in Definition 1.86.) We call $\|v\|$ the norm of v.

Notice that for $\mathbb{R}^n$, this norm agrees with the usual notion of the length of a vector.

From now on, when we say V is a complex inner product space, we assume (unless otherwise stated) that $\langle \cdot, \cdot \rangle$ denotes the inner product on V and $\| \cdot \|$ denotes the norm on V obtained from the inner product as in Definition 1.90.

Lemma 1.91   *(Cauchy-Schwarz inequality) Let V be a complex inner product space. Then for any $u, v \in V$,*

$$|\langle u, v \rangle| \leq \|u\| \|v\|.$$

Proof
Let $u, v \in V$. If $v = 0$, then $\langle u, v \rangle = \langle u, 0v \rangle = 0\langle u, v \rangle = 0$, so the result holds automatically. Now assume $v \neq 0$. For any $\lambda \in \mathbb{C}$, the positive definiteness of the inner product (property I4 in Definition 1.86) implies

$$0 \leq \langle u + \lambda v, u + \lambda v \rangle.$$

Expanding the right side (using the linearity properties I1, and I2, and equations (1.42) and (1.43)) gives

$$0 \leq \langle u, u \rangle + \langle u, \lambda v \rangle + \langle \lambda v, u \rangle + \langle \lambda v, \lambda v \rangle$$
$$= \|u\|^2 + \overline{\lambda}\langle u, v \rangle + \lambda \overline{\langle u, v \rangle} + |\lambda|^2 \|v\|^2,$$

where we used I3 in the last line. Now select

$$\lambda = -\frac{\langle u, v \rangle}{\|v\|^2}.$$

Substituting this into the preceding expression yields

$$0 \leq \|u\|^2 - 2\frac{|\langle u, v \rangle|^2}{\|v\|^2} + \frac{|\langle u, v \rangle|^2}{\|v\|^4}\|v\|^2 = \|u\|^2 - \frac{|\langle u, v \rangle|^2}{\|v\|^2},$$

which yields

$$|\langle u, v \rangle|^2 \leq \|u\|^2 \|v\|^2.$$

Taking the square root of both sides yields the result. ∎

A consequence of this is a general version of the triangle incquality.

Corollary 1.92   (Triangle inequality in an inner product space) *Let V be a complex inner product space. Then for $u, v \in V$,*

$$\|u + v\| \leq \|u\| + \|v\|.$$

Proof
Applying the definitions and the Cauchy-Schwarz inequality gives:

$$\|u + v\|^2 = \langle u + v, u + v \rangle = \langle u, u \rangle + \langle u, v \rangle + \langle v, u \rangle + \langle v, v \rangle$$
$$\leq \|u\|^2 + 2\|u\|\|v\| + \|v\|^2 = (\|u\| + \|v\|)^2.$$

Now take the square root of both sides. ∎

Recall from calculus that for x and y in $\mathbb{R}^2$ or $\mathbb{R}^3$, we have

$$x \cdot y = \|x\|\|y\| \cos \theta,$$

where θ is the angle between x and y. In particular, x is perpendicular to y if and only if $x \cdot y = 0$. In a general inner product space, which we may not be able to visualize, we use the inner product to define a generalized notion of perpendicularity.

Definition 1.93 *Suppose V is a complex inner product space. For $u, v \in V$, we say that u and v are* orthogonal *(written $u \perp v$) if $\langle u, v \rangle = 0$.*

Notice that if $u \perp v$, then expanding the inner product gives

$$\|u + v\|^2 = \|u\|^2 + \langle u, v \rangle + \langle v, u \rangle + \|v\|^2 = \|u\|^2 + \|v\|^2. \qquad (1.44)$$

This is a general version of the Pythagorean theorem.

Definition 1.94 *Suppose V is a complex inner product space. Let B be a collection of vectors in V. B is an* orthogonal *set if any two different elements of B are orthogonal. B is an* orthonormal *set if B is an orthogonal set and $\|v\| = 1$ for all $v \in B$.*

Orthogonal sets of nonzero vectors are linearly independent.

Lemma 1.95 *Suppose V is a complex inner product space. Suppose B is an orthogonal set of vectors in V and $0 \notin B$. Then B is a linearly independent set.*

Proof
Suppose $u_1, u_2, \ldots, u_k \in B$ and there exist scalars $\alpha_1, \alpha_2, \ldots, \alpha_k$ such that

$$\alpha_1 u_1 + \alpha_2 u_2 + \cdots + \alpha_k u_k = 0.$$

Take the inner product of both sides with u_j, where $j \in \{1, 2, \ldots, k\}$ is arbitrary. By the orthogonality assumption, $\langle u_l, u_j \rangle = 0$ for $l \neq j$. We obtain

$$\alpha_j \langle u_j, u_j \rangle = 0.$$

Since $u_j \neq 0$ (since $0 \notin B$ by assumption), we have $\langle u_j, u_j \rangle = \|u_j\|^2 \neq 0$ (by I4), hence $\alpha_j = 0$. Since j is arbitrary, this proves that B is linearly independent. ∎

Lemma 1.96 *Suppose V is a complex inner product space, and $B = \{u_1, u_2, \ldots, u_n\}$ is an orthogonal set in V with $u_j \neq 0$ for all j. If $v \in$ span B, then*

$$v = \sum_{j=1}^{n} \frac{\langle v, u_j \rangle}{\|u_j\|^2} u_j. \qquad (1.45)$$

Proof

Since $v \in \text{span } B$, there exist scalars $\alpha_1, \alpha_2, \ldots, \alpha_n$ such that

$$v = \alpha_1 u_1 + \alpha_2 u_2 + \cdots + \alpha_n u_n.$$

For each j, taking the inner product of both sides of this equation with u_j and using the orthogonality of the elements of B gives

$$\langle v, u_j \rangle = \alpha_j \langle u_j, u_j \rangle.$$

Solving for α_j gives equation (1.45). ∎

For a general element u belonging to the span of a finite set, one has to solve a system of linear equations to find the coefficients in the expansion of u. Lemma 1.96 demonstrates the basic advantage of orthogonality: for B orthogonal and $v \in \text{span } B$, it is easy to determine the coefficients in the expansion of the element v. Notice that if we assume that B is orthonormal, then equation (1.45) simplifies further to

$$v = \sum_{j=1}^{n} \langle v, u_j \rangle u_j. \tag{1.46}$$

Equation (1.45) suggests the notion of orthogonal projection.

Definition 1.97 *Suppose V is a complex inner product space, and $B = \{u_1, u_2, \ldots, u_n\}$ is an orthogonal set in V with $u_j \neq 0$ for all j. Let $S = \text{span } B$. (By Exercise 1.3.3, S is a subspace of V.) For $v \in V$, define the orthogonal projection $P_S(v)$ of v on S by*

$$P_S(v) = \sum_{j=1}^{n} \frac{\langle v, u_j \rangle}{\|u_j\|^2} u_j. \tag{1.47}$$

The orthogonal projection operator P_S has the following properties.

Lemma 1.98 *Let V, B, S, and P_S be as in Definition 1.97. Then*

  i. *P_S is a linear transformation.*
  ii. *For every $v \in V$, $P_S(v) \in S$.*
  iii. *If $s \in S$, then $P_S(s) = s$.*
  iv. *(Orthogonality property) For any $v \in V$ and $s \in S$,*

$$\left(v - P_S(v) \right) \perp s.$$

v. *(Best approximation property) For any $v \in V$ and $s \in S$,*

$$\|v - P_S(v)\| \leq \|v - s\|,$$

with equality if and only if $s = P_S(v)$.

Proof

Property i follows from the additivity of the inner product (I1 in Definition 1.86) and the relation (1.47). Also, expression (1.47) shows that $P_S(v) \in \mathrm{span}\, B = S$, so ii holds. Lemma 1.96 implies iii.

For iv, let $v \in V$ and first note that for each m, the orthogonality of B implies that

$$\langle P_S(v), u_m \rangle = \sum_{j=1}^{n} \frac{\langle v, u_j \rangle}{\|u_j\|^2} \langle u_j, u_m \rangle = \langle v, u_m \rangle.$$

Equivalently

$$\langle v - P_S(v), u_m \rangle = 0,$$

so $(v - P_S(v)) \perp u_m$ for $m = 1, 2, \ldots, n$. Since any element $s \in S$ is a linear combination of $u_1, \ldots, u_n$, it follows that $(v - P_S(v)) \perp s$.

To prove v, let $v \in V$ and $s \in S$. Then

$$\|v - s\|^2 = \|v - P_S(v) + P_S(v) - s\|^2 = \|v - P_S(v)\|^2 + \|P_S(v) - s\|^2$$
$$\geq \|v - P_S(v)\|^2,$$

where the next to last step follows from equation (1.44), since s and $P_S(v)$ belong to the subspace S, so $P_S(v) - s \in S$, and by part iv, $v - P_S(v)$ is orthogonal to everything in S. Taking the square root implies v.   ∎

Property v of Lemma 1.98 says that the closest element to v in the subspace S is the orthogonal projection $P_S(v)$. This corresponds to our geometric intuition in the case of $\mathbb{R}^2$ or $\mathbb{R}^3$. It also shows that, despite the definition, P_S does not depend on the choice of orthonormal basis B for S: for any two orthonormal bases of S, the resulting projections are the same. Exercise 1.6.8 gives a more direct proof of this fact.

Starting with a linearly independent set of vectors, we can obtain an orthonormal set with the same span by the Gram-Schmidt procedure, as shown by Lemma 1.99.

Lemma 1.99 (Gram–Schmidt procedure) *Suppose V is a complex inner product space, and $\{u_1, u_2, \ldots, u_n\}$ is a linearly independent set in V. Then there exists an orthonormal set $\{v_1, v_2, \ldots, v_n\}$ with the same span.*

Proof
For each $k = 1, 2, \ldots, n$, let $S_k = \operatorname{span}\{u_1, u_2, \ldots, u_k\}$. We define $w_1, w_2, \ldots, w_n$ inductively, so that at each stage the set $B_k = \{w_1, w_2, \ldots w_k\}$ is orthogonal (we normalize at the end) and $\operatorname{span} B_k = S_k$. To start, let $w_1 = u_1$. Then B_1 satisfies the requirements. By the induction hypothesis, suppose $B_{k-1} = \{w_1, w_2, \ldots w_{k-1}\}$ is orthogonal and $\operatorname{span} B_{k-1} = S_{k-1}$.

Let $P_{S_{k-1}}$ be the orthogonal projection operator onto S_{k-1} and set

$$w_k = u_k - P_{S_{k-1}} u_k = u_k - \sum_{j=1}^{k-1} \frac{\langle u_k, w_j \rangle}{\|w_j\|^2} w_j, \tag{1.48}$$

since by assumption B_{k-1} is an orthogonal set that spans S_{k-1}. Notice that $w_k \neq 0$ since if $w_k = 0$, equation (1.48) would imply that u_k belongs to S_{k-1}, contradicting (by Exercise 1.3.9(i)) the linear independence of $\{u_1, \ldots, u_n\}$. By property iv in Lemma 1.98, w_k is orthogonal to every element in S_{k-1}, in particular to $w_1, \ldots, w_{k-1}$. So B_k is orthogonal.

Now $w_1, w_2, \ldots w_{k-1}$ all belong to $B_{k-1} \subseteq S_{k-1} \subseteq S_k$. Also, $P_{S_{k-1}} u_k \in S_{k-1} \subseteq S_k$, and $u_k \in S_k$, so by equation (1.48), $w_k \in S_k$. It follows that $\operatorname{span} B_k \subseteq S_k$.

To prove the converse inclusion, first note that $u_1, u_2, \ldots, u_{k-1} \in S_{k-1} = \operatorname{span} B_{k-1} \subseteq \operatorname{span} B_k$. Also, $P_{S_{k-1}} u_k \in S_{k-1} = \operatorname{span} B_{k-1} \subseteq \operatorname{span} B_k$, and $w_k \in \operatorname{span} B_k$. So by equation (1.48), $u_k \in \operatorname{span} B_k$. It follows that $S_k \subseteq \operatorname{span} B_k$.

So altogether we obtain $S_k = \operatorname{span} B_k$, completing the induction step.

After n steps, this process gives an orthogonal set $\{w_1, \ldots, w_n\}$ with the same span as $\{u_1, \ldots, u_n\}$. The orthonormal set $\{v_1, \ldots, v_n\}$ in the statement of the theorem is obtained by normalization: set $v_j = w_j / \|w_j\|$ for each j. This does not change the orthogonality or the span. ∎

Definition 1.100   *Suppose V is a complex inner product space. An* orthonormal basis *for V is an orthonormal set in V that is also a basis.*

Every finite dimensional complex inner product space has an orthonormal basis: by definition, it has a basis, so by Lemma 1.99 there is an orthonormal set that spans V and has the same number of elements, and hence is also a basis. The standard basis in $\mathbb{C}^n$ (Definition 1.39) is an example of an orthonormal basis.

We can compute inner products and norms easily using the components with respect to an orthonormal basis.

Lemma 1.101   *Let V be a complex inner product space with (finite)* orthonormal basis $R = \{u_1, u_2, \ldots, u_n\}$.

   i. *For any $v \in V$,*

$$v = \sum_{j=1}^{n} \langle v, u_j \rangle u_j. \tag{1.49}$$

   ii. *(Parseval's relation) For any $v, w \in V$,*

$$\langle v, w \rangle = \sum_{j=1}^{n} \langle v, u_j \rangle \overline{\langle w, u_j \rangle}. \tag{1.50}$$

   iii. *(Plancherel's formula) For any $v \in V$,*

$$\|v\|^2 = \sum_{j=1}^{n} |\langle v, u_j \rangle|^2. \tag{1.51}$$

Proof
Part i is equation (1.46), which applies to any $v \in V$ since R is a basis for V. To prove ii, apply equation (1.49) to v and w and use the linearity of the inner product to write

$$\langle v, w \rangle = \left\langle \sum_{j=1}^{n} \langle v, u_j \rangle u_j, w \right\rangle = \sum_{j=1}^{n} \langle v, u_j \rangle \langle u_j, w \rangle = \sum_{j=1}^{n} \langle v, u_j \rangle \overline{\langle w, u_j \rangle}.$$

Part iii follows from part ii by letting $w = v$.   ∎

Properties i, ii, and iii make orthonormal bases simple to work with.

We will consider unitary matrices, which are closely connected with orthonormal bases.

Definition 1.102 *Let $A = [a_{ij}]$ be an $m \times n$ matrix over $\mathbb{C}$.*
  The transpose *A^t of A is the $n \times m$ matrix $B = [b_{ij}]$ defined by $b_{ij} = a_{ji}$, for all i, j.*
  The conjugate transpose *A^* of A is the $n \times m$ matrix $C = [c_{ij}]$ defined by $c_{ij} = \overline{a_{ji}}$, for all i, j.*

In other words, the transpose A^t is obtained by interchanging the rows and columns of A. The conjugate transpose A^* is obtained by taking the complex conjugates of all the entries of A^t.

The entries of an $m \times n$ complex matrix $A = [a_{ij}]$ can be expressed using inner products by the formula

$$a_{ij} = \langle Ae_j, e_i \rangle, \tag{1.52}$$

where $e_j \in \mathbb{C}^n$ and $e_i \in \mathbb{C}^m$ are the Euclidean basis vectors in Definition 1.39. To see this, note that the kth component of Ae_j is

$$(Ae_j)_k = \sum_{\ell=1}^{n} a_{k\ell}(e_j)_\ell = a_{kj},$$

since $(e_j)_\ell$, the ℓth component of e_j, is 1 if $\ell = j$ and 0 otherwise. Hence,

$$\langle Ae_j, e_i \rangle = \sum_{k=1}^{m} a_{kj}(e_i)_k = a_{ij}.$$

A more general formula is given in Exercise 1.6.18.

Lemma 1.103 *Let A be an $m \times n$ matrix over $\mathbb{C}$. Then*

$$\langle Az, w \rangle = \langle z, A^*w \rangle,$$

for every $z \in \mathbb{C}^n$ and $w \in \mathbb{C}^m$. Furthermore, A^* is the only matrix with this property.

Proof
Let $z \in \mathbb{C}^n$ and $w \in \mathbb{C}^m$ have components $z_1, z_2, \ldots, z_n$ and $w_1, w_2, \ldots w_m$, respectively. Let $A = [a_{ij}]$ be an $m \times n$ matrix, and let $B = [b_{ij}]$ be an $n \times m$ matrix. Note that the jth component $(Bw)_j$ of Bw is $\sum_{i=1}^{m} b_{ji}w_i$. Therefore,

$$\langle z, Bw \rangle = \sum_{j=1}^{n} z_j \overline{(Bw)_j} = \sum_{j=1}^{n} z_j \overline{\sum_{i=1}^{m} b_{ji}w_i} = \sum_{j=1}^{n} \sum_{i=1}^{m} \overline{b_{ji}} z_j \overline{w_i}.$$

On the other hand,

$$\langle Az, w \rangle = \sum_{i=1}^{m} (Az)_i \overline{w_i} = \sum_{i=1}^{m} \sum_{j=1}^{n} a_{ij} z_j \overline{w_i}.$$

Thus, if $B = A^*$, that is, $b_{ij} = \overline{a_{ji}}$ or equivalently $\overline{b_{ji}} = a_{ij}$ for each i, j, then $\langle Az, w \rangle = \langle z, Bw \rangle$.

To show uniqueness, suppose $B = [b_{ij}]$ satisfies $\langle Az, w \rangle = \langle z, Bw \rangle$ for all $z \in \mathbb{C}^n$ and $w \in \mathbb{C}^m$. By formula (1.52),

$$a_{ij} = \langle Ae_j, e_i \rangle = \langle e_j, Be_i \rangle = \overline{\langle Be_i, e_j \rangle} = \overline{b_{ji}},$$

as required. ∎

Definition 1.104 *Let A be an $n \times n$ matrix. A is* unitary *if A is invertible and $A^{-1} = A^*$.*

For a matrix over the real numbers, the conjugate transpose is the same as the transpose. So a real unitary matrix A is one that satisfies $A^{-1} = A^t$; such a matrix is called *orthogonal*.

Unitary matrices can be characterized in several interesting ways.

Lemma 1.105 *Let A be an $n \times n$ matrix over $\mathbb{C}$. Then the following statements are equivalent:*

   i. A is unitary.
   ii. The columns of A form an orthonormal basis for $\mathbb{C}^n$.
   iii. The rows of A form an orthonormal basis for $\mathbb{C}^n$.
   iv. A preserves inner products, that is, $\langle Az, Aw \rangle = \langle z, w \rangle$ for all $z, w \in \mathbb{C}^n$.
   v. $\|Az\| = \|z\|$, for all $z \in \mathbb{C}^n$.

Proof
We first prove that i is equivalent to ii. Let v_j be the jth column of A. By definition, the ith row of A^* is the vector $\overline{v_i}$, whose components $(\overline{v_i})_k$, $1 \le k \le n$, are the conjugates of the corresponding components $(v_i)_k$ of v_i. By the definition of matrix multiplication, the $(i, j)^{\text{th}}$ entry of A^*A is

$$(A^*A)_{ij} = \sum_{k=1}^{n} (\overline{v_i})_k (v_j)_k = \langle v_j, v_i \rangle.$$

By Exercise 1.4.12(ii), A is invertible with $A^{-1} = A^*$ if and only if $A^*A = I$. Hence A is unitary if and only if $\langle v_j, v_i \rangle = 1$ when $i = j$ and 0 otherwise; that is, if and only if the set $\{v_1, \ldots, v_n\}$ is orthonormal. If so, this set is automatically a basis for $\mathbb{C}^n$ since it is a linearly independent set (Lemma 1.95) with n elements in an n-dimensional space (Theorem 1.42). This proves that i and ii are equivalent.

Applying a similar argument to AA^* proves that i and iii are equivalent (Exercise 1.6.9).

Next, we show that i and iv are equivalent. By Lemma 1.103,

$$\langle Az, Aw \rangle = \langle z, A^*Aw \rangle.$$

If i holds, that is, $A^*A = I$, then iv holds. Conversely, if iv holds, then for any z and w,

$$\langle z, w - A^*Aw \rangle = 0.$$

Taking $z = w - A^*Aw$ shows that $A^*Aw = w$. Since this holds for all w, we obtain (by Exercise 1.4.6(i)) that $A^*A = I$.

We leave the proof that i is equivalent to v as Exercise 1.6.7(iv). ∎

If O is an orthonormal basis, then the change of basis matrices going from the O coordinates to the Euclidean coordinates or vice-versa are unitary.

Lemma 1.106 Let $E = \{e_1, e_2, \ldots, e_n\}$ be the standard basis for $\mathbb{C}^n$ (Definition 1.39) and suppose that $O = \{u_1, u_2, \ldots, u_n\}$ is an orthonormal basis for $\mathbb{C}^n$. Let U be the $n \times n$ matrix whose j^{th} column is the vector u_j.

    i. Then U is unitary, U is the O-to-E change of basis matrix, and U^* is the E-to-O change of basis matrix.

    ii. Suppose $T : \mathbb{C}^n \to \mathbb{C}^n$ is a linear transformation that is represented by the matrix A in the standard basis (i.e., $T(z) = Az$). Then T is represented in the basis O by the matrix

$$A_{T,O} = U^*AU.$$

Proof

    i. The fact that U is unitary follows from the orthonormality of O and Lemma 1.105, since any n orthonormal vectors are a basis for $\mathbb{C}^n$ (Lemma 1.95 and Theorem 1.42). Exercise 1.4.15

shows that U is the O-to-E change of basis matrix. By Exercise 1.4.14, the E-to-O change of basis matrix is U^{-1}, which is U^* since U is unitary.

  ii. This follows from part i and Lemma 1.61. ∎

Recall the main theme of section 1.5: we have a linear transformation T represented in some basis R by a matrix A. We want to select another basis S so that the matrix B representing T with respect to S is as simple as possible. We have our choice of B from among all matrices similar to A, that is, all matrices such that there exists an invertible matrix P so that $B = P^{-1}AP$. When R is the standard basis and S is an orthonormal basis, the matrix P is unitary. This case is particularly easy, because, for example, a unitary matrix is easy to invert. This prompts special definitions for the case when P is unitary.

Definition 1.107 *Let A and B be $n \times n$ matrices over $\mathbb{C}$. We say that A and B are* unitarily similar *if there exists a unitary matrix U such that $B = U^*AU$. If A is unitarily similar to a diagonal matrix, we say that A is* unitarily diagonalizable.

Remarkably, there is a simple characterization of the unitarily diagonalizable matrices. The proof is a bit involved, so we leave it as a series of exercises (Exercises 1.6.11 through 1.6.16) for the exceptionally motivated student.

Definition 1.108 *An $n \times n$ matrix A is* normal *if $A^*A = AA^*$.*

Note that unitary matrices are normal. The term "normal" is misleading because most matrices are not normal. The next theorem shows that normal matrices are unitarily diagonalizable. They should be called "exceptional" or "outstanding" rather than "normal."

Theorem 1.109 *(Spectral theorem for matrices) Let A be an $n \times n$ matrix over $\mathbb{C}$. Then the following statements are equivalent:*

  i. *A is unitarily diagonalizable.*

  ii. *A is normal.*

  iii. *There is an orthonormal basis for $\mathbb{C}^n$ consisting of eigenvectors of A.*

Proof

The fact that i implies ii is Exercise 1.6.11. That ii implies iii is difficult; the proof is sketched in Exercises 1.6.12 through 1.6.16. The fact that iii implies i follows from Lemma 1.74: the diagonalizing matrix can be taken to have columns equal to the elements of the orthonormal basis of eigenvectors of A, hence this matrix is unitary by Lemma 1.105. ■

The spectral theorem is a major theorem in linear algebra. Its generalization to infinite dimensional vector spaces is one of the key theorems in a subject known as functional analysis. This generalization is much more difficult because in the infinite dimensional case, the spectrum of T (the appropriate generalization of the set of eigenvalues) is not necessarily discrete, so one needs to introduce some measure on the spectrum. The first step in understanding this deep result is to understand the matrix case.

There is one more special type of matrix that we should discuss.

Definition 1.110   *Let A be an $n \times n$ matrix. A is Hermitian if $A^* = A$.*

In other words, $A = [a_{ij}]$ is Hermitian if the entries obtained when the matrix A is flipped over its diagonal are the complex conjugates of the original entries: that is, $a_{ji} = \overline{a_{ij}}$ for all i, j. Note that the diagonal entries of a Hermitian matrix must be real. If A has only real entries, then A is Hermitian if and only if $A = A^t$. Such a real matrix is called *symmetric*.

Hermitian matrices have the following characterizations.

Lemma 1.111   *Suppose A is an $n \times n$ matrix. Then the following are equivalent:*

  i. A is Hermitian.

  ii. A is normal and the eigenvalues of A are real.

  iii. There exists a unitary matrix U and a diagonal matrix D with only real entries such that $A = U^*DU$.

Proof

(i implies ii): Suppose A is Hermitian. Then A is normal: $A^*A = A^2 = AA^*$. Suppose λ is an eigenvalue of A with nonzero eigenvector v. Then, using Lemma 1.103,

$$\lambda\langle v, v\rangle = \langle \lambda v, v\rangle = \langle Av, v\rangle = \langle v, A^*v\rangle = \langle v, Av\rangle = \langle v, \lambda v\rangle = \overline{\lambda}\langle v, v\rangle.$$

Since $\langle v, v \rangle \neq 0$, we obtain $\lambda = \bar{\lambda}$, which implies that λ is real.

(ii implies iii): Since A is normal, by Theorem 1.109 there exists a unitary matrix U and a diagonal matrix D such that $A = U^*DU$. The diagonal entries of D are its eigenvalues, which are the same as the eigenvalues of A (Corollary 1.68), and hence are real by assumption.

(iii implies i): Suppose $A = U^*DU$ with U unitary and D diagonal with real entries. Then by Exercise 1.6.10(i),

$$A^* = (U^*DU)^* = U^*D^*(U^*)^* = U^*D^*U,$$

by definition of the conjugate transpose. But since D is diagonal with real entries, $D^* = D$. So

$$A^* = U^*DU = A.$$

Hence A is Hermitian. ∎

Unitary matrices have a characterization of a similar nature; see Exercise 1.6.19.

Exercises

1.6.1. Suppose V is a vector space over $\mathbb{C}$ and $\langle \cdot, \cdot \rangle$ is a (complex) inner product on V. Prove equations (1.42) and (1.43).

1.6.2.   i. Check that $\langle \cdot, \cdot \rangle$ as defined in Example 1.87 is a complex inner product on $\mathbb{C}^n$.

  ii. Check that the dot product is a real inner product on $\mathbb{R}^n$.

1.6.3.   i. For $z, w \in \mathbb{C}$, prove that $2|zw| \leq |z|^2 + |w|^2$, and deduce that $|z + w|^2 \leq 2(|z|^2 + |w|^2)$.

  ii. For $z = \{z_j\}_{j=1}^{\infty}, w = \{w_j\}_{j=1}^{\infty} \in \ell^2(\mathbb{N})$ (defined in Example 1.88), and $\lambda \in \mathbb{C}$, define $z + w = \{z_j + w_j\}_{j=1}^{\infty}$, and $\lambda z = \{\lambda z_j\}_{j=1}^{\infty}$. Prove that $\ell^2(\mathbb{N})$, with these operations, is a vector space over $\mathbb{C}$. (The only property in Definition 1.30 that is not obvious is A1, which follows from part i.)

  iii. For $z = \{z_j\}_{j=1}^{\infty} \in \ell^2(\mathbb{N})$ and $w = \{w_j\}_{j=1}^{\infty} \in \ell^2(\mathbb{N})$, prove that the series $\sum_{j=1}^{\infty} z_j \overline{w_j}$ converges absolutely.

  iv. Prove that $\langle \cdot, \cdot \rangle$ as defined in Example 1.88 is a complex inner product on $\ell^2(\mathbb{N})$.

1.6.4. Prove that $\langle \cdot, \cdot \rangle$ as defined in Example 1.89 is a complex inner product on $C([0, 1])$.

1.6.5. A *normed* vector space is a vector space V with a map $\| \cdot \|$ (called a *norm*) defined on V such that

N1. (Positive definiteness) For any $v \in V$, $\|v\|$ is a nonnegative real number.

N2. (Nondegeneracy) $\|v\| = 0$ if and only if $v = 0$.

N3. (Scalar compatibility) $\|\lambda v\| = |\lambda| \|v\|$, for every scalar λ and every $v \in V$.

N4. (Triangle inequality) $\|u + v\| \le \|u\| + \|v\|$ for every $u, v \in V$.

  i. Suppose V is a complex inner product space. Define $\| \cdot \|$ as in Definition 1.90. Show that V with $\| \cdot \|$ is a normed vector space.

    Remark: Exercise 1.6.6 shows that there are norms that do not come from inner products in this way.

  ii. Suppose V with $\|\cdot\|$ is a normed vector space. For $u, v \in V$, define $d(u, v) = \|u - v\|$. Prove that V with d is a metric space. (See Exercise 1.1.2 for the definition of a metric space.)

    This is an example of the different levels of structure in mathematics. A metric space is more general (i.e., less structure) than a normed vector space, that in turn is more general than an inner product space. Roughly speaking, an inner product space that is complete (called a *Hilbert space*) is about as close to having all the structure of $\mathbb{C}^n$ as possible without necessarily being finite dimensional.

1.6.6. i. Let V be a complex inner product space and $\| \cdot \|$ a norm obtained from the inner product as in Definition 1.90. Prove the *parallelogram identity*:

$$2\|z\|^2 + 2\|w\|^2 = \|z + w\|^2 + \|z - w\|^2$$

for every $z, w \in V$. To see why this is called the parallelogram identity, draw two nonzero vectors $z, w \in \mathbb{R}^2$ with a common base point and consider the parallelogram with sides z and w. Show that the sums of the squares of the lengths of the two diagonals equals the sums of the squares of the lengths of the four sides.

    ii. For $z \in \mathbb{C}^2$, define

$$\|z\|_1 = |z_1| + |z_2|,$$

    where z_1 and z_2 are the components of z. Prove that $\|\cdot\|_1$ is a norm on $\mathbb{C}^2$. This is called the ℓ^1 norm.

    iii. Find vectors $z, w \in \mathbb{C}^2$ such that

$$2\|z\|_1^2 + 2\|w\|_1^2 \neq \|z + w\|_1^2 + \|z - w\|_1^2.$$

    Deduce that $\|\cdot\|_1$ is a norm that does not come from an inner product.

1.6.7. Let V be a complex inner product space and $T : V \to V$ a linear transformation.

    i. (Polarization identity) Prove that for any $u, v \in V$,

$$4\langle T(u), v\rangle = \langle T(u + v), u + v\rangle - \langle T(u - v), u - v\rangle$$
$$+ i\langle T(u + iv), u + iv\rangle - i\langle T(u - iv), u - iv\rangle.$$

    ii. Prove that a complex inner product is determined by the norm it induces; that is, if two inner products on V result in the same norm, then these inner products must be the same. Hint: Take T in part i to be the identity operator.

    iii. Suppose that for all $u \in V$, $\langle T(u), u\rangle = 0$. Prove that T is the 0 operator, that is, $T(v) = 0$ for all $v \in V$. Hint: Apply part i to obtain that $\langle T(u), v\rangle = 0$ for all $u, v \in V$, and then let $v = T(u)$.

    iv. Prove that an $n \times n$ matrix A is unitary if and only if $\|Az\| = \|z\|$ for all $z \in \mathbb{C}^n$. Suggestion: Apply part iii to $A^*A - I$.

1.6.8. Let S be a subspace of a finite dimensional vector space V. Let $v \in V$. Suppose $w \in S$ and $(v - w) \perp s$ for every $s \in S$. Prove that $w = P_S(v)$ for P_S as in Definition 1.97. This gives a characterization of P_S that does not involve a choice of basis for S, and hence shows that P_S is independent of the choice of orthonormal basis used in its definition. Hint: Write $w - P_S(v) = (w - v) + (v - P_S(v))$ and show that $w - P_S(v)$ belongs to S but is also orthogonal to every element of S.

1.6.9. Prove the equivalence of i and iii in Lemma 1.105.

1.6.10. Let A and B be $n \times n$ matrices over $\mathbb{C}$.

    i. Prove that $(AB)^* = B^*A^*$. (Hint: There is a direct proof, but it is easier to use the uniqueness result in Lemma 1.103.)

    ii. Define $A \approx B$ if there exists a unitary matrix U such that $B = U^*AU$. Prove that $\approx$ is an equivalence relation; that is, (a) $A \approx A$ for any A, (b) if $A \approx B$ then $B \approx A$, and (c) if $A \approx B$ and $B \approx C$ then $A \approx C$.

1.6.11. In Theorem 1.109, prove that i implies ii.

1.6.12. Suppose $T : \mathbb{C}^n \to \mathbb{C}^n$ is a linear transformation. Let A be an $n \times n$ matrix over $\mathbb{C}$ which represents T in the standard basis, that is, such that $T(z) = Az$. We define the *adjoint* T^* of T to be the operator $T^* : \mathbb{C}^n \to \mathbb{C}^n$ defined by $T^*(z) = A^*z$. We say T is *normal* if $T^*T = TT^*$, that is, if $T^*T(z) = TT^*(z)$ for all $z \in \mathbb{C}$.

    i. Prove that T is normal if and only if A is normal. (Hint: Use Exercise 1.4.6(ii).)

    ii. Show that to prove statement ii implies statement iii in Theorem 1.109, it is enough to prove that if T is normal, then $\mathbb{C}^n$ has an orthonormal basis consisting of eigenvectors of T.

1.6.13. Suppose $T : \mathbb{C}^n \to \mathbb{C}^n$ is a linear transformation. Prove that T is normal (see Exercise 1.6.12) if and only if $\|T^*(z)\| = \|T(z)\|$ for all $z \in \mathbb{C}^n$. Hint: Show that the condition $\langle T^*(z), T^*(z) \rangle = \langle T(z), T(z) \rangle$ is equivalent to $\langle z, (T^*T - TT^*)(z) \rangle = 0$ (use Lemma 1.103) and apply Exercise 1.6.7(iii).

1.6.14. Suppose $T : \mathbb{C}^n \to \mathbb{C}^n$ is a normal linear transformation (see Exercise 1.6.12). Suppose $\lambda \in \mathbb{C}$.

    i. Prove that $(\lambda I - T)^* = \bar{\lambda} I - T^*$. (This part doesn't require the normality of T. Hint: The corresponding statement for matrices is easy.)

    ii. Prove that $\lambda I - T$ is also normal.

    iii. For $v \in \mathbb{C}^n$, prove that $T(v) = \lambda v$ if and only if $T^*(v) = \bar{\lambda} v$. Hint: Apply Exercise 1.6.13 to $\lambda I - T$.

1.6.15. Suppose $T : \mathbb{C}^n \to \mathbb{C}^n$ is a normal linear transformation (see Exercise 1.6.12). Suppose $\lambda, \mu \in \mathbb{C}$ and $\lambda \neq \mu$. Suppose u, $v \in \mathbb{C}^n$ are eigenvectors of T with eigenvalues λ and μ; that is, $T(u) = \lambda u$ and $T(v) = \mu v$. Prove that $u \perp v$. Hint: Note that by Exercise 1.6.14(iii),

$$\lambda \langle u, v \rangle = \langle T(u), v \rangle = \langle u, T^*(v) \rangle = \langle u, \overline{\mu} v \rangle = \mu \langle u, v \rangle.$$

1.6.16. Suppose $T : \mathbb{C}^n \to \mathbb{C}^n$ is a normal linear transformation (see Exercise 1.6.12). Denote the eigenvalues of T by $\lambda_1, \ldots \lambda_k$ and the corresponding eigenspaces by $E_{\lambda_1}, \ldots, E_{\lambda_k}$. Let m_i be the dimension of E_{λ_i}. Let $\{u_{ij}\}_{j=1}^{m_i}$ be an orthonormal basis for E_{λ_i}. By Exercise 1.6.15, the set

$$S = \{u_{ij}\}_{1 \le i \le k, 1 \le j \le m_i}$$

is orthonormal. Let

$$W = \operatorname{span} S.$$

Define

$$W^\perp = \left\{ z \in \mathbb{C}^n : \quad \text{for all} \quad w \in W, z \perp w \right\}.$$

i. Prove that $W^\perp$ is a subspace of $\mathbb{C}^n$.

ii. Prove that if $w \in W$, then $T^*(w) \in W$. Hint: Use Exercise 1.6.14(iii).

iii. Prove that if $y \in W^\perp$, then $T(y) \in W^\perp$. Hint: For $y \in W^\perp$ and $w \in W$,

$$\langle T(y), w \rangle = \langle y, T^*(w) \rangle.$$

Apply part ii.

iv. Let $T_{W^\perp}$ be the restriction of the transformation T to the subspace $W^\perp$. By part iii, $T_{W^\perp}$ is a linear transformation from $W^\perp$ to $W^\perp$. However, any nonzero eigenvector of $T_{W^\perp}$ would be a nonzero eigenvector of T in $W^\perp$. There cannot be any such eigenvector because all of the eigenvectors of T belong to W by its definition. (An element belonging to both W and $W^\perp$ must be 0 because it is orthogonal to itself.) So $T_{W^\perp}$ has no nonzero eigenvectors, hence no eigenvalues. But this is impossible unless $W^\perp = \{0\}$ (by Lemma 1.78, applied to a matrix representing $T_{W^\perp}$ in some basis, and the fundamental theorem of algebra). Deduce that $W = \mathbb{C}^n$. Hint: For a general $z \in \mathbb{C}^n$, write $z = z - P_W(z) + P_W(z)$; by Lemma 1.98, $z - P_W(z) \in W^\perp$.

v. Deduce that $\mathbb{C}^n$ has an orthonormal basis consisting of eigenvectors of T. With Exercise 1.6.12, this proves the difficult implication that statement ii implies statement iii in Theorem 1.109.

1.6.17. Suppose A is an $n \times n$ matrix such that A^* is a polynomial in A, that is, there exists $m \in \mathbb{N}$ and scalars $a_0, \ldots, a_m$ such that

$$A^* = a_m A^m + a_{m-1} A^{m-1} + \cdots + a_1 A + a_0 I.$$

Prove that A is unitarily diagonalizable.

1.6.18. Suppose that $R = \{v_1, v_2, \ldots, v_n\}$ is an orthonormal basis for a complex inner product space V and $T : V \to V$ is a linear transformation. Let $A_{T,R} = [a_{ij}]_{1 \le i,j \le n}$ be the matrix that represents T with respect to R. Prove that

$$a_{ij} = \langle T(v_j), v_i \rangle.$$

1.6.19. Let A be an $n \times n$ matrix over $\mathbb{C}$. Prove that A is unitary if and only if A is normal and all eigenvalues of A have magnitude 1.

1.6.20. Let $A = [a_{i,j}]$ be an $m \times n$ matrix over $\mathbb{C}$ with row vectors $u_1, u_2, \ldots, u_m \in \mathbb{C}^n$ and column vectors $v_1, v_2, \ldots, v_n \in \mathbb{C}^m$ (i.e., u_i is the vector with components $a_{i,1}, a_{i,2}, \ldots, a_{i,n}$ and v_j is the matrix with components $a_{1j}, a_{2j}, \ldots, a_{mj}$). The *row space* of A is

$$U = \operatorname{span} \{u_1, u_2, \ldots, u_m\},$$

and the *column space* of A is

$$V = \operatorname{span} \{v_1, v_2, \ldots, v_m\}.$$

i. Prove that $V = \{Az : z \in \mathbb{C}^n\}$. Hint: Show that $Az = z_1 v_1 + \cdots + z_n v_n$, where $z_1, z_2, \ldots, z_n$ are the components of z.

For $z \in \mathbb{C}^n$ with components $z_1, z_2, \ldots, z_n$, let $\bar{z} \in \mathbb{C}^n$ be the vector with components $\bar{z}_1, \bar{z}_2, \ldots, \bar{z}_n$.

ii. Suppose $z \in \mathbb{C}^n$. Prove that $A\bar{z} = 0$ if and only if $z \perp U$. (The statement $z \perp U$ means that z is orthogonal to every element of U).

Define $T : U \to V$ by $T(z) = A\bar{z}$. Note that $T(z) \in V$ by i.

iii. Prove that T is 1–1. Hint: Use Exercise 1.4.7 and part ii.

iv. Prove that T is onto. Hint: Let $y \in V$. By i, $A\bar{z} = y$ for some $z \in \mathbb{C}^n$. Let w be the orthogonal projection of z onto U. By definition, $(z - w) \perp U$. Apply ii.

v. Prove that the row space of A and the column space of A have the same dimension, equal to the rank of A. Hint: Use Exercise 1.4.8(iv).

vi. Prove that rank $A^t = $ rank A. (Here A^t is the transpose of A; see Definition 1.102.)

2 The Discrete Fourier Transform

CHAPTER

2.1 Definition and Basic Properties of the Discrete Fourier Transform

In chapter 1 we worked with vectors in $\mathbb{C}^N$, that is, sequences of N complex numbers. Here we change notation in several ways. First, for reasons that will be more clear later, we index these N numbers over $j \in \{0, 1, \ldots, N-1\}$ instead of $\{1, 2, \ldots, N\}$. Second, instead of writing the components of z as z_j, we write them as $z(j)$. This indicates a new point of view: we regard z as a function defined on the finite set

$$\mathbb{Z}_N = \{0, 1, \ldots, N-1\}.$$

(This is consistent with the formal definition of a sequence as a function on the set of indices.) To save space, we usually write such a z horizontally instead of vertically:

$$z = (z(0), z(1), \ldots z(N-1)).$$

However, when convenient, we still identify z with the column vector

$$z = \begin{bmatrix} z(0) \\ z(1) \\ z(2) \\ \cdot \\ \cdot \\ z(N-1) \end{bmatrix}. \tag{2.1}$$

This allows us to write the product of an $N \times N$ matrix A by z as Az. Finally, in order to be consistent with notation for functions used later in the infinite dimensional context, we write $\ell^2(\mathbb{Z}_N)$ in place of $\mathbb{C}^N$. So, formally,

$$\ell^2(\mathbb{Z}_N) = \left\{ z = (z(0), z(1), \ldots, z(N-1)) : z(j) \in \mathbb{C}, \ 0 \le j \le N-1 \right\}.$$

With the usual componentwise addition and scalar multiplication, $\ell^2(\mathbb{Z}_N)$ is an N-dimensional vector space over $\mathbb{C}$. One basis for $\ell^2(\mathbb{Z}_N)$ is the standard, or Euclidean, basis $E = \{e_0, e_1, \ldots, e_{N-1}\}$, where $e_j(n) = 1$ if $n = j$ and $e_j(n) = 0$ if $n \ne j$. In this notation, the complex inner product on $\ell^2(\mathbb{Z}_N)$ is

$$\langle z, w \rangle = \sum_{k=0}^{N-1} z(k)\overline{w(k)},$$

with the associated norm

$$\|z\| = \left(\sum_{k=0}^{N-1} |z(k)|^2 \right)^{1/2}$$

(called the ℓ^2 *norm*). We maintain the notion of orthogonality: $z \perp w$ if and only if $\langle z, w \rangle = 0$.

We make one more convention. Originally, for $z \in \ell^2(\mathbb{Z}_N)$, $z(j)$ is defined for $j = 0, 1, \ldots, N-1$. Now we extend z to be defined at all integers by requiring z to be periodic with period N:

$$z(j + N) = z(j), \quad \text{for all} \quad j \in \mathbb{Z}.$$

Hence, to find $z(j)$ for $j \ne \{0, 1, \ldots, N-1\}$, add some positive or negative integer multiple mN of N to j until $j + mN \in \{0, 1, \ldots, N-1\}$;

then define $z(j) = z(j + mN)$. For example, if $N = 12$, then

$$z(-21) = z(-9) = z(15) = z(27) = z(3).$$

Note that the value of $z(j)$ depends only on the residue of j modulo N; one can regard z as defined on the equivalence classes of $\mathbb{Z}$ mod N. In particular we can regard z as defined on any other set of N consecutive integers instead of $\{0, 1, \ldots, N - 1\}$.

The reader may be familiar with the theory of Fourier series, in which a function on an interval of real numbers is represented as a sum of sines and cosines (this is discussed in chapter 4). A similar phenomenon occurs with functions on $\mathbb{Z}_N$.

Definition 2.1 *Define $E_0, E_1, \ldots, E_{N-1} \in \ell^2(\mathbb{Z}_N)$ by*

$$E_0(n) = \tfrac{1}{\sqrt{N}} \quad \text{for} \quad n = 0, 1, \ldots, N - 1;$$

$$E_1(n) = \tfrac{1}{\sqrt{N}} e^{2\pi i n/N} \quad \text{for} \quad n = 0, 1, \ldots, N - 1;$$

$$E_2(n) = \tfrac{1}{\sqrt{N}} e^{2\pi i 2n/N} \quad \text{for} \quad n = 0, 1, \ldots, N - 1;$$

$$\vdots$$

and

$$E_{N-1}(n) = \tfrac{1}{\sqrt{N}} e^{2\pi i (N-1)n/N} \quad \text{for} \quad n = 0, 1, \ldots, N - 1.$$

More concisely,

$$E_m(n) = \frac{1}{\sqrt{N}} e^{2\pi i mn/N} \quad \text{for} \quad 0 \leq m, n \leq N - 1. \qquad (2.2)$$

We use the capital E notation to suggest the exponential function instead of using the lower-case e, which we reserve for the standard basis vectors noted earlier.

Lemma 2.2 *The set $\{E_0, \ldots, E_{N-1}\}$ is an orthonormal basis for $\ell^2(\mathbb{Z}_N)$.*

Proof

Suppose $j, k \in \{0, 1, \ldots, N - 1\}$. Then

$$\langle E_j, E_k \rangle = \sum_{n=0}^{N-1} E_j(n) \overline{E_k(n)} = \sum_{n=0}^{N-1} \frac{1}{\sqrt{N}} e^{2\pi i j n/N} \overline{\frac{1}{\sqrt{N}} e^{2\pi i k n/N}}$$

$$= \frac{1}{N} \sum_{n=0}^{N-1} e^{2\pi i j n/N} e^{-2\pi i k n/N} = \frac{1}{N} \sum_{n=0}^{N-1} e^{2\pi i (j-k)n/N}$$

$$= \frac{1}{N} \sum_{n=0}^{N-1} \left(e^{2\pi i(j-k)/N}\right)^n,$$

where we have used equations (1.14), (1.16), and (1.17). If $j = k$, the terms inside the last sum are all 1, so $\langle E_j, E_j \rangle = N^{-1} \sum_{j=0}^{N-1} 1 = 1$. Hence $\|E_j\|^2 = 1$ for each j, so each E_j has norm one. If $j \neq k$, then $e^{2\pi i(j-k)/N} \neq 1$ for $0 \leq j, k \leq N - 1$ since $-N < j - k < N$. Therefore the sum is the partial sum of a geometric series, so

$$\sum_{n=0}^{N-1} \left(e^{2\pi i(j-k)/N}\right)^n = \frac{1 - \left(e^{2\pi i(j-k)/N}\right)^N}{1 - e^{2\pi i(j-k)/N}},$$

by equation (1.5). But by equation (1.17),

$$\left(e^{2\pi i(j-k)/N}\right)^N = e^{2\pi i(j-k)} = 1,$$

since $j - k$ is an integer. So for $j \neq k$, $\langle E_j, E_k \rangle = 0$, that is, $E_j \perp E_k$. Thus $\{E_0, \ldots, E_{N-1}\}$ is an orthonormal set, and hence is linearly independent, by Lemma 1.95. Therefore $\{E_0, \ldots, E_{N-1}\}$ is a basis for $\ell^2(\mathbb{Z}_N)$ (by Theorem 1.42). ∎

Example 2.3
Let $N = 2$. Then

$$E_0 = \left(E_0(0), E_0(1)\right) = \frac{1}{\sqrt{2}}(1, 1),$$

and

$$E_1 = \left(E_1(0), E_1(1)\right) = \frac{1}{\sqrt{2}}(1, -1),$$

since $e^{2\pi i 0/2} = 1$ and $e^{2\pi i 1/2} = e^{i\pi} = -1$ by Euler's formula (Theorem 1.22). It is clear that $\{E_0, E_1\}$ is an orthonormal basis for $\ell^2(\mathbb{Z}_2)$.

The values in the case $N = 3$ do not work out as simply, so we pass on to $N = 4$.

Example 2.4
Let $N = 4$. Then (Exercise 2.1.1)

$$E_0 = \frac{1}{2}(1, 1, 1, 1),$$

$$E_1 = \frac{1}{2}(1, i, -1, -i),$$

$$E_2 = \frac{1}{2}(1, -1, 1, -1),$$

and

$$E_3 = \frac{1}{2}(1, -i, -1, i).$$

One can check directly (Exercise 2.1.2) that $\{E_0, E_1, E_2, E_3\}$ is an orthonormal basis for $\ell^2(\mathbb{Z}_4)$.

Since $\{E_0, E_1, \ldots, E_{N-1}\}$ is an orthonormal basis for $\ell^2(\mathbb{Z}_N)$, equations (1.49), (1.50), and (1.51) give us, for all $z, w \in \ell^2(\mathbb{Z}_N)$,

$$z = \sum_{m=0}^{N-1} \langle z, E_m \rangle E_m, \tag{2.3}$$

$$\langle z, w \rangle = \sum_{m=0}^{N-1} \langle z, E_m \rangle \overline{\langle w, E_m \rangle} \tag{2.4}$$

and

$$\|z\|^2 = \sum_{m=0}^{N-1} |\langle z, E_m \rangle|^2. \tag{2.5}$$

By definition,

$$\langle z, E_m \rangle = \sum_{n=0}^{N-1} z(n) \overline{\frac{1}{\sqrt{N}} e^{2\pi i m n/N}} = \frac{1}{\sqrt{N}} \sum_{n=0}^{N-1} z(n) e^{-2\pi i m n/N}. \tag{2.6}$$

Although equations (2.3) and (2.6) are the most natural equations from the standpoint of orthonormal bases, a renormalized version in which the factor of $1/\sqrt{N}$ in equation (2.6) is deleted is usually used in practice.

Definition 2.5 *Suppose $z = (z(0), \ldots, z(N-1)) \in \ell^2(\mathbb{Z}_N)$. For $m = 0, 1, \ldots, N-1$, define*

$$\hat{z}(m) = \sum_{n=0}^{N-1} z(n) e^{-2\pi i m n/N}. \tag{2.7}$$

Let

$$\hat{z} = (\hat{z}(0), \hat{z}(1), \ldots, \hat{z}(N-1)). \tag{2.8}$$

Then $\hat{z} \in \ell^2(\mathbb{Z}_N)$. The map $\hat{\ }: \ell^2(\mathbb{Z}_N) \to \ell^2(\mathbb{Z}_N)$, which takes z to $\hat{z}$, is called the discrete Fourier transform, *usually abbreviated DFT.*

Notice that if we use formula (2.7) to define $\hat{z}(m)$ for all $m \in \mathbb{Z}$, the result is periodic with period N:

$$\hat{z}(m+N) = \sum_{n=0}^{N-1} z(n)e^{-2\pi i(m+N)n/N}$$

$$= \sum_{n=0}^{N-1} z(n)e^{-2\pi imn/N}e^{-2\pi iNn/N} = \hat{z}(m),$$

since $e^{-2\pi iNn/N} = e^{-2\pi in} = 1$ for every $n \in \mathbb{Z}$. Thus using formula (2.7) for all m is consistent with regarding $\hat{z}$ as an element of $\ell^2(\mathbb{Z}_N)$, thought of as defined on $\mathbb{Z}$ and having period N.

There are a couple of advantages of formula (2.7) compared with formula (2.6). First, for numerical calculations, it is better to avoid computing $\sqrt{N}$. Second, we will see later that certain formulas (e.g., the formula for the DFT of a convolution, in Lemma 2.30 below) are simpler with the normalization in formula (2.7). Comparing formulae (2.7) and (2.6), note that

$$\hat{z}(m) = \sqrt{N}\langle z, E_m\rangle. \tag{2.9}$$

This leads to the following reformulation of formulae (2.3), (2.4), and (2.5).

Theorem 2.6   Let $z = (z(0), z(1), \ldots, z(N-1)), w = (w(0), w(1), \ldots, w(N-1)) \in \ell^2(\mathbb{Z}_N)$. Then
   i. (Fourier inversion formula)

$$z(n) = \frac{1}{N}\sum_{m=0}^{N-1} \hat{z}(m)e^{2\pi imn/N} \text{ for } n = 0, 1, \ldots, N-1. \tag{2.10}$$

   ii. (Parseval's relation)

$$\langle z, w\rangle = \frac{1}{N}\sum_{m=0}^{N-1} \hat{z}(m)\overline{\hat{w}(m)} = \frac{1}{N}\langle \hat{z}, \hat{w}\rangle. \tag{2.11}$$

   iii. (Plancherel's formula)

$$\|z\|^2 = \frac{1}{N}\sum_{m=0}^{N-1} |\hat{z}(m)|^2 = \frac{1}{N}\|\hat{z}\|^2. \tag{2.12}$$

Proof

By equations (2.3), (2.9), and (2.2), we have

$$z(n) = \sum_{m=0}^{N-1} \langle z, E_m \rangle E_m(n) = \sum_{m=0}^{N-1} N^{-1/2} \hat{z}(m) N^{-1/2} e^{2\pi i m n/N},$$

which gives equation (2.10). Similarly, by equation (2.4),

$$\langle z, w \rangle = \sum_{m=0}^{N-1} \langle z, E_m \rangle \overline{\langle w, E_m \rangle} = \sum_{m=0}^{N-1} N^{-1/2} \hat{z}(m) N^{-1/2} \overline{\hat{w}(m)},$$

yielding equation (2.11). Then equation (2.12) follows either by a similar argument or by letting $w = z$ in equation (2.11). ∎

To interpret the Fourier inversion formula (2.10), we make the following definition.

Definition 2.7   *For $m = 0, 1, \ldots, N-1$, define $F_m \in \ell^2(\mathbb{Z}_N)$ by*

$$F_m(n) = \frac{1}{N} e^{2\pi i m n/N}, \quad \text{for} \quad n = 0, 1, \ldots, N-1. \tag{2.13}$$

Let

$$F = \{F_0, F_1, \ldots, F_{N-1}\}. \tag{2.14}$$

We call F the Fourier basis *for $\ell^2(\mathbb{Z}_N)$.*

By equation (2.2), $F_m = N^{-1/2} E_m$. Hence Lemma 2.2 shows that F, as its name suggests, is a basis (in fact an orthogonal basis) for $\ell^2(\mathbb{Z}_N)$. With this notation, equation (2.10) becomes

$$z = \sum_{m=0}^{N-1} \hat{z}(m) F_m. \tag{2.15}$$

In other words, if we expand z in terms of the Fourier basis F, the coefficient of F_m is $\hat{z}(m)$. Therefore, the vector representing z with respect to the Fourier basis is $\hat{z}$; that is,

$$\hat{z} = [z]_F, \tag{2.16}$$

in the notation of Definition 1.43. Thus the Fourier inversion formula (2.10) is the change-of-basis formula for the Fourier basis. The DFT components $\hat{z}(m)$ are the components of z in the Fourier basis.

The DFT can be represented by a matrix (as we should expect because equation (2.7) shows that the map taking z to $\hat{z}$ is a linear transformation). To simplify notation, define

$$\omega_N = e^{-2\pi i/N}.$$

Then

$$e^{-2\pi imn/N} = \omega_N^{mn}$$

and

$$e^{2\pi imn/N} = \omega_N^{-mn}.$$

In this notation,

$$\hat{z}(m) = \sum_{n=0}^{N-1} z(n)\omega_N^{mn}. \tag{2.17}$$

To match up with our notation for vectors, we modify our notation for matrices by indexing the rows and columns from 0 to $N-1$ instead of from 1 to N.

Definition 2.8   *Let W_N be the matrix $[w_{mn}]_{0 \le m,n \le N-1}$ such that $w_{mn} = \omega_N^{mn}$. Written out, this is*

$$W_N = \begin{bmatrix} 1 & 1 & 1 & 1 & \cdot & \cdot & 1 \\ 1 & \omega_N & \omega_N^2 & \omega_N^3 & \cdot & \cdot & \omega_N^{N-1} \\ 1 & \omega_N^2 & \omega_N^4 & \omega_N^6 & \cdot & \cdot & \omega_N^{2(N-1)} \\ 1 & \omega_N^3 & \omega_N^6 & \omega_N^9 & \cdot & \cdot & \omega_N^{3(N-1)} \\ \cdot & \cdot & \cdot & \cdot & \cdot & \cdot & \cdot \\ \cdot & \cdot & \cdot & \cdot & \cdot & \cdot & \cdot \\ 1 & \omega_N^{N-1} & \omega_N^{2(N-1)} & \omega_N^{3(N-1)} & \cdot & \cdot & \omega_N^{(N-1)(N-1)} \end{bmatrix}. \tag{2.18}$$

Regarding z, $\hat{z} \in \ell^2(\mathbb{Z}_N)$ as column vectors, as in equation (2.1), the mth component ($0 \le m \le N-1$) of $W_N z$ is $\sum_{n=0}^{N-1} w_{mn} z(n) = \sum_{n=0}^{N-1} z(n)\omega_N^{mn}$, which is $\hat{z}(m)$, by equation (2.17). In other words,

$$\hat{z} = W_N z. \tag{2.19}$$

In section 2.3, we will see that there is a fast algorithm for computing $\hat{z}$. For now, we only compute a simple example in order to demonstrate the definitions. We could use equation (2.9) for this, but it is easier to use equation (2.19). For convenience, we record the values obtained in W_2 and W_4:

$$W_2 = \begin{bmatrix} 1 & 1 \\ 1 & -1 \end{bmatrix} \tag{2.20}$$

and

$$W_4 = \begin{bmatrix} 1 & 1 & 1 & 1 \\ 1 & -i & -1 & i \\ 1 & -1 & 1 & -1 \\ 1 & i & -1 & -i \end{bmatrix}. \tag{2.21}$$

Example 2.9
Let $z = (1, 0, -3, 4) \in \ell^2(\mathbb{Z}_4)$. Find $\hat{z}$.

Solution

$$\hat{z} = W_4 z = \begin{bmatrix} 1 & 1 & 1 & 1 \\ 1 & -i & -1 & i \\ 1 & -1 & 1 & -1 \\ 1 & i & -1 & -i \end{bmatrix} \begin{bmatrix} 1 \\ 0 \\ -3 \\ 4 \end{bmatrix} = \begin{bmatrix} 2 \\ 4 + 4i \\ -6 \\ 4 - 4i \end{bmatrix}.$$

Note that $\hat{z}$ has complex entries even though all entries of z are real. ∎

The Fourier inversion formula (2.10) shows that the linear transformation $\hat{\ } : \ell^2(\mathbb{Z}_N) \to \ell^2(\mathbb{Z}_N)$ is a 1-1 map: if $\hat{z} = \hat{w}$, then $z = w$. Therefore $\hat{\ }$ is invertible (Exercise 1.4.8(v)). More directly, equation (2.10) gives us a formula for the inverse of $\hat{\ }$, which we denote $\check{\ }$.

Definition 2.10 *For* $w = (w(0), \ldots, w(N-1)) \in \ell^2(\mathbb{Z}_N)$, *define*

$$\check{w}(n) = \frac{1}{N} \sum_{m=0}^{N-1} w(m) e^{2\pi i m n/N}, \text{ for } n = 0, 1, \ldots, N-1. \tag{2.22}$$

We set

$$\check{w} = \big(\check{w}(0), \check{w}(1), \ldots, \check{w}(N-1)\big).$$

The map $\check{\ } : \ell^2(\mathbb{Z}_N) \to \ell^2(\mathbb{Z}_N)$ *is the* inverse discrete Fourier transform, *or* IDFT.

In this notation, equation (2.10) states that for $z \in \ell^2(\mathbb{Z}_N)$,

$$(\hat{z})\check{\ }(n) = z(n), \quad \text{for } n = 0, 1, \ldots, N-1, \tag{2.23}$$

or just

$$(\hat{z})^{\vee} = z. \tag{2.24}$$

For a general $w \in \ell^2(\mathbb{Z}_N)$, there exists $z \in \ell^2(\mathbb{Z}_N)$ such that $\hat{z} = w$ (since $\hat{} : \ell^2(\mathbb{Z}_N) \to \ell^2(\mathbb{Z}_N)$ is onto). Taking the DFT of both sides of equation (2.24) and substituting w for $\hat{z}$ gives

$$(\check{w})^{\wedge} = w. \tag{2.25}$$

Since the DFT is an invertible linear transformation, the matrix W_N is invertible (Lemma 1.55), and we must have $z = W_N^{-1}\hat{z}$. Substituting $\hat{z} = w$ and (equivalently) $z = \check{w}$ in equation (2.19) gives

$$\check{w} = W_N^{-1}w. \tag{2.26}$$

Although one can determine W_N^{-1} directly (Exercise 2.1.9), it is easier to read what W_N^{-1} must be from formula (2.22). In the notation of formula (2.17), formula (2.22) becomes

$$\check{w}(n) = \sum_{m=0}^{N-1} w(m)\frac{1}{N}\omega_N^{-mn} = \sum_{m=0}^{N-1} \frac{1}{N}\overline{\omega_N^{nm}}w(m).$$

This shows that the (n, m) entry of W_N^{-1} is $\overline{\omega_N^{nm}}/N$, which is $1/N$ times the complex conjugate of the (n, m) entry of W_N. If we denote by $\overline{W_N}$ the matrix whose entries are the complex conjugates of the entries of W_N, we have

$$W_N^{-1} = \frac{1}{N}\overline{W_N}. \tag{2.27}$$

For future reference, we note that

$$W_2^{-1} = \frac{1}{2}\begin{bmatrix} 1 & 1 \\ 1 & -1 \end{bmatrix}, \tag{2.28}$$

and

$$W_4^{-1} = \frac{1}{4}\begin{bmatrix} 1 & 1 & 1 & 1 \\ 1 & i & -1 & -i \\ 1 & -1 & 1 & -1 \\ 1 & -i & -1 & i \end{bmatrix}. \tag{2.29}$$

Example 2.11
Let $w = (2, 4 + 4i, -6, 4 - 4i) \in \ell^2(\mathbb{Z}_4)$. Find $\check{w}$.

Solution

By formulas (2.26) and (2.29),

$$\check{w} = \frac{1}{4} \begin{bmatrix} 1 & 1 & 1 & 1 \\ 1 & i & -1 & -i \\ 1 & -1 & 1 & -1 \\ 1 & -i & -1 & i \end{bmatrix} \begin{bmatrix} 2 \\ 4+4i \\ -6 \\ 4-4i \end{bmatrix} = \begin{bmatrix} 1 \\ 0 \\ -3 \\ 4 \end{bmatrix}.$$

Note that for z as in Example 2.9, we obtained $\hat{z} = w$, so we have just verified in this example that $(\hat{z})^\vee = z$. ∎

By the same reasoning as for the DFT, if we regard $\check{w}$ as defined on $\mathbb{Z}$ by formula (2.22), then $\check{w}$ has period N: $\check{w}(n+N) = \check{w}(n)$ for all n. With this understanding, comparing formulas (2.7) and (2.22), we see that

$$\check{w}(n) = \frac{1}{N}\hat{w}(-n).$$

Since $\check{w}$ has period N, we can write this as

$$\check{w}(n) = \frac{1}{N}\hat{w}(N-n), \tag{2.30}$$

which is more convenient since $n \in \{1, \ldots, N-1\}$ if and only if $N - n \in \{1, \ldots, N-1\}$; the exceptional case is $n = 0$, for which $N - n = N$.

We summarize the basic facts about the DFT. The map $\hat{}$: $\ell^2(\mathbb{Z}_N) \to \ell^2(\mathbb{Z}_N)$ defined by equations (2.7) and (2.8) is an invertible linear transformation with inverse $\check{}$ defined by formula (2.22). We interpret the Fourier inversion formula (2.10) in the form of equation (2.15):

$$z = \sum_{m=0}^{N-1} \hat{z}(m)F_m,$$

where F_m is the mth element of the Fourier basis F:

$$F_m(n) = \frac{1}{N}e^{2\pi imn/N}.$$

Thus $\hat{z}(m)$ is the weight of the vector F_m used in making up z.

We consider a simple example. Let $N = 128$ and

$$z(n) = \cos\left(2\pi \cdot \frac{7n}{128}\right) + 4\cos\left(2\pi \cdot \frac{12n}{128}\right).$$

The vector z is plotted in Figure 4a. Its DFT $\hat{z}$ is plotted in Figure 4b, but in this case it is easy to determine $\hat{z}$. By Euler's formula (1.13),

$$z(n) = \frac{1}{2}\left(e^{2\pi i 7n/128} + e^{-2\pi i 7n/128}\right) + 4\frac{1}{2}\left(e^{2\pi i 12n/128} + e^{-2\pi i 12n/128}\right)$$

$$= \frac{1}{128}\left(64e^{2\pi i 7n/128} + 64e^{2\pi i 121/128}\right.$$

$$\left. + 256e^{2\pi i 12n/128} + 256e^{2\pi i 116n/128}\right).$$

By comparing this with equations (2.13) and (2.15), we see that

$$\hat{z}(7) = \hat{z}(121) = 64, \ \hat{z}(12) = \hat{z}(116) = 256,$$

and $\hat{z}(m) = 0$ for the other values of m in $\{0, 1, 2, \ldots, 127\}$. This is confirmed by Figure 4b, which was generated using the DFT program (called fft) in Matlab.

To get an intuition for this, we take a closer look at the vector $e^{2\pi i mn/N}$, for m fixed, as a function of $n = 0, 1, \ldots, N-1$. (The factor $1/N$ is a scale factor, which we temporarily drop for convenience.) By Euler's formula,

$$e^{2\pi i nm/N} = \cos(2\pi mn/N) + i\sin(2\pi mn/N).$$

For simplicity, consider only the real part $\cos(2\pi nm/N)$. For $m = 0$, this is just the constant function 1. For $m = 1$, this is the function $\cos(2\pi n/N)$. If we regard N as being large, and plot these values for $n = 0, 1, \ldots, N-1$, we trace out a set of N evenly spaced sample

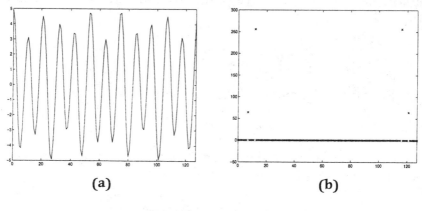

(a)   (b)

FIGURE 4   (a) z, (b) $\hat{z}$

points on the graph of one period of a cosine function. In particular, $\cos(2\pi n/N)$ carries out one full cycle as n runs from 0 to $N-1$ (if we continued it to N, we would obtain the value at 0 again). This is illustrated in Figure 5a, with $N = 16$. Now let $m = 2$, and consider the function $\cos(2\pi 2n/N)$. It is similar, only it makes two full cycles of the cosine as n runs from 0 to $N-1$. This continues similarly for a while as m increases; the function $\cos(2\pi mn/N)$ carries out m full cycles of the cosine wave as n goes from 0 to $N-1$ (see Figure 5b). So as m increases (up to a point, as we will see), the functions $\cos(2\pi mn/N)$ oscillate more and more times over the interval $0 \le n \le N-1$. See Figures 5c ($m = 3$, $N = 16$), 5d ($m = 7$, $N = 16$), and 5e ($m = 8$, $N = 16$). The imaginary part of $e^{2\pi inm/N}$ is a sine wave that behaves similarly. Terms that oscillate more are called higher frequency components, or just higher frequencies. Suppose $z(n)$ is a sound signal as a function of time (sampled at even time intervals). To the ear, the higher frequency signals sound higher pitched. Because the vectors $e^{2\pi imn/N}$ contain only one rate of oscillation, we regard them, and hence their rescaled versions F_m, as pure frequencies.

There is one technicality here that makes this a little confusing. The functions $e^{2\pi inm/N}$, defined only for integer values of n, have period N in the variable m as well as in n. Thus higher m cannot continue to mean higher frequency signals for all m because, for example, the function at $m = N$ is the same as the function when $m = 0$. A signal for $N = 16$ cannot oscillate faster than in Figure 5e ($m = 8$, $N = 16$). Note that the vector in Figure 5f ($m = 9$, $N = 16$) is the same as in Figure 5d ($m = 7$, $N = 16$). (However, it is not true that $F_7 = F_9$ when $N = 16$, just that their real parts are the same—their imaginary parts are negatives of each other.) This is a consequence of considering these functions only for values $n \in \mathbb{Z}$. If we consider $e^{2\pi imt/N}$ as a function of the real variable t, it does give a higher frequency as m increases (for $m \ge 0$). But when we consider only the integer values n, we cannot see the oscillation going on between the integers, which explains how $e^{2\pi in0/N}$ can be the same function of n as $e^{2\pi inN/N}$ (namely both are the constant function 1 on the integers, since $e^{2\pi in} = 1$, by Euler's formula). This is illustrated in Figure 6, which shows the graph of $\cos(2\pi 15n/16)$ (Figure 6a), the

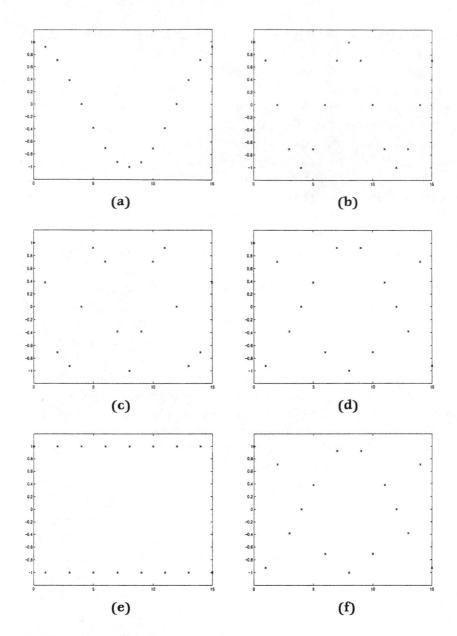

FIGURE 5 **(a)** $z(n) = \cos(2\pi n/16)$, **(b)** $z(n) = \cos(2\pi 2n/16)$, **(c)** $z(n) = \cos(2\pi 3n/16)$, **(d)** $z(n) = \cos(2\pi 7n/16)$, **(e)** $z(n) = \cos(2\pi 8n/16)$, **(f)** $z(n) = \cos(2\pi 9n/16)$

underlying function of a continuous variable $\cos(2\pi 15t/16)$ (Figure 6b), and their superposition (Figure 6c).

Looking at the graphs of the real and imaginary parts, we see that as m runs from 0 to $N/2$ (or the closest integer to $N/2$), the function $e^{2\pi i nm/N}$ oscillates more and more rapidly and so represents a higher frequency component. However, for $N/2 \leq m \leq N - 1$, we let $k = N - m$, and use periodicity to write

$$e^{2\pi i nm/N} = e^{2\pi i n(N-k)/N} = e^{-2\pi i nk/N} = \cos(2\pi nk/N) - i\sin(2\pi nk/N).$$

This vector oscillates k times over the interval $0 \leq n \leq N - 1$. So the number of oscillations over $0, 1, \ldots, N - 1$ of $e^{2\pi i nm/N}$ is m for $0 \leq m \leq N/2$, but $N - m$ for $N/2 \leq m \leq N - 1$. Note that when m is near $N/2$, so is $N - m$, but as m goes toward N, $N - m$ goes toward 0. Thus we regard $e^{2\pi i mn/N}$ as a high frequency vector for m near $N/2$, that is, in the middle of the range $0, 1, \ldots, N - 1$, and as a low frequency vector for m near 0 or $N - 1$.

Incidentally, this is why some prefer to regard the basic interval for $\mathbb{Z}_N$ as $-M + 1, -M + 2, \ldots, -1, 0, 1, \ldots, M$ when $N = 2M$ is even and as $-M, -M + 1, \ldots - 1, 0, 1, \ldots, M - 1, M$ when $N = 2M + 1$ is odd, instead of $0, 1, \ldots, N - 1$ as we have here. In the 0-centered version, the low frequencies are $e^{2\pi i mn/N}$ for m near 0 and the high frequencies are those for m near $\pm M$.

Returning to the basic formula

$$z = \sum_{m=0}^{N-1} \hat{z}(m)F_m,$$

we have interpreted each F_m as a pure frequency. Therefore, we regard $|\hat{z}(m)|$ as the strength of that frequency component in z. The argument of the complex number $\hat{z}(m)$ is more difficult to interpret (this is discussed later in this chapter), but its magnitude measures how much of the pure frequency F_m is needed to make up z. If z has the property that $|\hat{z}(m)|$ is large for values of m near $N/2$, then z has strong high-frequency components. If $|\hat{z}(m)|$ is large for m near 0 and near $N - 1$, then z has strong low-frequency components. For example, if z is the appropriately sampled audio signal of a person playing the drums, we will see relatively large values of $|\hat{z}(m)|$ when m is small if the drummer likes to pound on the bass drum a lot. If the drummer likes to bang on the cymbals, $|\hat{z}(m)|$ will be large

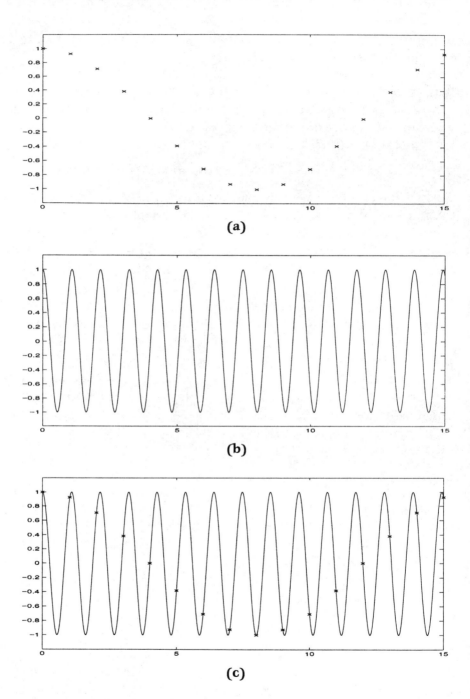

FIGURE 6 **(a)** $z(n) = \cos(2\pi 15n/16)$, **(b)** $y(t) = \cos(2\pi 15t/16)$, **(c)** Superposition of Figure 6a and Figure 6b

for values of m near the middle of the DFT range. Thus $\hat{z}$ gives us a frequency analysis of the signal z.

This phenomenon is illustrated in Figure 7, where $N = 128$. The graph of the vector $z(n) = \sin(\pi n^2/256)$ in Figure 7a oscillates more and more rapidly as n goes from 0 to 127. (Such a signal is called a "chirp.") Figure 7b shows the argument (phase) of $\hat{z}$, Figure 7c shows the real part of $\hat{z}$, and Figure 7d shows the imaginary part of $\hat{z}$. All three of these graphs seem to behave wildly and are difficult to interpret. However, the graph of $|\hat{z}|$, the magnitude of $\hat{z}$, in Figure 7e, shows significantly large values (between 5 and 9) of $|\hat{z}(n)|$ for all values of n. The interpretation is that a large range of frequencies is required to make up z, due to the large variation in its oscillation rate over different parts of its graph.

A comparison of Figures 7, 8, and 9 makes this behavior clearer. Figure 8a is the graph of $z(n) = \sin(\pi n^2/512)$. This is also a chirp, but the oscillation rate of this z is less at corresponding points than for the vector in Figure 7a. This suggests that much less in the way of high frequencies is required to synthesize z. This is demonstrated by the graph of $|\hat{z}|$ in Figure 8b. Note that only about half the values of $|\hat{z}(n)|$ (those corresponding to the lower frequencies) are large. Our interpretation is further demonstrated in Figure 9. In Figure 9a, the function $z(n) = \sin(\pi n^2/1024)$ is plotted. It is an even lower frequency chirp than in Figure 8a. The magnitude of its DFT, plotted in Figure 9b, shows that roughly only the lower fourth of the frequencies contribute substantially to z.

In Exercise 2.1.7 there is an expansion of z in terms of sines and cosines that is equivalent to formula (2.10). With this expansion, it is a little easier to see the high and low frequency interpretations just discussed. However, this expansion fails to have some of the key properties of formula (2.10), which we consider in the next two sections.

Next we consider how the DFT behaves under a few important operations. The first of these is translation.

Definition 2.12 *Suppose* $z \in \ell^2(\mathbb{Z}_N)$ *and* $k \in \mathbb{Z}$. *Define*

$$(R_k z)(n) = z(n - k) \quad \text{for} \quad n \in \mathbb{Z}.$$

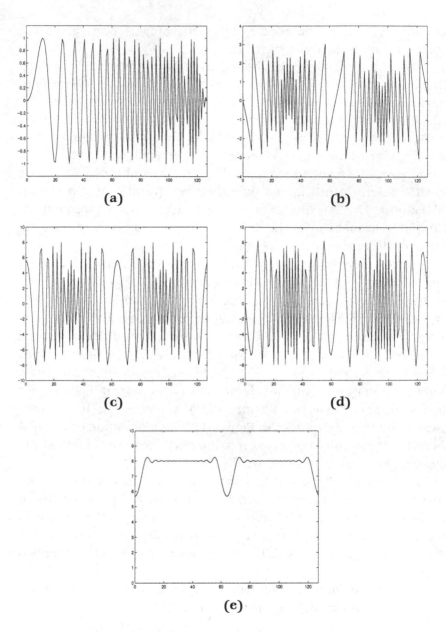

FIGURE 7 **(a)** $z(n) = \sin(\pi n^2/256)$, **(b)** Phase (angle) of $\hat{z}$, **(c)** Real part of $\hat{z}$, **(d)** Imaginary part of $\hat{z}$, **(e)** Magnitude of $\hat{z}$

We call $R_k z$ the translate *of z by k. We call R_k the* translation by k operator.

Perhaps we should write $R_k(z)$, but this leads to too many parentheses, so we write only $R_k z$. Definition 2.12 requires some interpretation. For $n, k \in \mathbb{Z}_N$, it may be that $n - k \notin \mathbb{Z}_N$, so it may appear that $z(n - k)$ is not defined. However, recall that we regard z as extended to all of $\mathbb{Z}$ in such a way that z has period N. With this understanding, Definition 2.12 makes sense. As an example, suppose $N = 6$, $k = 2$, and $z = (2, 3 - i, 2i, 4 + i, 0, 1)$. Then, for example,

$$(R_2 z)(0) = z(0 - 2) = z(-2) = z(4) = 0.$$

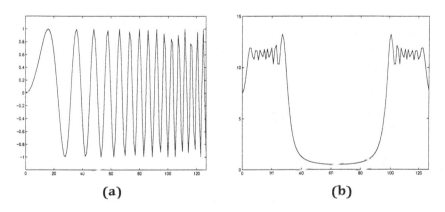

(a)   (b)

FIGURE 8 **(a)** $z(n) = \sin(\pi n^2 / 512)$, **(b)** Magnitude of $\hat{z}$

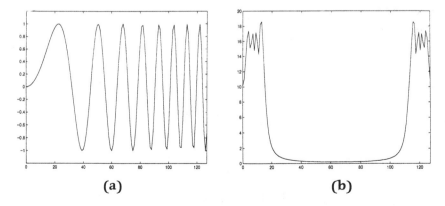

(a)   (b)

FIGURE 9 **(a)** $z(n) = \sin(\pi n^2 / 1024)$, **(b)** Magnitude of $\hat{z}$

Similarly, $(R_2z)(1) = z(-1) = z(5) = 1, (R_2z)(2) = z(0) = 2$, etc. We obtain

$$R_2z = (0, 1, 2, 3 - i, 2i, 4 + i).$$

Thus the effect of R_2 on z is to move the components two positions to the right, except for the last two, which were rotated around into the first two positions, in the same order as originally. This can be visualized as a rotation by two if the positions $0, 1, 2, 3, 4, 5$ are marked off on a circle. For this reason, this operation is sometimes known as *circular translation* or *rotation*, which explains the notation R_k.

How is the DFT affected by translation? It is intuitive that translation should not affect the magnitudes of the different frequencies making up the signal, but it might change their phase (i.e., the angle in the polar representation $re^{i\theta}$ of $\hat{z}(n)$). This is verified by the next result.

Lemma 2.13   *Suppose $z \in \ell^2(\mathbb{Z}_N)$ and $k \in \mathbb{Z}$. Then for any $m \in \mathbb{Z}$,*

$$(R_kz)\hat{\ }(m) = e^{-2\pi imk/N}\hat{z}(m).$$

Proof

By definition,

$$(R_kz)\hat{\ }(m) = \sum_{n=0}^{N-1}(R_kz)(n)e^{-2\pi imn/N} = \sum_{n=0}^{N-1}z(n-k)e^{-2\pi imn/N}.$$

In this last sum, we change variables by letting $\ell = n - k$ (recall k is fixed, and n is the summation variable). If $n = 0$, then $\ell = -k$, whereas for $n = N - 1$, we have $\ell = N - k - 1$. Since $n = \ell + k$, we obtain

$$(R_kz)\hat{\ }(m) = \sum_{\ell=-k}^{N-k-1}z(\ell)e^{-2\pi im(\ell+k)/N} = e^{-2\pi imk/N}\sum_{\ell=-k}^{N-k-1}z(\ell)e^{-2\pi im\ell/N}.$$

However, we claim that

$$\sum_{\ell=-k}^{N-k-1}z(\ell)e^{-2\pi im\ell/N} = \sum_{n=0}^{N-1}z(n)e^{-2\pi imn/N} = \hat{z}(m). \qquad (2.31)$$

If so, substituting this gives the final result. To see equation (2.31), note that both $z(\ell)$ and $e^{-2\pi im\ell/N}$ are periodic functions in the variable

ℓ with period N. If $k = 0$, there is nothing to prove, so suppose $0 < k \leq N - 1$. Then

$$\sum_{\ell=-k}^{N-k-1} z(\ell)e^{-2\pi i m \ell/N} = \sum_{\ell=-k}^{-1} z(\ell+N)e^{-2\pi i m(\ell+N)/N} + \sum_{\ell=0}^{N-k-1} z(\ell)e^{-2\pi i m \ell/N}.$$

In the first of these last two sums, we let $n = \ell + N$, whereas in the second we just let $n = \ell$. This gives us

$$\sum_{\ell=-k}^{N-k-1} z(\ell)e^{-2\pi i m \ell/N} = \sum_{n=N-k}^{N-1} z(n)e^{-2\pi i m n/N} + \sum_{n=0}^{N-k-1} z(n)e^{-2\pi i m n/N}$$

$$= \sum_{n=0}^{N-1} z(n)e^{-2\pi i m n/N},$$

as desired. Now let $k \in \mathbb{Z}$ be arbitrary. Then there is some integer r such that $k' = k + rN \in \{0, 1, 2, \ldots, N-1\}$. Then, by changing the summation variable by setting $\ell' = \ell - rN$ we get

$$\sum_{\ell=-k}^{N-k-1} z(\ell)e^{-2\pi i m \ell/N} = \sum_{\ell'=-k-rN}^{N-k-rN-1} z(\ell'+rN)e^{-2\pi i m(\ell'+rN)/N}$$

$$= \sum_{\ell'=-k'}^{N-k'-1} z(\ell')e^{-2\pi i m \ell'/N},$$

by the N-periodicity of both z and the exponential. So by the case $k' \in \{0, 1, 2, \ldots, N-1\}$ considered above, the last sum is $\hat{z}(m)$. ∎

This last summation trick is important enough to be made into a general principle (see Exercise 2.1.8).

This shows a limitation of the DFT. By Lemma 2.13 and equation (1.15), we have $|(R_k z)\hat{\ }(m)| = |\hat{z}(m)|$ for each m. Thus, by looking only at the magnitude of the DFT, we cannot distinguish z from any circular translation $R_k z$. The locations of the particular features of z are not determined by $|\hat{z}|$. That information is contained in the phase (or argument) of $\hat{z}$, but in this form it is very difficult to interpret, as we noted in connection with Figure 7b. We will see later that wavelets have an advantage over the DFT in this regard.

The next operation we consider is complex conjugation

Definition 2.14   *For* $z = (z(0), z(1), \ldots, z(N-1)) \in \ell^2(\mathbb{Z}_N)$, *let* $\bar{z}$ *be the vector*

$$\bar{z} = (\overline{z(0)}, \overline{z(1)}, \ldots, \overline{z(N-1)}),$$

that is, $\bar{z}(n) = \overline{z(n)}$.

Lemma 2.15   *For* $z \in \ell^2(\mathbb{Z}_N)$,

$$(\bar{z})\hat{}(m) = \overline{\hat{z}(-m)} = \overline{\hat{z}(N-m)},$$

for all m.

Proof

By the properties of complex conjugation (Lemma 1.7),

$$(\bar{z})\hat{}(m) = \sum_{n=0}^{N-1} \bar{z}(n) e^{-2\pi i m n/N} = \overline{\sum_{n=0}^{N-1} z(n) e^{2\pi i m n/N}} = \overline{\hat{z}(-m)}.$$

∎

Corollary 2.16   *Suppose* $z \in \ell^2(\mathbb{Z}_N)$. *Then* z *is real (i.e., every component of* z *is a real number) if and only if* $\hat{z}(m) = \overline{\hat{z}(N-m)}$ *for all* m.

Proof

Note that z is real if and only if $z = \bar{z}$. By the invertibility of the DFT, this holds if and only if $\hat{z} = (\bar{z})\hat{}$. By Lemma 2.15, this is equivalent to $\hat{z}(m) = \overline{\hat{z}(N-m)}$ for all m. ∎

The DFT is the change-of-basis operator that converts from the Euclidean basis to the Fourier basis. We have interpreted the elements of the Fourier basis as pure frequencies. The Fourier inversion formula (2.15) shows that a general signal consists of a superposition of pure frequencies. The DFT component $\hat{z}(m)$ measures the strength of the pure frequency F_m in the signal z. In the next two sections, we see two strong reasons for using the Fourier basis.

Exercises

2.1.1. Use Definition 2.1 and Euler's formula to check the values given in Example 2.4.

2.1.2. Check directly (not using Lemma 2.2) that the set $\{E_0, E_1, E_2, E_3\}$ in Example 2.4 is an orthonormal basis for $\ell^2(\mathbb{Z}_4)$.

2.1.3. Let $z = (1, i, 2 + i, -3) \in \ell^2(\mathbb{Z}_4)$.
  i. Compute $\hat{z}$.
  ii. Compute $(\hat{z})\check{}$ directly and check that you get z.

2.1.4. Verify formula (2.10) by direct computation (not using Lemma 1.101), that is, substitute the definition of $\hat{z}(m)$ and calculate directly using equation (1.5). Warning: Change one summation index from n to something else, to avoid using the same letter for two different variables.

2.1.5. Verify relation (2.11) by direct computation (as in Exercise 2.1.4).

2.1.6. Define $z \in \ell^2(\mathbb{Z}_{512})$ by

$$z(n) = 3\sin(2\pi 7n/512) - 4\cos(2\pi 8n/512).$$

Find $\hat{z}$.

2.1.7. (DFT in real notation) Suppose N is even, say $N = 2M$. Define

$$c_0(n) = N^{-1/2} \text{ for } n = 0, 1, \ldots, N - 1; \qquad (2.32)$$
$$c_M(n) = N^{-1/2}\cos(2\pi(N/2)n/N) = N^{-1/2}(-1)^n \quad (2.33)$$

for $n = 0, 1, \ldots, N - 1$; and

$$c_m(n) = (\sqrt{2/N})\cos(2\pi mn/N), \text{ for } n = 0, 1, \ldots, N - 1 \qquad (2.34)$$

for $m = 1, 2, \ldots, M - 1$. Also, for $m = 1, 2, \ldots, M - 1$, define

$$s_m(n) = (\sqrt{2/N})\sin(2\pi mn/N) \text{ for } n = 0, 1, \ldots, N - 1. \qquad (2.35)$$

  i. Prove that

$$\{c_0, c_1, \ldots, c_{M-1}, c_M, s_1, \ldots, s_{M-1}\}$$

is an orthonormal basis for $\ell^2(\mathbb{Z}_N)$.
  By Lemma 1.101, this implies that

$$z = \sum_{m=0}^{M} \langle z, c_m \rangle c_m + \sum_{m=1}^{M-1} \langle z, s_m \rangle s_m. \qquad (2.36)$$

This is a variant of the DFT written in real notation. It has the advantage that if z is real, then all computations involve only real numbers. However, we see that this form does not have some of the key advantages discussed in the next two sections. In any case, there are simple transformations relating these coefficients to the DFT coefficients and vice-versa.

ii. Prove that

$$\langle z, c_0 \rangle = N^{-1/2} \hat{z}(0),$$
$$\langle z, c_m \rangle = (2N)^{-1/2} \left(\hat{z}(m) + \hat{z}(N - m) \right)$$

for $m = 1, 2, \ldots, M - 1$,

$$\langle z, c_M \rangle = N^{-1/2} \hat{z}(M),$$

and

$$\langle z, s_m \rangle = -i(2N)^{-1/2} \left(\hat{z}(m) - \hat{z}(N - m) \right)$$

for $m = 1, 2, \ldots, M - 1$.

iii. Conversely, prove that

$$\hat{z}(0) = \sqrt{N} \langle z, c_0 \rangle,$$
$$\hat{z}(m) = \sqrt{N/2} \left(\langle z, c_m \rangle - i \langle z, s_m \rangle \right)$$

for $m = 1, 2, \ldots, M - 1$,

$$\hat{z}(M) = \sqrt{N} \langle z, c_M \rangle,$$

and

$$\hat{z}(m) = \sqrt{N/2} \left(\langle z, c_{N-m} \rangle + i \langle z, s_{N-m} \rangle \right)$$

for $m = M + 1, M + 2, \ldots, N - 1$.

iv. Derive relation (2.36) from formula (2.10) and Euler's formula, without using part i. (Suggestion: Break up the sum in formula (2.10) into four parts: the term for $m = 0$; the term for $m = M$; the sum on $m = 1, 2, \ldots, M - 1$; and the sum on $m = M + 1, M + 2, \ldots, N - 1$. Then substitute the relations from part iii. Change the summation index in the last sum to run over $1, 2, \ldots, M - 1$, combine the c_m terms, and use Euler's formula.)

Remark: If N is odd, say $N = 2J + 1$, then we define c_0 by equation (2.32); c_m by equation (2.34) for $m = 1, 2, \ldots, J$;

and s_m by equation (2.35) for $m = 1, 2, \ldots, J$. This yields an orthonormal basis. We have the same formulae relating the coefficients in this orthonormal expansion to the DFT coefficients as above, except that the exceptional case $m = N/2$ does not occur.

2.1.8. Suppose h is a function on $\mathbb{Z}$ that is periodic with period N, that is, $h(n + N) = h(n)$ for all n. Prove: for any $m \in \mathbb{Z}$,

$$\sum_{n=m}^{m+N-1} h(n) = \sum_{n=0}^{N-1} h(n).$$

In other words, any sum over an interval of length N yields the same result.

2.1.9.   i. Prove that $N^{-1/2} W_N$ is a unitary matrix. (Suggestion: Use Lemma 2.2 and Lemma 1.105 ii.)

ii. Use part i to give a direct proof that $W_N^{-1} = \overline{W_N}/N$.

2.1.10. Let $z = (z(0), z(1), \ldots, z(N - 1))$, $w = (w(0), w(1), \ldots, w(N - 1)) \in \ell^2(\mathbb{Z}_N)$. Prove that

$$\langle \check{z}, \check{w} \rangle = \frac{1}{N} \langle z, w \rangle$$

and

$$\|\check{z}\|^2 = \frac{1}{N} \|z\|^2.$$

2.1.11. Suppose $z \in \ell^2(\mathbb{Z}_N)$. We say z is pure imaginary if $z = iw$ for some w that is real, in other words, if every component of z is pure imaginary. Prove that z is pure imaginary if and only if

$$\hat{z}(m) = -\overline{\hat{z}(N - m)}$$

for all m.

2.1.12. Suppose $z \in \ell^2(\mathbb{Z}_N)$.

i. Prove that $\hat{z}$ is real if and only if $z(m) = \overline{z(N - m)}$ for every m.

ii. Prove that $\hat{z}$ is pure imaginary (see Exercise 2.1.11 for the definition) if and only if $z(m) = -\overline{z(N - m)}$ for every m.

2.1.13. Suppose $z \in \ell^2(\mathbb{Z}_N)$. Define $\tilde{z} \in \ell^2(\mathbb{Z}_N)$ by $\tilde{z}(n) = \overline{z(N-n)}$, for $n = 0, 1, \ldots, N-1$. Prove that

$$(\tilde{z})\hat{}(n) = \overline{\hat{z}(n)},$$

for all n.

2.1.14. Let N and k be positive integers with $k < N$, with $(k, N) = 1$ (this means k and N are relatively prime, i.e., they have no integer factors in common other than ± 1). Let $\omega = e^{2\pi i k/N}$. (Such an ω is called a *primitive N^{th} root of unity*.) Prove that

$$1, \omega, \omega^2, \omega^3, \ldots, \omega^{N-1}$$

are distinct N^{th} roots of unity. Hint: Use the fact from number theory that if $q|ab$ and $(q, b) = 1$, then $q|a$. This is easy to see by considering prime factors.

Remark: This shows that we can define the DFT starting with any primitive N^{th} root of unity ω instead of $e^{2\pi i/N}$; the result is only a permutation of the entries in the DFT matrix.

2.1.15. Let N_1 and N_2 be positive integers. Let

$$\ell^2(\mathbb{Z}_{N_1} \times \mathbb{Z}_{N_2}) = \left\{ z : \mathbb{Z}_{N_1} \times \mathbb{Z}_{N_2} \to \mathbb{C} \right\}.$$

In other words, if $z \in \ell^2(\mathbb{Z}_{N_1} \times \mathbb{Z}_{N_2})$, then for every n_1 and n_2 with $0 \leq n_1 \leq N_1 - 1$ and $0 \leq n_2 \leq N_2 - 1$, $z(n_1, n_2)$ is defined and is a complex number. With the usual addition and scalar multiplication, $\ell^2(\mathbb{Z}_{N_1} \times \mathbb{Z}_{N_2})$ is a vector space over $\mathbb{C}$ (assume this). For $z, w \in \ell^2(\mathbb{Z}_{N_1} \times \mathbb{Z}_{N_2})$, define

$$\langle z, w \rangle = \sum_{n_1=0}^{N_1-1} \sum_{n_2=0}^{N_2-1} z(n_1, n_2) \overline{w(n_1, n_2)}.$$

Then $\langle \cdot, \cdot \rangle$ is a complex inner product on $\ell^2(\mathbb{Z}_{N_1} \times \mathbb{Z}_{N_2})$ (assume this).

Prove that $\ell^2(\mathbb{Z}_{N_1} \times \mathbb{Z}_{N_2})$ is $N_1 N_2$ dimensional. Hint: What plays the role of the standard basis here?

2.1.16. Let $\ell^2(\mathbb{Z}_{N_1} \times \mathbb{Z}_{N_2})$ be as in Exercise 2.1.15. Suppose $\{B_0, B_1, \ldots, B_{N_1-1}\}$ is an orthonormal basis for $\ell^2(\mathbb{Z}_{N_1})$ and $\{C_0, C_1, \ldots, C_{N_2-1}\}$ is an orthonormal basis for $\ell^2(\mathbb{Z}_{N_2})$. For $0 \leq m_1 \leq N_1 - 1$ and $0 \leq m_2 \leq N_2 - 1$, define $D_{m_1, m_2} \in \ell^2(\mathbb{Z}_{N_1} \times \mathbb{Z}_{N_2})$

by

$$D_{m_1,m_2}(n_1, n_2) = B_{m_1}(n_1)C_{m_2}(n_2).$$

Prove that

$$\{D_{m_1,m_2}\}_{0 \le m_1 \le N_1 - 1, 0 \le m_2 \le N_2 - 1}$$

is an orthonormal basis for $\ell^2(\mathbb{Z}_{N_1} \times \mathbb{Z}_{N_2})$. Hint: Prove that

$$\langle D_{m_1,m_2}, D_{k_1,k_2} \rangle = \langle B_{m_1}, B_{k_1} \rangle \langle C_{m_2}, C_{k_2} \rangle,$$

where the inner product on the left is in $\ell^2(\mathbb{Z}_{N_1} \times \mathbb{Z}_{N_2})$, whereas those on the right are in $\ell^2(\mathbb{Z}_{N_1})$ and $\ell^2(\mathbb{Z}_{N_2})$, respectively.

2.1.17.   i. (Two-dimensional Fourier basis) For $m_1 \in \mathbb{Z}_{N_1}$ and $m_2 \in \mathbb{Z}_{N_2}$, define $E_{m_1,m_2} \in \ell^2(\mathbb{Z}_{N_1} \times \mathbb{Z}_{N_2})$ (see Exercise 2.1.15 for the definition) by

$$E_{m_1,m_2}(n_1, n_2) = \frac{1}{\sqrt{N_1 N_2}} e^{2\pi i m_1 n_1 / N_1} e^{2\pi i m_2 n_2 / N_2}.$$

Prove that

$$\{E_{m_1,m_2}\}_{0 \le m_1 \le N_1 - 1, \, 0 \le m_2 \le N_2 - 1}$$

is an orthonormal basis for $\ell^2(\mathbb{Z}_{N_1} \times \mathbb{Z}_{N_2})$. As in the one-dimensional case, we usually renormalize by setting

$$F_{m_1,m_2}(n_1, n_2) = \frac{1}{N_1 N_2} e^{2\pi i m_1 n_1 / N_1} e^{2\pi i m_2 n_2 / N_2}.$$

We call $F = \{F_{m_1,m_2}\}_{0 \le m_1 \le N_1 - 1, \, 0 \le m_2 \le N_2 - 1}$ the Fourier basis for $\ell^2(\mathbb{Z}_{N_1} \times \mathbb{Z}_{N_2})$.

  ii. (Two-dimensional DFT) For $z \in \ell^2(\mathbb{Z}_{N_1} \times \mathbb{Z}_{N_2})$, define $\hat{z} \in \ell^2(\mathbb{Z}_{N_1} \times \mathbb{Z}_{N_2})$ by

$$\hat{z}(m_1, m_2) = \sum_{n_1=0}^{N_1-1} \sum_{n_2=0}^{N_2-1} z(n_1, n_2) e^{-2\pi i m_1 n_1 / N_1} e^{-2\pi i m_2 n_2 / N_2}.$$

Also, for $w \in \ell^2(\mathbb{Z}_{N_1} \times \mathbb{Z}_{N_2})$, define $\check{w} \in \ell^2(\mathbb{Z}_{N_1} \times \mathbb{Z}_{N_2})$ by

$$\check{w}(n_1, n_2) = \frac{1}{N_1 N_2} \sum_{m_1=0}^{N_1-1} \sum_{m_2=0}^{N_2-1} w(m_1, m_2) e^{2\pi i m_1 n_1 / N_1} e^{2\pi i m_2 n_2 / N_2}.$$

Prove that

$$z = (\hat{z})^{\check{}},$$

for all $z \in \ell^2(\mathbb{Z}_{N_1} \times \mathbb{Z}_{N_2})$. Deduce that $\hat{} : \ell^2(\mathbb{Z}_{N_1} \times \mathbb{Z}_{N_2}) \to \ell^2(\mathbb{Z}_{N_1} \times \mathbb{Z}_{N_2})$ is 1–1, hence invertible, with inverse $\check{}$. Hint: Use Exercise 2.1.16, and follow the reasoning in the text for the one-dimensional case.

2.1.18. With definitions as in Exercises 2.1.16 and 2.1.17, and for $z, w \in \ell^2(\mathbb{Z}_{N_1} \times \mathbb{Z}_{N_2})$, prove Parseval's formula:

$$\langle z, w \rangle = \frac{1}{N_1 N_2} \langle \hat{z}, \hat{w} \rangle,$$

and Plancherel's formula:

$$\|z\|^2 = \frac{1}{N_1 N_2} \|\hat{z}\|^2.$$

Remark: Because visual images are two-dimensional, to do image processing one needs to work with the two-dimensional DFT discussed in Problems 2.1.15 through 2.1.18.

2.1.19. Formulate the analogue of Exercises 2.1.15–2.1.18 for d dimensions and carry out the proofs. It will probably be helpful to use vector notation; for example, for $n = (n_1, n_2, \ldots, n_d)$ and with m defined similarly, let $n \cdot m = \sum_{k=1}^d n_k m_k$.

2.2 Translation-Invariant Linear Transformations

Most electrical engineering departments have a course titled "Signals and Systems" or something similar. From our point of view, a "signal" is just a function. It could be a function on an interval of real numbers (a continuous, or analog, signal) or a function on a finite set of points or on an infinite discrete set such as $\mathbb{Z}$ (a discrete, or digital, signal). As shown in section 2.1, a vector in $\mathbb{C}^N$ can be thought of as a function on N points, hence as a signal. Physically, we can think of an audio signal, for example, a piece of music. Examples of "systems" include amplifiers, graphic equalizers, and other audio equipment. A system is something that transforms an input signal into an output signal. Mathematically, a system is a transformation.

What assumptions are reasonable for a system, which we model as a transformation T? Ideally, amplifiers and most other audio equipment should be linear. First, the effect of an amplifier on two signals together should be the sum of its effects on each signal separately, that is, $T(u + v) = T(u) + T(v)$. Second, if we multiply the volume of the input signal by some amount, the output volume should be multiplied by that amount, that is, $T(\alpha u) = \alpha T(u)$. These are the ideal characteristics of an amplifier. In reality this is not possible. If a signal is multiplied by a large enough factor, it will blow out the system and the output becomes 0. However, within the usual range of operation, a good amplifier is close to linear. So this is a reasonable assumption to make on our mathematical model of a system. The conditions we have just considered are that T is a linear transformation in the sense of Definition 1.44.

Another natural assumption is that if we delay our input signal by a certain amount, the only effect on the output is to delay it by the same amount. In other words, the system does not behave differently at different times of the day. Such a system is called *time-invariant* or *shift-invariant*. The linear transformation associated with such a system is called *translation invariant*. To formulate the translation invariance property mathematically, recall the translation operator R_k defined by (Definition 2.12)

$$(R_k z)(n) = z(n - k) \quad \text{for} \quad n \in \mathbb{Z},$$

for $z \in \ell^2(\mathbb{Z}_N)$. Shifting a signal to the right by k units gives $R_k(z)$. (Recall that we are working with periodic signals defined on all of $\mathbb{Z}$, so the translated signal does not actually start earlier or later. But all the values are shifted to the right by k.) Suppose our input signal is z and the resulting output is $w = T(z)$. Our hypothesis of shift invariance says that if the new input signal is $R_k(z)$, the new output should be the shift $R_k w = R_k(T(z))$. But by definition of T, the output is $T(R_k z)$. Thus our formal definition of translation invariance is as follows.

Definition 2.17 Let $T : \ell^2(\mathbb{Z}_N) \to \ell^2(\mathbb{Z}_N)$ be a linear transformation. T is translation invariant if

$$T(R_k z) = R_k T(z), \tag{2.37}$$

for all $z \in \ell^2(\mathbb{Z}_N)$ and all $k \in \mathbb{Z}$.

Note that equation (2.37) states that T commutes with the translation operator R_k.

Recall the Fourier basis F from Definition 2.7. Probably the single most important fact about the Fourier basis F is that all translation-invariant linear transformations (from $\ell^2(\mathbb{Z}_N)$ to $\ell^2(\mathbb{Z}_N)$) are diagonalized by F. Before proving this, we should pause to appreciate this fact. We noted above that translation invariant linear transformations are the most natural. In chapter 1 we learned that diagonalizable linear transformations are the easiest to use. Now we find that translation-invariant linear transformations are all diagonalizable, in fact by the same basis! Moreover, this basis is orthogonal (by Lemma 2.2).

We begin with a direct proof of this main result. Recall Definition 1.71: a linear transformation T is diagonalizable if its domain has a basis consisting of eigenvectors of T.

Theorem 2.18 Let $T : \ell^2(\mathbb{Z}_N) \to \ell^2(\mathbb{Z}_N)$ be a translation-invariant linear transformation. Then each element of the Fourier basis F is an eigenvector of T. In particular, T is diagonalizable.

Proof
Fix $m \in \{0, 1, \ldots, N-1\}$. Let F_m be the mth element of the Fourier basis (defined in relation (2.13)). Then there exist complex scalars $a_0, a_1, \ldots, a_{N-1}$ such that

$$T(F_m)(n) = \sum_{k=0}^{N-1} a_k F_k(n) = \frac{1}{N} \sum_{k=0}^{N-1} a_k e^{2\pi i k n/N}, \tag{2.38}$$

for all n, because F is a basis for $\ell^2(\mathbb{Z}_N)$. Notice that

$$(R_1 F_m)(n) = F_m(n-1) = \frac{1}{N} e^{2\pi i m(n-1)/N}$$

$$= \frac{1}{N} e^{-2\pi i m/N} e^{2\pi i m n/N} = e^{-2\pi i m/N} F_m(n).$$

Since $e^{-2\pi i m/N}$ is independent of n, the linearity of T implies that

$$T(R_1 F_m)(n) = e^{-2\pi i m/N} T(F_m)(n)$$

$$= e^{-2\pi i m/N} \sum_{k=0}^{N-1} a_k F_k(n) = \sum_{k=0}^{N-1} a_k e^{-2\pi i m/N} F_k(n),$$

by equation (2.38). On the other hand, equation (2.38) also implies

$$(R_1 T(F_m))(n) = (T(F_m))(n-1) = \frac{1}{N} \sum_{k=0}^{N-1} a_k e^{2\pi i k(n-1)/N}$$

$$= \frac{1}{N} \sum_{k=0}^{N-1} a_k e^{-2\pi i k/N} e^{2\pi i k n/N} = \sum_{k=0}^{N-1} a_k e^{-2\pi i k/N} F_k(n).$$

But $T(R_1 F_m)(n) = (R_1 T(F_m))(n)$ for all n, by the assumption that T is translation invariant. Comparing the expressions above for these two quantities, and using the uniqueness of the coefficients of the representation of a vector in terms of a basis, we obtain that for each $k = 0, 1, \ldots, N-1$,

$$a_k e^{-2\pi i m/N} = a_k e^{-2\pi i k/N}. \qquad (2.39)$$

If $k \neq m$, we have $e^{-2\pi i m/N} \neq e^{-2\pi i k/N}$ since $0 \leq k, m \leq N-1$. Thus the only way equation (2.39) can hold is if $a_k = 0$. Hence we have proved that $a_k = 0$ whenever $k \neq m$. Therefore in equation (2.38), all terms disappear except the term with $k = m$, leaving

$$T(F_m)(n) = a_m F_m(n),$$

that is, $T(F_m) = a_m F_m$. So F_m is an eigenvector of T with eigenvalue a_m.

Since m was arbitrary, this shows that every element of the Fourier basis F is an eigenvector of T. So T is diagonalizable. ∎

Although this proves our main point, this result is so important that we will consider it again in more detail and approach it through a different route. Along the way we will encounter a variety of key concepts such as circulant matrices, convolutions, and Fourier multipliers. Our goal is to prove the following theorem, which we state now to map out our objectives, even though we have not defined some of the terms yet.

Theorem 2.19 *Let $T : \ell^2(\mathbb{Z}_N) \to \ell^2(\mathbb{Z}_N)$ be a linear transformation. Then the following statements are equivalent:*

    i. T is translation invariant.

    ii. The matrix $A_{T,E}$ representing T in the standard basis E is circulant.

    iii. T is a convolution operator.

    iv. T is a Fourier multiplier operator.

    v. The matrix $A_{T,F}$ representing T in the Fourier basis F is diagonal.

Statement v is another way of saying that T is diagonalized by the Fourier basis. The strategy of the proof of Theorem 2.19 is to prove i $\Rightarrow$ ii $\Rightarrow$ iii $\Rightarrow$ i, and then to prove iii $\Leftrightarrow$ iv and finally iv $\Leftrightarrow$ v. Note that Theorem 2.19 gives not only Theorem 2.18 but its converse, which states that a linear transformation that is diagonalized by the Fourier basis is translation invariant.

As in Definition 2.8, the indices for matrices will now run from 0 to $N-1$, to match our notation for vectors. That is, we write an $n \times n$ matrix A as $[a_{mn}]_{0 \le m,n \le N-1}$. When we multiply the matrix A by a vector z as in expression (2.1), the result is the vector Az whose m^{th} component $(Az)(m)$ $(0 \le m \le N-1)$ is

$$(Az)(m) = \sum_{n=0}^{N-1} a_{mn} z(n). \tag{2.40}$$

Also we will adapt the same periodicity convention for matrices that we have adapted for vectors. For $[a_{mn}]_{0 \le m,n \le N-1}$ given, we define a_{mn} for all $m, n \in \mathbb{Z}$ by assuming periodicity with period N in each index:

$$a_{m+N,n} = a_{mn} \quad \text{and} \quad a_{m,n+N} = a_{mn}$$

for all m and n (the extra comma has been added for clarity). Thus for example $a_{-1,N+2} = a_{N-1,N+2} = a_{N-1,2}$, where only the indices in the last expression are in the original range $0 \le m, n \le N-1$.

Definition 2.20 *A matrix $A = [a_{mn}]_{0 \le m,n \le N-1}$, periodized as above, is circulant if*

$$a_{m+k,n+k} = a_{m,n} \tag{2.41}$$

for all $m, n, k \in \mathbb{Z}$.

Note that it is equivalent to require only $a_{m+1,n+1} = a_{m,n}$ for all $m, n \in \mathbb{Z}$, since one can repeat this k times to obtain equation (2.41).

Definition 2.20 states that we obtain column $m + 1$ in a circulant matrix by shifting the mth column down (circular translation) by 1. Equivalently, we obtain row $m + 1$ by shifting row m to the right by 1. Thus circulant matrices are easy to identify by sight.

Example 2.21
The matrix

$$\begin{bmatrix} 3 & 2+i & -1 & 4i \\ 4i & 3 & 2+i & -1 \\ -1 & 4i & 3 & 2+i \\ 2+i & -1 & 4i & 3 \end{bmatrix}$$

is circulant. The matrix

$$\begin{bmatrix} 2 & i & 3 \\ 3 & 2 & i \\ i & 2 & 3 \end{bmatrix}$$

is not circulant, but if the last row were $i\ 3\ 2$, it would be.

Recall that the standard, or Euclidean, basis E for $\ell^2(\mathbb{Z}_N)$ is $E = \{e_0, e_1, \ldots, e_{N-1}\}$, defined by $e_n(m) = 1$ if $m = n$, and $e_n(m) = 0$ if $m \neq n$ and $0 \leq m \leq N - 1$. It is useful to define e_n for all $n \in \mathbb{Z}$, although some redundancy is introduced in the process. Observe that because we regard e_n as defined on all of $\mathbb{Z}$, with period N, we see that $e_n(m) = 1$ if and only if $m = n + kN$ for some $k \in \mathbb{Z}$. With this in mind, for each $n \in \mathbb{Z}$ we define $e_n \in \ell^2(\mathbb{Z}_N)$ by periodicity, that is, by letting

$$e_{n+N} = e_n$$

for any n. It follows that

$$R_1 e_n = e_{n+1} \tag{2.42}$$

for any n. For example, when $n = N - 1$, the definition of circular translation shows that $R_1 e_{N-1} = e_0 = e_N$, which is consistent with equation (2.42).

If we multiply A by e_n, we obtain the nth column of A. The formal proof is:

$$(Ae_n)(m) = \sum_{k=0}^{N-1} a_{mk}e_n(k) = a_{mn}, \qquad (2.43)$$

using relation (2.40) and the fact that $e_n(k) = 0$ unless $k = n$, in which case it is 1. However, this is probably easier to see by writing it out. By the periodicity in our definitions, this is true for all $n, m \in \mathbb{Z}$.

Recall that if the matrix $A_{T,E}$ represents a linear transformation T in the standard basis, then $T(z) = A_{T,E}z$ (since $z = [z]_E$, by equation (1.24)). We are now prepared to prove the implication i $\Rightarrow$ ii in Theorem 2.19.

Lemma 2.22 *Suppose $T : \ell^2(\mathbb{Z}_N) \rightarrow \ell^2(\mathbb{Z}_N)$ is a linear transformation. Let $A_{T,E}$ be the matrix representing T in the standard basis E. If T is translation invariant, then $A_{T,E}$ is circulant.*

Proof
For each m, n we have, by equation (2.43) and then equation (2.42),

$$
\begin{aligned}
a_{m+1,n+1} &= (A_{T,E}e_{n+1})(m+1) = (T(e_{n+1}))(m+1) = (T(R_1e_n))(m+1) \\
&= (R_1T(e_n))(m+1) = T(e_n)(m+1-1) = T(e_n)(m) \\
&= (A_{T,E}e_n)(m) = a_{m,n},
\end{aligned}
$$

by the assumed translation invariance of T. As noted earlier, iterating this k times gives $a_{m+k,n+k} = a_{m,n}$ for any k. ∎

For the next step in Theorem 2.19, we define the convolution of two vectors.

Definition 2.23 *For $z, w \in \ell^2(\mathbb{Z}_N)$, the convolution $z * w \in \ell^2(\mathbb{Z}_N)$ is the vector with components*

$$z * w(m) = \sum_{n=0}^{N-1} z(m-n)w(n),$$

for all m.

Example 2.24
Let $z = (1, 1, 0, 2)$ and $w = (i, 0, 1, i)$ be vectors in $\ell^2(\mathbb{Z}_4)$. Then, using the periodicity of z,

$$
\begin{aligned}
z * w(0) &= \sum_{n=0}^{3} z(-n)w(n) \\
&= z(0)w(0) + z(-1)w(1) + z(-2)w(2) + z(-3)w(3) \\
&= z(0)w(0) + z(3)w(1) + z(2)w(2) + z(1)w(3) \\
&= 1 \cdot i + 2 \cdot 0 + 0 \cdot 1 + 1 \cdot i = 2i.
\end{aligned}
$$

Similarly,

$$
z * w(1) = \sum_{n=0}^{3} z(1 - n)w(n) = 1 \cdot i + 1 \cdot 0 + 2 \cdot 1 + 0 \cdot i = 2 + i,
$$

$$
z * w(2) = \sum_{n=0}^{3} z(2 - n)w(n) = 0 \cdot i + 1 \cdot 0 + 1 \cdot 1 + 2 \cdot i = 1 + 2i,
$$

and

$$
z * w(3) = \sum_{n=0}^{3} z(3 - n)w(n) = 2 \cdot i + 0 \cdot 0 + 1 \cdot 1 + 1 \cdot i = 1 + 3i.
$$

Therefore

$$
z * w = (2i, 2 + i, 1 + 2i, 1 + 3i).
$$

If we fix one vector in the convolution, we can regard convolution with this fixed vector as a linear transformation.

Definition 2.25 *Suppose $b \in \ell^2(\mathbb{Z}_N)$. Define $T_b : \ell^2(\mathbb{Z}_N) \to \ell^2(\mathbb{Z}_N)$ by*

$$
T_b(z) = b * z,
$$

for all $z \in \ell^2(\mathbb{Z}_N)$. Any transformation T of the form $T = T_b$, for some $b \in \ell^2(\mathbb{Z}_N)$, is called a convolution operator.

It is not difficult to see that a convolution operator T_b is a linear transformation. The implication ii $\Rightarrow$ iii in Theorem 2.19 states that a circulant matrix gives rise to a convolution operator. This explains why convolution is of interest to us.

Lemma 2.26   *Let A be an $N \times N$ matrix, $A = [a_{mn}]_{0 \leq m,n \leq N-1}$. Suppose A is circulant. Define $b \in \ell^2(\mathbb{Z}_N)$ by*

$$b(n) = a_{n,0}$$

for all n. (In other words, b is the first column of A, regarded as a vector. By equation (2.43), this means that $b = Ae_0$.) Then for all $z \in \ell^2(\mathbb{Z}_N)$,

$$Az = b * z = T_b(z).$$

Proof

Since A is circulant, we have

$$a_{mn} = a_{m-n,0} = b(m - n),$$

for any $m, n \in \mathbb{Z}$. Hence, by the definition of matrix multiplication (equation (2.40)),

$$(Az)(m) = \sum_{n=0}^{N-1} a_{mn} z(n) = \sum_{n=0}^{N-1} b(m - n) z(n) = b * z(m). \qquad \blacksquare$$

We have proved i $\Rightarrow$ ii $\Rightarrow$ iii in Theorem 2.19, so we now know that any translation-invariant linear transformation T is a convolution operator T_b. The converse, that a convolution operator is translation invariant, is relatively easy.

Lemma 2.27   *Let $b \in \ell^2(\mathbb{Z}_N)$, and let T_b be the convolution operator associated with b as in Definition 2.25. Then T_b is translation invariant.*

Proof

Let $z \in \ell^2(\mathbb{Z}_N)$. Let $k \in \mathbb{Z}$. Then for any m,

$$T_b(R_k z)(m) = b * (R_k z)(m) = \sum_{n=0}^{N-1} b(m - n)(R_k z)(n)$$

$$= \sum_{n=0}^{N-1} b(m - n) z(n - k).$$

In this last sum we make the change of index $\ell = n - k$. Using Exercise 2.1.8, we obtain

$$T_b(R_k z)(m) = \sum_{\ell=-k}^{N-1-k} b(m - k - \ell) z(\ell) = \sum_{\ell=0}^{N-1} b(m - k - \ell) z(\ell)$$

$$= (b * z)(m - k) = R_k(b * z)(m) = R_k T_b(z)(m),$$

for all m. In other words, $T_b(R_k z) = R_k T_b(z)$; that is, T_b is translation invariant. ∎

Thus we have proved that statements i, ii, and iii in Theorem 2.19 are equivalent. We pause to consider an application of this result.

Definition 2.28   *Define $\delta \in \ell^2(\mathbb{Z}_N)$ by*

$$\delta(n) = \begin{cases} 1 & \text{if } n = 0 \\ 0 & \text{if } n = 1, 2, \ldots, N-1. \end{cases}$$

This is the discrete version of what is sometimes called the *Dirac delta function*. It is also known as the unit impulse. Note that δ is just e_0, but because the notation δ is standard, we accept this redundancy.

Lemma 2.29   *For any $w \in \ell^2(\mathbb{Z}_N)$,*

$$w * \delta = w.$$

Proof
For each $m \in \mathbb{Z}_N$,

$$(w * \delta)(m) = \sum_{n=0}^{N-1} w(m-n)\delta(n) = w(m),$$

since $\delta(n) = 0$ unless $n = 0$, in which case $\delta(0) = 1$. ∎

This simple result has the following interpretation. Suppose we have a system, say an amplifier or some other audio equipment. By our general discussion above, we model the system as a translation-invariant linear transformation T on $\ell^2(\mathbb{Z})$. We have proved that T is a convolution operator T_b for some $b \in \ell^2(\mathbb{Z}_N)$. If we knew b, we would know the action of our system on any signal z since $T(z) = T_b(z) = b * z$. Thus the system is completely determined by b. How can we find b? By Lemma 2.29, this is easy:

$$T(\delta) = T_b(\delta) = b * \delta = b.$$

Thus to recover b we only have to measure the output of the system when the input is δ. Since δ is called the unit impulse, b is often called the *impulse response* of the system.

We now consider how the DFT interacts with convolution.

Lemma 2.30   *Suppose $z, w \in \ell^2(\mathbb{Z}_N)$. Then for each m,*

$$(z * w)\hat{\ }(m) = \hat{z}(m)\hat{w}(m).$$

Proof

By definition,

$$(z * w)\hat{}(m) = \sum_{n=0}^{N-1} (z * w)(n)e^{-2\pi imn/N} = \sum_{n=0}^{N-1}\sum_{k=0}^{N-1} z(n-k)w(k)e^{-2\pi imn/N}$$

$$= \sum_{n=0}^{N-1}\sum_{k=0}^{N-1} z(n-k)w(k)e^{-2\pi im(n-k)/N}e^{-2\pi imk/N}$$

$$= \sum_{k=0}^{N-1} w(k)e^{-2\pi imk/N} \sum_{n=0}^{N-1} z(n-k)e^{-2\pi im(n-k)/N}.$$

In the last sum we change index, letting $\ell = n - k$, to obtain (by Exercise 2.1.8)

$$\sum_{n=0}^{N-1} z(n-k)e^{-2\pi im(n-k)/N} = \sum_{\ell=-k}^{N-1-k} z(\ell)e^{-2\pi im\ell/N} = \sum_{\ell=0}^{N-1} z(\ell)e^{-2\pi im\ell/N}.$$

Substituting this gives

$$(z * w)\hat{}(m) = \sum_{k=0}^{N-1} w(k)e^{-2\pi imk/N} \sum_{\ell=0}^{N-1} z(\ell)e^{-2\pi im\ell/N} = \hat{z}(m)\hat{w}(m).$$

∎

Thus, the DFT transforms the relatively complicated operation of convolution into the simple operation of multiplication.

Example 2.31

In Example 2.24, for $z = (1, 1, 0, 2)$ and $w = (i, 0, 1, i)$, we calculated that $z * w = (2i, 2 + i, 1 + 2i, 1 + 3i)$. Proceeding as in Example 2.9, we find that

$$\hat{z} = (4, 1 + i, -2, 1 - i).$$

and

$$\hat{w} = (1 + 2i, -2 + i, 1, i).$$

Similarly, we compute

$$(z * w)\hat{} = (4 + 8i, -3 - i, -2, 1 + i).$$

Now one can check that $(z * w)\hat{}(n) = \hat{z}(n)\hat{w}(n)$ for $n = 0, 1, 2, 3$ (e.g., $\hat{z}(1)\hat{w}(1) = (1 + i)(-2 + i) = -3 - i = (z * w)\hat{}(1)$).

Now we consider linear transformations obtained by taking the DFT, multiplying the resulting components by some numbers, and taking the inverse DFT (IDFT) of the result.

Definition 2.32 *Let* $m \in \ell^2(\mathbb{Z}_N)$. *Define* $T_{(m)} : \ell^2(\mathbb{Z}_N) \to \ell^2(\mathbb{Z}_N)$ *by*

$$T_{(m)}(z) = (m\hat{z})^{\check{}}, \tag{2.44}$$

where $m\hat{z}$ *is the vector obtained from multiplying* m *and* $\hat{z}$ *component-wise; that is,* $(m\hat{z})(n) = m(n)\hat{z}(n)$ *for each* n. *Any transformation of this form is called a* Fourier multiplier operator.

Note that we write $T_{(m)}$ to distinguish Fourier multiplier operators from convolution operators (denoted T_b) in Definition 2.25. It is not difficult to see that any Fourier multiplier operator is a linear transformation. Another way to describe $T_{(m)}$ in Definition 2.32 is to note that for each k,

$$(T_{(m)}(z))^{\hat{}}(k) = m(k)\hat{z}(k), \tag{2.45}$$

by applying the DFT to both sides of equation (2.44). To understand this, recall, by Fourier inversion (2.15), that

$$z = \sum_{k=0}^{N-1} \hat{z}(k)F_k.$$

By applying inversion (2.15) to $T_{(m)}(z)$ and using equation (2.45), we get

$$T_{(m)}(z) = \sum_{k=0}^{N-1} (T_{(m)}(z))^{\hat{}}(k)F_k = \sum_{k=0}^{N-1} m(k)\hat{z}(k)F_k.$$

Thus the effect of $T_{(m)}$ on z is to multiply the kth DFT coefficient $\hat{z}(k)$ by $m(k)$. This explains the name Fourier multiplier operator.

A piece of audio equipment known as a graphic equalizer is modeled by a Fourier multiplier operator. The purpose of a graphic equalizer is to allow one to boost or lower separate frequency components of an audio signal. Recall that the strength of the frequency $e^{2\pi i k n/N}$ in a signal is proportional to the magnitude of its kth DFT coefficient. By equation (2.45), the operator $T_{(m)}$ multiplies the kth DFT coefficient by $m(k)$. The settings on a graphic equalizer correspond to different frequency components of a signal. Each can be manually raised or lowered, corresponding to an increased or decreased multiplier factor for that frequency. In doing so, one is choosing a value of the Fourier multiplier $m(k)$. A graphic equalizer allows one to tune the frequency response to one's satisfaction.

The equivalence of iii and iv in Theorem 2.19 follows easily from Lemma 2.30.

Lemma 2.33   Let $T : \ell^2(\mathbb{Z}_N) \to \ell^2(\mathbb{Z}_N)$ be a linear transformation. Then T is a convolution operator if and only if T is a Fourier multiplier operator. More precisely, for a given convolution operator T_b, let $m = \hat{b}$; then $T_b = T_{(m)}$. Conversely, given a Fourier multiplier operator $T_{(m)}$, let $b = \check{m}$. Then $T_{(m)} = T_b$.

Proof

In either case, we have $m = \hat{b}$ (by Fourier inversion in the second case). Then by Fourier inversion and Lemma 2.30,

$$T_b(z) = b * z = ((b * z)\hat{)}\check{} = (\hat{b}\hat{z})\check{} = (m\hat{z})\check{} = T_{(m)}(z),$$

for any $z \in \ell^2(\mathbb{Z}_N)$. ∎

The last step in the proof of Theorem 2.19 is to show that iv and v are equivalent. So far we know that i, ii, iii, and iv are equivalent, so we could apply Theorem 2.18, which shows that i implies v. However, we require the converse also, and we prefer to show the direct connection between iv and v in the next lemma. Recall by equation (2.16) that $\hat{z}$ is the vector representing z in the Fourier basis: $\hat{z} = [z]_F$. Then equation (2.45) shows that $T_{(m)}$ behaves like multiplication by a diagonal matrix in the Fourier basis. The proof of the equivalence of iv and v just requires writing this observation out.

Lemma 2.34   Let $T : \ell^2(\mathbb{Z}_N) \to \ell^2(\mathbb{Z}_N)$ be a linear transformation. Then T is a Fourier multiplier operator $T_{(m)}$ for some $m \in \ell^2(\mathbb{Z}_N)$ if and only if the matrix representing T in the Fourier basis $F = \{F_0, F_1, \ldots, F_{N-1}\}$ is a diagonal matrix D.

Moreover, if $T = T_{(m)}$ is a Fourier multiplier operator, then the diagonal matrix $D = [d_{mn}]_{0 \le m,n \le N-1}$ satisfies $d_{nn} = m(n)$ for $n = 0, 1, \ldots, N-1$.

Proof

Let $T_{(m)}$ be a Fourier multiplier operator. Define a diagonal matrix $D = [d_{mn}]_{0 \leq m,n \leq N-1}$ by setting $d_{nn} = m(n)$, for $0 \leq m \leq N-1$. By equation (2.45), $(T_{(m)}(z))\hat{} = m\hat{z}$ and therefore

$$
[T_{(m)}(z)]_F =
\begin{bmatrix}
m(0)\hat{z}(0) \\
m(1)\hat{z}(1) \\
\cdot \\
\cdot \\
m(N-1)\hat{z}(N-1)
\end{bmatrix}
=
\begin{bmatrix}
d_{00}\hat{z}(0) \\
d_{11}\hat{z}(1) \\
\cdot \\
\cdot \\
d_{N-1,N-1}\hat{z}(N-1)
\end{bmatrix}
$$

$$
=
\begin{bmatrix}
d_{00} & 0 & \cdot & \cdot & 0 \\
0 & d_{11} & 0 & \cdot & 0 \\
\cdot & \cdot & \cdot & \cdot & \cdot \\
\cdot & \cdot & \cdot & \cdot & \cdot \\
0 & \cdot & \cdot & 0 & d_{N-1,N-1}
\end{bmatrix}
\begin{bmatrix}
\hat{z}(0) \\
\hat{z}(1) \\
\cdot \\
\cdot \\
\hat{z}(N-1)
\end{bmatrix}
$$

$$
= D\hat{z} = D[z]_F,
$$

by using equation (2.16). Hence a Fourier multiplier opertor $T_{(m)}$ is represented by the diagonal matrix D with respect to the Fourier basis.

Conversely, suppose the diagonal matrix $D = [d_{mn}]_{0 \leq m,n \leq N-1}$ represents T with respect to the Fourier basis F. Set $m(n) = d_{nn}$, for $0 \leq n \leq N-1$, and let $T_{(m)}$ be the corresponding Fourier multiplier operator. Then by the calculation above,

$$
[T(z)]_F = D[z]_F = [T_{(m)}(z)]_F.
$$

Therefore, $T = T_{(m)}$. ∎

This completes the proof of Theorem 2.19. These results can be used in a practical way in computation. Suppose T is a translation-invariant linear transformation. From its definition, we can write down the matrix $A = [a_{mn}]_{0 \leq m,n \leq N-1}$ representing T in Euclidean coordinates (that is, A such that $T(z) = Az$). This matrix must be circulant. We let $b \in \ell^2(\mathbb{Z}_N)$ be the first column of A. Then T is the convolution operator T_b, by Lemma 2.26. Let $m = \hat{b}$. By Lemma 2.33, T is the Fourier multiplier operator $T_{(m)}$. Form the diagonal matrix D with n^{th} diagonal entry $d_{nn} = m(n)$. Then D represents T

with respect to the Fourier basis F; that is,

$$[T(z)]_F = D[z]_F.$$

To understand this on the matrix level, recall from equation (2.19) that $\hat{z} = W_N z$, where W_N is the matrix (2.18). Thus

$$W_N Az = (Az)\hat{} = [Az]_F = [T(z)]_F = D[z]_F = D\hat{z} = DW_N z.$$

Multiplying on the left by W_N^{-1} gives

$$Az = W_N^{-1} DW_N z,$$

and hence that

$$A = W_N^{-1} DW_N, \quad \text{or} \quad W_N A W_N^{-1} = D. \tag{2.46}$$

This is an explicit diagonalization of A. Notice that the diagonalizing matrix W_N is the same for any circulant matrix. Recall that the diagonal entries of D are the eigenvalues of A; we see here that if A is a circulant matrix, they are just the components of the vector m determined above. For circulant matrices, this is a much easier way to find the eigenvalues than trying to factor the characteristic polynomial.

Example 2.35
Define $T : \ell^2(\mathbb{Z}_4) \to \ell^2(\mathbb{Z}_4)$ by

$$T(z)(n) = z(n) + 2z(n+1) + z(n+3).$$

Find the eigenvalues and eigenvectors of T, and diagonalize the matrix A representing T in the standard basis, if possible.

Solution
One can check that T is translation invariant:

$$\begin{aligned}
T(R_k z) &= (R_k z)(n) + 2(R_k z)(n+1) + (R_k z)(n+3) \\
&= z(n-k) + 2z(n+1-k) + z(n+3-k) \\
&= R_k(z(n) + 2z(n+1) + z(n+3)) = R_k T(z).
\end{aligned}$$

Alternatively, one can write the matrix A that represents T in the standard basis (i.e., satisfying $T(z) = Az$) by considering

$T(z)(0)$, $T(z)(1)$, $T(z)(2)$, and $T(z)(3)$, obtaining

$$
A = \begin{bmatrix} 1 & 2 & 0 & 1 \\ 1 & 1 & 2 & 0 \\ 0 & 1 & 1 & 2 \\ 2 & 0 & 1 & 1 \end{bmatrix},
$$

which is circulant. Then $b = (1, 1, 0, 2)$.

In Example 2.31 we calculated $\hat{b}$ (b was called z there), where we obtained $m = \hat{b} = (4, 1+i, -2, 1-i)$. These components are the eigenvalues of A, and the eigenvectors are the Fourier basis vectors in F. In particular,

$$
D = \begin{bmatrix} 4 & 0 & 0 & 0 \\ 0 & 1+i & 0 & 0 \\ 0 & 0 & -2 & 0 \\ 0 & 0 & 0 & 1-i \end{bmatrix}
$$

satisfies $W_4 A W_4^{-1} = D$. ∎

The transformation considered in the next example is the second difference operator, which is used to approximate the second derivative f'' when doing numerical solutions of differential equations in the periodic setting.

Example 2.36
Define $\Delta : \ell^2(\mathbb{Z}_N) \to \ell^2(\mathbb{Z}_N)$ by

$$
(\Delta(z))(n) = z(n+1) - 2z(n) + z(n-1).
$$

Find the eigenvalues of Δ.

Solution
As in Example 2.35, we can check that Δ is translation invariant. By definition,

$$
(\Delta(z))(0) = z(1) - 2z(0) + z(-1) = z(1) - 2z(0) + z(N-1).
$$

Therefore the first row of the matrix A of Δ in the standard basis is $(-2, 1, 0, \ldots, 0, 1)$. Since A must be circulant,

$$
A = \begin{bmatrix}
-2 & 1 & 0 & \cdot & \cdot & 0 & 1 \\
1 & -2 & 1 & 0 & \cdot & \cdot & 0 \\
0 & 1 & -2 & 1 & 0 & \cdot & 0 \\
\cdot & \cdot & \cdot & \cdot & \cdot & \cdot & \cdot \\
\cdot & \cdot & \cdot & \cdot & \cdot & \cdot & \cdot \\
\cdot & \cdot & \cdot & \cdot & \cdot & \cdot & \cdot \\
1 & 0 & \cdot & \cdot & 0 & 1 & -2
\end{bmatrix}. \tag{2.47}
$$

Therefore b is the first column of A, that is, $b = (-2, 1, 0, \ldots, 0, 1)$.

The eigenvalues of Δ are the components of $m = \hat{b}$. We obtain

$$
\hat{b}(k) = \sum_{n=0}^{N-1} b(n) e^{-2\pi i k n / N} = (-2) \cdot 1 + 1 \cdot e^{-2\pi i k / N} + 1 \cdot e^{-2\pi i k (N-1)/N}
$$

$$
= -2 + e^{-2\pi i k / N} + e^{2\pi i k / N} = -2 + 2\cos(\frac{2\pi k}{N})
$$

$$
= -4\left(\frac{1}{2} - \frac{1}{2}\cos(\frac{2\pi k}{N})\right) = -4\sin^2(\frac{\pi k}{N}).
$$

These are the eigenvalues of Δ. The matrix D with diagonal entries $d_{kk} = -4\sin^2(\pi k/N)$ satisfies $W_N A W_N^{-1} = D$. ∎

Exercises

2.2.1. For $z \in \ell^2(\mathbb{Z}_N)$, define $T(z) \in (\mathbb{Z}_N)$ by

$$
(T(z))(n) = z(n-1),
$$

for all n.

    i. Prove that T is translation invariant.

    ii. Let $w(n) = \cos(2\pi n/N)$, for $n \in \mathbb{Z}_N$. For each $N \geq 3$, show that w is not an eigenvector of T.

       Remark: This shows that the orthonormal basis of sines and cosines in Exercise 2.1.7 does not diagonalize T.

2.2.2. Define $T : \ell^2(\mathbb{Z}_N) \to \ell^2(\mathbb{Z}_N)$ by

$$
(T(z))(n) = 3z(n-2) + iz(n) - (2+i)z(n+1),
$$

for all n.

    i. Prove that T is translation invariant.

    ii. Write the matrix that represents T with respect to the standard (Euclidean) basis for the case $N = 4$.

    iii. For the case $N = 4$, show by direct computation that the vectors $E_0, E_1, E_2,$ and E_3 from Example 2.4 are eigenvectors of T.

2.2.3. Define $T : \ell^2(\mathbb{Z}_4) \to \ell^2(\mathbb{Z}_4)$ by

$$T(z) = (2z(0) - z(1), iz(1) + 2z(2), z(1), 0).$$

    i. Let $z = (1, 0, -2, i)$. Compute $T(R_1 z)$ and $R_1 T(z)$. Observe that they are not equal. Hence T is not translation invariant.

    ii. Find the matrix that represents T with respect to the standard basis. Observe that it is not circulant, as we expect from part i.

    iii. Show that $(1, i, -1, -i)$ is not an eigenvector of T. (Recall by Example 2.4 that $(1, i, -1, -i)$ is a multiple of the Fourier basis element F_1.)

2.2.4. Let $z = (2, i, 1, 0)$ and $w = (1, 0, 2i, 3)$.

    i. Compute $\hat{z}$ and $\hat{w}$.

    ii. Compute $z * w$ directly.

    iii. Compute $(z * w)\hat{}$ directly and check that it agrees with $\hat{z}\hat{w}$.

2.2.5. Let $z, w \in \ell^2(\mathbb{Z}_N)$.

    i. Prove that

$$z * w = w * z$$

directly from the definition of convolution.

    ii. Prove that $z * w = w * z$ by using Lemma 2.30 and the Fourier inversion formula.

2.2.6. Prove that convolution is associative, that is,

$$(x * y) * z = x * (y * z),$$

for $x, y, z \in \ell^2(\mathbb{Z}_N)$. Suggestion: Use the easier of the two methods in Exercise 2.2.5.

2.2.7. Define $T : \ell^2(\mathbb{Z}_4) \to \ell^2(\mathbb{Z}_4)$ by

$$(T(z))(n) = 3z(n - 1) + z(n).$$

    i. Write the matrix $A_{T,E}$ that represents T with respect to the standard basis. Observe that it is circulant.

    ii. Find $b \in \ell^2(\mathbb{Z}_4)$ such that $T(z) = b * z$.

    iii. Find $m \in \ell^2(\mathbb{Z}_4)$ such that $T = T_{(m)}$, that is, such that $(T(z))\hat{}(n) = m(n)\hat{z}(n)$ for each n.

    iv. Find the matrix $A_{T,F}$ representing T in the Fourier basis F.

    v. By direct computation, check that $A_{T,E} = W_4^{-1} A_{T,F} W_4$, where W_4 is the matrix in equation (2.21).

2.2.8. Let $A = [a_{mn}]_{0 \leq m, n \leq N-1}$ be an $N \times N$ circulant matrix. Define

$$\lambda_m = \sum_{n=0}^{N-1} a_{0,n} e^{2\pi i m n / N}$$

for $m = 0, 1, \ldots, N - 1$. Prove directly, without using Theorem 2.19, that the eigenvalues of A are $\lambda_0, \lambda_1, \ldots, \lambda_{N-1}$, which may be repeated according to multiplicity (algebraic or geometric, which are the same here because A is diagonalizable).

    Hint: For $m = 0, 1, \ldots, N-1$, define vectors $B_m \in \ell^2(\mathbb{Z}_N)$ by

$$B_m(n) = e^{2\pi i m n / N} \quad \text{for} \quad n = 0, 1, \ldots, N - 1.$$

These are multiples of the Fourier basis elements, hence by Theorem 2.18, each B_m is an eigenvector of A. Note that $B_m(0) = 1$ for each m, and observe that λ_m as defined above is the 0th component of AB_m.

    Remark: Note that if we change summation index in the expression for λ_m above by setting $k = -n$, and use the fact that A is circulant, we obtain

$$\lambda_m = \sum_{k=0}^{N-1} a_{0,-k} e^{-2\pi i m k / N} = \sum_{k=0}^{N-1} a_{k,0} e^{-2\pi i m k / N} = \hat{b}(m),$$

for b as in Lemma 2.26. Thus the expression above is consistent with our earlier results. The proof above is a little more direct than the proof of Theorem 2.19 in the text, but we preferred to show the connections with circulant matrices, convolution operators, and Fourier multiplier operators.

2.2.9. Define $T : \ell^2(\mathbb{Z}_N) \to \ell^2(\mathbb{Z}_N)$ by

$$(T(z))(n) = z(n+1) - z(n).$$

Find all eigenvalues of T.

2.2.10. Let $T_{(m)} : \ell^2(\mathbb{Z}_4) \to \ell^2(\mathbb{Z}_4)$ be the Fourier multiplier operator defined by $T_{(m)}(z) = (m\hat{z})^\vee$ where $m = (1, 0, i, -2)$.

    i. Find $b \in \ell^2(\mathbb{Z}_4)$ such that $T_{(m)}$ is the convolution operator T_b (defined by $T_b(z) = b * z$).

    ii. Find the matrix that represents $T_{(m)}$ with respect to the standard basis.

2.2.11. i. Suppose $T_1, T_2 : \ell^2(\mathbb{Z}_N) \to \ell^2(\mathbb{Z}_N)$ are translation-invariant linear transformations. Prove that the composition $T_2 \circ T_1$ is translation invariant.

    ii. Suppose A and B are circulant $N \times N$ matrices. Prove directly (i.e., just using the definition of a circulant matrix, not using Theorem 2.19) that AB is circulant. Show that this result and Theorem 2.19 imply part i. Hint: Write out the $(m+1, n+1)$ entry of AB using the definition of matrix multiplication; compare with the hint to Exercise 2.2.12 (i).

    iii. Suppose $b_1, b_2 \in \ell^2(\mathbb{Z}_N)$. Prove that the composition $T_{b_2} \circ T_{b_1}$ of the convolution operators T_{b_2} and T_{b_1} is the convolution operator T_b with $b = b_2 * b_1$. Hint: Use Exercise 2.2.6.

    iv. Suppose $m_1, m_2 \in \ell^2(\mathbb{Z}_N)$. Prove that the composition $T_{(m_2)} \circ T_{(m_1)}$ of the Fourier multiplier operators $T_{(m_2)}$ and $T_{(m_1)}$ is the Fourier multiplier operator $T_{(m)}$ where $m(n) = m_2(n)m_1(n)$ for all n.

    v. Suppose $T_1, T_2 : \ell^2(\mathbb{Z}_N) \to \ell^2(\mathbb{Z}_N)$ are linear transformations. Prove that if T_1 is represented by a matrix A_1 with respect to the Fourier basis F (i.e., $[T_1(z)]_F = A_1[z]_F$) and T_2 is represented by a matrix A_2 with respect to F, then the composition $T_2 \circ T_1$ is represented by the matrix $A_2 A_1$ with respect to F. Deduce part i again.

Remark: By Theorem 2.19, we have just proved the same thing five times. This may not seem very intelligent, but at least we have seen how to interpret the given information in each of these five formulations. In practice, it is useful to have these five ways because for any given problem one can select the formulation that seems simplest. Exercise 2.2.12 is another example of this phenomenon.

2.2.12. Suppose $T_1, T_2 : \ell^2(\mathbb{Z}_N) \to \ell^2(\mathbb{Z}_N)$ are translation-invariant linear transformations. In this problem we prove, in four different ways, that T_1 and T_2 commute. This means, by definition, that for any $z \in \ell^2(\mathbb{Z}_N)$,

$$T_2(T_1(z)) = T_1(T_2(z)).$$

i. Suppose A and B are circulant $N \times N$ matrices. Prove directly from the definitions of matrix multiplication and circulant matrices that $AB = BA$. Deduce (from Theorem 2.19) that T_1 and T_2 commute. Hint: For $A = [a_{mn}]_{0 \le m,n \le N-1}$, and $B = [b_{mn}]_{0 \le m,n \le N-1}$, the (m, n) entry of AB is

$$\sum_{k=0}^{N-1} a_{mk} b_{kn},$$

by the definition of matrix multiplication. But $a_{mk} = a_{m+n-k,n}$ and $b_{kn} = b_{m,m+n-k}$, since A and B are circulant. Now change summation index and apply Exercise 2.1.8.

ii. Prove that T_1 and T_2 commute by using Exercises 2.2.5 and 2.2.11 (iii).

iii. Prove that T_1 and T_2 commute by using Exercise 2.2.11 (iv).

iv. Prove that T_1 and T_2 commute by using Exercise 2.2.11 (v).

Remark: This exercise and Exercise 1.5.13 explain why all circulant matrices are diagonalized by the same basis.

2.2.13. Suppose $T : \ell^2(\mathbb{Z}_N) \to \ell^2(\mathbb{Z}_N)$ is a translation-invariant linear transformation, and $z, w \in \ell^2(\mathbb{Z}_N)$.

i. Prove that

$$T(z * w) = T(z) * w = w * T(z).$$

ii. Prove that $T(z) = T(e_0) * z$, where e_0 is the first element of the standard basis. Hint: Note that $e_0 = \delta$, then apply

Lemma 2.29 and part i. Note that by equation (2.43), $T(e_0) = b$, for b as in Lemma 2.26. This gives another proof of Lemma 2.26.

2.2.14. Let $T : \ell^2(\mathbb{Z}_N) \to \ell^2(\mathbb{Z}_N)$ be a linear transformation. Prove that T is translation invariant if and only if $T(z) = \sum_{k=0}^{N-1} a_k R_k(z)$ for some $a_0, a_1, \ldots, a_{N-1} \in \mathbb{C}$. Since R_k is the same as R_1^k, the kth iterate of R_1, this states that T is a polynomial in R_1. Hint: Consider the convolution operator T_b, where $b(k) = a_k$.

2.2.15. Let A be an $N \times N$ circulant matrix. Prove that A is normal (Definition 1.108). This can be proved directly, or it follows from Theorem 1.109, Lemma 2.2, and Theorem 2.18.

2.2.16. Let $T : \ell^2(\mathbb{Z}_N) \to \ell^2(\mathbb{Z}_N)$ be a translation-invariant linear transformation.

    i. Suppose u is an eigenvector of T with eigenvalue λ. Prove that for each $k \in \mathbb{Z}_N$, $R_k u$ is also an eigenvector of T with eigenvalue λ.

    ii. By Theorem 2.18, each Fourier basis element F_m is an eigenvector of T. By part i, so is every $R_k F_m$, for $0 \le k, m \le N - 1$. Explain how T can have so many eigenvectors.

2.2.17. Show that there exist $z, w \in \ell^2(\mathbb{Z}_4)$ such that $z \ne 0$ and $w \ne 0$, but $z * w = 0$, where 0 is the zero vector $(0, 0, 0, 0)$.

2.2.18. Recall the definitions in Exercises 2.1.15 and 2.1.17. For $z \in \ell^2(\mathbb{Z}_{N_1} \times \mathbb{Z}_{N_2})$, define $z(n_1, n_2)$ as for all $(n_1, n_2) \in \mathbb{Z} \times \mathbb{Z}$ by requiring z to be periodic with period N_1 in the first variable and period N_2 in the second variable; that is,

$$z(n_1 + j_1 N_1, n_2 + j_2 N_2) = z(n_1, n_2)$$

for all $n_1, n_2, j_1, j_2 \in \mathbb{Z}$. For $k_1, k_2 \in \mathbb{Z}$, define the translation-invariant linear transformation $R_{k_1, k_2} : \ell^2(\mathbb{Z}_{N_1} \times \mathbb{Z}_{N_2}) \to \ell^2(\mathbb{Z}_{N_1} \times \mathbb{Z}_{N_2})$ by

$$(R_{k_1, k_2} z)(n_1, n_2) = z(n_1 - k_1, n_2 - k_2).$$

We say $T : \ell^2(\mathbb{Z}_{N_1} \times \mathbb{Z}_{N_2}) \to \ell^2(\mathbb{Z}_{N_1} \times \mathbb{Z}_{N_2})$ is translation invariant if, for all $k_1, k_2 \in \mathbb{Z}$ and all $z \in \ell^2(\mathbb{Z}_{N_1} \times \mathbb{Z}_{N_2})$,

$$T(R_{k_1, k_2} z) = R_{k_1, k_2} T(z).$$

If $T : \ell^2(\mathbb{Z}_{N_1} \times \mathbb{Z}_{N_2}) \to \ell^2(\mathbb{Z}_{N_1} \times \mathbb{Z}_{N_2})$ is translation invariant, prove that each F_{m_1,m_2} (defined in Exercise 2.1.17) is an eigenvector of T. Hint: Follow the proof of Theorem 2.18.

2.2.19. For $z, w \in \ell^2(\mathbb{Z}_{N_1} \times \mathbb{Z}_{N_2})$ (see Exercise 2.2.18), define $z * w \in \ell^2(\mathbb{Z}_{N_1} \times \mathbb{Z}_{N_2})$ by

$$z * w(m_1, m_2) = \sum_{n_1=0}^{N_1-1} \sum_{n_2=0}^{N_2-1} z(m_1 - n_1, m_2 - n_2)w(n_1, n_2),$$

for all m_1, m_2.

i. Prove that

$$(z * w)\hat{}(m_1, m_2) = \hat{z}(m_1, m_2)\hat{w}(m_1, m_2),$$

for all m_1, m_2.

ii. For $b \in \ell^2(\mathbb{Z}_{N_1} \times \mathbb{Z}_{N_2})$, define $T_b : \ell^2(\mathbb{Z}_{N_1} \times \mathbb{Z}_{N_2}) \to \ell^2(\mathbb{Z}_{N_1} \times \mathbb{Z}_{N_2})$ by

$$T_b(z) = b * z.$$

Any linear transformation of this form is called *a convolution operator*. Prove that T_b is translation invariant.

iii. For $m \in \ell^2(\mathbb{Z}_{N_1} \times \mathbb{Z}_{N_2})$, define $T_{(m)} : \ell^2(\mathbb{Z}_{N_1} \times \mathbb{Z}_{N_2}) \to \ell^2(\mathbb{Z}_{N_1} \times \mathbb{Z}_{N_2})$ by

$$T_{(m)}(z) = (m\hat{z})^{\vee},$$

where $(m\hat{z})(n_1, n_2) = m(n_1, n_2)\hat{z}(n_1, n_2)$ for each (n_1, n_2). Any linear transformation of this type is called a *Fourier multiplier operator*. Prove that any convolution operator T_b is a Fourier multiplier operator $T_{(m)}$ with $m = \hat{b}$.

2.2.20. Suppose $T : \ell^2(\mathbb{Z}_{N_1} \times \mathbb{Z}_{N_2}) \to \ell^2(\mathbb{Z}_{N_1} \times \mathbb{Z}_{N_2})$ is a linear transformation. Prove that the following are equivalent:

i. T is translation invariant.

ii. T is a Fourier multiplier operator.

iii. T is a convolution operator.

Hint: The implication i $\Rightarrow$ ii follows from Exercise 2.2.18, ii $\Rightarrow$ iii follows from Exercise 2.2.19 (iii) and Fourier inversion, and iii $\Rightarrow$ i was proved in Exercise 2.2.19 (ii).

Remark: This approach avoids matrices, which are more difficult to write out in the two-dimensional case, although they can be represented by writing

$$(Az)(n_1, n_2) = \sum_{m_1=0}^{N_1-1} \sum_{m_2=0}^{N_2-1} A(n_1, n_2; m_1, m_2) z(m_1, m_2).$$

We could have followed the sequence of steps in Exercise 2.2.20 for the one-dimensional case presented in the text, but we preferred to make the matrix description explicit in that case.

2.3 The Fast Fourier Transform

In section 2.2, we saw the main advantage of the Fourier basis F (Definition 2.7): all translation-invariant linear transformations are diagonalized by F. In this section we discuss a second key feature of F: the DFT can be computed by a fast algorithm, known as the *fast Fourier transform*, or *FFT*. Without the FFT, use of the DFT in analyzing real speech or video signals would be dramatically limited.

Consider the amount of computation required for a general change of basis. Suppose $z \in \ell^2(\mathbb{Z}_N)$. If B is a basis for $\ell^2(\mathbb{Z}_N)$, one can obtain the components $[z]_B$ of z with respect to the basis B from the Euclidean components $z = [z]_E$, by multiplying z by the E to B change-of-basis matrix, which we call A. That is,

$$[z]_B = A[z]_E = Az.$$

Since the mth component of Az is $\sum_{n=0}^{N-1} a_{mn} z(n)$, it takes N complex multiplications to compute each component of Az. Since Az has N components, it takes N^2 multiplications to compute the entire vector $Az = [z]_B$.

For the Fourier basis, the situation does not appear any different: we have $\hat{z} = [z]_F = W_N z$, where W_N is the matrix in equation (2.18). So direct computation of $\hat{z}$ takes N^2 complex mutliplications. To be more precise, we could also count the number of additions. However, because multiplication is much slower on a computer than addition, we get a good idea of the speed of computation by just considering the number of complex multiplications required.

When we say *complex multiplication*, we mean the multiplication of two complex numbers. This would appear to require four real multiplications, but by a trick (Exercise 2.3.1), it requires only three real multiplications.

In signal and image processing, the vectors under consideration can be very large. A television signal, for example, requires roughly 10,000,000 pixel values per second to preserve all relevant information (Proakis and Manolakis, 1996, pp. 29–30). Thus one second of the sampled signal is a vector of length 10,000,000. A fingerprint image (see the Prologue) is represented digitally by breaking each square inch of the image into a 500 by 500 grid of pixels, each of which is assigned a gray-scale value (a darkness). These values are the components of a large vector. For a video image, one may have 20–30 vectors of comparable size every second. Computation of the DFTs of these vectors in real time by direct means may be beyond the capacity of one's computational hardware. So a fast algorithm is needed.

We begin with the simplest version of the FFT, in which the length N of the vector is assumed to be even. This case demonstrates the basic idea behind the FFT.

Lemma 2.37 *Suppose $M \in \mathbb{N}$, and $N = 2M$. Let $z \in \ell^2(\mathbb{Z}_N)$. Define $u, v \in \ell^2(\mathbb{Z}_M)$ by*

$$u(k) = z(2k) \quad for \quad k = 0, 1, \ldots, M - 1,$$

and

$$v(k) = z(2k + 1) \quad for \quad k = 0, 1, \ldots, M - 1.$$

In other words,

$$u = \big(z(0), z(2), z(4), \ldots, z(N - 4), z(N - 2)\big)$$

and

$$v = \big(z(1), z(3), z(5), \ldots, z(N - 3), z(N - 1)\big).$$

Let $\hat{z}$ denote the DFT of z defined on N points, that is, $\hat{z} = W_N z$. Let $\hat{u}$ and $\hat{v}$ denote the DFTs of u and v respectively, defined on $M = N/2$ points, that is, $\hat{u} = W_M u$ and $\hat{v} = W_M v$. Then for $m = 0, 1, \ldots, M - 1$,

$$\hat{z}(m) = \hat{u}(m) + e^{-2\pi i m/N} \hat{v}(m). \tag{2.48}$$

Also, for $m = M, M + 1, M + 2, \ldots, N - 1$, *let* $\ell = m - M$. *Note that the corresponding values of* ℓ *are* $\ell = 0, 1, \ldots, M - 1$. *Then*

$$\hat{z}(m) = \hat{z}(\ell + M) = \hat{u}(\ell) - e^{-2\pi i \ell / N} \hat{v}(\ell). \tag{2.49}$$

Proof

For any $m = 0, 1, \ldots, N - 1$,

$$\hat{z}(m) = \sum_{n=0}^{N-1} z(n) e^{-2\pi i m n / N},$$

by definition. The sum over $n = 0, 1, \ldots, N - 1$ can be broken up into the sum over the even values $n = 2k$, $k = 0, 1, \ldots, M - 1$, plus the sum over the odd values $n = 2k + 1$, for $k = 0, 1, \ldots M - 1$:

$$
\begin{aligned}
\hat{z}(m) &= \sum_{k=0}^{M-1} z(2k) e^{-2\pi i 2 k m / N} + \sum_{k=0}^{M-1} z(2k+1) e^{-2\pi i (2k+1) m / N} \\
&= \sum_{k=0}^{M-1} u(k) e^{-2\pi i k m / (N/2)} + e^{-2\pi i m / N} \sum_{k=0}^{M-1} v(k) e^{-2\pi i k m / (N/2)} \\
&= \sum_{k=0}^{M-1} u(k) e^{-2\pi i k m / M} + e^{-2\pi i m / N} \sum_{k=0}^{M-1} v(k) e^{-2\pi i k m / M}.
\end{aligned}
$$

In the case $m = 0, 1, \ldots, M - 1$, the last expression is $\hat{u}(m) + e^{-2\pi i m / N} \hat{v}(m)$, so we have equation (2.48). Now suppose $m = M, M + 1, \ldots, N - 1$. By writing $m = \ell + M$ as in the statement of the theorem and substituting this for m above, we get

$$
\begin{aligned}
\hat{z}(m) &= \sum_{k=0}^{M-1} u(k) e^{-2\pi i k (\ell + M) / M} + e^{-2\pi i (\ell + M) / N} \sum_{k=0}^{M-1} v(k) e^{-2\pi i k (\ell + M) / M} \\
&= \sum_{k=0}^{M-1} u(k) e^{-2\pi i k \ell / M} - e^{-2\pi i \ell / N} \sum_{k=0}^{M-1} v(k) e^{-2\pi i k \ell / M},
\end{aligned}
$$

since the exponentials $e^{-2\pi i k l / M}$ are periodic with period M, and $e^{-2\pi i M / N} = e^{-\pi i} = -1$ for $N = 2M$. This yields equation (2.49). ∎

Example 2.38

Let $z = (1, 1, 1, i, 1, -1, 1, -i)$. Find $\hat{z}$.

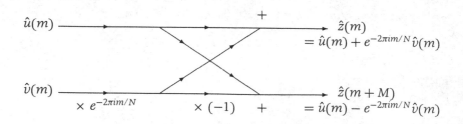

FIGURE 10

Solution

Following Lemma 2.37, we obtain

$$u = (1, 1, 1, 1) \quad \text{and} \quad v = (1, i, -1, -i).$$

Note that $u = 4F_0$ and $v = 4F_1$ (e.g., by Example 2.4 and the fact that $F_m = N^{-1/2}E_m$). Hence by Exercise 2.3.3 (or by direct computation using equations (2.19) and (2.21)),

$$\hat{u} = (4, 0, 0, 0) \quad \text{and} \quad \hat{v} = (0, 4, 0, 0).$$

Hence by equation (2.48),

$$\hat{z}(0) = \hat{u}(0) + 1\hat{v}(0) = 4 + 0 = 4,$$
$$\hat{z}(1) = \hat{u}(1) + e^{-2\pi i 1/8}\hat{v}(1) = 0 + 4e^{-\pi i/4} = 2\sqrt{2} - 2\sqrt{2}i,$$
$$\hat{z}(2) = \hat{u}(2) + e^{-2\pi i 2/8}\hat{v}(2) = 0 + 0 = 0,$$

and

$$\hat{z}(3) = \hat{u}(3) + e^{-2\pi i 3/8}\hat{v}(3) = 0 + 0 = 0.$$

Then by equation (2.49),

$$\hat{z}(4) = \hat{u}(0) - 1\hat{v}(0) = 4 - 0 = 4,$$
$$\hat{z}(5) = \hat{u}(1) - e^{-2\pi i 1/8}\hat{v}(1) = 0 - 4e^{-\pi i/4} = -2\sqrt{2} + 2\sqrt{2}i,$$
$$\hat{z}(6) = \hat{u}(2) - e^{-2\pi i 2/8}\hat{v}(2) = 0 - 0 = 0,$$

and

$$\hat{z}(7) = \hat{u}(3) - e^{-2\pi i 3/8}\hat{v}(3) = 0 - 0 = 0.$$

Hence

$$\hat{z} = (4, 2\sqrt{2} - 2\sqrt{2}i, 0, 0, 4, -2\sqrt{2} + 2\sqrt{2}i, 0, 0).$$

■

The basic step of this procedure starts with the values $\hat{u}(m)$ and $\hat{v}(m)$ and gives $\hat{z}(m)$ and $\hat{z}(m+M)$ according to the diagram in Figure 10, called a *butterfly*. This computation is so basic that computer hardware is sometimes evaluated by how many butterflies can be computed per second.

Notice that the same values are used in equations (2.48) and (2.49), namely $\hat{u}(\ell)$ and $\hat{v}(\ell)$ for $0 \le \ell \le M - 1$. To apply equations (2.48) and (2.49), we first compute $\hat{u}$ and $\hat{v}$. Because each of these is a vector of length $M = N/2$, each can be computed directly with M^2 complex multiplications. We then compute the products $e^{-2\pi i m/N}\hat{v}(m)$ for $m = 0, 1, \ldots, M - 1$. This requires an additional M multiplications. The rest is done using only additions and subtractions of these quantities, which we do not count. So the total number of complex multiplications required to compute $\hat{z}$ by equations (2.48) and (2.49) is at most

$$2M^2 + M = 2\left(\frac{N}{2}\right)^2 + \frac{N}{2} = \frac{1}{2}\left(N^2 + N\right).$$

For N large, this is essentially $N^2/2$, whereas the number of complex multiplications required to compute $\hat{z}$ directly is N^2. So Lemma 2.37 already cuts our computation time nearly in half.

If N is divisible by 4 instead of just 2, we can go further. Because u and v have even order, we can then apply the same method to reduce the time required to compute them. If N is divisible by 8, we can carry this one step further, and so on. A more general way to describe this is to define $\#_N$, for any positive integer N, to be the least number of complex multiplications required to compute the DFT of a vector of length N. If $N = 2M$, then equations (2.48) and (2.49) reduce the computation of $\hat{z}$ to the computation of two DFTs of size M, plus M additional complex multiplications. Hence

$$\#_N \le 2\#_M + M. \tag{2.50}$$

The most favorable case is when N is a power of 2.

Lemma 2.39 *Suppose $N = 2^n$ for some $n \in \mathbb{N}$. Then*

$$\#_N \le \frac{1}{2}N \log_2 N.$$

Proof

The proof is by induction on n. When $n = 1$, a vector of length 2^1 is of the form $z = (a, b)$. Then from the definition (see equations (2.19) and (2.20)),

$$\hat{z} = (a + b, a - b).$$

Note that this computation does not require any complex multiplications, so $\#_2 = 0 < 1 = (2 \log_2 2)/2$. So the result holds in this case. By induction, suppose it holds for $n = k - 1$. Then for $n = k$, we have by equation (2.50) and the induction hypothesis that

$$\#_{2^k} \leq 2\#_{2^{k-1}} + 2^{k-1} \leq 2\frac{1}{2}2^{k-1}(k-1) + 2^{k-1}$$

$$= k2^{k-1} = \frac{1}{2}k2^k = \frac{1}{2}N \log_2 N.$$

This completes the induction step, and hence establishes the result. ∎

As an example, for a vector of size $262{,}144 = 2^{18}$, the FFT reduces the number of complex multiplications needed to compute the DFT from 6.87×10^{10} to $2{,}359{,}296$, making the computation more than 29,000 times faster! Thus if it takes 8 hours to do this DFT directly, it would take about 1 second to do it via the FFT. This ratio becomes more extreme as N increases, to the point that some computations that can be done by the FFT in a reasonable length of time could not be done directly in an entire lifetime. This radical difference in speed has been essential for modern-day digital signal processing.

What if N is not even? If N is prime, the method of the FFT does not apply. However, if N is composite, say $N = pq$, a generalization of Lemma 2.37 can be applied.

Lemma 2.40 *Suppose $p, q \in \mathbb{N}$ and $N = pq$. Let $z \in \ell^2(\mathbb{Z}_N)$. Define $w_0, w_1, \ldots, w_{p-1} \in \ell^2(\mathbb{Z}_q)$ by*

$$w_\ell(k) = z(kp + \ell) \quad \text{for} \quad k = 0, 1, \ldots, q - 1.$$

For $b = 0, 1, \ldots, q - 1$, define $v_b \in \ell^2(\mathbb{Z}_p)$ by

$$v_b(\ell) = e^{-2\pi i b\ell/N} \hat{w}_\ell(b) \quad \text{for} \quad \ell = 0, 1, \ldots, p - 1.$$

Then for $a = 0, 1, \ldots, p - 1$ and $b = 0, 1, \ldots, q - 1$,

$$\hat{z}(aq + b) = \hat{v}_b(a). \tag{2.51}$$

Note that by the division algorithm, every $m = 0, 1, \ldots, N - 1$ is of the form $aq + b$ for some $a \in \{0, 1, \ldots, p - 1\}$ and $b \in \{0, 1, \ldots, q - 1\}$, so equation (2.51) gives the full DFT of z.

Proof

We can write each $n = 0, 1, \ldots, N - 1$ uniquely in the form $kp + \ell$ for some $k \in \{0, 1, \ldots, q - 1\}$ and $\ell \in \{0, 1, \ldots, p - 1\}$. Hence

$$\hat{z}(aq+b) = \sum_{n=0}^{N-1} z(n)e^{-2\pi i(aq+b)n/N} = \sum_{\ell=0}^{p-1}\sum_{k=0}^{q-1} z(kp+\ell)e^{-2\pi i(aq+b)(kp+\ell)/(pq)}.$$

Note that

$$e^{-2\pi i(aq+b)(kp+\ell)/(pq)} = e^{-2\pi iak}e^{-2\pi ia\ell/p}e^{-2\pi ibk/q}e^{-2\pi ib\ell/(pq)}.$$

Since $e^{-2\pi iak} = 1$ and $pq = N$, using the definition of $w_\ell(k)$ we obtain

$$\hat{z}(aq + b) = \sum_{\ell=0}^{p-1} e^{-2\pi ia\ell/p} e^{-2\pi ib\ell/N} \sum_{k=0}^{q-1} w_\ell(k)e^{-2\pi ibk/q}$$

$$= \sum_{\ell=0}^{p-1} e^{-2\pi ia\ell/p} e^{-2\pi ib\ell/N} \hat{w}_\ell(b) = \sum_{\ell=0}^{p-1} e^{-2\pi ia\ell/p} v_b(\ell) = \hat{v}_b(a). \quad\blacksquare$$

This proof shows the basic principle behind the FFT. In computing $\hat{z}(aq + b)$, the same quantities $v_b(\ell), 0 \le \ell \le p - 1$, arise for each value of a. The FFT algorithm recognizes this and computes these values only once. Direct computation of $\hat{z}$ involves implicitly recomputing these intermediate values each time they arise.

Consider the number of multiplications required for the algorithm in Lemma 2.40. We first compute the vectors $\hat{w}_\ell$, for $\ell = 0, 1, \ldots, p-1$. Each of these is a vector of length q, so computing each $\hat{w}_\ell$ requires $\#_q$ complex multiplications. So this step requires a total of $p\#_q$ complex multiplications. The next step is to multiply each $\hat{w}_\ell(b)$ by $e^{-2\pi ib\ell/N}$ to obtain the vectors $v_b(\ell)$. This requires a total of pq complex multiplications, one for each of the q values of b and p values of ℓ. Finally we compute the vectors $\hat{v}_b$ for $b = 0, 1, \ldots, q-1$.

Each v_b is a vector of length p, so each of the q vectors $\hat{v}_b$ requires $\#_p$ complex multiplications, for a total of $q\#_p$ muliplications. Adding up, we have an estimate for the number of multiplications required to compute a DFT of size $N = pq$, namely

$$\#_{pq} \le p\#_q + q\#_p + pq. \tag{2.52}$$

This estimate can be used inductively to make various estimates on the time required to compute the FFT (see Exercises 2.3.7 and 2.3.8). The advantage of using the FFT is greater the more composite N is. In many applications we can segment our data stream into pieces of any size we choose. In this case we usually take N to be a power of 2 so we can apply Lemma 2.39. If we cannot choose the length of the vector, sometimes it is harmless to pad it with some extra zeros at the end until it has length that is highly composite.

Although equations (2.50) and (2.52) can be used to estimate the total number of complex multiplications needed to compute $\hat{z}$, they don't show how to set up the full computation. Lemmas 2.37 and 2.40 show how to do each step but not how to organize them iteratively. We discuss one more FFT algorithm to show how the computation can be arranged. For simplicity, we restrict ourselves here to the case where N is a power of 2, say $N = 2^n$. Then we can expand any $m \in \{0, 1, \dots, N - 1\}$ in base 2 in the form

$$m = m_0 + 2m_1 + 2^2 m_2 + \cdots + 2^{n-1} m_{n-1},$$

where $m_0, m_1, \dots, m_{n-1} \in \{0, 1\}$. For $z \in \ell^2(\mathbb{Z}_N)$, denote

$$z(m) = z(m_{n-1}, m_{n-2}, \dots, m_1, m_0).$$

For any $k = k_0 + 2k_1 + 2^2 k_2 + \cdots + 2^{n-1} k_{n-1}$ with $k_0, k_1, \dots, k_{n-1} \in \{0, 1\}$,

$$\hat{z}(k) = \sum_{m=0}^{N-1} z(m) e^{-2\pi i k m / N}$$

$$= \sum_{m_0=0}^{1} \sum_{m_1=0}^{1} \cdots \sum_{m_{n-1}=0}^{1} z(m_{n-1}, m_{n-2}, \dots, m_1, m_0)$$

$$\times \exp\left(\frac{-2\pi i (k_0 + 2k_1 + \cdots + 2^{n-1} k_{n-1})(m_0 + 2m_1 + \cdots + 2^{n-1} m_{n-1})}{2^n} \right),$$

where $\exp(t)$ denotes e^t. Now

$$\exp\left(\frac{-2\pi i(k_0 + 2k_1 + \cdots + 2^{n-1}k_{n-1})(m_0 + 2m_1 + \cdots + 2^{n-1}m_{n-1})}{2^n}\right)$$

$$= \exp\left(\frac{-2\pi i(k_0 + 2k_1 + \cdots + 2^{n-1}k_{n-1})2^{n-1}m_{n-1}}{2^n}\right)\cdots$$

$$\times \exp\left(\frac{-2\pi i(k_0 + 2k_1 + \cdots + 2^{n-1}k_{n-1})2m_1}{2^n}\right)$$

$$\times \exp\left(\frac{-2\pi i(k_0 + 2k_1 + \cdots + 2^{n-1}k_{n-1})m_0}{2^n}\right).$$

In each exponent we can delete all products that give an integer multiple of $2\pi i 2^n$ in the numerator because after division by 2^n, the argument is an integer multiple of 2π. Thus

$$\exp\left(\frac{-2\pi i(k_0 + 2k_1 + \cdots + 2^{n-1}k_{n-1})(m_0 + 2m_1 + \cdots + 2^{n-1}m_{n-1})}{2^n}\right)$$

$$= \exp\left(\frac{-2\pi i k_0 2^{n-1}m_{n-1}}{2^n}\right)\exp\left(\frac{-2\pi i(k_0 + 2k_1)2^{n-2}m_{n-2}}{2^n}\right)$$

$$\times \exp\left(\frac{-2\pi i(k_0 + 2k_1 + 2^2 k_2)2^{n-3}m_{n-3}}{2^n}\right)\cdots$$

$$\times \exp\left(\frac{-2\pi i(k_0 + 2k_1 + 2^2 k_2 + \cdots + 2^{n-1}k_{n-1})m_0}{2^n}\right).$$

Substituting this into the preceding equation gives

$$\hat{z}(k) = \sum_{m_0=0}^{1}\sum_{m_1=0}^{1}\cdots\sum_{m_{n-1}=0}^{1} z(m_{n-1}, \ldots, m_1, m_0)\exp\left(\frac{-2\pi i k_0 2^{n-1}m_{n-1}}{2^n}\right)$$

$$\times \exp\left(\frac{-2\pi i(k_0 + 2k_1)2^{n-2}m_{n-2}}{2^n}\right)\cdots \qquad (2.53)$$

$$\times \exp\left(\frac{-2\pi i(k_0 + 2k_1 + \cdots + 2^{n-1}k_{n-1})m_0}{2^n}\right).$$

Notice that the inside sum depends on the outside summation variables $m_0, m_1, \ldots, m_{n-2}$ and on k_0 but not on $k_1, \ldots, k_{n-1}$. So define

$$y_1(k_0, m_{n-2}, m_{n-3}, \ldots, m_0)$$

$$= \sum_{m_{n-1}=0}^{1} z(m_{n-1}, m_{n-2}, \ldots, m_1, m_0) \exp(-2\pi i k_0 2^{n-1} m_{n-1}/2^n)$$

$$= z(0, m_{n-2}, \ldots, m_1, m_0) \cdot 1$$

$$+ z(1, m_{n-2}, \ldots, m_1, m_0) \exp\left(\frac{-2\pi i k_0 2^{n-1}}{2^n}\right).$$

Computing $y_1(k_0, m_{n-2}, m_{n-3}, \ldots, m_0)$ requires only one complex multiplication for each of the 2^n choices of $k_0, m_{n-2}, m_{n-3}, \ldots, m_0 \in \{0, 1\}$, for a total of 2^n complex multiplications to compute all 2^n possible values of y_1. At the next step, define

$$y_2(k_0, k_1, m_{n-3}, \ldots, m_0)$$

$$= \sum_{m_{n-2}=0}^{1} y_1(k_0, m_{n-2}, m_{n-3}, \ldots, m_0)$$

$$\times \exp\left(\frac{-2\pi i (k_0 + 2k_1) 2^{n-2} m_{n-2}}{2^n}\right).$$

It takes just one complex multiplication to compute each one of these, hence 2^n total to compute all possible values of y_2. We continue in this way, each time replacing the highest remaining m index by the next k index. Thus the scheme is to make the sequence of transformations

$$z(m_{n-1}, m_{n-2}, \ldots, m_1, m_0) \to y_1(k_0, m_{n-2}, m_{n-3}, \ldots, m_0),$$

$$y_1(k_0, m_{n-2}, m_{n-3}, \ldots, m_0) \to y_2(k_0, k_1, m_{n-3}, \ldots, m_0),$$

$$\vdots$$

$$y_{n-1}(k_0, k_1, k_2, \ldots, k_{n-2}, m_0) \to y_n(k_0, k_1, k_2, \ldots, k_{n-2}, k_{n-1}).$$

By equation (2.53), the final vector $y_n(k_0, k_1, k_2, \ldots, k_{n-2}, k_{n-1})$ is $\hat{z}(k)$. At each step we compute the next vector y_j at all the 2^n possible choices of its variables. So each step requires at most 2^n complex multiplications, and there are a total of n steps, for at most $n2^n = N \log_2 N$ complex multiplications. (By Lemma 2.39, this must be an overcounting by a factor of 2. Exercise 2.3.9 explains the discrepancy.) Note that once the vector y_j has been computed, y_{j-1} is no longer needed, and hence can be discarded. Thus the computation can be done "in place," that is, at each stage the previous data can be replaced by the new data. This reduces the amount of memory needed to run the computation.

There are many variations on the FFT algorithm, sometimes leading to slight advantages over the basic one given here. But the main point is that the DFT of a vector of length $N = 2^n$ can be computed with at most $n2^{n-1} = (N/2)\log_2 N$ complex multiplications as opposed to $N^2 = 2^{2n}$ if done directly.

What about the inverse DFT, the IDFT? By equation (2.30),

$$\check{w}(n) = \frac{1}{N}\hat{w}(N - n),$$

so the FFT algorithm can be used to compute the IDFT quickly also, in at most $(N/2)\log_2 N$ steps if N is a power of 2. (We don't count division by N because integer division is relatively fast.)

Given that the DFT and IDFT can be computed rapidly, it follows that we can compute convolutions quickly also. Namely, we can write

$$z * w = (\hat{z}\hat{w})^{\check{}}$$

(by applying the inverse transform to the result of Lemma 2.30). If $z, w \in \ell^2(\mathbb{Z}_N)$, for N a power of 2, it takes at most $N\log_2 N$ multiplications to compute $\hat{z}$ and $\hat{w}$, N multiplications to compute $\hat{z}\hat{w}$ (the componentwise product of these two vectors), and at most $(N/2)\log_2 N$ multiplications to take the IDFT of $\hat{z}\hat{w}$. Thus overall it takes no more than $N + (3N/2)\log_2 N$ multiplications to compute $z * w$.

In section 2.2, we saw that any translation-invariant linear transformation on $\ell^2(\mathbb{Z}_N)$ can be written as a convolution operator. By Theorem 2.19, this includes the operation of multiplication by a circulant matrix. Thus the product of an $N \times N$ circulant matrix times a vector of length N can be computed using at most $N + (3N/2)\log_2 N$ multiplications, instead of the usual N^2. In other words, when T is a translation-invariant linear transformation, the DFT not only diagonalizes T, it gives (via the FFT) a fast, practical way to compute T.

Exercises

2.3.1. Observe that

$$(a+ib)(c+id) = (a-b)d+(c-d)a+\left[(a-b)d+(c+d)b\right]i.$$

This means that to compute the product of two complex numbers, we need to compute only three real multiplications, namely $(a-b)d, (c-d)a,$ and $(c+d)b$.

2.3.2. Let $u = (1,3)$, $v = (0,4)$, and $z = (1,0,3,4)$.
  i. Compute $\hat{u}$ and $\hat{v}$.
  ii. Use part i and equations (2.48) and (2.49) to compute $\hat{z}$.
  iii. Compute $\hat{z}$ directly and compare the answer with your answer to part ii.
  iv. Let $w = (0,1,4,3)$. Use equations (2.48) and (2.49) to compute $\hat{w}$.

2.3.3. Let $\{e_0, e_1, \ldots, e_{N-1}\}$ be the Euclidean basis for $\ell^2(\mathbb{Z}_N)$, and let $\{F_0, F_1, \ldots, F_{N-1}\}$ be the Fourier basis.
  i. Show that $\hat{e}_m(k) = e^{-2\pi imk/N}$ for all k. Notice that $\hat{e}_m$ is very nearly (up to a reflection and a normalization) an element of the Fourier basis.
  ii. Show that $\hat{F}_m = e_m$.

2.3.4. Let $w = (1,0,1,0,1,0,1,0)$. Compute $\hat{w}$. Hint: consider $z = (1,1,1,1)$.

2.3.5. Let $u = (1, i, -1, -i)$, $v = (1, -1, 1, -1)$, and $z = (1, 1, i, -1, -1, 1, -i, -1)$.
  i. Compute $\hat{u}$ and $\hat{v}$. (Suggestion: Use Exercise 2.3.3.)
  ii. Compute $\hat{z}$.

2.3.6. Suppose $u = (a, b, c, d)$, $v = (\alpha, \beta, \gamma, \delta)$, and $z = (a, \alpha, b, \beta, c, \gamma, d, \delta)$. If

$$\hat{u} = (2, i, -1, 0) \quad \text{and} \quad \hat{v} = (3, -2, 0, 4i),$$

find $\hat{z}$.

2.3.7. Suppose p is a prime number and $N = p^n$ for some positive integer n. Prove that

$$\#_{p^n} \le np^{n+2} = p^2 N \log_p N.$$

Hint: Use induction and apply equation (2.52) with $q = p^n$ when going from n to $n + 1$.

2.3.8. Suppose $N = p_1^{m_1} p_2^{m_2} p_3^{m_3} \cdots p_n^{m_n}$, for some positive integers $p_1, p_2, \ldots, p_n$ and $m_1, \ldots, m_n$. Prove that there exist constants $C(p_1, p_2, \ldots, p_m)$ depending on $p_1, p_2, \ldots, p_n$ but not on $m_1, m_2, \ldots, m_n$, such that

$$\#_N \le C(p_1, p_2, \ldots, p_m) N \log_2 N.$$

Hint: Recall that $\log_b(x) = \log_a(x) / \log_a(b)$; use Exercise 2.3.7, equation (2.52), and induction on n.

2.3.9. Show that the number of complex multiplications required to compute equation (2.53) is at most $(N/2) \log_2 N$. Hint: At the general step going from y_j to y_{j+1}, we must compute

$$y_{j+1}(k_0, \ldots, k_j, m_{n-j-2}, \ldots, m_0)$$

$$= \sum_{m_{n-j-1}=0}^{1} y_j(k_0, \ldots, k_{j-1}, m_{n-j-1}, \ldots, m_0)$$

$$\times \exp\left(\frac{-2\pi i(k_0 + \cdots + 2^{j-1}k_{j-1} + 2^j k_j)2^{n-j-1}m_{n-j-1}}{2^n}\right).$$

This seems to require one multiplication (only one because there is no multiplication when $m_{n-j-1} = 0$) for each possible value of $k_0, \ldots, k_j$ and $m_{n-j-2}, \ldots, m_0$. Separate out the term $e^{-2\pi i 2^j k_j 2^{n-j-1} m_{n-j-1}/2^n}$ and show that it is always either $+1$ or -1. Thus the two choices of k_j are done via the same multiplication.

2.3.10. (Operation count for the two-dimensional DFT) Let $z \in \ell^2(\mathbb{Z}_{N_1} \times \mathbb{Z}_{N_2})$ (defined in Exercise 2.1.15). Direct computation of the two-dimensional DFT $\hat{z}(m_1, m_2)$ (defined in Exercise 2.1.17) would at first appear to require $N_1 N_2$ complex multiplications for each of the $N_1 N_2$ values of (m_1, m_2), for a total of $N_1^2 N_2^2$ complex multiplications. However, define

$$y(n_1, m_2) = \sum_{n_2=0}^{N_2-1} z(n_1, n_2)e^{-2\pi i m_2 n_2 / N_2}.$$

Then by the definition,

$$\hat{z}(m_1, m_2) = \sum_{n_1=0}^{N_1-1} y(n_1, m_2)e^{-2\pi i m_1 n_1 / N_1}.$$

i. Show that the direct computation of all values of $y(n_1, m_2)$ takes $N_1 N_2^2$ complex multiplications, after which direct computation of all values of $\hat{z}(m_1, m_2)$ requires $N_1^2 N_2$ multiplications. So the full computation can be done using $N_1 N_2 (N_1 + N_2)$ multiplications.

Remark: The reason for this advantage is that the computation of the two-dimensional DFT consists, by its definition, of computing one transform of size N_2 (the inner sum) followed by a second sum of size N_1. Any computation that can be broken down into a sequence of smaller summations in this way is called *parallelizable*. The principle behind the FFT is that the DFT is parallelizable, as shown by equation (2.53).

ii. Suppose N_1 and N_2 are powers of 2. Using the FFT at each of the two stages outlined above, instead of direct computation, show that $\hat{z}$ can be computed using at most $\frac{1}{2} N_1 N_2 \log_2(N_1 N_2)$ complex multiplications.

3

Wavelets on $\mathbb{Z}_N$

3.1 Construction of Wavelets on $\mathbb{Z}_N$: The First Stage

In this chapter we continue to work with discrete signals $z \in \ell^2(\mathbb{Z}_N)$. In chapter 2 we noted the two key advantages of the Fourier basis F: (1) translation-invariant linear transformations are diagonalized by F, and (2) the coordinates in the Fourier basis can be computed quickly using the FFT. However, for many purposes in signal analysis and other fields, the Fourier basis has serious limitations. Many of these come from the fact that the Fourier basis elements are not localized in space, in the following sense.

We say that a vector $z \in \ell^2(\mathbb{Z}_N)$ is localized in space near n_0 if most of the components $z(n)$ of z are 0 or at least relatively small, except for a few values of n close to n_0. A Fourier basis element F_m is not localized in space because its components $F_m(n) = \frac{1}{N}e^{2\pi imn/N}$ have the same magnitude (namely $1/N$) for every $n \in \mathbb{Z}_N$. This is the opposite of being localized: the Fourier basis vectors are as evenly dispersed as possible.

Suppose $B = \{v_0, v_1, \ldots, v_{N-1}\}$ is a basis for $\ell^2(\mathbb{Z}_N)$ such that all the basis elements of B are localized in space. For a vector z, we can write

$$z = \sum_{n=0}^{N-1} a_n v_n, \tag{3.1}$$

for some scalars $a_0, \ldots, a_{N-1}$. Suppose that we wish to focus on the portion of z near some particular point n_0. Terms involving basis vectors that are 0 or negligibly small near n_0 can be deleted from relation (3.1) without changing the behavior near n_0 significantly. Thus we may be able to replace a full sum over N terms by a much smaller sum when considering only the portion of z near n_0.

More generally, a spatially localized basis is useful because it provides a local analysis of a signal: if a certain coefficient in the expansion of z is large, we can identify the location with which this large coefficient is associated. We could then, for example, focus on this location and analyze it in more detail. One example for which this is useful is medical image processing, for example to look closely at a potential tumor. Another is radar or sonar imaging, for example in oil prospecting to identify the boundary of an oil pocket, or in archaeology to locate artifacts.

Another example comes from video image analysis. Currently, the use of video telephones is not widespread because a high-quality sequence of video images cannot be transmitted along a phone line in real time, because such a sequence of images exceeds the phone line's capacity. If video images can be represented with a smaller data set without serious degradation of the image, video telephones may become practical. There is an entire field called *data compression* devoted to this and similar problems.

Having a localized basis would help compress video images for the following reason. For television video images, it is likely that one frame differs only slightly from the previous frame: for example, the background may be the same but perhaps a person's hand is moving. Instead of transmitting the entire new frame, only the difference between one frame and the next has to be transmitted. For a localized basis, the coefficients of the basis vectors that were concentrated far away from the moving hand would be almost unaffected by that movement. These coefficients would not change appreciably from one frame to the next, and so would not require updating. The only updating required would be for a small number of coefficients of basis vectors localized near the hand. The updating could be done with relatively few data bits, thus achieving a high rate of compression.

This cannot be done if we are using the Fourier basis to represent our data. Since

$$\hat{z}(m) = \sum_{n=0}^{N-1} z(n) e^{-2\pi i m n / N},$$

and $e^{-2\pi i m n / N}$ has magnitude 1 at every n, a change in $z(n)$ for a single value of n can affect all values of $\hat{z}(m)$ significantly. Similarly, the spatially localized hand movement can affect nearly all the DFT values substantially. Updating the image in the Fourier basis may require a large number of data bits, even though the image changed only locally. (For the example of a video image, we should consider the two-dimensional DFT from Exercise 2.1.17, but the principles are the same.)

We have one example of a localized basis, namely the standard, or Euclidean, basis. It is as localized as possible; each basis vector has only one nonzero component. However, we would also like to obtain the advantages of the Fourier basis discussed in chapter 2, in particular the fast computation of translation-invariant linear transformations. For this we would like our basis to be *frequency localized*. By this we mean that we would like the DFTs of our basis vectors to be negligibly small except near one particular region. This means that the basis vectors should consist of a very small group of frequencies. Note that a standard basis vector e_m is not frequency localized: by Exercise 2.3.3 (i), $|\hat{e}_m(k)| = 1$ for all k. The Fourier basis vectors are perfectly localized in frequency: by Exercise 2.3.3 (ii), $\hat{F}_m(n)$ is nonzero only at one n, namely $n = m$. Because the Fourier basis, which is perfectly frequency localized, exactly diagonalizes translation-invariant linear transformations, we expect that a basis that is somewhat frequency localized will nearly diagonalize these transformations, in some sense.

Working with a frequency localized basis also allows us to mimic common filtering techniques. For example, it may be that the high-frequency components of the signal have very small coefficients, so that these values can be deleted without altering the signal in a serious way. Or it may be that these high-frequency components are not humanly perceptible, so deleting them does not affect our perception of the image. (This is the case, for example, for audio signals.) It may even be that some high-frequencies come from noise added to the signal, so that the signal becomes more clear when these terms are removed. With a frequency localized expansion, we know which terms in our expansion to delete to remove the high-frequency components of the signal. If the result is that the signal is satisfactorily represented by a reduced number of data bits, then we have obtained compression.

Thus, our ultimate goal is to obtain a basis whose elements are both spatially and frequency localized. Then a vector's expansion coefficients in this basis will provide both spatial and frequency information. Hence, we would obtain a simultaneous space/frequency analysis of this vector. Wavelets will provide such a basis.

When we talk about an audio signal, we regard the original variable as time (i.e., $z(n)$ is the amplitude of the signal at time n), whereas if we are talking about a two-dimensional video image, we regard the original variable as position (i.e., $z(n_1, n_2)$ is the darkness of the image at position (n_1, n_2); see Exercises 2.1.15–2.1.18). In either case the DFT variable is considered the frequency (e.g., in one dimension, $\hat{z}(m)$ is the coefficient of the frequency component F_m in the sum making up z). Thus we can talk of time/frequency analysis or space/frequency analysis, depending on the physical context; the mathematical meaning is the same.

We would also like the change of basis from the standard basis E to the new basis B to be computable by a fast algorithm, because otherwise B will be useless for audio and video signals of a realistic size. We focus on the issue of rapid computation for the moment. We can compute the DFT quickly via the FFT but the Fourier basis is not spatially localized. However, we noted in section 2.3 that we can use the FFT to compute convolutions quickly also, via the formula

$$z * w = (\hat{z}\hat{w})^{\vee}.$$

Can we use this to compute a change of basis quickly? To answer this, we first note the connection between convolution and inner products.

Definition 3.1 *For any $w \in \ell^2(\mathbb{Z}_N)$, define $\tilde{w} \in \ell^2(\mathbb{Z}_N)$ by*

$$\tilde{w}(n) = \overline{w(-n)} = \overline{w(N-n)} \quad \text{for all} \quad n. \tag{3.2}$$

We call $\tilde{w}$ the conjugate reflection *of w.*

(This will be standard notation for us: from now on, $\tilde{w}, \tilde{z}$, etc., are assumed without comment to be defined as in equation (3.2).) By Exercise 2.1.13,

$$(\tilde{w})\hat{\ }(n) = \overline{\hat{w}(n)} \tag{3.3}$$

for all n. Recall the definition of the circular translate $R_k z$ of a vector z: $(R_k z)(n) = z(n-k)$ (Definition 2.12).

Lemma 3.2 *Suppose $z, w \in \ell^2(\mathbb{Z}_N)$. For any $k \in \mathbb{Z}$,*

$$z * \tilde{w}(k) = \langle z, R_k w \rangle \tag{3.4}$$

and

$$z * w(k) = \langle z, R_k \tilde{w} \rangle. \tag{3.5}$$

Proof
By definition,

$$\langle z, R_k w \rangle = \sum_{n=0}^{N-1} z(n)\overline{R_k w(n)} = \sum_{n=0}^{N-1} z(n)\overline{w(n-k)}$$

$$= \sum_{n=0}^{N-1} z(n)\tilde{w}(k-n) = \tilde{w} * z(k) = z * \tilde{w}(k),$$

by the commutativity of convolution (Exercise 2.2.5). This proves equation (3.4). Then equation (3.5) follows by replacing w with $\tilde{w}$ in equation (3.4), and noting that $\tilde{\tilde{w}} = w$. ∎

Can Lemma 3.2 be used to obtain a basis that can be computed quickly? Suppose $w \in \ell^2(\mathbb{Z}_N)$ is such that $B = \{R_k w\}_{k=0}^{N-1}$ is an orthonormal basis for $\ell^2(\mathbb{Z}_N)$. Then the coefficients of the expansion of a vector z in terms of B are the inner products $\langle z, R_k w \rangle$ (Lemma 1.101 i). By equation (3.4), these coefficients are just the components of $z * \tilde{w}$, that is,

$$[z]_B = z * \tilde{w}.$$

Using the FFT, this convolution can be computed rapidly. So for an orthonormal basis B generated by translates of a single vector w, the E to B change of basis can be computed quickly (where E is the Euclidean basis).

The standard basis E is the only obvious example of an orthonormal basis of the form $\{R_k w\}_{k=0}^{N-1}$. Remarkably, there is a simple condition, in terms of the DFT of w, that characterizes all such bases.

Lemma 3.3   Let $w \in \ell^2(\mathbb{Z}_N)$. Then $\{R_k w\}_{k=0}^{N-1}$ is an orthonormal basis for $\ell^2(\mathbb{Z}_N)$ if and only if $|\hat{w}(n)| = 1$ for all $n \in \mathbb{Z}_N$.

Proof

Recall the Dirac function $\delta \in \ell^2(\mathbb{Z}_N)$ defined by $\delta(n) = 1$ if $n = 0$ and $\delta(n) = 0$ if $n = 1, 2, \ldots, N - 1$. By Exercise 2.3.3 and the fact that $\delta = e_0$ (or a simple computation), $\hat{\delta}(n) = 1$ for all n. By Exercise 3.1.1(ii), $\{R_k w\}_{k=0}^{N-1}$ is an orthonormal basis for $\ell^2(\mathbb{Z}_N)$ if and only if

$$\langle w, R_k w \rangle = \begin{cases} 1 & \text{if } k = 0 \\ 0 & \text{if } k = 1, 2, \ldots, N - 1. \end{cases} \tag{3.6}$$

By equation (3.4), $\langle w, R_k w \rangle = w * \tilde{w}(k)$, so equation (3.6) is equivalent to

$$w * \tilde{w} = \delta.$$

By Fourier inversion, Lemma 2.30, and equation (3.3), this is the same as

$$1 = \hat{\delta}(n) = (w * \tilde{w})\hat{\ }(n) = \hat{w}(n)(\tilde{w})\hat{\ }(n) = \hat{w}(n)\overline{\hat{w}(n)} = |\hat{w}(n)|^2,$$

for all n.   ∎

Although it is gratifying to get such a simple condition for B to be orthonormal, Lemma 3.3 is a disappointment from our point of view. It says that we cannot obtain a frequency localized orthonormal basis of the form $\{R_k w\}_{k=0}^{N-1}$, since $\hat{w}(n)$ will have magnitude 1 for all n. By Lemma 2.13, $|(R_k w)\hat{\ }(n)| = |\hat{w}(n)|$, so every element $R_k w$ has the same property. Thus the situation for any orthonormal basis of this form is similar to the case of the Euclidean basis.

This observation is not as devastating to our plans as it appears because it turns out that a slight modification of the original idea leads to the key results. Instead of looking for one vector w whose full set of translates form an orthonormal basis, we look for two vectors u and v such that the set of their translates by even integers forms an orthonormal basis. For this result, we must restrict ourselves to even values of N.

Definition 3.4 *Suppose N is an even integer, say $N = 2M$ for some $M \in \mathbb{N}$. An orthonormal basis for $\ell^2(\mathbb{Z}_N)$ of the form*

$$\{R_{2k}u\}_{k=0}^{M-1} \cup \{R_{2k}v\}_{k=0}^{M-1}$$

for some $u, v \in \ell^2(\mathbb{Z}_N)$, is called a first-stage wavelet basis *for $\ell^2(\mathbb{Z}_N)$. We call u and v the* generators *of the first-stage wavelet basis. We sometimes also call u the* father wavelet *and v the* mother wavelet.

Our goal is to determine when a pair u, v generates a first-stage wavelet basis. In Theorem 3.8, we characterize such pairs in terms of conditions on $\hat{u}$ and $\hat{v}$. First we build up some necessary background.

Lemma 3.5 *Suppose $M \in \mathbb{N}$, $N = 2M$, and $z \in \ell^2(\mathbb{Z}_N)$. Define $z^* \in \ell^2(\mathbb{Z}_N)$ by*

$$z^*(n) = (-1)^n z(n) \quad \text{for all} \quad n. \tag{3.7}$$

Then

$$(z^*)\hat{}(n) = \hat{z}(n + M) \quad \text{for all} \quad n. \tag{3.8}$$

Proof
By definition,

$$(z^*)\hat{}(n) = \sum_{k=0}^{N-1} z^*(k)e^{-2\pi ikn/N} = \sum_{k=0}^{N-1} (-1)^k z(k)e^{-2\pi ikn/N}$$

$$= \sum_{k=0}^{N-1} z(k)e^{-i\pi k}e^{-2\pi ikn/N} = \sum_{k=0}^{N-1} z(k)e^{-2\pi ik(n+M)/N} = \hat{z}(n + M).$$ ∎

Observe that for any $z \in \ell^2(\mathbb{Z}_N)$, with N even,

$$(z + z^*)(n) = z(n)(1 + (-1)^n) = \begin{cases} 2z(n) & \text{if } n \text{ is even} \\ 0 & \text{if } n \text{ is odd.} \end{cases} \tag{3.9}$$

From equation (3.9), we see the utility of z^*: it provides a means to restrict to even values of n. This is exhibited in the proof of Lemma 3.6.

Lemma 3.6   *Suppose* $M \in \mathbb{N}$, $N = 2M$, *and* $w \in \ell^2(\mathbb{Z}_N)$. *Then* $\{R_{2k}w\}_{k=0}^{M-1}$ *is an orthonormal set with* M *elements if and only if*

$$|\hat{w}(n)|^2 + |\hat{w}(n+M)|^2 = 2 \quad \text{for} \quad n = 0, 1, \dots, M-1. \tag{3.10}$$

Proof

By equation (3.4) and Exercise 3.1.1 (iii), $\{R_{2k}w\}_{k=0}^{M-1}$ is an orthonormal set with M elements if and only if

$$w * \tilde{w}(2k) = \langle w, R_{2k}w \rangle = \begin{cases} 1 & \text{if } k = 0 \\ 0 & \text{if } k = 1, 2, \dots, M-1. \end{cases} \tag{3.11}$$

By equation (3.9),

$$(w * \tilde{w} + (w * \tilde{w})^*)(n) = \begin{cases} 2\, w * \tilde{w}(n) & \text{if } n \text{ is even} \\ 0 & \text{if } n \text{ is odd.} \end{cases} \tag{3.12}$$

Hence, for even values of n, say $n = 2k$, equation (3.11) holds if and only if

$$(w * \tilde{w} + (w * \tilde{w})^*)(2k) = 2\, w * \tilde{w}(2k) = \begin{cases} 2 & \text{if } k = 0 \\ 0 & \text{if } k = 1, 2, \dots, M-1. \end{cases}$$

For odd values of n, $(w * \tilde{w} + (w * \tilde{w})^*)(n)$ is automatically 0, by equation (3.12). Hence equation (3.11) holds if and only if

$$w * \tilde{w} + (w * \tilde{w})^* = 2\delta.$$

Since $\hat{\delta}(n) = 1$ for all n, Fourier inversion shows that equation (3.11) holds if and only if

$$(w * \tilde{w})\hat{}(n) + ((w * \tilde{w})^*)\hat{}(n) = 2 \quad \text{for} \quad n = 0, 1, \dots, N-1. \tag{3.13}$$

By Lemma 2.30 and equation (3.3),

$$(w * \tilde{w})\hat{}(n) = \hat{w}(n)(\tilde{w})\hat{}(n) = \hat{w}(n)\overline{\hat{w}(n)} = |\hat{w}(n)|^2.$$

Using this equation and equation (3.8),

$$((w * \tilde{w})^*)\hat{}(n) = (w * \tilde{w})\hat{}(n+M) = |\hat{w}(n+M)|^2.$$

By substituting these last two identities, we see that the left side of equation (3.13) is $|\hat{w}(n)|^2 + |\hat{w}(n+M)|^2$. Note that this expression is periodic with period M:

$$|\hat{w}(n+M)|^2 + |\hat{w}(n+M+M)|^2 = |\hat{w}(n+M)|^2 + |\hat{w}(n)|^2,$$

because $\hat{w}$ has period $N = 2M$. Hence $|\hat{w}(n)|^2 + |\hat{w}(n+M)|^2$ is 2 for $n = 0, 1, \ldots, M - 1$ if and only if it is always 2. Therefore equation (3.10) is equivalent to equation (3.13), which we noted is equivalent to equation (3.11), and hence to the orthonormality of $\{R_{2k}w\}_{k=0}^{M-1}$. ∎

The phrase "with M elements" in the statement of Lemma 3.6 is included to guarantee that the elements $R_{2k}w$ are distinct for $k = 0, 1, \ldots, M - 1$. For example, if $w = N^{-1/2}(1, 1, , , , 1)$, then technically the set $\{R_{2k}w\}_{k=0}^{M-1}$ is orthonormal because it has only one element. We remark that Lemma 3.6 has an alternate proof based on Parseval's formula and Fourier inversion (Exercise 3.1.1 (v)).

Definition 3.7 *Suppose $M \in \mathbb{N}$, $N = 2M$, and $u, v \in \ell^2(\mathbb{Z}_N)$. For $n \in \mathbb{Z}$, define $A(n)$, the system matrix of u and v, by*

$$A(n) = \frac{1}{\sqrt{2}} \begin{bmatrix} \hat{u}(n) & \hat{v}(n) \\ \hat{u}(n+M) & \hat{v}(n+M) \end{bmatrix}. \tag{3.14}$$

Now we can characterize orthonormal bases generated by the even integer translates of two vectors.

Theorem 3.8 *Suppose $M \in \mathbb{N}$ and $N = 2M$. Let $u, v \in \ell^2(\mathbb{Z}_N)$. Then*

$$B = \{R_{2k}v\}_{k=0}^{M-1} \cup \{R_{2k}u\}_{k=0}^{M-1}$$
$$= \{v, R_2v, R_4v, \ldots, R_{N-2}v, u, R_2u, R_4u, \ldots, R_{N-2}u\}$$

is an orthonormal basis for $\ell^2(\mathbb{Z}_N)$ if and only if the system matrix $A(n)$ of u and v is unitary for each $n = 0, 1, \ldots, M - 1$. Equivalently, B is a first-stage wavelet basis for $\ell^2(\mathbb{Z}_N)$ if and only if

$$|\hat{u}(n)|^2 + |\hat{u}(n+M)|^2 = 2, \tag{3.15}$$

$$|\hat{v}(n)|^2 + |\hat{v}(n+M)|^2 = 2, \tag{3.16}$$

and

$$\hat{u}(n)\overline{\hat{v}(n)} + \hat{u}(n+M)\overline{\hat{v}(n+M)} = 0, \tag{3.17}$$

for all $n = 0, 1, \ldots M - 1$.

Proof
Recall (Lemma 1.105) that a 2×2 matrix is unitary if and only if its columns form an orthonormal basis for $\mathbb{C}^2$. Applying Lemma 3.6,

$\{R_{2k}u\}_{k=0}^{M-1}$ is orthonormal if and only if equation (3.15) holds, that is, the first column of $A(n)$ has length 1 for every $n = 0, 1, \ldots M - 1$. Similarly, $\{R_{2k}v\}_{k=0}^{M-1}$ is orthonormal if and only if equation (3.16) holds, which states that the second column of $A(n)$ has length 1 for $n = 0, 1, \ldots, M - 1$. Next, we claim that

$$\langle R_{2k}u, R_{2j}v \rangle = 0 \quad \text{for all} \quad j, k = 0, 1, \ldots, M - 1 \tag{3.18}$$

if and only if equation (3.17) holds. Note that equation (3.17) is the statement that the columns of $A(n)$ are orthogonal. Assuming this claim momentarily, it follows that B is an orthonormal set, hence an orthonormal basis for $\ell^2(\mathbb{Z}_N)$ (because it has N elements), if and only if $A(n)$ is unitary for each $n = 0, 1, \ldots, M - 1$.

To prove the equivalence of equations (3.17) and (3.18), first note that equation (3.18) is equivalent to

$$u * \tilde{v}(2k) = \langle u, R_{2k}v \rangle = 0 \quad \text{for all} \quad k = 0, 1, \ldots, M - 1,$$

by equation (3.4) and Exercise 3.1.1 (iv). By equation (3.9) with $z = u * \tilde{v}$, this is equivalent to

$$u * \tilde{v} + (u * \tilde{v})^* = 0,$$

because the values at odd indices are automatically 0. By DFT inversion, this is equivalent to

$$(u * \tilde{v})\hat{} + (u * \tilde{v})^*)\hat{} = 0.$$

By Lemma 2.30 and relation (3.3),

$$(u * \tilde{v})\hat{}(n) = \hat{u}(n)\overline{\hat{v}(n)}.$$

Hence by equation (3.8),

$$((u * \tilde{v})^*)\hat{}(n) = \hat{u}(n + M)\overline{\hat{v}(n + M)}.$$

Note that the left side of equation (3.17) is periodic with period M, so it is 0 for $n = 0, 1, \ldots, M - 1$ if and only if it is 0 for all n. Therefore equation (3.18) is equivalent to equation (3.17). ∎

There is a generalization of Theorem 3.8 to the case of ℓ functions (Exercise 3.1.11), which includes Lemma 3.3 as the special case $\ell = 1$.

It is generally not easy to see directly that $\{R_{2k}v\}_{k=0}^{M-1} \cup \{R_{2k}u\}_{k=0}^{M-1}$ is an orthonormal basis for $\ell^2(\mathbb{Z}_N)$ (although one case for which this is possible is the discrete Haar basis, defined in Exercise 3.1.2). However, it is not difficult to construct $\hat{u}$ and $\hat{v}$ such that the system matrix $A(n)$ is unitary for all $n = 0, 1, \ldots, M-1$. By Theorem 3.8, once we do so, we can take the IDFT of $\hat{u}$ and $\hat{v}$ to obtain an example of a first-stage wavelet basis.

Before constructing an example, we compare the conditions in Lemma 3.3 and Theorem 3.8. In Lemma 3.3, $|\hat{w}(n)|^2$ is constrained to be 1 for every n. In Theorem 3.8, the only constraint on $|\hat{u}(n)|^2$ and $|\hat{u}(n+M)|^2$ is that their average is 1. This allows, for example, $|\hat{u}(n)|^2 = 2$ and $|\hat{u}(n+M)|^2 = 0$ for some n. Then equation (3.17) forces $\hat{v}(n) = 0$, which, by equation (3.16), forces $|\hat{v}(n+M)|^2 = 2$. In this case the component $\hat{v}(n)$ of F_n in $v = \sum_{n=0}^{N-1} \hat{v}(n)F_n$ is 0, that is, v has no component in the direction F_n. This permits us, for example, to select u to contain only low-frequency components and v to contain only high-frequency components (see Example 3.10 below). We remark that when the high and low frequencies are partitioned to some extent between the generators of a first-stage wavelet basis, it is standard notation to take u to be the vector containing the low frequencies (the *low pass filter*) and v to be the one containing the high frequencies (the *high pass filter*). By convention, the basis B in Theorem 3.8 is ordered with the translates of v coming first for reasons related to the iteration step in section 3.2.

This shows the appropriateness of the terms mother wavelet and father wavelet. The advantage of generating first-stage wavelet bases via two parents, as compared with bases generated by one parent as in Lemma 3.3, is (as in biology) that more diversity is allowed in the result.

Example 3.9
Let $\hat{u} = (\sqrt{2}, 1, 0, 1)$ and $\hat{v} = (0, 1, \sqrt{2}, -1)$. Then

$$A(0) = \frac{1}{\sqrt{2}} \begin{bmatrix} \sqrt{2} & 0 \\ 0 & \sqrt{2} \end{bmatrix} = I,$$

and

$$A(1) = \frac{1}{\sqrt{2}} \begin{bmatrix} 1 & 1 \\ 1 & -1 \end{bmatrix}.$$

Clearly $A(0)$ and $A(1)$ are unitary. By Theorem 3.8, $\{v, R_2v, u, R_2u\}$ is an orthonormal basis for $\ell^2(\mathbb{Z}_4)$. Computing as in Example 2.11,

$$u = (\hat{u})^\vee = W_4^{-1}\hat{u} = \frac{1}{4}(2 + \sqrt{2}, \sqrt{2}, -2 + \sqrt{2}, \sqrt{2}),$$

and similarly

$$v = \frac{1}{4}(\sqrt{2}, -\sqrt{2} + 2i, \sqrt{2}, -\sqrt{2} - 2i).$$

Of course,

$$R_2u = \frac{1}{4}(-2 + \sqrt{2}, \sqrt{2}, 2 + \sqrt{2}, \sqrt{2}),$$

and

$$R_2v = \frac{1}{4}(\sqrt{2}, -\sqrt{2} - 2i, \sqrt{2}, -\sqrt{2} + 2i).$$

One can check the orthonormality of $\{u, R_2u, v, R_2v\}$ directly.

Next we consider an example defined for general N, which is designed to partition the high and low frequencies. Recall that the high frequencies are the vectors F_m in the Fourier basis with m in the middle of $0, 1, \ldots, N - 1$, that is, near $N/2$, whereas the low frequencies are those with m near 0 or $N - 1$.

Example 3.10

(First-stage Shannon basis) Suppose N is divisible by 4. Define $\hat{u}, \hat{v} \in \ell^2(\mathbb{Z}_N)$ by

$$\hat{u}(n) = \begin{cases} \sqrt{2} & \text{if } n = 0, 1, \ldots, \frac{N}{4} - 1 \text{ or } n = \frac{3N}{4}, \frac{3N}{4} + 1, \ldots, N - 1 \\ 0 & \text{if } n = \frac{N}{4}, \frac{N}{4} + 1, \ldots, \frac{3N}{4} - 2, \frac{3N}{4} - 1, \end{cases}$$

and

$$\hat{v}(n) = \begin{cases} 0 & \text{if } n = 0, 1, \ldots, \frac{N}{4} - 1 \text{ or } n = \frac{3N}{4}, \frac{3N}{4} + 1, \ldots, N - 1 \\ \sqrt{2} & \text{if } n = \frac{N}{4}, \frac{N}{4} + 1, \ldots, \frac{3N}{4} - 2, \frac{3N}{4} - 1. \end{cases}$$

Notice that at every n, either $\hat{u}(n) = 0$ or $\hat{v}(n) = 0$, so equation (3.17) holds, that is, the columns of the system matrix $A(n)$ are orthogonal. Also at each n, either $\hat{u}(n) = \sqrt{2}$ and $\hat{u}(n + N/2) = 0$ or vice versa, so (3.15) holds, that is, the first column of $A(n)$ has length one for each n. The same observation holds for $\hat{v}$, so the second column of $A(n)$ has length one. Thus $A(n)$ is unitary for all n, so by Theorem 3.8,

$\{R_{2k}v\}_{k=0}^{(N/2)-1} \cup \{R_{2k}u\}_{k=0}^{(N/2)-1}$ is a first-stage wavelet basis. We call this the first-stage *Shannon wavelet basis* because it is similar to a basis arising from something known as the Shannon sampling theorem (see Exercises 5.3.2 and 5.4.17). Note that we have prescribed u and v by their DFTs; to obtain the actual values of u and v we must compute the IDFTs of $\hat{u}$ and $\hat{v}$. The resulting sums can be evaluated in closed form (Exercise 3.1.8). We obtain

$$u(0) = v(0) = \frac{1}{\sqrt{2}}, \qquad (3.19)$$

and

$$u(n) = \frac{\sqrt{2}}{N} e^{-i\pi n/N} \frac{\sin(\frac{\pi n}{2})}{\sin(\frac{\pi n}{N})}, \qquad (3.20)$$

and

$$v(n) = \frac{\sqrt{2}}{N} (-1)^n e^{-i\pi n/N} \frac{\sin(\frac{\pi n}{2})}{\sin(\frac{\pi n}{N})}, \qquad (3.21)$$

for $n = 1, 2, \ldots, N - 1$.

Observe that v has $|\hat{v}(m)| = \sqrt{2}$ for the $N/2$ high frequencies $N/4 \leq m \leq (3N/4) - 1$, and $|\hat{v}(m)| = 0$ for the remaining $N/2$ low frequencies. Since $v = N^{-1} \sum_{m=0}^{N-1} \hat{v}(m) F_m$, this means that v contains no frequencies in the lower half of the frequency scale. The translates $R_{2k}v$ have the same property because of Lemma 2.13. In the same way, u and its translates have have no frequencies in the upper half of the scale. Thus in the representation

$$z = \sum_{k=0}^{(N/2)-1} \langle z, R_{2k}v \rangle R_{2k}v + \sum_{k=0}^{(N/2)-1} \langle z, R_{2k}u \rangle R_{2k}u,$$

for $z \in \ell^2(\mathbb{Z}_N)$, the higher half of the frequencies in z are contained in the first sum whereas the lower half are contained in the second sum.

Sometimes it is advantageous to have an orthonormal basis consisting entirely of real-valued vectors. For example, suppose the signals z that we want to expand are all real-valued, as is often the case in applications (e.g., audio or visual signals). If the basis elements are real also, then the coefficients in the expansion of z

will be real because they are the inner products of z with the basis elements. Thus in this case we would know that we could store the components of z in this basis as a real vector. This simplifies computation and saves space in the computer memory because complex vectors are stored as pairs of real vectors.

Note that the Shannon basis in Example 3.10 is not real valued. By Corollary 2.16, a vector z is real valued if and only if its DFT satisfies the symmetry condition $\hat{z}(m) = \overline{\hat{z}(N - m)}$ for every m. Looking at $\hat{u}$, we see that this condition is satisfied for $m = 0, 1, \ldots, (N/4) - 1$ and for $m = (3N/4) + 1, \ldots, N - 1$, where both values are $\sqrt{2}$, and for $m = (N/4) + 1, \ldots, (3N/4) - 1$, where both values are 0. However, the condition fails for $m = N/4$ (equivalently, $m = 3N/4$) because $\hat{u}(N/4) = 0$ and $\hat{u}(3N/4) = \sqrt{2}$. A similar observation holds for v. However, we can modify $\hat{u}$ and $\hat{v}$ at these values in such as way as to obtain the symmetry condition (hence u and v will be real) while still satisfying the conditions of Theorem 3.8.

Example 3.11

(First-stage real Shannon basis) Suppose N is divisible by 4. Define $\hat{u}, \hat{v} \in \ell^2(\mathbb{Z}_N)$ by

$$\hat{u}(n) = \begin{cases} \sqrt{2} & \text{if } n = 0, 1, \ldots, \frac{N}{4} - 1 \text{ or } n = \frac{3N}{4} + 1, \ldots, N - 1 \\ i & \text{if } n = \frac{N}{4} \\ -i & \text{if } n = \frac{3N}{4} \\ 0 & \text{if } n = \frac{N}{4} + 1, \ldots, \frac{3N}{4} - 1, \end{cases}$$

and

$$\hat{v}(n) = \begin{cases} 0 & \text{if } n = 0, 1, \ldots, \frac{N}{4} - 1 \text{ or } n = \frac{3N}{4} + 1, \ldots, N - 1 \\ 1 & \text{if } n = \frac{N}{4} \text{ or } n = \frac{3N}{4} \\ \sqrt{2} & \text{if } n = \frac{N}{4} + 1, \ldots, \frac{3N}{4} - 1. \end{cases}$$

Note that at $N/4$ or $3N/4$, $\hat{v}(N/4) = \overline{\hat{v}(3N/4)}$ because both are 1, whereas $\hat{u}(N/4) = i = \overline{-i} = \overline{\hat{u}(3N/4)}$. At the other values, $\hat{u}$ and $\hat{v}$ agree with the Shannon basis and hence satisfy the symmetry condition. Thus u and v are real-valued vectors. Also at $n = N/4$, the system matrix is

$$A(N/4) = \frac{1}{\sqrt{2}} \begin{bmatrix} i & 1 \\ -i & 1 \end{bmatrix},$$

which is unitary. (Note that, given our choice of $\hat{v}$, we were forced to take $\hat{u}$ pure imaginary at $N/4$ and $3N/4$ to obtain both the symmetry condition and the property that the matrix is unitary.) At other values, the system matrix is the same as for the Shannon basis, and hence is unitary. Thus by Theorem 3.8, $\{R_{2k}v\}_{k=0}^{(N/2)-1} \cup \{R_{2k}u\}_{k=0}^{(N/2)-1}$ is a first-stage wavelet basis such that u and v are real valued. It follows that all the basis elements are real valued because they are translates of u and v.

Note that the high and low frequencies are still partitioned between u and v, as in Example 3.10, except that there is an overlap when $n = N/4$ or $3N/4$.

Here it is relatively difficult to write closed-form expressions for u and v. In practice, for a specific N, one would use an IDFT program to compute u and v and store them for future use. The graphs of $R_{32}u$ and $R_{32}v$ in the case $N = 64$ are shown in Figure 11 (we used translates by 32 to center the graph in the range $0 \le n \le 63$).

Similarly Figure 12 shows the graphs of $R_{256}u$ and $R_{256}v$ in the case $N = 512$. Note that these functions are relatively localized around their center points (32 in the case $N = 64$ and 256 in the case $N = 512$). This may be surprising, because we did not specifically arrange for this. However, we should expect v to have its maximum at 0 (and hence $R_{256}v$, in the case $N = 512$, to have its maximum at

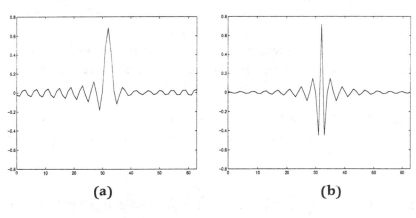

(a)   **(b)**

FIGURE 11   (a) $R_{32}u$, (b) $R_{32}v$

256) because, by Fourier inversion,

$$v(n) = \frac{1}{N} \sum_{m=0}^{N-1} \hat{v}(m) e^{2\pi i m n / N},$$

so

$$v(0) = \frac{1}{N} \sum_{m=0}^{N-1} \hat{v}(m).$$

Since $\hat{v}(m) \geq 0$ for all m in this case, the sum giving $v(0)$ has no cancellation. For $n \neq 0$, the sum giving $v(n)$ has cancellation, and so will give a smaller value than $v(0)$. The functions $e^{2\pi i m n / N}$ are lined up so as to agree at $n = 0$. As n moves away from 0, the exponentials become less aligned, until eventually they are more or less independent. This means the terms in the sum tend to cancel out, resulting in small values of $v(n)$ for n far from the center point.

Notice also that $\sum_{n=0}^{N-1} v(n) = \hat{v}(0) = 0$ by construction, which explains why v seems to have equal positive and negative mass. However, $\sum_{n=0}^{N-1} u(n) = \hat{u}(0) = \sqrt{2}$, so u has more positive than negative mass.

Note that $R_{32}v$ in Figure 11b appears to be symmetric around $n = 32$. This can be deduced from Exercise 2.1.12 (i): v is symmetric around 0 because $\hat{v}$ is real. Similarly, u is not symmetric around 0 because $\hat{u}$ is not real, hence (as we can see in Figure 11b), $R_{32}u$ is not symmetric around $n = 32$. Similar remarks hold for Figure 12.

Because 2×2 unitary matrices are easy to characterize, Theorem 3.8 can be used to describe all first-stage wavelet bases explicitly (see Exercise 3.1.6). Later we will find the following result very useful. It says that every potential father wavelet u has a companion mother wavelet v such that u and v generate a first-stage wavelet basis.

Lemma 3.12   *Suppose $M \in \mathbb{N}$, $N = 2M$, and $u \in \ell^2(\mathbb{Z}_N)$ is such that $\{R_{2k}u\}_{k=0}^{M-1}$ is an orthonormal set with M elements. Define $v \in \ell^2(\mathbb{Z}_N)$ by*

$$v(k) = (-1)^{k-1}\overline{u(1-k)} \tag{3.22}$$

for all k. Then $\{R_{2k}v\}_{k=0}^{M-1} \cup \{R_{2k}u\}_{k=0}^{M-1}$ is a first-stage wavelet basis for $\ell^2(\mathbb{Z}_N)$.

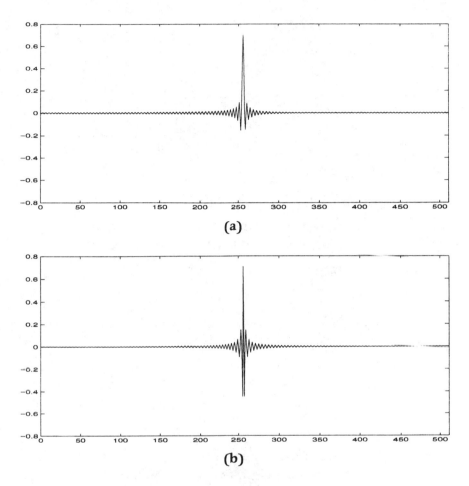

FIGURE 12 (a) $R_{256}u$, (b) $R_{256}v$

Proof

Using relation (3.22) and the substitution $k = 1 - n$,

$$\hat{v}(m) = \sum_{n=0}^{N-1} v(n)e^{-2\pi imn/N} = \sum_{n=0}^{N-1}(-1)^{n-1}\overline{u(1-n)}e^{-2\pi imn/N}$$

$$= \sum_{k=0}^{N-1} \overline{u(k)}(-1)^{-k}e^{-2\pi im(1-k)/N}$$

$$= e^{-2\pi im/N}\sum_{k=0}^{N-1} \overline{u(k)}(e^{-i\pi})^{-k}e^{2\pi imk/N}$$

$$= e^{-2\pi i m/N} \sum_{k=0}^{N-1} u(k) e^{-2\pi i (m+M)k/N} = e^{-2\pi i m/N} \overline{\hat{u}(m+M)}.$$

Therefore

$$\hat{v}(m+M) = e^{-2\pi i (m+M)/N} \overline{\hat{u}(m+2M)}$$
$$= e^{-2\pi i M/N} e^{-2\pi i m/N} \overline{\hat{u}(m)} = -e^{-2\pi i m/N} \overline{\hat{u}(m)},$$

since $2M = N$, so that $\hat{u}(m+2M) = \hat{u}(m+N) = \hat{u}(m)$ and $e^{-2\pi i M/N} = e^{-i\pi} = -1$. Hence

$$|\hat{v}(m)|^2 + |\hat{v}(m+M)|^2 = |\hat{u}(m+M)|^2 + |\hat{u}(m)|^2 = 2,$$

for $m = 0, 1, \ldots, M-1$, by Lemma 3.6 and the orthonormality of $\{R_{2k}u\}_{k=0}^{M-1}$. Thus equations (3.15) and (3.16) hold. Finally

$$\hat{u}(m)\overline{\hat{v}(m)} + \hat{u}(m+M)\overline{\hat{v}(m+M)}$$
$$= \hat{u}(m)e^{2\pi i m/N}\hat{u}(m+M) - \hat{u}(m+M)e^{2\pi i m/N}\hat{u}(m) = 0,$$

which is equation (3.17). By Theorem 3.8, the pair u, v generates a first-stage wavelet basis for $\ell^2(\mathbb{Z}_N)$. ∎

Suppose $B = \{R_{2k}v\}_{k=0}^{M-1} \cup \{R_{2k}u\}_{k=0}^{M-1}$ is a first-stage wavelet basis. By Lemma 1.106 i, the B to E change-of-basis matrix is the matrix U with column vectors $v, R_2v, \ldots R_{N-2}v, u, R_2u, \ldots R_{N-2}u$, in that order. Since B is orthonormal, U is unitary (Lemma 1.105), so the E to B change-of-basis matrix is $U^{-1} = U^*$. However, computing $[z]_B$ by multiplying out U^*z directly is slow, requiring N^2 multiplications. To compute this change of basis quickly, we should use the fact that the coefficient of $R_{2k}v$ in the expansion of z is $\langle z, R_{2k}v \rangle = z * \tilde{v}(2k)$ and similarly for u. Thus

$$[z]_B = \begin{bmatrix} z * \tilde{v}(0) \\ z * \tilde{v}(2) \\ \cdot \\ \cdot \\ z * \tilde{v}(N-2) \\ z * \tilde{u}(0) \\ z * \tilde{u}(2) \\ \cdot \\ \cdot \\ z * \tilde{u}(N-2) \end{bmatrix}. \tag{3.23}$$

We can represent the calculation of this vector as the result of two convolutions of z followed in each case by the operation of throwing out the odd-indexed entries.

Definition 3.13 *Suppose $M \in \mathbb{N}$ and $N = 2M$. Define $D : \ell^2(\mathbb{Z}_N) \to \ell^2(\mathbb{Z}_M)$ by setting, for $z \in \ell^2(\mathbb{Z}_N)$,*

$$D(z)(n) = z(2n), \quad for \quad n = 0, 1, \ldots, M - 1.$$

The operator D is called the downsampling *or* decimation *operator.*

In other words, if $z = (z(0), z(1), z(2), z(3), \ldots, z(N - 1))$, then

$$D(z) = (z(0), z(2), z(4), \ldots, z(N - 2)).$$

The downsampling operator is often denoted $\downarrow 2$ in diagrams. The calculation of $[z]_B$ is represented in Figure 13.

This is a simple example of a *filter bank*. A general filter bank is any sequence of convolutions and other operations. The study of filter banks is an entire subject in engineering called *multirate signal analysis*, or *subband coding*. The term "filter" is used to denote a convolution operator because such an operator can cut out various frequencies if the associated Fourier multiplier is 0 (or sufficiently small) at those frequencies.

We have seen that we can compute the E to B change of basis quickly. What about going the opposite way, computing the B to E change of basis? Of course, this can be obtained by multiplying by the matrix U, but multiplying by an $N \times N$ matrix is slow. There is a fast procedure based on the filter bank approach.

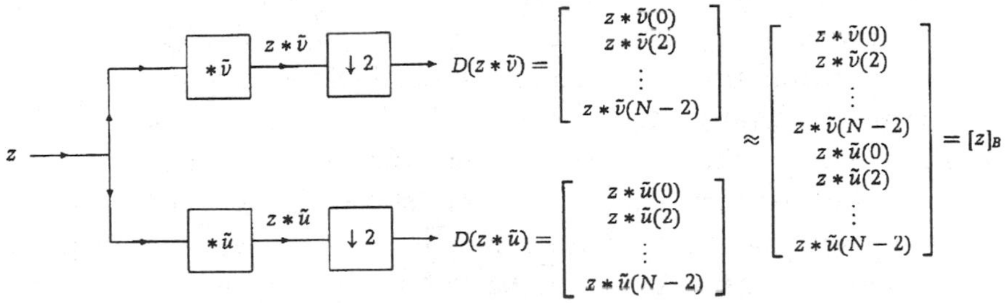

FIGURE 13

Definition 3.14   *Suppose $M \in \mathbb{N}$ and $N = 2M$. Define $U : \ell^2(\mathbb{Z}_M) \rightarrow \ell^2(\mathbb{Z}_N)$ by setting, for $z \in \ell^2(\mathbb{Z}_M)$,*

$$U(z)(n) = \begin{cases} z(n/2) & \text{if } n \text{ is even} \\ 0 & \text{if } n \text{ is odd.} \end{cases}$$

The operator U is called the upsampling *operator. It is denoted $\uparrow 2$.*

The upsampling operator doubles the size of a vector by inserting a 0 between any two adjacent values. For example, if

$$z = (2, 5, -1, i)$$

then

$$U(z) = (2, 0, 5, 0, -1, 0, i, 0).$$

Note that if we upsample and then downsample, we get back what we started with, that is, $D(U(z)) = z$ for any z. However, if we first downsample and then upsample, we throw away the odd-index values and then put them back in as 0s. Thus the composition $U \circ D$ has the effect of zeroing out all the odd-index values, and hence is not the identity. By comparing $U \circ D$ with equation (3.9), we see that

$$U \circ D(z) = \frac{1}{2}(z + z^*). \tag{3.24}$$

This example shows that, unlike the case of square matrices in Exercise 1.4.12 (ii), a one-sided inverse is not necessarily a two-sided inverse.

To regain z from the output of the left filter bank in Figure 13, we follow up with a right filter bank as in the right half of Figure 14. Here $s, t \in \ell^2(\mathbb{Z}_N)$ are unknown.

The output of the top branch of Figure 14 is $\tilde{t} * U(D(z * \tilde{v}))$ and the output of the lower branch is $\tilde{s} * U(D(z * \tilde{u}))$. Lemma 3.15 gives conditions under which the sum of these outputs is always the original input z. When this happens we say we have *perfect reconstruction* in the filter bank. Note that we do not necessarily assume the conditions of Theorem 3.8. Thus we have a more general result about filter banks that do not necessarily correspond to orthonormal bases. Although we will not pursue this further here, this more general case leads to a generalization of orthonormal wavelets called *biorthogonal wavelets*.

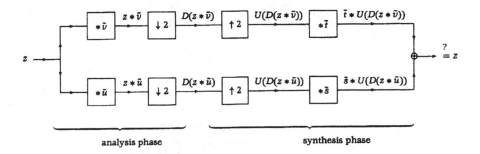

FIGURE 14

Lemma 3.15 *Suppose $M \in \mathbb{N}$, $N = 2M$, and $u, v, s, t \in \ell^2(\mathbb{Z}_N)$. For $n = 0, 1, \ldots, N - 1$, let $A(n)$ be the system matrix (Definition 3.7) for u and v. Then we have perfect reconstruction in Figure 14, that is,*

$$\tilde{t} * U(D(z * \tilde{v})) + \tilde{s} * U(D(z * \tilde{u})) = z$$

for all $z \in \ell^2(\mathbb{Z}_N)$, if and only if

$$A(n) \begin{bmatrix} \hat{s}(n) \\ \hat{t}(n) \end{bmatrix} = \begin{bmatrix} \sqrt{2} \\ 0 \end{bmatrix} \tag{3.25}$$

for all $n = 0, 1, \ldots, N - 1$. In the case that $A(n)$ is unitary, this simplifies to $\hat{t}(n) = \overline{\hat{v}(n)}$ and $\hat{s}(n) = \overline{\hat{u}(n)}$. If $A(n)$ is unitary for all n (equivalently, by Theorem 3.8, if $\{R_{2k}v\}_{k=0}^{M-1} \cup \{R_{2k}u\}_{k=0}^{M-1}$ is an orthonormal basis for $\ell^2(\mathbb{Z}_N)$), then $t = \tilde{v}$ and $s = \tilde{u}$.

Proof
By equation (3.24),

$$U(D(z * \tilde{v})) = \frac{1}{2} \left(z * \tilde{v} + (z * \tilde{v})^* \right)$$

and similarly with v replaced by u. Hence, as in the proof of the equivalence of equations (3.17) and (3.18),

$$\left(U(D(z * \tilde{v})) \right)\hat{}(n) = \frac{1}{2} \left(\hat{z}(n)\overline{\hat{v}(n)} + \hat{z}(n + M)\overline{\hat{v}(n + M)} \right),$$

for any n, and similarly with v replaced by u. Therefore,

$$\left[\tilde{t} * U(D(z * \tilde{v})) + \tilde{s} * U(D(z * \tilde{u})) \right]\hat{}(n)$$

$$= \overline{\hat{t}(n)} \frac{1}{2} \left(\hat{z}(n)\overline{\hat{v}(n)} + \hat{z}(n + M)\overline{\hat{v}(n + M)} \right)$$

$$+ \overline{\hat{s}(n)} \frac{1}{2} \left(\hat{z}(n)\overline{\hat{u}(n)} + \hat{z}(n + M)\overline{\hat{u}(n + M)} \right)$$

$$= \frac{1}{2} \left(\overline{\hat{t}(n)\hat{v}(n)} + \overline{\hat{s}(n)\hat{u}(n)} \right) \hat{z}(n)$$

$$+ \frac{1}{2} \left(\overline{\hat{t}(n)\hat{v}(n + M)} + \overline{\hat{s}(n)\hat{u}(n + M)} \right) \hat{z}(n + M). \quad (3.26)$$

By Fourier inversion, we have perfect reconstruction if and only if, for all n and $z \in \ell^2(\mathbb{Z}_N)$, each expression in equation (3.26) agrees with $\hat{z}(n)$. We claim that this holds if and only if

$$\hat{s}(n)\hat{u}(n) + \hat{t}(n)\hat{v}(n) = 2 \quad (3.27)$$

and

$$\hat{s}(n)\hat{u}(n + M) + \hat{t}(n)\hat{v}(n + M) = 0. \quad (3.28)$$

To prove this, substituting in equation (3.26) shows that equations (3.27) and (3.28) are sufficient conditions for perfect reconstruction. Conversely, if we assume perfect reconstruction, fix n and pick z such that $\hat{z}(n) = 1$ and $\hat{z}(n+M) = 0$. Using the perfect reconstruction condition for this z shows that equation (3.27) holds, whereas considering another z with $\hat{z}(n) = 0$ and $\hat{z}(n + M) = 1$ implies equation (3.28).

By dividing by $\sqrt{2}$ and rewriting equations (3.27) and (3.28) in matrix notation, we get equation (3.25).

In the case that $A(n)$ is unitary, $A(n)$ is invertible and $A(n)^{-1} = A(n)^*$, so solving equation (3.25) gives $\hat{s}(n) = \overline{\hat{u}(n)}$ and $\hat{t}(n) = \overline{\hat{v}(n)}$. If $A(n)$ is unitary for all n, then Fourier inversion and relation (3.3) imply that $s = \tilde{u}$ and $t = \tilde{v}$.   ∎

In the case that $\{R_{2k}v\}_{k=0}^{M-1} \cup \{R_{2k}u\}_{k=0}^{M-1}$ is an orthonormal basis for $\ell^2(\mathbb{Z}_N)$, there is a simpler proof (Exercise 3.1.13) that $s = \tilde{u}$ and $t = \tilde{v}$, but we have presented this more general result because of its independent interest.

This tells us how we can reconstruct z using Figure 14, in the case where u and v generate a first-stage wavelet basis. We take $s = \tilde{u}$ and $t = \tilde{v}$ (or equivalently, $\tilde{s} = u$ and $\tilde{t} = v$). In particular, the reconstruction step involves only two more convolutions and hence can be computed rapidly.

To put this another way, to compute the B to E change of basis, input the first half of $[z]_B$ in the top portion of the right-hand part of Figure 11 and input the bottom half in the bottom portion, with $\tilde{t} = v$ and $\tilde{s} = u$. The output will be $[z]_E = z$.

So far we have found conditions that allow us to create orthonormal bases for which the change of basis and its inverse can be computed fast via convolutions using the filter bank diagram. From the examples of the Shannon and real Shannon bases, we see that we can get some degree of frequency localization with such a basis, because v and its translates carry the high frequencies whereas u and its translates carry the low frequencies. We have also seen by experiment that we can obtain a fair degree of spatial localization as well.

Next, we iterate this type of splitting. This gives us a basis that naturally reflects different scales. The easiest way to understand this iteration is in terms of the filter bank diagram, as we see in section 3.2.

Exercises

3.1.1. Let $z, w, u, v \in \ell^2(\mathbb{Z}_N)$.

    i. Prove that

$$\langle R_k z, R_j w \rangle = \langle z, R_{j-k} w \rangle = \langle R_{k-j} z, w \rangle,$$

    for any $k, j \in \mathbb{Z}$.

    ii. Prove that $\{R_k w\}_{k=0}^{N-1}$ is an orthonormal basis for $\ell^2(\mathbb{Z}_N)$ if and only if equation (3.6) holds.

    iii. Suppose $M \in \mathbb{N}$ and $N = 2M$. Prove that $\{R_{2k}u\}_{k=0}^{M-1}$ is an orthonormal set with M elements if and only if equation (3.11) holds.

    iv. Suppose $M \in \mathbb{N}$ and $N = 2M$. Prove that equation (3.18) holds if and only if $\langle u, R_{2k}v \rangle = 0$ for all $k = 0, 1, \ldots, M-1$.

    v. Complete the following alternate proof of Lemma 3.6: By Parseval's relation (2.11) and Lemma 2.13,

$$\langle w, R_{2k}w \rangle = \frac{1}{N} \sum_{m=0}^{N-1} |\hat{w}(m)|^2 \, e^{2\pi imk/(N/2)}.$$

Write $\sum_{m=0}^{N-1}$ as $\sum_{m=0}^{M-1} + \sum_{m=M}^{N-1}$ and replace m in the second sum by $m - M$ to write

$$\langle w, R_{2k}w \rangle = \frac{1}{2M} \sum_{m=0}^{M-1} \left(|\hat{w}(m)|^2 + |\hat{w}(m+M)|^2 \right) e^{2\pi imk/M}.$$

Regard $|\hat{w}(m)|^2 + |\hat{w}(m+M)|^2$ as a vector in $\ell^2(\mathbb{Z}_M)$ and apply Fourier inversion.

3.1.2. (The first-stage Haar basis) Suppose $N = 2M$, for $M \in \mathbb{N}$. Define $u, v \in \ell^2(\mathbb{Z}_N)$ by

$$u = \left(\frac{1}{\sqrt{2}}, \frac{1}{\sqrt{2}}, 0, 0, \dots, 0 \right)$$

and

$$v = \left(\frac{1}{\sqrt{2}}, -\frac{1}{\sqrt{2}}, 0, 0, \dots, 0 \right).$$

i. Prove that $\{R_{2k}v\}_{k=0}^{M-1} \cup \{R_{2k}u\}_{k=0}^{M-1}$ is an orthonormal basis for $\ell^2(\mathbb{Z}_N)$ directly from the definitions of u and v, that is, not using the DFT or Theorem 3.8.

ii. Compute $\hat{u}$ and $\hat{v}$. Check that the system matrix $A(n)$ (Definition 3.7) is unitary for all n.

iii. For $z \in \ell^2(\mathbb{Z}_N)$, define

$$P(z) = \sum_{k=0}^{M-1} \langle z, R_{2k}u \rangle R_{2k}u,$$

and

$$Q(z) = \sum_{k=0}^{M-1} \langle z, R_{2k}v \rangle R_{2k}v.$$

By part i and Lemma 1.101 i, $z = P(z) + Q(z)$. Prove that for $m = 0, 1, \dots, M - 1$,

$$P(z)(2m) = P(z)(2m+1) = \left(z(2m) + z(2m+1) \right)/2.$$

In other words, $P(z)$ is obtained from z by replacing the values of z at $2m$ and $2m+1$ by their average. This can be regarded as the vector z seen at a resolution of 2. Then $Q(z)$ is the "detail" needed to pass from a resolution of 2 to a resolution of 1.

iv. For $N = 8$, suppose

$$z = (4, 2, 3, 7, 10, 8, 10, 14).$$

Find $P(z)$ and $Q(z)$. Then graph z, $P(z)$, and $Q(z)$.

3.1.3. Let $u \in \ell^2(\mathbb{Z}_4)$ be such that $\hat{u} = (1, \sqrt{2}, i, 0)$. Find some $\hat{v}$ such that $\{v, R_2v, u, R_2u\}$ is an orthonormal basis for $\ell^2(\mathbb{Z}_4)$.

3.1.4. Suppose $u, v \in \ell^2(\mathbb{Z}_N)$.

i. Prove that

$$\langle \tilde{u}, R_{2k}\tilde{v} \rangle = \langle v, R_{2k}u \rangle.$$

ii. Suppose $M \in \mathbb{N}$ and $N = 2M$. Deduce from part i and Exercise 3.1.1 that u and v generate a first-stage wavelet basis for $\ell^2(\mathbb{Z}_N)$ if and only if $\tilde{u}$ and $\tilde{v}$ do.

iii. Obtain the result in part ii from Theorem 3.8 instead.

3.1.5. Suppose $\hat{u} = (\sqrt{2}, \sqrt{2}, 0, 0)$ and $\hat{v} = (0, 0, \sqrt{2}, \sqrt{2})$.

i. Check that the system matrix $A(n)$ (Definition 3.7) is the identity, and hence is unitary, for $n = 0, 1$. Deduce that $\{v, R_2v, u, R_2u\}$ is an orthonormal basis for $\ell^2(\mathbb{Z}_4)$.

ii. Use the IDFT to compute u, v.

iii. Check directly (i.e., without using Theorem 3.8) that $\{v, R_2v, u, R_2u\}$ is an orthonormal set in $\ell^2(\mathbb{Z}_4)$.

3.1.6. Suppose $M \in \mathbb{N}$ and $N = 2M$.

i. Let $\{r(n)\}_{n=0}^{M-1}$ be real numbers such that

$$0 \leq r(n) \leq \sqrt{2}, \quad \text{for all} \quad n = 0, 1, \ldots, M - 1.$$

Let $\{\theta(n)\}_{n=0}^{M-1}$, $\{\varphi(n)\}_{n=0}^{M-1}$, $\{\sigma(n)\}_{n=0}^{M-1}$, $\{\rho(n)\}_{n=0}^{M-1}$ be real numbers such that, if $n \in \{0, 1, \ldots, M-1\}$ and $0 < r(n) < \sqrt{2}$, then

$$\theta(n) + \rho(n) - \varphi(n) - \sigma(n) = (2k+1)\pi \quad \text{for some } k = k(n) \in \mathbb{Z}.$$

(If $r(n) = 0$ or $r(n) = \sqrt{2}$, then $\theta(n), \varphi(n), \sigma(n)$, and $\rho(n)$ are unconstrained.) Define $\hat{u}, \hat{v} \in \ell^2(\mathbb{Z}_N)$ by setting

$$\hat{u}(n) = r(n)e^{i\theta(n)}, \quad \hat{u}(n + N/2) = \sqrt{2 - (r(n))^2}\, e^{i\varphi(n)},$$

$$\hat{v}(n) = \sqrt{2 - (r(n))^2}\, e^{i\sigma(n)}, \quad \text{and} \quad \hat{v}(n + N/2) = r(n)e^{i\rho(n)}$$

for $n = 0, 1, \ldots M - 1$. Define $u, v \in \ell^2(\mathbb{Z}_N)$ by $u = (\hat{u})^\vee$ and $v = (\hat{v})^\vee$. Prove that $\{R_{2k}v\}_{k=0}^{M-1} \cup \{R_{2k}u\}_{k=0}^{M-1}$ is an orthonormal basis for $\ell^2(\mathbb{Z}_N)$.

ii. Prove that for any first-stage wavelet basis $\{R_{2k}v\}_{k=0}^{M-1} \cup \{R_{2k}u\}_{k=0}^{M-1}$, $\hat{u}$ and $\hat{v}$ are of the form stated in part i for some real-valued $r(n)$, $\theta(n)$, $\varphi(n)$, $\sigma(n)$, $\rho(n)$, $n = 0, 1, \ldots, M-1$, satisfying $0 \le r(n) \le \sqrt{2}$ and $\theta + \rho - \varphi - \sigma = (2k+1)\pi$ for some $k = k(n) \in \mathbb{Z}$, for each $n = 0, 1, \ldots, N-1$. Hint: By Theorem 3.8, this just comes down to parameterizing 2×2 unitary matrices.

3.1.7. Suppose $M \in \mathbb{N}$, $N = 2M$, $z \in \ell^2(\mathbb{Z}_N)$, and $w \in \ell^2(\mathbb{Z}_M)$. Prove that

$$\langle D(z), w \rangle = \langle z, U(w) \rangle.$$

Note that the inner product on the left side is in $\ell^2(\mathbb{Z}_M)$, whereas the inner product on the right is in $\ell^2(\mathbb{Z}_N)$.

3.1.8. Verify equations (3.19), (3.20), and (3.21). Hint: Change summation indices so that each sum begins at 0, and apply equation (1.5).

3.1.9. In the case $N = 2M$, with $M \in \mathbb{N}$, show that Theorem 3.8 implies Lemma 3.3 by applying Theorem 3.8 with $v = R_1 u$.

3.1.10. Suppose $M \in \mathbb{N}$ and $N = 2M$. Let $u, s \in \ell^2(\mathbb{Z}_N)$. Consider the filter bank with only one branch shown in Figure 15. Prove that no matter how u and s are chosen, this filter bank cannot give perfect reconstruction.

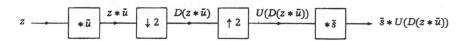

FIGURE 15

3.1.11. (Generalization of Theorem 3.8 to ℓ functions) Suppose $\ell \in \mathbb{N}$ and $\ell | N$ (recall this means that there exists $q \in \mathbb{Z}$ such that $N = q\ell$).

i. Suppose $z, v, w \in \ell^2(\mathbb{Z}_N)$. Prove that

$$\{R_{\ell k}z\}_{k=0}^{(N/\ell)-1}$$

is an orthonormal set with N/ℓ elements if and only if

$$\sum_{k=0}^{\ell-1} |\hat{z}(n + kN/\ell)|^2 = \ell \text{ for all } n.$$

Also prove that

$$\langle R_{\ell k}v, R_{\ell j}w \rangle = 0 \text{ for all } j, k$$

if and only if

$$\sum_{k=0}^{\ell-1} \hat{v}(n + kN/\ell)\overline{\hat{w}(n + kN/\ell)} = 0 \text{ for all } n.$$

Hint: Method 1: Prove that

$$\sum_{m=0}^{\ell-1} e^{-2\pi i n m/\ell} = \begin{cases} \ell & \text{if } \ell | n \\ 0 & \text{if } \ell \nmid n. \end{cases}$$

Hence

$$\langle v, R_{\ell k}w \rangle = v * \tilde{w}(\ell k) = \begin{cases} 1 & \text{if } k = 0 \text{ and } v = w \\ 0 & \text{if } k = 1, 2, \ldots, (N/\ell) - 1 \\ & \text{or } v \neq w \end{cases}$$

if and only if

$$\frac{1}{\ell} \sum_{m=0}^{\ell-1} e^{-2\pi i n m/\ell} v * \tilde{w}(n) = \begin{cases} \delta(n) & \text{for all } n, \text{ if } v = w \\ 0 & \text{for all } n, \text{ if } v \neq w. \end{cases}$$

Take the DFT of both sides. Prove that

$$(e^{-2\pi i n m/\ell} v * \tilde{w})\hat{}(k) = \hat{v}(k + mN/\ell)\overline{\hat{w}(k + mN/\ell)}.$$

Hint: Method 2: (compare to Exercise 3.1.1 (v)) By Parseval's relation (2.11) and Lemma 2.13,

$$\langle v, R_{\ell k}w \rangle$$

$$= \frac{1}{N} \sum_{m=0}^{N-1} \hat{v}(m) e^{2\pi i \ell k m/N} \overline{\hat{w}(m)}$$

$$= \frac{1}{N} \sum_{j=0}^{\ell-1} \sum_{n=0}^{(N/\ell)-1} e^{2\pi i \ell k(n+jN/\ell)/N} \hat{v}(n + jN/\ell) \overline{\hat{w}(n + jN/\ell)}$$

$$= \frac{1}{(N/\ell)} \sum_{n=0}^{(N/\ell)-1} e^{2\pi i k n/(N/\ell)} \frac{1}{\ell} \sum_{j=0}^{\ell-1} \hat{v}(n + jN/\ell) \overline{\hat{w}(n + jN/\ell)},$$

and apply Fourier inversion in $\ell^2(\mathbb{Z}_{N/\ell})$.

ii. Suppose $u_0, u_1, \ldots, u_{\ell-1} \in \ell^2(\mathbb{Z}_N)$. Prove that

$$\{R_{\ell k}u_0\}_{k=0}^{(N/\ell)-1} \cup \{R_{\ell k}u_1\}_{k=0}^{(N/\ell)-1} \cup \cdots \cup \{R_{\ell k}u_{\ell-1}\}_{k=0}^{(N/\ell)-1}$$

is an orthonormal basis for $\ell^2(\mathbb{Z}_N)$ if and only if the matrix

$$\frac{1}{\sqrt{\ell}} \begin{bmatrix} \hat{u}_0(n) & \hat{u}_1(n) & \cdots & \hat{u}_{\ell-1}(n) \\ \hat{u}_0(n+\frac{N}{\ell}) & \hat{u}_1(n+\frac{N}{\ell}) & \cdots & \hat{u}_{\ell-1}(n+\frac{N}{\ell}) \\ \hat{u}_0(n+\frac{2N}{\ell}) & \hat{u}_1(n+\frac{2N}{\ell}) & \cdots & \hat{u}_{\ell-1}(n+\frac{2N}{\ell}) \\ \vdots & \vdots & \ddots & \vdots \\ \hat{u}_0(n+\frac{(\ell-1)N}{\ell}) & \hat{u}_1(n+\frac{(\ell-1)N}{\ell}) & \cdots & \hat{u}_{\ell-1}(n+\frac{(\ell-1)N}{\ell}) \end{bmatrix}$$

is unitary for all $n = 0, 1, \ldots, (N/\ell) - 1$.

3.1.12. (Two-dimensional case of Theorem 3.8) Suppose $M_1, M_2 \in \mathbb{N}$, $N_1 = 2M_1$ and $N_2 = 2M_2$. Recall Exercises 2.1.15, 2.1.17, 2.2.18, and 2.2.19 as background for this problem.

i. Suppose $z \in \ell^2(\mathbb{Z}_{N_1} \times \mathbb{Z}_{N_2})$. Define z^*, z^{**}, and $z^{***} \in \ell^2(\mathbb{Z}_{N_1} \times \mathbb{Z}_{N_2})$ by

$$z^*(n_1, n_2) = (-1)^{n_1} z(n_1, n_2),$$
$$z^{**}(n_1, n_2) = (-1)^{n_2} z(n_1, n_2),$$

and

$$z^{***}(n_1, n_2) = (-1)^{n_1+n_2} z(n_1, n_2).$$

Prove that

$$z(k_1, k_2) + z^*(k_1, k_2) + z^{**}(k_1, k_2) + z^{***}(k_1, k_2)$$
$$= \begin{cases} 4z(k_1, k_2) & \text{if } k_1 \text{ and } k_2 \text{ are even} \\ 0 & \text{if } k_1 \text{ or } k_2 \text{ is odd.} \end{cases}$$

ii. Let $u_0, u_1, u_2, u_3 \in \ell^2(\mathbb{Z}_{N_1} \times \mathbb{Z}_{N_2})$. Let $A(n_1, n_2)$ be the matrix whose m^{th} column ($m = 0, 1, 2, 3$) is

$$\frac{1}{2} \begin{bmatrix} \hat{u}_m(n_1, n_2) \\ \hat{u}_m(n_1 + \frac{N_1}{2}, n_2) \\ \hat{u}_m(n_1, n_2 + \frac{N_2}{2}) \\ \hat{u}_m(n_1 + \frac{N_1}{2}, n_2 + \frac{N_2}{2}) \end{bmatrix}$$

Prove that

$$\{R_{2k_1, 2k_2} u_0\}_{0 \le k_1 \le M_1-1, 0 \le k_2 \le M_2-1}$$

$$\cup \left\{ R_{2k_1,2k_2} u_1 \right\}_{0 \le k_1 \le M_1 - 1, 0 \le k_2 \le M_2 - 1}$$

$$\cup \left\{ R_{2k_1,2k_2} u_2 \right\}_{0 \le k_1 \le M_1 - 1, 0 \le k_2 \le M_2 - 1}$$

$$\cup \left\{ R_{2k_1,2k_2} u_3 \right\}_{0 \le k_1 \le M_1 - 1, 0 \le k_2 \le M_2 - 1}$$

is an orthonormal basis for $\ell^2(\mathbb{Z}_{N_1} \times \mathbb{Z}_{N_2})$ if and only if $A(n_1, n_2)$ is unitary for all $n_1 = 0, 1, \ldots, (N_1/2) - 1$ and $n_2 = 0, 1, \ldots, (N_2/2) - 1$.

iii. (First-stage product wavelets for the two-dimensional case) The easiest way to generate a basis of the type in part ii is to form product wavelets. Suppose $\{R_{2k} v_1\}_{k=0}^{N_1-1} \cup \{R_{2k} u_1\}_{k=0}^{N_1-1}\}$ is a first-stage wavelet basis for $\ell^2(\mathbb{Z}_{N_1})$ and $\{R_{2k} v_2\}_{k=0}^{N_2-1} \cup \{R_{2k} u_2\}_{k=0}^{N_2-1}$ is a first-stage wavelet basis for $\ell^2(\mathbb{Z}_{N_2})$. Define

$$w_0(n_1, n_2) = v_1(n_1) v_2(n_2),$$
$$w_1(n_1, n_2) = u_1(n_1) v_2(n_2),$$
$$w_2(n_1, n_2) = v_1(n_1) u_2(n_2),$$

and

$$w_3(n_1, n_2) = u_1(n_1) u_2(n_2).$$

Prove that

$$\left\{ R_{2k_1,2k_2} w_0 \right\}_{k_1,k_2} \cup \left\{ R_{2k_1,2k_2} w_1 \right\}_{k_1,k_2}$$
$$\cup \left\{ R_{2k_1,2k_2} w_2 \right\}_{k_1,k_2} \cup \left\{ R_{2k_1,2k_2} w_3 \right\}_{k_1,k_2}$$

is an orthonormal basis for $\ell^2(\mathbb{Z}_{N_1} \times \mathbb{Z}_{N_2})$, where in every case k_1 runs from 0 to $M_1 - 1$ and k_2 runs from 0 to $M_2 - 1$. Hint: The easy way to do this is to notice that

$$R_{2k_1,2k_2} w_0(n_1, n_2) = R_{2k_1} v_1(n_1) R_{2k_2} v_2(n_2),$$

and similarly for w_1, w_2, and w_3. Then the result follows from the general result about product bases in Exercise 2.1.16. Another way, which is more difficult, but also instructive, is to use Theorem 3.8 and part ii. Let $A_1(n_1)$ be the system matrix with $u = u_1, v = v_1, n = n_1$, and $M = M_1$, and let $A_2(n_2)$ be the system matrix for u_2, v_2, n_2, and M_2. By assumption and Theorem 3.8, $A_1(n_1)$ and $A_2(n_2)$ are unitary for all n_1 and n_2. Use this to show that the matrix $A(n_1, n_2)$ defined as in part ii, but with w_j in place of u_j for $j = 0, 1, 2, 3$, is unitary for all n_1, n_2.

We remark that not all orthonormal bases of the form in part ii are product bases.

3.1.13. Suppose $M \in \mathbb{N}$, $N = 2M$, and $u, v, s, t, z \in \ell^2(\mathbb{Z}_N)$.

i. Prove that

$$\tilde{t} * U(D(z * \tilde{v})) = \sum_{k=0}^{M-1} \langle z, R_{2k}v \rangle R_{2k}\tilde{t}$$

and

$$\tilde{s} * U(D(z * \tilde{u})) = \sum_{k=0}^{M-1} \langle z, R_{2k}u \rangle R_{2k}\tilde{s}.$$

ii. Suppose $\{R_{2k}v\}_{k=0}^{M-1} \cup \{R_{2k}u\}_{k=0}^{M-1}$ is an orthonormal basis for $\ell^2(\mathbb{Z}_N)$. Use part i to give a simple proof that we have perfect reconstruction in Figure 14 if and only if $s = \tilde{u}$ and $t = \tilde{v}$. (Hint: Consider $z = u$ or v.)

3.1.14. (Perfect reconstruction for two-dimensional filter banks) Suppose $M_1, M_2 \in \mathbb{N}$, $N_1 = 2M_1$ and $N_2 = 2M_2$. Define the two-dimensional downsampling operator $D : \ell^2(\mathbb{Z}_{N_1} \times \mathbb{Z}_{N_2}) \rightarrow \ell^2(\mathbb{Z}_{M_1} \times \mathbb{Z}_{M_2})$ by

$$D(z)(k_1, k_2) = z(2k_1, 2k_2)$$

for $k_1 = 0, 1, \ldots, M_1 - 1, k_2 = 0, 1, \ldots, M_2 - 1$, for any $z \in \ell^2(\mathbb{Z}_{M_1} \times \mathbb{Z}_{M_2})$. Define the two-dimensional upsampling operator $U : \ell^2(\mathbb{Z}_{M_1} \times \mathbb{Z}_{M_2}) \rightarrow \ell^2(\mathbb{Z}_{N_1} \times \mathbb{Z}_{N_2})$ by

$$U(z)(k_1, k_2) = \begin{cases} z(k_1/2, k_2/2) & \text{if } k_1 \text{ and } k_2 \text{ are even} \\ 0 & \text{if } k_1 \text{ or } k_2 \text{ is odd.} \end{cases}$$

Note that U inserts three zeros for every nonzero component that it retains.

i. Suppose $z \in \ell^2(\mathbb{Z}_{N_1} \times \mathbb{Z}_{N_2})$. Prove that

$$U(D(z)) = \frac{1}{4}\left(z + z^* + z^{**} + z^{***}\right),$$

with definitions as in Exercise 3.1.12 (i).

ii. The basic filter bank in the two-dimensional case has four branches, just as our first-stage wavelet basis as in Exercise 3.1.12 (ii) has four generators. Suppose $u_0, u_1, u_2, u_3, s_0, s_1, s_2, s_3 \in \ell^2(\mathbb{Z}_{N_1} \times \mathbb{Z}_{N_2})$. In the jth branch of the filter bank, with input $z \in \ell^2(\mathbb{Z}_{N_1} \times \mathbb{Z}_{N_2})$, we compute $z * \tilde{u}_j$. Then we apply D. This is the decomposition stage of the process. In the reconstruction stage, we take the output of the jth branch so far, namely $D(z*\tilde{u}_j)$, apply U, and convolve with $\tilde{s}_j$, giving $\tilde{s}_j * U(D(z* \tilde{u}_j))$. We have perfect reconstruction if the sum of these always equals z, that is, if

$$\sum_{j=0}^{3} \tilde{s}_j * U(D(z * \tilde{u}_j)) = z$$

for all $z \in \ell^2(\mathbb{Z}_{N_1} \times \mathbb{Z}_{N_2})$. Let $A(n_1, n_2)$ be the matrix defined in Exercise 3.1.12 (ii). Prove that we have perfect reconstruction if and only if

$$A(n_1, n_2) \begin{bmatrix} \hat{s}_0 \\ \hat{s}_1 \\ \hat{s}_2 \\ \hat{s}_3 \end{bmatrix} = \begin{bmatrix} 2 \\ 0 \\ 0 \\ 0 \end{bmatrix}$$

for all n_1, n_2.

iii. The output of the decomposition stage in part ii is the vector whose components are of the form

$$(z * \tilde{u}_j)(2k_1, 2k_2) = \langle z, R_{2k_1, 2k_2} u_j \rangle$$

for $k_1 = 0, 1, \ldots, M_1 - 1$ and $k_2 = 0, 1, \ldots, M_2 - 1$. Suppose that $A(n_1, n_2)$ is unitary for all n_1, n_2, so that the set in Exercise 3.1.12 (ii) is an orthonormal basis for $\ell^2(\mathbb{Z}_{N_1} \times \mathbb{Z}_{N_2})$. Then the output of the decomposition stage is the vector whose components are the coefficients in the orthonormal expansion of z with respect to this orthonormal basis. Prove that in this case we have perfect reconstruction if and only if $s_j = \tilde{u}_j$ for $j = 0, 1, 2, 3$.

3.2 Construction of Wavelets on $\mathbb{Z}_N$: The Iteration Step

So far we have constructed orthonormal bases for $\ell^2(\mathbb{Z}_N)$ of the form

$$\{R_{2k}v\}_{k=0}^{(N/2)-1} \cup \{R_{2k}u\}_{k=0}^{(N/2)-1}, \tag{3.29}$$

which we call first-stage wavelet bases. Theorem 3.8 gives necessary and sufficient conditions on $\hat{u}$ and $\hat{v}$ for such a collection to form an orthonormal basis. As we see in Examples 3.10 and 3.11, we can concentrate the high frequencies in the terms involving v in relation (3.29) and the low frequencies in the terms involving u. Thus some degree of frequency localization is obtained. As shown by Figures 11 and 12, we also obtained some degree of spatial localization. We saw that for a first-stage wavelet basis, the change of basis can be computed via the filter bank scheme in Figure 13. The inverse change of basis can also be computed by a filter bank arrangement, namely the right half of Figure 14. These basis changes can be computed quickly, because they are accomplished by a pair of convolutions, which can be done via the FFT.

The filter bank arrangement in Figure 14 suggests a possibility for iteration. Namely, on either or both of the outputs of the left side, we can apply the same procedure again. We can pass the output of either branch through two more filters, and downsample again in each new branch. Similarly on the right side, we can pass the signal from each of these two new branches through an upsampler and a new filter. If the filters at this second stage are compatible, as at the first stage, we will still have perfect reconstruction. Then we can iterate. In principle we can iterate in any branch, but in standard wavelet analysis we iterate in only one of the two branches from the previous stage, normally the branch coming from the convolution with the low pass filter u. (More complicated, adaptive iteration procedures occur in something known as *wavelet-packet* theory.) In this section, we study this iteration procedure, and in particular we see that it corresponds to a certain type of orthonormal basis, which we call a wavelet basis for $\ell^2(\mathbb{Z}_N)$. First we comment on why we want to iterate.

Consider the first-stage Shannon wavelet basis in Example 3.10. Recall that u and v are chosen in this case to split the frequency scale in half. But from music we know that it is more natural to consider frequencies on a logarithmic scale, in octaves. This suggests that we should leave the terms carrying the top half of the frequencies, but that we should subdivide the frequencies in the lower half into two equal parts, the lowest quarter of frequencies and the next quarter. Then we could split the lowest quarter again, and so on. In this way our basis decomposition could give us a more refined frequency analysis of a signal.

Another motivation comes from considering the first-generation discrete Haar basis in Exercise 3.1.2. In part iii of that exercise, we split z into $P(z)$ and $Q(z)$, where $P(z)$ corresponded to the terms coming from the translates of u and $Q(z)$ corresponded to the terms coming from the translates of v. In part iv we saw that $P(z)$ is obtained by replacing z at adjacent values $2k$ and $2k+1$ by the average of these values. This can be regarded as viewing z evened out to a scale of 2 instead of at a scale of 1. This may be enough to determine the general large-scale behavior of z, if the details at scale 1 that are contained in $Q(z)$ are not essential in our considerations. If we need to consider only $P(z)$, we have compressed our original data by a factor of 2, because $P(z)$ is determined by $N/2$ coefficients.

However, it may be that we can go further. Perhaps the type of behavior with which we are concerned can be understood at a scale of 4 or 8 or some larger number. Then we should be able to take our approximation $P(z)$ at scale 2 and split it into two parts, one being the approximation at scale 4, the other being the detail needed to pass from scale 4 to scale 2. We should be able to continue in such a way that the data at scale 2^ℓ are determined by $N/2^\ell$ numbers. This may, for example, allow us to transmit a rough approximation to a signal very quickly, and then to add detail gradually as needed. If we can determine early on that we do not need to look at the full detail, we have saved time, energy, or both. This could be relevant to a radar or sonar search, where our first concern is whether there is anything there; if there is, we look more closely at details to identify it.

This approach also gives us a natural notion of different scales of behavior. This is important in many applications, for example in studying fluid flow such as ocean waves. Something that only shows up when looking at the fine-scale terms of the expansion indicates small-scale behavior, such as small ripples in a wave; the large-scale terms might indicate the main wave.

To describe the iteration step, consider the filter bank diagram in Figure 14, now denoting the filters in the left half as $\tilde{u}_1$ and $\tilde{v}_1$ instead of $\tilde{u}$ and $\tilde{v}$. Assume that u_1, v_1 generate a first-stage wavelet basis, that is, that the system matrix $A(n)$ of u_1 and v_1 (Definition 3.7) is unitary for all n. Lemma 3.15 states that for perfect reconstruction, the filters in the right half of Figure 14 should be u_1 and v_1.

Let the input of the diagram be some vector $z \in \ell^2(\mathbb{Z}_N)$, where N is even. For the second stage that we are about to describe, N must be divisible by 4. The output of the left half of Figure 10, the analysis phase, is the pair of vectors $D(z * \tilde{v}_1), D(z * \tilde{u}_1) \in \ell^2(\mathbb{Z}_{N/2})$. We think of v_1 as corresponding to the high-frequency part, although strictly speaking this does not have to be the case (in fact at this point u and v are interchangeable). As suggested by our examples above, we leave the vector $D(z * \tilde{v}_1)$ alone. However, we operate on the other vector $D(z * \tilde{u}_1)$ in the same way we have done so far with z. Namely, we pick two vectors $u_2, v_2 \in \ell^2(\mathbb{Z}_{N/2})$ whose system matrix (Definition 3.7 with N replaced by $N/2$) is unitary for all n. We pass $D(z * \tilde{u}_1)$ through the filters corresponding to $\tilde{v}_2$ and $\tilde{u}_2$, followed in each case by a downsampling operator. These, plus the output of the top branch, become the output of our second-stage analysis procedure. In other words the ouput is the set of vectors $D(z * \tilde{v}_1), D(D(z * \tilde{u}_1) * \tilde{v}_2)$, and $D(D(z * \tilde{u}_1) * \tilde{u}_2)$. This is shown in the left side of Figure 16.

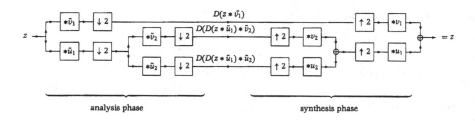

analysis phase   synthesis phase

FIGURE 16

For the synthesis, or reconstruction, phase, we need only to introduce a new pair of branches on the right side corresponding to the two branches on the left introduced at the second stage. Each branch consists of an upsampling operator followed by a convolution, namely with v_2 in one branch and with u_2 in the lowest branch, as in Figure 16. We then add the outputs of these two convolutions. By Theorem 3.15, the net effect of all the branches introduced at the second stage is the identity. Thus by the results for the first stage, the total effect of the entire diagram is the identity, and we have perfect reconstruction.

If N is divisible by 2^p, we can repeat this process up to p times. Each time we subdivide only the lowest branch, each time using filters u_ℓ, v_ℓ satisfying the conditions of Theorem 3.8, to guarantee perfect reconstruction. Figure 17 shows this process at stage 3. In the analysis phase of this figure (the left half), each box represents convolution of the incoming signal with the two filters, followed by downsampling each result. For example, the output of the first box is the pair $D(z * \tilde{v}_1), D(z * \tilde{u}_1)$, similar to Figure 13. In the synthesis phase, each box represents upsampling of the two incoming signals, followed by convolutions with the two filters, as in the right half of Figure 14.

We formalize this process with the following definition.

Definition 3.16 *Suppose N is divisible by 2^p. A pth-stage wavelet filter sequence is a sequence of vectors $u_1, v_1, u_2, v_2, \ldots, u_p, v_p$ such that, for each $\ell = 1, 2, \ldots, p$,*

$$u_\ell, v_\ell \in \ell^2(\mathbb{Z}_{N/2^{\ell-1}}),$$

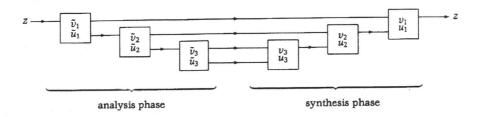

analysis phase   synthesis phase

FIGURE 17

and the system matrix

$$A_\ell(n) = \frac{1}{\sqrt{2}} \begin{bmatrix} \hat{u}_\ell(n) & \hat{v}_\ell(n) \\ \hat{u}_\ell(n + \frac{N}{2^\ell}) & \hat{v}_\ell(n + \frac{N}{2^\ell}) \end{bmatrix} \tag{3.30}$$

is unitary for all $n = 0, 1, \ldots, (N/2^\ell) - 1$.

For an input $z \in \ell^2(\mathbb{Z}_N)$, *define*

$$x_1 = D(z * \tilde{v}_1) \in \ell^2(\mathbb{Z}_{N/2}), \tag{3.31}$$

and

$$y_1 = D(z * \tilde{u}_1) \in \ell^2(\mathbb{Z}_{N/2}). \tag{3.32}$$

Define $x_2, y_2, \ldots, x_p, y_p$ *inductively by*

$$x_\ell = D(y_{\ell-1} * \tilde{v}_\ell) \in \ell^2(\mathbb{Z}_{N/2^\ell}) \tag{3.33}$$

and

$$y_\ell = D(y_{\ell-1} * \tilde{u}_\ell) \in \ell^2(\mathbb{Z}_{N/2^\ell}) \tag{3.34}$$

for $\ell = 2, \ldots, p$.

The output of the analysis phase *of the* p^{th}-stage wavelet filter bank *is the set of vectors* $\{x_1, x_2, \ldots, x_p, y_p\}$.

The recursion formulae (3.33) and (3.34) can be best understood by considering the ℓth stage of the filtering sequence, as in the left half of Figure 18.

Note that $y_1, y_2, \ldots, y_{p-1}$ do not occur in the output of the analysis phase. For $\ell < p$, y_ℓ is used only to define $x_{\ell+1}$ and $y_{\ell+1}$. If we write

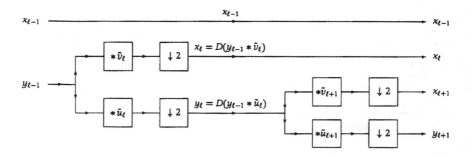

FIGURE 18

out x_ℓ and y_ℓ explicitly, we obtain

$$x_\ell = D(D(\cdots D(D(z * \tilde{u}_1) * \tilde{u}_2)\cdots *\tilde{u}_{\ell-1}) * \tilde{v}_\ell) \tag{3.35}$$

and

$$y_\ell = D(D(\cdots D(D(z * \tilde{u}_1) * \tilde{u}_2)\cdots *\tilde{u}_{\ell-1}) * \tilde{u}_\ell). \tag{3.36}$$

Notice that the sum of the number of components of all the output vectors of the analysis stage is

$$\frac{N}{2} + \frac{N}{4} + \cdots + \frac{N}{2^{p-1}} + \frac{N}{2^p} + \frac{N}{2^p} = N,$$

as expected.

The reconstruction phase can be described by the following sequence of steps. At the first stage of the reconstruction, we compute $(U(y_p)) * u_p$ and $(U(x_p)) * v_p$. By the perfect reconstruction property of the lowest block of the filter bank (the analysis and synthesis phases corresponding to the filters v_p and u_p), we have

$$(U(y_p)) * u_p + (U(x_p)) * v_p = y_{p-1}.$$

Continue similarly with y_{p-1} and x_{p-1} to obtain y_{p-2}. Then use y_{p-2} and x_{p-2} to obtain y_{p-3}, etc., as suggested by Figure 19.

Formally,

$$y_{p-2} = (U(y_{p-1})) * u_{p-1} + (U(x_{p-1})) * v_{p-1},$$
$$y_{p-3} = (U(y_{p-2})) * u_{p-2} + (U(x_{p-2})) * v_{p-2},$$

and so on until

$$y_1 = (U(y_2)) * u_2 + (U(x_2)) * v_2.$$

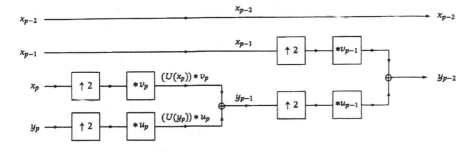

FIGURE 19

Then one more step yields the original vector z, that is,

$$z = (U(y_1)) * u_1 + (U(x_1)) * v_1.$$

We now consider the number of multiplications needed to compute the output of the analysis phase of the wavelet filter bank sequence. The reconstruction phase is computable in the same number of multiplications (Exercise 3.2.4).

Lemma 3.17   *Suppose $N = 2^n, 1 \le p \le n$, and $u_1, v_1, u_2, v_2, \ldots,$ u_p, v_p form a pth stage wavelet filter sequence. Suppose $z \in \ell^2(\mathbb{Z}_N)$. Then the output $\{x_1, x_2, x_3, \ldots, x_p, y_p\}$ of the analysis phase of the corresponding p^{th} stage wavelet filter bank can be computed using no more than*

$$4N + N \log_2 N$$

complex multiplications. (We are assuming that the vectors $\hat{u}_1, \hat{v}_1, \ldots,$ $\hat{u}_p, \hat{v}_p$ have been precomputed and stored. We are determining only the subsequent computation required for each z.)

Proof
We use the fact (Lemma 2.39) that the DFT of a vector of length 2^k can be computed using at most $k2^{k-1}$ complex multiplications, via the FFT. We first compute $\hat{z}$, which requires

$$n2^{n-1} = \frac{1}{2}N \log_2 N$$

multiplications. From now until the final step, we stay in the frequency domain, that is, we work with the DFT at every stage.

At the first stage, we compute $(z * \tilde{v}_1)\hat{} = \hat{z}\overline{\tilde{v}_1}$ (by Lemma 2.30 and equation (3.3)) via N multiplications, and similarly for $(z * \tilde{u}_1)\hat{}$. Now, instead of applying the IDFT to compute $z * \tilde{v}_1$ and then downsampling, we apply Exercise 3.2.1 (i) to obtain $\hat{x}_1 = (D(z * \tilde{v}_1))\hat{}$ from $(z * \tilde{v}_1)\hat{}$ without any multiplications. Also define $y_1 = D(z * \tilde{u}_1)$ and compute $\hat{y}_1$ in the same way. So we have obtained $\hat{x}_1$ and $\hat{y}_1$ with $2N$ additional multiplications.

At the second stage, recall that $x_2 = D(y_1 * \tilde{v}_2)$ and $y_2 = D(y_1 * \tilde{u}_2)$. Because we have already computed $\hat{y}_1$, we can compute $(y_1 * \tilde{v}_2)\hat{} = \hat{y}_1\overline{\tilde{v}_2}$ in $N/2$ multiplications, since these vectors have length $N/2$. Then we use Exercise 3.2.1 (i) to do the downsampling on the transform side, yielding $\hat{x}_2$. We compute $\hat{y}_2$ with $N/2$ multiplications similarly.

We continue in this way. Each step ends with a downsampling, which reduces the size of the vectors by a factor of 2. The jth step yields a pair of vectors, x_j and y_j; x_j is kept to the end, and y_j is used at the next stage to yield x_{j+1} and y_{j+1}. The jth stage requires a total of $2(N/2^{j-1})$ multiplications. After p stages, we have $\hat{x}_1, \hat{x}_2, \ldots, \hat{x}_p$, and $\hat{y}_p$, and altogether we have used

$$2\left(N + \frac{N}{2} + \frac{N}{4} + \cdots + \frac{N}{2^{p-1}}\right) < 4N$$

multiplications for these stages.

To compute $\{x_1, x_2, x_3, \ldots, x_p, y_p\}$, we need to compute the IDFTs of $\hat{x}_1, \hat{x}_2, \ldots, \hat{x}_p$, and $\hat{y}_p$. Since $x_1 \in \ell^2(\mathbb{Z}_{N/2})$; $x_2 \in \ell^2(\mathbb{Z}_{N/4}), \ldots$; $x_p \subseteq \ell^2(\mathbb{Z}_{N/2}^p)$; and $y_p \in \ell^2(\mathbb{Z}_{N/2}^p)$, this requires at most

$$\frac{1}{2}\left((n-1)2^{n-1} + (n-2)2^{n-2} + \cdots + (n-p)2^{n-p} + (n-p)2^{n-p}\right)$$

$$\leq n2^{n-1} = \frac{1}{2}N\log_2 N$$

multiplications (see Exercise 3.2.3 for the proof of the inequality).

Thus the total number of multiplications is at most

$$2 \cdot \frac{1}{2}N\log_2 N + 4N = 4N + N\log_2 N. \qquad \blacksquare$$

Thus this recursive filter bank procedure can be computed rapidly, requiring only about two to three times as many multiplications as the FFT for large N. If the filters $u_1, \ldots, u_p, v_1, \ldots, v_p$ all have at most K nonzero components, we can compute the output of the analysis phase using at most $4KN$ multiplications (Exercise 3.2.12).

Although the recursive description in Definition 3.16 is useful for computational purposes, it is not clear what it has to do with our original intent of constructing orthonormal bases for $\ell^2(\mathbb{Z}_N)$. There is an equivalent nonrecursive reformulation of our filter bank structure, which gives us more insight and leads to orthonormal bases. We begin with a lemma. To be clear about our notation, we point out that when we write $D(z) * w$, we mean $(D(z)) * w$, not $D(z * w)$.

Lemma 3.18   *Suppose N is even, say $N = 2M$, $z \in \ell^2(\mathbb{Z}_N)$, and $x, y, w \in \ell^2(\mathbb{Z}_{N/2})$. Then*

$$D(z) * w = D(z * U(w)), \tag{3.37}$$

and

$$U(x) * U(y) = U(x * y). \tag{3.38}$$

Proof

To prove equation (3.37), we note that $w(m) = U(w)(2m)$ for each m; hence,

$$D(z) * w(n) = \sum_{m=0}^{(N/2)-1} D(z)(n-m)w(m)$$

$$= \sum_{m=0}^{(N/2)-1} z(2n-2m)U(w)(2m)$$

$$= \sum_{k=0}^{N-1} z(2n-k)U(w)(k) = z * U(w)(2n)$$

$$= D(z * U(w))(n),$$

since $U(w)(k) = 0$ at the odd values of k, so they can be added to the sum on m with no change.

To prove equation (3.38), the fact that $U(y)(m) = 0$ if m is odd and $U(y)(m) = y(m/2)$ if m is even, shows that

$$U(x) * U(y)(n) = \sum_{m=0}^{N-1} U(x)(n-m)U(y)(m) = \sum_{k=0}^{(N/2)-1} U(x)(n-2k)y(k).$$

If n is odd, then so is $n - 2k$, hence $U(x)(n-2k) = 0$ for all k. So in this case,

$$U(x) * U(y)(n) = 0 = U(x * y)(n),$$

by definition. If n is even, say $n = 2\ell$, then $U(x)(n - 2k) = U(x)(2\ell - 2k) = x(\ell - k)$, so, from above,

$$U(x) * U(y)(n) = \sum_{k=0}^{(N/2)-1} x(\ell-k)y(k) = x * y(\ell) = (U(x * y))(n). \qquad \blacksquare$$

When we write $D^{\ell}(z)$, we mean the composition of D with itself ℓ times, applied to z. More formally, $D^1 = D$ and, by induction, we define $D^{\ell}(z) = (D \circ D^{\ell-1})(z)$ for $\ell > 1$. We define $U^{\ell}(z) = (U \circ U^{\ell-1})(z)$ in the same way. Note that $D^{\ell} : \ell^2(\mathbb{Z}_N) \to \ell^2(\mathbb{Z}_{N/2}^{\ell})$ is given by

$$D^{\ell}(z)(n) = z(2^{\ell}n),$$

whereas $U^{\ell} : \ell^2(\mathbb{Z}_{N/2}^{\ell}) \to \ell^2(\mathbb{Z}_N)$ is given by

$$U^{\ell}(w)(n) = \begin{cases} w(n/2^{\ell}) & \text{if } 2^{\ell} | n \\ 0 & \text{if } 2^{\ell} \nmid n. \end{cases}$$

Corollary 3.19 *Suppose N is divisible by 2^{ℓ}, $x, y, w \in \ell^2(\mathbb{Z}_{N/2^{\ell}})$, and $z \in \ell^2(\mathbb{Z}_N)$. Then*

$$D^{\ell}(z) * w = D^{\ell}(z * U^{\ell}(w)), \tag{3.39}$$

and

$$U^{\ell}(x * y) = U^{\ell}(x) * U^{\ell}(y). \tag{3.40}$$

Proof
Exercise 3.2.5. ∎

We now introduce a nonrecursive notation that will be seen to be equivalent to the recursive notation in Definition 3.16.

Definition 3.20 *Suppose N is divisible by 2^p. Suppose vectors $u_1, v_1, u_2, v_2, \ldots, u_p, v_p$ are given, such that, for each $\ell = 1, 2, \ldots, p$,*

$$u_{\ell}, v_{\ell} \in \ell^2(\mathbb{Z}_{N/2^{\ell-1}}).$$

Define $f_1 = v_1$, and $g_1 = u_1$. Then inductively define $f_{\ell}, g_{\ell} \in \ell^2(\mathbb{Z}_N)$, for $\ell = 2, 3, \ldots, p$, by

$$f_{\ell} = g_{\ell-1} * U^{\ell-1}(v_{\ell}) \tag{3.41}$$

and

$$g_{\ell} = g_{\ell-1} * U^{\ell-1}(u_{\ell}). \tag{3.42}$$

Writing a few of these out, we see that

$$f_2 = u_1 * U(v_2), \quad g_2 = u_1 * U(u_2),$$
$$f_3 = u_1 * U(u_2) * U^2(v_3), \quad g_3 = u_1 * U(u_2) * U^2(u_3),$$

and so on, with general terms

$$f_\ell = u_1 * U(u_2) * U^2(u_3) * \cdots U^{\ell-2}(u_{\ell-1}) * U^{\ell-1}(v_\ell) \qquad (3.43)$$

and

$$g_\ell = u_1 * U(u_2) * U^2(u_3) * \cdots U^{\ell-2}(u_{\ell-1}) * U^{\ell-1}(u_\ell). \qquad (3.44)$$

Note that all the convolution operations in the definitions of f_ℓ and g_ℓ involve only u_j filters, except for the last convolution in the case of f_ℓ, which involves v_ℓ.

For future reference, we note that

$$\tilde{f}_\ell = (g_{\ell-1} * U^{\ell-1}(v_\ell))\tilde{} = \tilde{g}_{\ell-1} * (U^{\ell-1}(v_\ell))\tilde{} = \tilde{g}_{\ell-1} * U^{\ell-1}(\tilde{v}_\ell), \quad (3.45)$$

by Exercise 3.2.2. Similarly,

$$\tilde{g}_\ell = \tilde{g}_{\ell-1} * U^{\ell-1}(\tilde{u}_\ell). \qquad (3.46)$$

The next lemma allows us to describe the output of the analysis phase of a pth stage recursive wavelet filter bank as a set of single (nonrecursive) convolutions. It also allows us to describe the reconstruction phase in a similar way.

Lemma 3.21　*Suppose N is divisible by 2^p, $z \in \ell^2(\mathbb{Z}_N)$, and $u_1, v_1, \ldots, u_p, v_p$ are such that*

$$u_\ell, v_\ell \in \ell^2(\mathbb{Z}_{N/2^{\ell-1}}),$$

for each $\ell = 1, 2, \ldots, p$. Define $x_1, x_2, \ldots, x_p, y_1, y_2, \ldots, y_p$ as in equations (3.31)–(3.34), and $f_1, f_2, \ldots f_p, g_1, g_2, \ldots, g_p$ as in Definition 3.20. Then for $\ell = 1, 2, \ldots, p$,

$$x_\ell = D^\ell(z * \tilde{f}_\ell), \qquad (3.47)$$

and

$$y_\ell = D^\ell(z * \tilde{g}_\ell). \qquad (3.48)$$

Proof

We prove equations (3.47) and (3.48) together by induction on ℓ. When $\ell = 1$, equations (3.47) and (3.48) hold by equations (3.31) and (3.32) and the definitions of f_1 and g_1. Now suppose equations (3.47) and (3.48) hold for $\ell - 1$. By equation (3.33), induction, and equation (3.39),

$$
\begin{aligned}
x_\ell &= D(y_{\ell-1} * \tilde{v}_\ell) = D(D^{\ell-1}(z * \tilde{g}_{\ell-1}) * \tilde{v}_\ell) \\
&= D \circ D^{\ell-1}(z * \tilde{g}_{\ell-1} * U^{\ell-1}(\tilde{v}_\ell)) \\
&= D^\ell(z * \tilde{g}_{\ell-1} * U^{\ell-1}(\tilde{v}_\ell)) = D^\ell(z * \tilde{f}_\ell),
\end{aligned}
$$

by equation (3.45). Similarly, using equation (3.34) in place of equation (3.33),

$$
\begin{aligned}
y_\ell &= D(y_{\ell-1} * \tilde{u}_\ell) = D(D^{\ell-1}(z * \tilde{g}_{\ell-1}) * \tilde{u}_\ell) \\
&= D \circ D^{\ell-1}(z * \tilde{g}_{\ell-1} * U^{\ell-1}(\tilde{u}_\ell)) \\
&= D^\ell(z * \tilde{g}_{\ell-1} * U^{\ell-1}(\tilde{u}_\ell)) = D^\ell(z * \tilde{g}_\ell),
\end{aligned}
$$

by equation (3.46). This completes the induction step and hence the proof. ∎

Thus the output of the ℓth branch of the analysis phase of the filter bank sequence is $D^\ell(z * \tilde{f}_\ell)$, for $\ell = 1, 2, \ldots, p$. The output of the final branch is $D^p(z * \tilde{g}_p)$. This is exhibited in the left half of Figure 20.

There is a similar description of the reconstruction phase of the filter bank sequence.

Lemma 3.22 *Suppose N is divisible by 2^p. Consider a pth-stage filter bank sequence $u_1, v_1, \ldots, u_p, v_p$ as in Definition 3.16 (except that we do not require that the system matrix in equation (3.30) be unitary for this result). Define $f_1, \ldots, f_p, g_p$ as in Definition 3.20. If the input to the ℓth branch $(1 \le \ell \le p)$ of the reconstruction phase (i.e., the branch for which the next operation is convolution with v_ℓ) is x_ℓ, and all other inputs are zero, then the output of the reconstruction phase is*

$$
f_\ell * U^\ell(x_\ell).
$$

If the input to the final branch (for which the next operation is convolution with u_p) is y_p, and all other inputs are zero, then the output of the reconstruction phase is

$$
g_p * U^p(y_p).
$$

Proof
Exercise 3.2.6. ∎

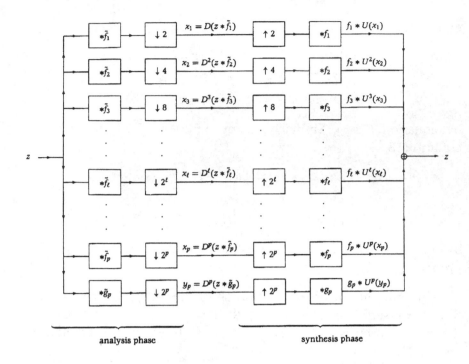

FIGURE 20

Thus the full recursive p^{th}-stage wavelet filter sequence can be represented by the nonrecursive structure shown in Figure 20.

Recall our original goal of constructing orthonormal bases for $\ell^2(\mathbb{Z}_N)$. Recall also from Definition 3.16 that the output of the analysis phase of our filter bank is the set of vectors $x_1, x_2, \ldots x_{p-1}, x_p, y_p$. By Lemma 3.21, for each $\ell = 1, 2, \ldots, p$,

$$x_\ell(k) = D^\ell(z * \tilde{f}_\ell)(k) = z * \tilde{f}_\ell(2^\ell k) = \langle z, R_{2^\ell k}f_\ell \rangle, \qquad (3.49)$$

for $k = 0, 1, \ldots, (N/2^\ell) - 1$, by equation (3.4). Similarly,

$$y_p(k) = D^p(z * \tilde{g}_p)(k) = z * \tilde{g}_p(2^p k) = \langle z, R_{2^p k}g_p \rangle, \qquad (3.50)$$

for $k = 0, 1, \ldots, (N/2^p) - 1$. As we noted above, the total number of components of $x_1, x_2, \ldots x_{p-1}, x_p, y_p$ is N. As one might hope, it turns out (by Theorem 3.27) that these are the components of the expansion of z with respect to an orthonormal basis.

Definition 3.23   *Suppose N is divisible by 2^p, where p is a positive integer. Let B be a set of the form*

$$\{R_{2k}f_1\}_{k=0}^{(N/2)-1} \cup \{R_{4k}f_2\}_{k=0}^{(N/4)-1} \cup \cdots \cup \{R_{2^p k}f_p\}_{k=0}^{(N/2^p)-1} \cup \{R_{2^p k}g_p\}_{k=0}^{(N/2^p)-1}$$

for some $f_1, f_2, \ldots, f_p, g_p \in \ell^2(\mathbb{Z}_N)$. If B forms an orthonormal basis for $\ell^2(\mathbb{Z}_N)$, we call B a pth stage wavelet basis for $\ell^2(\mathbb{Z}_N)$. We say that $f_1, f_2, \ldots, f_p, g_p$ generate B.

Our goal is to show that $f_1, f_2, \ldots, f_p, g_p$, obtained by Definitions 3.16 and 3.20, generate a pth stage wavelet basis. The key step is contained in the next lemma.

Lemma 3.24   *Suppose N is divisible by 2^ℓ, $g_{\ell-1} \in \ell^2(\mathbb{Z}_N)$, and the set*

$$\{R_{2^{\ell-1}k}g_{\ell-1}\}_{k=0}^{(N/2^{\ell-1})-1} \tag{3.51}$$

is orthonormal with $N/2^{\ell-1}$ elements. Suppose $u_\ell, v_\ell \in \ell^2(\mathbb{Z}_{N/2^{\ell-1}})$, and the system matrix $A_\ell(n)$ in equation (3.30) is unitary for all $n = 0, 1, \ldots, (N/2^\ell) - 1$. Define

$$f_\ell = g_{\ell-1} * U^{\ell-1}(v_\ell) \quad \text{and} \quad g_\ell = g_{\ell-1} * U^{\ell-1}(u_\ell).$$

Then

$$\{R_{2^\ell k}f_\ell\}_{k=0}^{(N/2^\ell)-1} \cup \{R_{2^\ell k}g_\ell\}_{k=0}^{(N/2^\ell)-1} \tag{3.52}$$

is an orthonormal set with $N/2^{\ell-1}$ elements.

Proof
By equation (3.4) and the assumed orthonormality of the set (3.51),

$$g_{\ell-1} * \tilde{g}_{\ell-1}(2^{\ell-1}k) = \langle g_{\ell-1}, R_{2^{\ell-1}k}g_{\ell-1}\rangle$$

$$= \begin{cases} 1 & \text{if } k = 0 \\ 0 & \text{if } k = 1, 2, \ldots, (N/2^{\ell-1}) - 1. \end{cases} \tag{3.53}$$

By Theorem 3.8, applied to $\ell^2(\mathbb{Z}_{N/2}^{\ell-1})$, our assumption that $A_\ell(n)$ is unitary guarantees that the set

$$\{R_{2k}v_\ell\}_{k=0}^{(N/2^\ell)-1} \cup \{R_{2k}u_\ell\}_{k=0}^{(N/2^\ell)-1} \tag{3.54}$$

is an orthonormal basis for $\ell^2(\mathbb{Z}_{N/2}^{\ell-1})$. In particular, using equation (3.4),

$$v_\ell * \tilde{v}_\ell(2k) = \langle v_\ell, R_{2k}v_\ell \rangle = \begin{cases} 1 & \text{if } k = 0 \\ 0 & \text{if } k = 1, 2, \ldots, (N/2^\ell) - 1, \end{cases} \quad (3.55)$$

$$v_\ell * \tilde{u}_\ell(2k) = \langle v_\ell, R_{2k}u_\ell \rangle = 0 \quad \text{for all } k, \quad (3.56)$$

and

$$u_\ell * \tilde{u}_\ell(2k) = \langle u_\ell, R_{2k}u_\ell \rangle = \begin{cases} 1 & \text{if } k = 0 \\ 0 & \text{if } k = 1, 2, \ldots, (N/2^\ell) - 1. \end{cases} \quad (3.57)$$

To prove the orthonormality of the set (3.52), use equation (3.4) to write, for $k = 0, 1, \ldots, (N/2^{\ell+1}) - 1$,

$$\begin{aligned} \langle f_\ell, R_{2^\ell k}f_\ell \rangle &= f_\ell * \tilde{f}_\ell(2^\ell k) \\ &= g_{\ell-1} * U^{\ell-1}(v_\ell) * \tilde{g}_{\ell-1} * U^{\ell-1}(\tilde{v}_\ell)(2^\ell k) \\ &= (g_{\ell-1} * \tilde{g}_{\ell-1}) * \left(U^{\ell-1}(v_\ell * \tilde{v}_\ell) \right)(2^\ell k), \end{aligned}$$

where we have used equation (3.45), the commutativity and associativity of convolution, and equation (3.40). By writing out the last convolution as bracketed, we have

$$\langle f_\ell, R_{2^\ell k}f_\ell \rangle = \sum_{n=0}^{N-1} (g_{\ell-1} * \tilde{g}_{\ell-1})(2^\ell k - n)U^{\ell-1}(v_\ell * \tilde{v}_\ell)(n).$$

Observe that $U^{\ell-1}(v_\ell * \tilde{v}_\ell)(n) = (v_\ell * \tilde{v}_\ell)(j)$ when $n = 2^{\ell-1}j$ and 0 otherwise. Hence the sum over all n reduces to a sum over n of the form $2^{\ell-1}j$, hence to a sum over j after a substitution. We obtain

$$\langle f_\ell, R_{2^\ell k}f_\ell \rangle = \sum_{j=0}^{(N/2^{\ell-1})-1} g_{\ell-1} * \tilde{g}_{\ell-1}(2^\ell k - 2^{\ell-1}j)(v_\ell * \tilde{v}_\ell)(j).$$

By equation (3.53),

$$g_{\ell-1} * \tilde{g}_{\ell-1}(2^\ell k - 2^{\ell-1}j)$$

$$= g_{\ell-1} * \tilde{g}_{\ell-1}(2^{\ell-1}(2k - j)) = \begin{cases} 1 & \text{if } j = 2k \\ 0 & \text{if } j \neq 2k, j \in \mathbb{Z}_{N/2^{\ell-1}}. \end{cases}$$

Therefore

$$\langle f_\ell, R_{2^\ell k}f_\ell \rangle = (v_\ell * \tilde{v}_\ell)(2k) = \begin{cases} 1 & \text{if } k = 0 \\ 0 & \text{if } k = 1, 2, \ldots, (N/2^\ell) - 1, \end{cases}$$

by equation (3.55). It follows that the set $\{R_{2^{\ell}k}f_{\ell}\}_{k=0}^{(N/2^{\ell})-1}$ is orthonormal, as in Exercise 3.1.1.

Following the same procedure, but with g_{ℓ} in place of f_{ℓ}, leads to

$$\langle g_{\ell}, R_{2^{\ell}k}g_{\ell}\rangle = (u_{\ell} * \tilde{u}_{\ell})(2k) = \begin{cases} 1 & \text{if } k = 0 \\ 0 & \text{if } k = 1, 2, \ldots, (N/2^{\ell}) - 1, \end{cases}$$

by equation (3.57). This proves that the set $\{R_{2^{\ell}k}g_{\ell}\}_{k=0}^{(N/2^{\ell})-1}$ is orthonormal.

Similarly, by equation (3.56), we obtain

$$\langle f_{\ell}, R_{2^{\ell}k}g_{\ell}\rangle = (v_{\ell} * \tilde{u}_{\ell})(2k) = 0$$

for all k. As in Exercise 3.1.1, this proves that $\langle R_{2^{\ell}k}f_{\ell}, R_{2^{\ell}j}g_{\ell}\rangle = 0$ for all j, k. Hence the set in relation (3.52) is orthonormal. ■

This result can be proved by a DFT argument (Exercise 3.2.7). Lemma 3.24 shows that we can break a subspace generated by translates by $2^{\ell-1}$ of one vector into two orthogonal subspaces, each generated by the translates by 2^{ℓ} of another vector. This is a generalization of Theorem 3.8, which shows how to do this when the original subspace is the whole space $\ell^2(\mathbb{Z}_N)$, regarded as generated by all translates of δ. The more general result in Lemma 3.24 allows us to iterate this splitting. To describe this, the following terminology is convenient.

Definition 3.25 *Suppose X is an inner product space and U and V are subspaces of X. Suppose $U \perp V$ (i.e., for all $u \in U$ and all $v \in V$, $\langle u, v \rangle = 0$). Define*

$$U \oplus V = \{u + v : u \in U, v \in V\}. \tag{3.58}$$

We call $U \oplus V$ the orthogonal direct sum of U and V. In particular, if we say $U \oplus V = X$, we mean that U and V are subspaces of X, $U \perp V$, and every element $x \in X$ can be written as $x = u + v$ for some $u \in U$ and $v \in V$.

Lemma 3.26 *Suppose N is divisible by 2^{ℓ}, $g_{\ell-1} \in \ell^2(\mathbb{Z}_N)$, and the set*

$$\{R_{2^{\ell-1}k}g_{\ell-1}\}_{k=0}^{(N/2^{\ell-1})-1}$$

is orthonormal and has $N/2^{\ell-1}$ elements. Suppose $u_{\ell}, v_{\ell} \in \ell^2(\mathbb{Z}_{N/2^{\ell-1}})$, and the system matrix $A_{\ell}(n)$ in equation (3.30) is unitary for all $n =$

$0, 1, \ldots, (N/2^\ell) - 1$. *Define*

$$f_\ell = g_{\ell-1} * U^{\ell-1}(v_\ell) \quad and \quad g_\ell = g_{\ell-1} * U^{\ell-1}(u_\ell).$$

Define spaces

$$V_{-\ell+1} = \text{span } \{R_{2^{\ell-1}k}g_{\ell-1}\}_{k=0}^{(N/2^{\ell-1})-1}, \tag{3.59}$$

$$W_{-\ell} = \text{span } \{R_{2^\ell k}f_\ell\}_{k=0}^{(N/2^\ell)-1}, \tag{3.60}$$

and

$$V_{-\ell} = \text{span } \{R_{2^\ell k}g_\ell\}_{k=0}^{(N/2^\ell)-1}. \tag{3.61}$$

Then

$$V_{-\ell} \oplus W_{-\ell} = V_{-\ell+1}. \tag{3.62}$$

Proof

By Lemma 3.24, every basis element $R_{2^\ell k}g_\ell$ of $V_{-\ell}$ is orthogonal to every basis element $R_{2^\ell j}f_\ell$ of $W_{-\ell}$. It follows by linearity that every element of $V_{-\ell}$ is orthogonal to every element of $W_{-\ell}$. This proves that $V_{-\ell} \perp W_{-\ell}$. Next we claim that $V_{-\ell}$ and $W_{-\ell}$ are subspaces of $V_{-\ell+1}$. To see this, note that, for $k = 0, 1, \ldots, (N/2^\ell) - 1$,

$$R_{2^\ell k}g_\ell(n) = g_\ell(n - 2^\ell k) = g_{\ell-1} * U^{\ell-1}(u_\ell)(n - 2^\ell k)$$

$$= \sum_{m=0}^{N-1} g_{\ell-1}(n - 2^\ell k - m)U^{\ell-1}(u_\ell)(m).$$

Since $U^{\ell-1}(u_\ell)(m) = u_\ell(m/2^{\ell-1})$ if $2^{\ell-1} \mid m$ and 0 otherwise, the sum over m reduces to a sum over m of the form $2^{\ell-1}j$, and

$$R_{2^\ell k}g_\ell(n) = \sum_{j=0}^{(N/2^{\ell-1})-1} g_{\ell-1}(n - 2^\ell k - 2^{\ell-1}j)u_\ell(j)$$

$$= \sum_{j=0}^{(N/2^{\ell-1})-1} u_\ell(j)R_{2^{\ell-1}(j+2k)}g_{\ell-1}(n).$$

Since this is true for any n, we have

$$R_{2^\ell k}g_\ell = \sum_{j=0}^{(N/2^{\ell-1})-1} u_\ell(j)R_{2^{\ell-1}(j+2k)}g_{\ell-1}. \tag{3.63}$$

In the same way, we obtain

$$R_{2^\ell k} f_\ell = \sum_{j=0}^{(N/2^{\ell-1})-1} v_\ell(j) R_{2^{\ell-1}(j+2k)} g_{\ell-1}. \tag{3.64}$$

Therefore $R_{2^\ell k} g_\ell$ and $R_{2^\ell k} f_\ell$ belong to $V_{-\ell+1}$, because the right-hand sides of equations (3.63) and (3.64) are linear combinations of translates of $g_{\ell-1}$ by integer multiples of $2^{\ell-1}$, that is, linear combinations of basis elements of $V_{-\ell+1}$. Thus the basis elements $R_{2^\ell k} g_\ell$ of $V_{-\ell}$ and $R_{2^\ell k} f_\ell$ of $W_{-\ell}$ belong to $V_{-\ell+1}$, and hence the same is true for all elements of their spans. So $V_{-\ell}$ and $W_{-\ell}$ are subspaces of $V_{-\ell+1}$. However, we have seen that $V_{-\ell}$ and $W_{-\ell}$ each have dimension $N/2^\ell$, so $V_{-\ell} \oplus W_{-\ell}$ has dimension $N/2^{\ell-1}$, which is the dimension of $V_{-\ell+1}$. It follows that $V_{-\ell} \oplus W_{-\ell} = V_{-\ell+1}$. ∎

It may seem strange that we define the spaces $V_{-\ell}$ with negative indices. This is done partly so that the spaces will increase with the index (i.e., $V_{-\ell} \subseteq V_{-\ell+1}$) and partly to be consistent with the notation we use later when considering wavelets on $\mathbb{R}$.

Lemma 3.26 contains the main effort required to prove that the output of the analysis phase of a pth-stage wavelet filter bank system with input z yields the coefficients of z with respect to a pth-stage wavelet basis.

Theorem 3.27   *Suppose N is divisible by 2^p, and $u_1, v_1, u_2, v_2, \ldots,$ u_p, v_p is a pth-stage wavelet filter sequence (Definition 3.16). Define $f_1,$ $f_2, \ldots$ $f_p, g_1, g_2, \ldots, g_p$ as in Definition 3.20. Then $f_1, f_2, \ldots, f_p, g_p$ generate a pth-stage wavelet basis (Definition 3.23) for $\ell^2(\mathbb{Z}_N)$.*

Proof
Our goal is to prove the orthonormality of the set in Definition 3.23. Given this, the fact that this set has N elements implies that it is an orthonormal basis for $\ell^2(\mathbb{Z}_N)$. Since $f_1 = v_1$ and $g_1 = u_1$, Theorem 3.8 guarantees that the set $\{R_{2k} f_1\}_{k=0}^{(N/2)-1} \cup \{R_{2k} g_1\}_{k=0}^{(N/2)-1}$ is orthonormal. Then an inductive argument and Lemma 3.24 show that the set $\{R_{2^\ell k} f_\ell\}_{k=0}^{(N/2^\ell)-1}$ is orthonormal for each $\ell = 1, 2, \ldots, p$, and $\{R_{2^p k} g_p\}_{k=0}^{(N/2^p)-1}$ is orthonormal. Therefore, to obtain the orthonormality of the full set, all that remains to be proved is the orthogonality of elements in the different subsets. Consider first

some $R_{2^\ell k} f_\ell$ and some $R_{2^m j} f_m$, where we may assume $m < \ell$. Lemma 3.26 implies (with the spaces $V_{-\ell}$ and $W_{-\ell}$ defined there) that

$$R_{2^\ell k} f_\ell \in W_{-\ell} \subseteq V_{-\ell+1} \subseteq \cdots \subseteq V_{-m},$$

and $R_{2^m j} f_m \in W_{-m}$. Also by Lemma 3.26, $V_{-m} \perp W_{-m}$, so $R_{2^\ell k} f_\ell$ is orthogonal to $R_{2^m j} f_m$. Similarly, for any $\ell \le p$, any $R_{2^p k} g_p$ belongs to $V_{-p} \subseteq V_{-\ell}$ and hence is orthogonal to any $R_{2^\ell k} f_\ell \in W_{-\ell}$.   ∎

The best way to understand what we have done is in terms of Figure 21, which shows the subspaces in Lemma 3.26. The arrows represent containment. Beginning at the right, we break $\ell^2(\mathbb{Z}_N)$ into orthogonal subspaces V_{-1} and W_{-1}. We keep W_{-1}, but we break V_{-1} into orthogonal subspaces V_{-2} and W_{-2}. We keep W_{-2}, and continue with V_{-2}. We keep going in this way until the pth stage, where we keep both W_{-p} and V_{-p}. We see in chapters 4 and 5 that this point of view can be applied to develop wavelets on $\mathbb{Z}$ and $\mathbb{R}$.

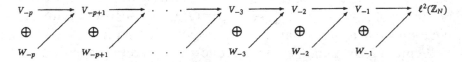

FIGURE 21

By equations (3.49) and (3.50), the output of the analysis phase of the pth-stage wavelet filter bank in Figure 20 with input z is a set of vectors whose components are the components of the expansion of z with respect to the pth-stage wavelet basis in Theorem 3.27. In particular, by Lemma 3.17, the wavelet coefficients are computable by a fast algorithm.

For comparison with later work, we set the following notation.

Definition 3.28   *Suppose N is divisible by 2^p. Let $u_1, v_1, \ldots, u_p, v_p$ be a pth-stage wavelet filter sequence (Definition 3.16). Let $f_1, g_1, \ldots, f_p, g_p$ be as in Definition 3.20. For $j = 1, 2, \ldots, p$ and $k = 0, 1, \ldots, (N/2^j) - 1$, define*

$$\psi_{-j,k} = R_{2^j k} f_j, \tag{3.65}$$

and

$$\varphi_{-j,k} = R_{2^j k} g_j. \tag{3.66}$$

Thus, in this notation, the pth-stage wavelet basis generated by $f_1, f_2, \ldots, f_p, g_p$ (Theorem 3.27) has the form

$$\{\psi_{-1,k}\}_{k=0}^{(N/2)-1} \cup \{\psi_{-2,k}\}_{k=0}^{(N/4)-1} \cup \cdots \cup \{\psi_{-p,k}\}_{k=0}^{(N/2^p)-1} \cup \{\varphi_{-p,k}\}_{k=0}^{(N/2^p)-1}.$$
$$(3.67)$$

The elements of this orthonormal basis are called wavelets on $\mathbb{Z}_N$.

Note that in this terminology,

$$V_{-j} = \text{span}\{\varphi_{-j,k}\}_{k=0}^{(N/2^j)-1} \tag{3.68}$$

and

$$W_{-j} = \text{span}\{\psi_{-j,k}\}_{k=0}^{(N/2^j)-1}. \tag{3.69}$$

We warn the reader that the term "wavelets" in general usage is reserved for wavelets on $\mathbb{R}$, which we consider in chapter 5. The version here, which we have called wavelets on $\mathbb{Z}_N$, is our analog for the finite-dimensional case. This case may have independent interest, and it serves as the easiest introduction to the train of thought in chapter 5.

Exercise 3.2.10 (ii) shows that the definition of the term "wavelet" in Definition 3.28 is not very restrictive, but we let it stand for heuristic reasons.

Summarizing our results, we have the following recipe for creating a wavelet basis for $\ell^2(\mathbb{Z}_N)$.

Recipe 3.29 *Suppose $2^p | N$. Let $u_1, v_1, \ldots, u_p, v_p$ be a pth-stage wavelet filter sequence (Definition 3.16). Define $f_1, f_2, \ldots, f_p, g_1, g_2, \ldots, g_p$ as in Definition 3.20, and $\psi_{-j,k}, \varphi_{-p,k}$ as in equations (3.65) and (3.66). Then the set in equation (3.67) is a pth-stage wavelet basis for $\ell^2(\mathbb{Z}_N)$.*

It turns out that all wavelet bases for $\ell^2(\mathbb{Z}_N)$ are obtained from some wavelet filter sequence by this recipe (Exercise 3.2.9).

We have seen that for any such wavelet basis, the components of some vector $z \in \ell^2(\mathbb{Z}_N)$ in this basis can be computed quickly (in roughly $N \log_2 N$ multiplications, if N is a power of 2) via the analysis phase of the filter bank diagram. The inverse transformation can also be computed with the same speed via the reconstruction phase of the filter bank diagram (Exercise 3.2.4).

It is sometimes useful to look at the wavelets on the DFT side. By taking the DFT of both sides of equation (3.43) and (3.44), and applying Exercise 3.2.1 (ii), we obtain

$$\hat{\psi}_{-j,0}(n) = \hat{f}_j(n) = \hat{u}_1(n)\hat{u}_2(n) \cdots \hat{u}_{j-1}(n)\hat{v}_j(n) \qquad (3.70)$$

and

$$\hat{\varphi}_{-j,0}(n) = \hat{g}_j(n) = \hat{u}_1(n)\hat{u}_2(n) \cdots \hat{u}_{j-1}(n)\hat{u}_j(n). \qquad (3.71)$$

Note also that

$$\psi_{-j,k} = R_{2^j k}f_j = R_{2^j k}\psi_{-j,0} \qquad (3.72)$$

and

$$\varphi_{-p,k} = R_{2^p k}g_p = R_{2^p k}\varphi_{-p,0}. \qquad (3.73)$$

By Lemma 2.13, the DFTs of these translates are given by

$$\hat{\psi}_{-j,k}(m) = e^{-2\pi i m 2^j k/N}\hat{\psi}_{-j,0}(m) \qquad (3.74)$$

and

$$\hat{\varphi}_{-j,k}(m) = e^{-2\pi i m 2^j k/N}\hat{\varphi}_{-j,0}(m) \qquad (3.75)$$

for all j, k. To clarify our notation, we remark that $\hat{\psi}_{-j,k}$ denotes $(\psi_{-j,k})\hat{}$, and similarly $\hat{\varphi}_{-j,k}$ denotes $(\varphi_{-j,k})\hat{}$.

So far we have not required any relationship between the filters u_ℓ, v_ℓ at different stages. It may even seem that no relationship is possible, because they are vectors of different lengths. However, the next lemma provides a way of obtaining filters u_ℓ, v_ℓ satisfying the criterion that the system matrix A_ℓ in Definition 3.16 is unitary for all n, directly from filters u_1, v_1 satisfying this property for A_1.

Lemma 3.30   (The folding lemma) *Suppose N is divisible by 2, and $u_1 \in \ell^2(\mathbb{Z}_N)$.*

   i. Define $u_2 \in \ell^2(\mathbb{Z}_{\frac{N}{2}})$ by

$$u_2(n) = u_1(n) + u_1\left(n + \frac{N}{2}\right). \qquad (3.76)$$

   (Note that the right side of equation (3.76) is periodic with period $N/2$.) Then for all m

$$\hat{u}_2(m) = \hat{u}_1(2m). \qquad (3.77)$$

ii. *Suppose N is divisible by 2^ℓ. Define $u_\ell \in \ell^2(\mathbb{Z}_{N/2^{\ell-1}})$ by*

$$u_\ell(n) = \sum_{k=0}^{2^{\ell-1}-1} u_1\left(n + \frac{kN}{2^{\ell-1}}\right).\qquad(3.78)$$

Then

$$\hat{u}_\ell(m) = \hat{u}_1(2^{\ell-1}m).\qquad(3.79)$$

Proof

To prove part i, write

$$\hat{u}_2(m) = \sum_{n=0}^{(N/2)-1} u_2(n)e^{-2\pi inm/(N/2)}$$

$$= \sum_{n=0}^{(N/2)-1} u_1(n)e^{-2\pi inm/(N/2)} + \sum_{n=0}^{(N/2)-1} u_1\left(n + \frac{N}{2}\right)e^{-2\pi inm/(N/2)},$$

by definition of u_2. In the first of the sums on the last line, let $k = n$; in the second, let $k = n + N/2$. We obtain

$$\hat{u}_2(m) = \sum_{k=0}^{(N/2)-1} u_1(k)e^{-2\pi ik(2m)/N} + \sum_{k=N/2}^{N-1} u_1(k)e^{-2\pi ik(2m)/N} = \hat{u}_1(2m),$$

as required. Part ii follows from part i by an induction argument (Exercise 3.2.11). ∎

We call this the folding lemma because u_2 is obtained from u_1 by cutting u_1 just before $N/2$, folding that part over the first part, and summing. It has the following corollary.

Corollary 3.31 *Suppose N is divisible by 2^p. Suppose $u, v \in \ell^2(\mathbb{Z}_N)$ are such that the system matrix $A(n)$ in Definition 3.7 is unitary for all n. Let $u_1 = u$ and $v_1 = v$, and, for all $\ell = 2, 3, \ldots p$, define u_ℓ by equation (3.78) and v_ℓ similarly with v_1 in place of u_1. Then $u_1, v_1, u_2, v_2, \ldots, u_p, v_p$ is a pth-stage wavelet filter sequence.*

Proof

By equation (3.79), the ℓth system matrix is

$$A_\ell(n) = \frac{1}{\sqrt{2}}\left[\begin{array}{cc} \hat{u}_\ell(n) & \hat{v}_\ell(n) \\ \hat{u}_\ell\left(n + \frac{N}{2^\ell}\right) & \hat{v}_\ell\left(n + \frac{N}{2^\ell}\right) \end{array}\right]$$

$$= \frac{1}{\sqrt{2}} \left[\begin{array}{cc} \hat{u}_1(2^{\ell-1}n) & \hat{v}_1(2^{\ell-1}n) \\ \hat{u}_1\left(2^{\ell-1}n + \frac{N}{2}\right) & \hat{v}_1\left(2^{\ell-1}n + \frac{N}{2}\right) \end{array} \right] = A_1(2^{\ell-1}n);$$

hence it is automatically unitary for all n. ∎

Thus if we are given u and v satisfying the condition in Theorem 3.8, we can obtain $u_1, v_1, \ldots, u_p, v_p$ in an automatic way by Corollary 3.31, and then obtain a wavelet basis by Recipe 3.29. In this case we say we have a wavelet basis with repeated filters. Note that for repeated filters, equations (3.70) and (3.71) become, respectively,

$$\hat{\psi}_{-j,0}(n) = \hat{u}(n)\hat{u}(2n)\hat{u}(4n)\cdots\hat{u}(2^{j-2}n)\hat{v}(2^{j-1}n), \tag{3.80}$$

and

$$\hat{\varphi}_{-j,0}(n) = \hat{u}(n)\hat{u}(2n)\hat{u}(4n)\cdots\hat{u}(2^{j-2}n)\hat{u}(2^{j-1}n). \tag{3.81}$$

Wavelet bases of this type are particularly easy to construct because they require constructing only one pair u, v such that the system matrix is unitary.

Exercises

3.2.1. Suppose $z \in \ell^2(\mathbb{Z}_N)$.
  i. If N is even, prove that

$$(D(z))\hat{\ }(n) = \frac{1}{2}\left(\hat{z}(n) + \hat{z}\left(n + \frac{N}{2}\right)\right),$$

  for all n. Hint: Show that the DFT on $N/2$ points of $D(z)$ agrees with the DFT on N points of $(z + z^*)/2$, for $z^*(n) = (-1)^n z(n)$.
  ii. Prove that

$$(U(z))\hat{\ }(n) = \hat{z}(n)$$

  for all n. Note that the DFT on the left side is in $\ell^2(\mathbb{Z}_{2N})$, whereas the one on the right is in $\ell^2(\mathbb{Z}_N)$. Nevertheless, the assertion is that they (or their periodic extensions) agree at every n. Hint: See equations (2.48) and (2.49).
3.2.2. Suppose $z, w \in \ell^2(\mathbb{Z}_N)$.
  i. Prove that $(z * w)\tilde{\ } = \tilde{z} * \tilde{w}$.

ii. Suppose that N is even. Prove that $(D(z))\tilde{} = D(\tilde{z})$.

iii. Prove that $(U(z))\tilde{} = U(\tilde{z})$.

3.2.3. Suppose $1 \le p \le n$. Prove that

$$(n-1)2^{n-1} + (n-2)2^{n-2} + \cdots + (n-p+1)2^{n-p+1}$$
$$+(n-p)2^{n-p} + (n-p)2^{n-p} \le n2^n.$$

Suggestion: Use induction on n.

3.2.4. Suppose $N = 2^n, 1 \le p \le n$. Prove that the reconstruction phase of the pth stage recursive wavelet filter bank can be computed using at most $4N + N \log_2 N$ complex multiplications.

Hint: As in the analysis stage, do everything in the Fourier domain. To start, one needs to compute $(U(x_1))\hat{}, (U(x_2))\hat{}, \ldots,$ $(U(x_p))\hat{}$ and $(U(y_p))\hat{}$. But since $x_1 \in \ell^2(\mathbb{Z}_{N/2})$, we can use Exercise 3.2.1 (ii) to compute $(U(x_1))\hat{}$ with at most $((n-1)2^{n-1})/2$ multiplications, and similarly for the other vectors. Thus these require the same number of multiplications as computing the IDFTs of $\hat{x}_1, \hat{x}_2, \ldots, \hat{x}_p$, and $\hat{y}_p$ in Lemma 3.17. The multiplications required to compute the DFTs of all the filtering operations is the same as for the analysis phase. The DFT of each $U(y_j)$ is obtained from $\hat{y}_j$ with no multiplications via Exercise 3.2.1 (ii). After this, we have $\hat{z}$, so to find z we perform the IDFT. So the total computation count is the same as in Lemma 3.17.

3.2.5. Prove equations (3.39) and (3.40). Suggestion: Use induction.

3.2.6. Suppose N is divisible by 2^p, and $u_1, v_1, \ldots, u_p, v_p$ are such that, for each $\ell = 1, 2, \ldots, p$,

$$u_\ell, v_\ell \in \ell^2(\mathbb{Z}_{N/2^{\ell-1}}).$$

Define $f_1, f_2, \ldots f_p, g_1, g_2, \ldots, g_p$ as in Definition 3.20. For $\ell = 1, 2, \ldots, p$, and $w \in \ell^2(\mathbb{Z}_{N/2}^\ell)$, define

$$A_\ell(w) = (U(\cdots(U(U(w) * v_\ell) * u_{\ell-1})\cdots) * u_2) * u_1$$

and

$$B_\ell(w) = (U(\cdots(U(U(w) * u_\ell) * u_{\ell-1})\cdots) * u_2) * u_1.$$

i. Prove that for $\ell = 1, 2, \ldots, p$,

$$A_\ell(w) = U^\ell(w) * f_\ell$$

and

$$B_\ell(w) = U^\ell(w) * g_\ell.$$

Hint: Note that $A_\ell(w) = B_{\ell-1}(U(w) * v_\ell)$ and $B_\ell(w) = B_{\ell-1}(U(w) * u_\ell)$, and use induction.

ii. Prove Lemma 3.22. Hint: If x_ℓ is the input to the ℓth branch of the reconstruction phase of the filter bank, and all other inputs are zero, check that the output is $A_\ell(x_\ell)$. Similarly, if the input in the last branch is y_p, and the other inputs are zero, check that the output is $B_p(y_p)$.

3.2.7. Prove Lemma 3.24 by taking DFTs and using the criteria in Exercise 3.1.11 (i). Hint: Write

$$\sum_{k=0}^{2^\ell-1} |\hat{f}_\ell(n + kN/2^\ell)|^2 = \sum_{k=0}^{2^\ell-1} |\hat{g}_{\ell-1}(n + kN/2^\ell)|^2 |\hat{v}_\ell(n + kN/2^\ell)|^2,$$

using Exercise 3.2.1 (ii). Break up the sum into a sum over k even and a sum over k odd. Note that because $\hat{v}_\ell$ has period $N/2^{\ell-1}$, the terms involving $\hat{v}$ are constant in each of these two sums. Take the constant out and apply the assumptions. This will give the orthonormality of the set $\{R_{2^\ell k} f_\ell\}_{k=0}^{(N/2^\ell)-1}$. Similar arguments deal with the remaining cases.

3.2.8. Suppose X is an inner product space, with subspaces U, V, and W. Suppose $U \perp V, U \perp W$, and $U \oplus V = U \oplus W$. Prove that $V = W$.

3.2.9. Suppose N is divisible by 2^p, and

$$\{R_{2k} f_1\}_{k=0}^{(N/2)-1} \cup \{R_{4k} f_2\}_{k=0}^{(N/4)-1} \cup \cdots \{R_{2^p k} f_p\}_{k=0}^{(N/2^p)-1} \cup \{R_{2^p k} g_p\}_{k=0}^{(N/2^p)-1}$$

is a wavelet basis for $\ell^2(\mathbb{Z}_N)$. Define

$$W_{-\ell} = \text{span}\{R_{2^\ell k} f_\ell\}_{k=0}^{(N/2^\ell)-1}$$

for $\ell = 1, 2, \ldots, p$ and

$$V_{-p} = \text{span}\{R_{2^p k} g_p\}_{k=0}^{(N/2^p)-1}.$$

Also define, for $\ell = 1, 2, \ldots, p-1$,

$$V_{-\ell} = W_{-\ell-1} \oplus W_{-\ell-2} \oplus \cdots \oplus W_{-p} \oplus V_{-p}.$$

Prove inductively that there exist g_ℓ for $\ell = 1, 2, \ldots, p-1$, and u_ℓ, v_ℓ for $\ell = 1, 2, \ldots, p$, such that

i. $u_\ell, v_\ell \in \ell^2(\mathbb{Z}_{N/2}^{\ell-1})$ for each ℓ.

ii. The system matrix $A_\ell(n)$ in equation (3.30) is unitary for each n and ℓ.

iii. $f_1 = v_1, g_1 = u_1$.

iv. $f_\ell = g_{\ell-1} * U^{\ell-1}(v_\ell)$, $g_\ell = g_{\ell-1} * U^{\ell-1}(u_\ell)$ for $2 \le \ell \le p$.

v. $V_{-\ell} = \text{span}\{R_{2^\ell k} g_\ell\}_{k=0}^{(N/2^\ell)-1}$ for $\ell = 1, 2, \ldots, p-1$.

Remark: This proves the assertion in the text that every wavelet basis is obtained from a wavelet filter sequence via Recipe 3.29.

Hint: For $\ell = 1$, define $v_1 = f_1$, and obtain u_1 as in Lemma 3.12, but with u and v interchanged. Then part i holds for $\ell = 1$, as does part ii, by Theorem 3.8. Then define $g_1 = u_1$. If we set

$$X_{-1} = \text{span}\{R_{2k} g_1\}_{k=0}^{(N/2)-1},$$

then by Theorem 3.8, $W_{-1} \perp X_{-1}$ and

$$W_{-1} \oplus X_{-1} = \ell^2(\mathbb{Z}_N) = W_{-1} \oplus V_{-1}.$$

By Exercise 3.2.8, then, $X_{-1} = V_{-1}$, and part v holds for $\ell = 1$. Now suppose the result is true for $\ell - 1$. To prove it for ℓ, note that $f_\ell \in W_{-\ell} \subseteq V_{-\ell+1}$. So by the induction hypothesis (part v) for $\ell-1$, there exist coefficients $v_\ell(k)$, $k = 0, 1, \ldots, (N/2^{\ell-1}) - 1$ such that

$$f_\ell = \sum_{k=0}^{(N/2^{\ell-1})-1} v_\ell(k) R_{2^{\ell-1}k} g_{\ell-1}.$$

This defines $v_\ell \in \ell^2(\mathbb{Z}_{N/2}^{\ell-1})$ such that $f_\ell = g_{\ell-1} * U^{\ell-1}(v_\ell)$. From this and Exercises 3.2.1 (ii) and 3.1.11 (i), we get

$$2^\ell = \sum_{k=0}^{2^\ell-1} \left| \hat{g}_{\ell-1}\left(n + \frac{kN}{2^\ell}\right) \right|^2 \left| \hat{v}_\ell\left(n + \frac{kN}{2^\ell}\right) \right|^2.$$

Break this sum into the terms with $k = 2j$ and the terms with $k = 2j + 1$. Since $\hat{v}_\ell$ has period $2^{\ell-1}$, the term involving $\hat{v}_\ell$ is constant in each sum, with value $|\hat{v}_\ell(n)|^2$ in the first sum, and value $|\hat{v}_\ell(n + N/2^\ell)|^2$ in the second. Factoring these out and applying Exercise 3.1.11 (i) to $g_{\ell-1}$ leads to

$$|\hat{v}_\ell(n)|^2 + \left|\hat{v}_\ell\left(n + \frac{N}{2^\ell}\right)\right|^2 = 2.$$

This allows us to pick u_ℓ so that the system matrix $A_\ell(n)$ is unitary for all n, by Lemma 3.12 with u and v reversed. For $\ell < p$, define $g_\ell = g_{\ell-1} * U^{\ell-1}(u_\ell)$. Define $X_{-\ell} = \text{span}\{R_{2^\ell k}g_\ell\}_{k=0}^{(N/2^\ell)-1}$. (When $\ell = p$, g_p is already given. Then u_p can be obtained from g_p in the same way that v_ℓ was obtained from f_ℓ.) Then by Lemma 3.26, $V_{-\ell+1} = X_{-\ell} \oplus W_{-\ell}$. However, by definition, $V_{-\ell+1} = V_{-\ell} \oplus W_{-\ell}$. By Exercise 3.2.8, it follows that $X_{-\ell} = V_{-\ell}$. This completes the induction.

3.2.10. i. Suppose N is a multiple of 2^p and $1 \le \ell \le p$. Suppose $f \in \ell^2(\mathbb{Z}_N)$. Prove that there is a pth-stage wavelet basis such that $f_\ell = f$ (notation as in Definition 3.23) if and only if the set $\{R_{2^\ell k}f\}_{k=0}^{(N/2^\ell)-1}$ is orthonormal. Hint: The "only if" direction is immediate. To prove the "only if" direction, let $\Sigma_0 = \hat{f}$ and for $j = 1, \ldots, \ell$, define

$$\Sigma_j(n) = \left(\frac{1}{2^j} \sum_{k=0}^{2^j-1} \left|\hat{f}\left(n + \frac{kN}{2^j}\right)\right|^2\right)^{1/2}.$$

Note that Σ_j has period $N/2^j$. If $\{R_{2^\ell k}f\}_{k=0}^{(N/2^\ell)-1}$ is orthonormal, then by Exercise 3.1.11 (i), $\Sigma_\ell(n) = 1$ for all n. Define $u_1, \ldots, u_{\ell-1}$ and v_ℓ so that $\hat{u}_j = \Sigma_{j-1}/\Sigma_j, j = 1, 2, \ldots, \ell-1$, and $\hat{v}_\ell = \Sigma_{\ell-1}/\Sigma_\ell$, when the denominators are not 0. (Note that when the denominator is 0, the numerator must be 0 also. In this case define the fraction to be 1.) Note $\hat{u}_j$ is periodic with period $N/2^{j-1}$ and $\hat{v}_\ell$ has period $N/2^{\ell-1}$. Check that

$$|\hat{u}_j(n)|^2 + |\hat{u}_j(n + N/2^j)|^2 = 2$$

for all n, and similarly for v_ℓ. This allows us to select $v_1, v_2, \ldots, v_{\ell-1}, u_\ell$ such that the system matrix $A_j(n)$ is unitary for all $j = 1, 2, \ldots, \ell$, by Lemma 3.12. We can select any admissable filters $u_{\ell+1}, v_{\ell+1}, \ldots, u_p, v_p$ and obtain a wavelet basis as in Theorem 3.27. Note that our definitions give us, using equation (3.70) and Exercise 3.2.1 (ii), that

$$\hat{f}_\ell = \hat{u}_1 \hat{u}_2 \cdots \hat{u}_{\ell-1} \hat{v}_\ell = \frac{\Sigma_0}{\Sigma_\ell},$$

by cancellation. But $\Sigma_0 = \hat{f}$ by definition, and $\Sigma_\ell = 1$ as noted above. (Note that if $\Sigma_j(n) = 0$ for some j and n, then $\hat{f}(n) = 0$, and at least one $\Sigma_{k-1}(n)/\Sigma_k(n)$ is 0, so we still have the equality in this case.)

ii. Suppose $N = 2^n$. Prove that if $f \in \ell^2(\mathbb{Z}_N)$ has $\|f\| = 1$, then there exists an nth-stage wavelet basis such that $f_n = f$. This means that, by our rather weak definition of the term, every vector of length one is a wavelet.

3.2.11. Prove Lemma 3.30, part ii.

3.2.12. Suppose N is divisible by 2^p, and $u_1, v_1, u_2, v_2, \ldots, u_p, v_p$ is a pth-stage wavelet filter sequence such that, for each $\ell = 1, 2, \ldots, p$, both u_ℓ and v_ℓ have at most K nonzero components. Prove that components of $z \in \ell^2(\mathbb{Z}_N)$ with respect to the corresponding wavelet basis can be computed using at most $4KN$ complex multiplications. Hint: Compute everything directly, not using the DFT. Each component of each convolution requires at most K multiplications to compute because the filter has at most K nonzero entries. In the recursive filter bank structure, one needs to compute two convolutions of length N, two of length $N/2$, two of length $N/4$, down to two of length $N/2^p$.

Remark: This shows that the wavelet transform for the case of bounded filter length can be computed in at most a fixed multiple of N multiplications, which is faster than the FFT! However, comparing to Lemma 3.17, this is an advantage only if $4KN < 4N + N\log_2 N$, that is, if $K < 1 + (1/4)\log_2 N$. For example, if all the filters have at most four nonzero entries, this requires $N > 2^{12}$ before this yields an advantage. However, if all the filters are real valued and z is real valued, then all these multiplications are real multiplications, which correspond to $1/3$ of a complex multiplication (Exercise 2.3.1).

3.2.13. (Iteration in the two-dimensional case) Suppose $2^{\ell}|N_1$ and $2^{\ell}|N_2$. Suppose that for each $\ell = 1, 2, \ldots, p$, we are given $u_{\ell,0}, u_{\ell,1}, u_{\ell,2}, u_{\ell,3} \in \ell^2(\mathbb{Z}_{N_1/2^{\ell-1}} \times \mathbb{Z}_{N_2/2^{\ell-1}})$. For $\ell = 1, 2, \ldots, p$, define $A_{\ell}(n_1, n_2)$ as in Exercise 3.1.12 (ii) except with $u_{\ell,m}$ in place of u_m, $m = 0, 1, 2, 3$, $N_1/2^{\ell}$ in place of N_1, and $N_2/2^{\ell}$ in place of N_2. Suppose that for each ℓ, n_1, and n_2, $A_{\ell}(n_1, n_2)$ is unitary. (Such a sequence of filters is called a two-dimensional wavelet filter sequence). Define $f_{1,0} = u_{1,0}, f_{1,1} = u_{1,1}, f_{1,2} = u_{1,2}$, and $g_1 = u_{1,3}$. Inductively define $f_{\ell,j} = g_{\ell-1} * U^{\ell-1}(u_{\ell,j})$, for $j = 0, 1, 2$ and $g_{\ell} = g_{\ell-1} * U^{\ell-1}(u_{\ell,4})$. Define, for $\ell = 1, 2, \ldots, p$,

$$B_{\ell} = \bigcup_{j=0}^{2} \{R_{2^{\ell}k_1, 2^{\ell}k_2} f_{\ell,j}\}_{k_1=0,1,\ldots,(N_1/2^{\ell})-1, k_2=0,1,\ldots,(N_2/2^{\ell})-1}.$$

Also define

$$C_p = \{R_{2^p k_1, 2^p k_2} g_p\}_{k_1=0,1,\ldots,(N_1/2^p)-1, k_2=0,1,\ldots,(N_2/2^p)-1}.$$

Prove that

$$B_1 \cup B_2 \cup \cdots \cup B_p \cup C_p$$

is an orthonormal basis for $\ell^2(\mathbb{Z}_{N_1} \times \mathbb{Z}_{N_2})$. Such a basis is called a two-dimensional (discrete) wavelet basis.

3.2.14. (Folding lemma in two-dimensions)

i. Suppose N_1 and N_2 are even, and $u_1 \in \ell^2(\mathbb{Z}_{N_1} \times \mathbb{Z}_{N_2})$. Define $u_2 \in \ell^2(\mathbb{Z}_{N_1/2} \times \mathbb{Z}_{N_2/2})$ by

$$u_2(n_1, n_2) = u_1(n_1, n_2) + u_1\left(n_1 + \frac{N_1}{2}, n_2\right)$$

$$+u_1\left(n_1, n_2 + \frac{N_2}{2}\right)$$

$$+u_1\left(n_1 + \frac{N_1}{2}, n_2 + \frac{N_2}{2}\right).$$

More generally, suppose $2^p|N_1$ and $2^p|N_2$. Define $u_\ell \in \ell^2(\mathbb{Z}_{N_1/2^{\ell-1}} \times \mathbb{Z}_{N_2/2^{\ell-1}})$ for $1 \le \ell \le p$ by

$$u_\ell(n_1, n_2) = \sum_{k_1=0}^{2^{\ell-1}-1} \sum_{k_2=0}^{2^{\ell-1}-1} u_1\left(n_1 + \frac{k_1 N_1}{2^{\ell-1}}, n_2 + \frac{k_2 N_2}{2^{\ell-1}}\right).$$

Prove that

$$\hat{u}_\ell(m_1, m_2) = \hat{u}_1(2^{\ell-1}m_1, 2^{\ell-1}m_2),$$

for all $\ell = 1, 2, \ldots, p$.

ii. Suppose $2^p|N_1$ and $2^p|N_2$. Suppose $u_{1,0}, u_{1,1}, u_{1,2}$, and $u_{1,3}$ are such that the matrix $A_1(n_1, n_2)$ (defined as in Exercise 3.2.13) is unitary for all $(n_1, n_2) \in \mathbb{Z}_{N_1} \times \mathbb{Z}_{N_2}$. For $\ell = 2, \ldots, p$, define $u_{\ell,0}, u_{\ell,1}, u_{\ell,2}$, and $u_{\ell,3}$ as in part i but with $u_{\ell j}$ in place of u_ℓ, for each $j = 0, 1, 2, 3$. Prove that $A_\ell(n_1, n_2)$ is unitary for all ℓ, n_1, and n_2; that is, the resulting $u_{\ell j}$ for $j = 0, 1, 2, 3$ and $\ell = 1, 2, \ldots, p$ form a two-dimensional wavelet filter sequence.

3.3 Examples and Applications

We summarize our algorithm for constructing wavelet bases for $\ell^2(\mathbb{Z}_N)$. Suppose N is divisible by 2^p. We begin with a wavelet filter sequence $u_1, v_1, u_2, v_2, \ldots, u_p, v_p$, that is, (Definition 3.16) a sequence such that for each $\ell = 1, 2, \ldots, p$,

$$u_\ell, v_\ell \in \ell^2(\mathbb{Z}_{N/2^{\ell-1}}),$$

and the system matrix

$$A_\ell(n) = \frac{1}{\sqrt{2}} \begin{bmatrix} \hat{u}_\ell(n) & \hat{v}_\ell(n) \\ \hat{u}_\ell\left(n + \frac{N}{2^\ell}\right) & \hat{v}_\ell\left(n + \frac{N}{2^\ell}\right) \end{bmatrix}$$

is unitary for $n = 0, 1, \ldots, (N/2^\ell) - 1$ (equivalently, for all n). If we have $u_1, v_1 \in \ell^2(\mathbb{Z}_N)$ such that $A_1(n)$ is unitary for all n, we can obtain a wavelet filter sequence with repeated filters by defining

$$u_\ell(n) = \sum_{k=0}^{2^{\ell-1}-1} u_1\left(n + \frac{kN}{2^{\ell-1}}\right) \quad \text{and} \quad v_\ell(n) = \sum_{k=0}^{2^{\ell-1}-1} v_1\left(n + \frac{kN}{2^{\ell-1}}\right).$$
$$(3.82)$$

We then define (Definition 3.20) $f_1 = v_1$, $g_1 = u_1$, and, by induction, for $2 \le \ell \le p$,

$$f_\ell = g_{\ell-1} * U^{\ell-1}(v_\ell), \quad \text{and} \quad g_\ell = g_{\ell-1} * U^{\ell-1}(u_\ell). \qquad (3.83)$$

More explicitly, this gives equations (3.43) and (3.44). We define

$$\psi_{-j,k} = R_{2^j k} f_j \quad \text{and} \quad \varphi_{-j,k} = R_{2^j k} g_j, \qquad (3.84)$$

for $j = 1, 2, \ldots, p$. Then

$$\{\psi_{-1,k}\}_{k=0}^{(N/2)-1} \cup \{\psi_{-2,k}\}_{k=0}^{(N/4)-1} \cup \cdots \cup \{\psi_{-p,k}\}_{k=0}^{(N/2^p)-1} \cup \{\varphi_{-p,k}\}_{k=0}^{(N/2^p)-1}$$

is an orthonormal basis for $\ell^2(\mathbb{Z}_N)$, called a pth-stage wavelet basis.

Notice that if we have a pth-stage wavelet basis and $1 \le j \le p$, then

$$\left(\bigcup_{l=1}^{j} \{\psi_{-\ell,k}\}_{0 \le k \le (N/2^\ell)-1}\right) \cup \{\varphi_{-j,k}\}_{0 \le k \le (N/2^j)-1} \qquad (3.85)$$

forms a jth-stage wavelet basis. In fact this basis is exactly what we would obtain from the jth-stage wavelet filter sequence $u_1, v_1, \ldots, u_j, v_j$ by the algorithm given above.

Suppose $j \in \{1, 2, \ldots, p\}$. Observe that

$$\{\varphi_{-j,k}\}_{k=0}^{(N/2^j)-1} = \{R_{2^j k} g_j\}_{k=0}^{(N/2^j)-1}$$

is an orthonormal basis for V_{-j} (orthonormality follows because the set (3.85) is orthonormal, and $V_{-j} = \text{span}\{R_{2^j k} g_j\}_{k=0}^{(N/2^j)-1}$ by definition). Hence the orthogonal projection $P_{-j}(z)$ of $z \in \ell^2(\mathbb{Z}_N)$ onto V_{-j} (see Definition 1.97) is

$$P_{-j}(z) = \sum_{k=0}^{(N/2^j)-1} \langle z, \varphi_{-j,k} \rangle \varphi_{-j,k}. \qquad (3.86)$$

We call $P_{-j}(z)$ the *partial reconstruction at level* $-j$ of z. It represents an approximation to z using only $N/2^j$ terms in the full expansion of z in terms of a jth-stage wavelet basis. As we would expect, this approximation is more and more coarse for larger values of j. We think of $P_{-j}(z)$ as the approximation to z "at level $-j$."

The orthogonal projection $Q_{-j}(z)$ of z onto

$$W_{-j} = \text{span}\{\psi_{-j,k}\}_{k=0}^{(N/2^j)-1} = \text{span}\{R_{2^j k} g_j\}_{k=0}^{(N/2^j)-1}$$

is defined by

$$Q_{-j}(z) = \sum_{k=0}^{(N/2^j)-1} \langle z, \psi_{-j,k} \rangle \psi_{-j,k}. \tag{3.87}$$

Recall (Lemma 3.26) that $V_{-j} \oplus W_{-j} = V_{-j+1}$, so $\{\varphi_{-j,k}\}_{k=0}^{(N/2^j)-1} \cup \{\psi_{-j,k}\}_{k=0}^{(N/2^j)-1}$ is an orthonormal basis for V_{-j+1} (as is $\{\varphi_{-j+1,k}\}_{k=0}^{(N/2^{j-1})-1}$); hence,

$$P_{-j+1}(z) = \sum_{k=0}^{(N/2^j)-1} \langle z, \varphi_{-j,k} \rangle \varphi_{-j,k} + \sum_{k=0}^{(N/2^j)-1} \langle z, \psi_{-j,k} \rangle \psi_{-j,k},$$

for any $z \in \ell^2(\mathbb{Z}_N)$. Therefore

$$P_{-j+1}(z) = P_{-j}(z) + Q_{-j}(z), \tag{3.88}$$

for $j = 2, 3, \ldots, p$. For convenience, set $V_0 = \ell^2(\mathbb{Z}_N)$ and let P_0 be the identity operator (i.e., $P_0(z) = z$ for all z). Then equation (3.88) is true even for $j = 1$, since $\{\varphi_{-1,k}\}_{k=0}^{N/2-1} \cup \{\psi_{-1,k}\}_{k=0}^{N/2-1} = \{R_{2k}v_1\}_{k=0}^{N/2-1} \cup \{R_{2k}u_1\}_{k=0}^{N/2-1}$ is an orthonormal basis for $\ell^2(\mathbb{Z}_N)$, by Theorem 3.8. We regard $Q_{-j}(z)$ as containing the "details at level $-j+1$" needed to pass from $P_{-j}(z)$, the level $-j$ approximation of z, to $P_{-j+1}(z)$, the level $-j+1$ approximation.

Let

$$R_{-j}(z) = \sum_{\ell=1}^{j} Q_{-\ell}(z).$$

Then applying equation (3.88) inductively, we obtain $z = P_0(z) = P_{-1}(z) + Q_{-1}(z) = P_{-2}(z) + Q_{-2}(z) + Q_{-1}(z)$ and so on, until

$$z = P_{-j}(z) + \sum_{\ell=1}^{j} Q_{-\ell}(z) = P_{-j}(z) + R_{-j}(z),$$

for each $j = 1, \ldots, p$ (this can also be seen from the fact that the set (3.85) is an orthonormal basis for $\ell^2(\mathbb{Z}_N)$). Thus $R_{-j}(z)$ is the error made in approximating z by $P_{-j}(z)$.

In this section, we first consider some examples of wavelet bases. Later we describe some basic compression examples.

Example 3.32

(The Haar system) In Exercise 3.1.2, we considered the first-stage Haar system. Here we consider the general p^{th}-stage Haar system, obtained from the first stage by the procedure just described, using repeated filters. Suppose N is divisible by 2^p. Define

$$u_1 = \left(\frac{1}{\sqrt{2}}, \frac{1}{\sqrt{2}}, 0, 0, \ldots, 0 \right)$$

and

$$v_1 = \left(\frac{1}{\sqrt{2}}, -\frac{1}{\sqrt{2}}, 0, 0, \ldots, 0 \right).$$

We saw in Exercise 3.1.2, that u_1, v_1 form a first-stage wavelet basis. Define u_ℓ, v_ℓ by equation (3.82) for $2 \leq \ell \leq p$. We can show (Exercise 3.3.1 (i)) that

$$u_\ell(0) = \frac{1}{\sqrt{2}}, u_\ell(1) = \frac{1}{\sqrt{2}}, \text{ and } u_\ell(n) = 0 \text{ for } 2 \leq n \leq \left(\frac{N}{2^{\ell-1}} \right) - 1,$$
$$(3.89)$$

$$v_\ell(0) = \frac{1}{\sqrt{2}}, v_\ell(1) = -\frac{1}{\sqrt{2}}, \text{ and } v_\ell(n) = 0 \text{ for } 2 \leq n \leq \left(\frac{N}{2^{\ell-1}} \right) - 1.$$
$$(3.90)$$

Then equations (3.83), (3.89), (3.90), and an induction argument (Exercise 3.3.1 (iii)) lead to

$$f_\ell(n) = \begin{cases} 2^{-\ell/2}, & n = 0, 1, \ldots, 2^{\ell-1} - 1 \\ -2^{-\ell/2}, & n = 2^{\ell-1}, 2^{\ell-1} + 1, \ldots, 2^\ell - 1 \\ 0, & n = 2^\ell, 2^\ell + 1, \ldots, N - 1 \end{cases} \qquad (3.91)$$

and

$$g_\ell(n) = \begin{cases} 2^{-\ell/2}, & n = 0, 1, \ldots, 2^\ell - 1 \\ 0, & n = 2^\ell, 2^\ell + 1, \ldots, N - 1, \end{cases} \qquad (3.92)$$

for $\ell = 1, 2, \ldots, p$. Hence (Exercise 3.3.1 (iv)) for $k = 0, 1, \ldots, (N/2^\ell) - 1$,

$$\psi_{-\ell,k}(n) = \begin{cases} 2^{-\ell/2}, & n = 2^\ell k, 2^\ell k + 1, \ldots, 2^\ell k + 2^{\ell-1} - 1 \\ -2^{-\ell/2}, & n = 2^\ell k + 2^{\ell-1}, 2^\ell k + 2^{\ell-1} + 1, \ldots, 2^\ell k + 2^\ell - 1 \\ 0, & n = 0, 1, \ldots, 2^\ell k - 1; 2^\ell k + 2^\ell, \ldots, N - 1 \end{cases}$$
(3.93)

and

$$\varphi_{-\ell,k}(n) = \begin{cases} 2^{-\ell/2}, & n = 2^\ell k, 2^\ell k + 1, \ldots, 2^\ell k + 2^\ell - 1 \\ 0, & n = 0, 1, \ldots, 2^\ell k - 1; 2^\ell k + 2^\ell, 2^\ell + 1, \ldots, N - 1. \end{cases}$$
(3.94)

For the Haar system, the reconstructions at level ℓ have a simple interpretation. One can show (Exercise 3.3.1 (v)) that for $2^\ell k \leq n \leq 2^\ell k + 2^\ell - 1$,

$$P_{-\ell}(z)(n) = \frac{1}{2^\ell} \left[z(2^\ell k) + z(2^\ell k + 1) + \cdots + z(2^\ell k + 2^\ell - 1) \right].$$
(3.95)

In other words, $P_{-\ell}(z)$ is obtained from z by replacing the 2^ℓ consecutive values of z on the segment $n = 2^\ell k, 2^\ell k + 1, \ldots, 2^\ell k + 2^\ell - 1$ by their average. We regard $P_{-\ell}(z)$ as z seen at a resolution of 2^ℓ. By equation (3.88), $Q_{-\ell}$ contains the information needed to upgrade the approximation from a resolution of 2^ℓ to one of $2^{\ell-1}$.

Example 3.33
(Shannon wavelets) In Example 3.10, we considered the first-stage Shannon basis. Calling the u and v there u_1 and v_1, respectively, we define $u_\ell, v_\ell \in \ell^2(\mathbb{Z}_{N/2}^{\ell-1})$ by equation (3.82). Then equation (3.79) and its analog for $\hat{v}_\ell$ imply that

$$\hat{v}_\ell(n) = \begin{cases} \sqrt{2}, & \frac{N}{2^{\ell+1}} \leq n \leq \frac{3N}{2^{\ell+1}} - 1 \\ 0, & 0 \leq n \leq \frac{N}{2^{\ell+1}} - 1; \frac{3N}{2^{\ell+1}} \leq n \leq \frac{N}{2^{\ell-1}} - 1 \end{cases}$$

and

$$\hat{u}_\ell(n) = \begin{cases} \sqrt{2}, & 0 \leq n \leq \frac{N}{2^{\ell+1}} - 1; \frac{3N}{2^{\ell+1}} \leq n \leq \frac{N}{2^{\ell-1}} - 1 \\ 0, & \frac{N}{2^{\ell+1}} \leq n \leq \frac{3N}{2^{\ell+1}} - 1. \end{cases}$$

Since $\hat{u}_\ell$ and $\hat{v}_\ell$ have period $N/2^{\ell-1}$, these formulae define them for all n. Define f_ℓ and g_ℓ by equation (3.83), and $\psi_{-j,k}$ and $\varphi_{-j,k}$ by equation (3.84), for $j = 1, 2, \ldots, p$ (where N is divisible by 2^{p+1}). Then $\hat{\psi}_{-1,0} = \hat{v}_1$ and $\hat{\varphi}_{-1,0} = \hat{u}_1$. For $\ell \geq 2$, Exercise 3.2.1 (ii) leads

(Exercise 3.3.2) to

$$
\hat{\psi}_{-\ell,0}(n) = \begin{cases} 2^{\ell/2}, & \frac{N}{2^{\ell+1}} \le n \le \frac{N}{2^\ell} - 1; \\ & N - \frac{N}{2^\ell} \le n \le N - \frac{N}{2^{\ell+1}} - 1 \\ 0, & 0 \le n \le \frac{N}{2^{\ell+1}} - 1; \frac{N}{2^\ell} \le n \le N - \frac{N}{2^{\ell-1}}; \\ & N - \frac{N}{2^{\ell+1}} \le n \le N - 1 \end{cases} \tag{3.96}
$$

and

$$
\hat{\varphi}_{-\ell,0}(n) = \begin{cases} 2^{\ell/2}, & 0 \le n \le \frac{N}{2^{\ell+1}} - 1; N - \frac{N}{2^{\ell+1}} \le n \le N - 1 \\ 0, & \frac{N}{2^{\ell+1}} \le n \le N - \frac{N}{2^{\ell+1}} - 1. \end{cases} \tag{3.97}
$$

Thus $\hat{\psi}_{-1,0}$ is nonzero only on the highest $N/2$ frequencies, $\hat{\psi}_{-2,0}$ is nonzero only on the next highest $N/4$ frequencies, and so on, down to $\hat{\varphi}_{-p,0}$, which is 0 except at the lowest $N/2^p$ frequencies. By Lemma 2.13, similar remarks hold for the translates $\psi_{-j,k}$ and $\varphi_{-p,k}$. Thus the partial reconstruction $P_{-p}(z)$ consists exactly of the lowest $N/2^p$ frequencies of z, $P_{-p+1}(z)$ the lowest $N/2^{p-1}$ frequencies, and so on. In this case the partial reconstructions give a filtering out of the high frequencies, to varying degrees.

Example 3.34

(Real Shannon wavelets) In Example 3.11 we constructed the first-stage real Shannon wavelet basis. It was a minor modification of the first stage Shannon basis that had the advantage that the basis vectors were all real valued. By applying the iteration procedure previously described to this basis, we can obtain a real-valued p^{th}-stage wavelet basis with structure very similar to the Shannon basis. Let u_1 and v_1 be the u and v of Example 3.11. Define u_ℓ and v_ℓ by equation (3.82), f_ℓ and g_ℓ by equation (3.83), and $\psi_{-j,k}$ and $\varphi_{-j,k}$ by equation (3.84). Then $\hat{\psi}_{-1,0} = \hat{v}$ and $\hat{\varphi}_{-1,0} = \hat{u}$ are as given in Example 3.11. For $\ell > 1$, a calculation (Exercise 3.3.3) similar to the one for the

Shannon wavelets in Example 3.33 shows that

$$
\hat{\psi}_{-\ell,0}(n) = \begin{cases}
2^{(\ell-1)/2}, & n = \frac{N}{2^{\ell+1}}, N - \frac{N}{2^{\ell+1}} \\
2^{\ell/2}, & \frac{N}{2^{\ell+1}} + 1 \leq n \leq \frac{N}{2^{\ell}} - 1; \\
& N - \frac{N}{2^{\ell}} + 1 \leq n \leq N - \frac{N}{2^{\ell+1}} - 1 \\
2^{(\ell-1)/2}i, & n = \frac{N}{2^{\ell}} \\
-2^{(\ell-1)/2}i, & n = N - \frac{N}{2^{\ell}} \\
0, & 0 \leq n \leq \frac{N}{2^{\ell+1}} - 1; \\
& \frac{N}{2^{\ell}} + 1 \leq n \leq N - \frac{N}{2^{\ell}} - 1; \\
& N - \frac{N}{2^{\ell+1}} + 1 \leq n \leq N - 1
\end{cases}
\tag{3.98}
$$

and

$$
\hat{\varphi}_{-\ell,0}(n) = \begin{cases}
2^{\ell/2}, & 0 \leq n \leq \frac{N}{2^{\ell+1}} - 1; \\
& N - \frac{N}{2^{\ell+1}} + 1 \leq n \leq N - 1 \\
2^{(\ell-1)/2}i, & n = \frac{N}{2^{\ell+1}} \\
-2^{(\ell-1)/2}i, & n = N - \frac{N}{2^{\ell+1}} \\
0, & \frac{N}{2^{\ell+1}} + 1 \leq n \leq N - \frac{N}{2^{\ell+1}} - 1.
\end{cases}
\tag{3.99}
$$

Figure 22 shows the graphs of a few 4th level real Shannon wavelet basis functions $\psi_{-j,k}$, $j = 1, 2, 3, 4$, and $\varphi_{-4,k}$, for $N = 512$. At each level, the translation parameter k has been picked so that the basis function is centered in the middle of the range. Note how well spatially localized the high-frequency basis function $\psi_{-1,128}$ is (Figure 22a). Thereafter $\psi_{-2,64}$ is relatively localized (Figure 22b), $\psi_{-3,32}$ less localized (Figure 22c), and $\psi_{-4,16}$ (Figure 22d) and $\varphi_{-4,16}$ (Figure 22e) even less so. Note that the 4th-level wavelet basis consists of 256 translates by 2 of $\psi_{-1,0}$; 128 translates by 4 of $\psi_{-2,0}$; 64 translates by 8 of $\psi_{-3,0}$; 32 translates by 16 of $\psi_{-4,0}$; and 32 translates by 16 of $\varphi_{-4,0}$. In Figure 22f, we have plotted the magnitude of the DFT of the first-generation wavelets $\psi_{-1,k}$ (because they are related to each other by translation, their DFTs have the same magnitude at every point, by Lemma 2.13). We already knew that the graph should look like this, by the definition of $\hat{v} = \hat{\psi}_{-1,0}$ in Example 3.11, but we have included the graph for comparison with Example 3.35.

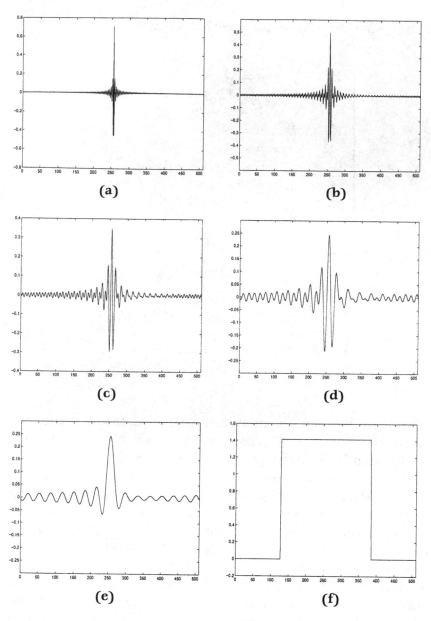

FIGURE 22   **(a)** $\psi_{-1,128}$,   **(b)** $\psi_{-2,64}$,   **(c)** $\psi_{-3,32}$,   **(d)** $\psi_{-4,16}$,   **(e)** $\varphi_{-4,16}$, **(f)** $|\hat{\psi}_{-1,k}|$

Figure 23 shows an example of a signal z and the magnitude of its DFT. The signal (Figure 23a) is

$$z(n) = \begin{cases} 0, & 0 \leq n \leq 127 \\ \sin\left(\frac{|n-128|^{1.7}}{128}\right), & 128 \leq n \leq 255 \\ 0, & 256 \leq n \leq 383 \\ \sin\left(\frac{|n-128|^{2}}{128}\right), & 384 \leq n \leq 447 \\ 0, & 448 \leq n \leq 511. \end{cases} \qquad (3.100)$$

For the first interval in which z is not identically 0 (n such that $128 \leq n \leq 255$), the rate of oscillation of z is increasing as we move from left to right. Also, the oscillation of z is more rapid in the second nontrivial interval (n such that $384 \leq n \leq 447$). So we expect that there is a substantial range of frequencies present in z. This is reflected in the magnitude of the DFT of z (Figure 23b): the two outer humps show the existence of a low-frequency component (from the first interval, although the magnitude of the DFT does not tell us which portion of the signal contributes this low-frequency part), and the two inner humps show that there is a higher frequency component (from the second interval).

Figure 23 should be compared with the graphs in Figure 24, which show the 4th level real Shannon wavelet coefficients of z in equation (3.100) (z is plotted again in Figure 24a). In Figure 24f, we plot the value of $\langle z, \varphi_{-4,k} \rangle$ at the point $2^4 k$, since $\varphi_{-4,k} = R_{2^4 k} g_4$ is centered around the point $2^4 k$. This graph shows the coefficients of the lowest frequency wavelets; note that we get fairly large values near the left portion of the first interval $128 \leq n \leq 255$, where the lower frequency part of z occurs. Figures 24 b–e show the other levels of wavelet coefficients, each time with the value of the wavelet coefficient $\langle z, \psi_{-j,k} \rangle$ plotted above the point $2^j k$ where the corresponding wavelet basis vector is centered. In Figure 24e (plotting $\langle z, \psi_{-4,k} \rangle$), we see that the next higher frequency wavelet coefficients are still mostly concentrated on the left portion of the first interval. Figure 24d shows somewhat higher frequency coefficients $\langle z, \psi_{-3,k} \rangle$, which are concentrated more on the right half of the interval containing the lower frequency component. This is due to the increase in the frequency of z from the left to right of the first interval. The two highest frequency sets of coefficients $\langle z, \psi_{-1,k} \rangle$

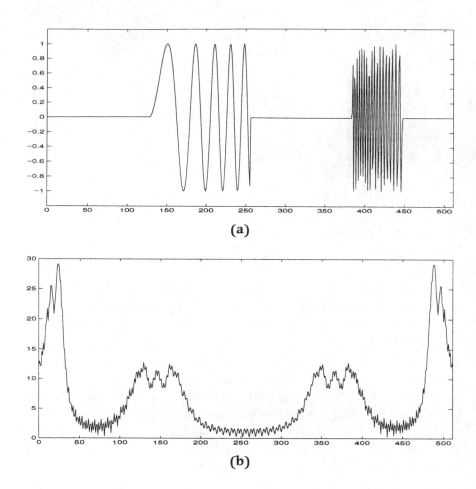

FIGURE 23   **(a)** z,   **(b)** Magnitude of $\hat{z}$

and $\langle z, \psi_{-2,k} \rangle$ in Figures 24b and 24c are mostly large on or near the second interval $384 \le n \le 511$. The exception is that there are a few large high-frequency wavelet coefficients near the endpoint $n = 255$ of the first interval; this is due to the more rapid jump of z there due to the cutoff of the graph at 255. The main point is that the wavelet coefficients give a sense not only of which frequency levels make up the signal, but which locations in the signal require the different frequencies. Thus the wavelet coefficients give a simultaneous space and frequency analysis of a signal.

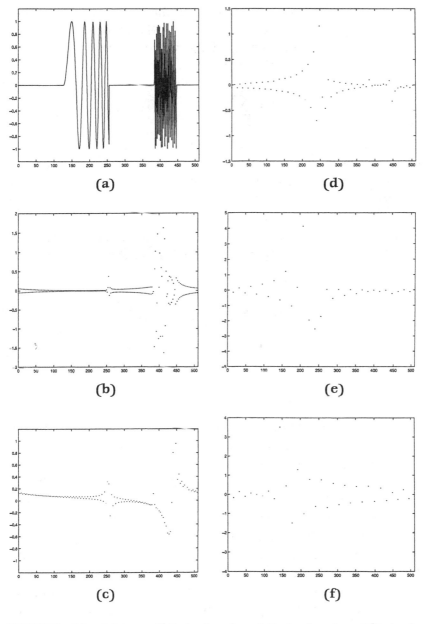

FIGURE 24 **(a)** z, **(b)** $\langle z, \psi_{-1,k} \rangle$, **(c)** $\langle z, \psi_{-2,k} \rangle$, **(d)** $\langle z, \psi_{-3,k} \rangle$, **(e)** $\langle z, \psi_{-4,k} \rangle$, **(f)** $\langle z, \varphi_{-4,k} \rangle$

The role of the wavelet coefficients at different levels is further clarified by Figure 25, which shows the partial reconstructions, in the real Shannon wavelet basis, of the vector z from Figure 24a. Figure 25a is the graph of the vector

$$P_{-4}(z) = \sum_{k=0}^{31} \langle z, \varphi_{-4,k} \rangle \, \varphi_{-4,k},$$

as in equation (3.86). Thus Figure 25a exhibits the portion of the wavelet expansion of z corresponding to the coefficients plotted in Figure 24f. Roughly speaking, these are the lowest frequency terms in the wavelet expansion. This is why the portion of the vector z that is best reconstructed by this part of the expansion is the portion from about $n = 128$ to about $n = 170$, where z is oscillating relatively slowly. The partial expansion corresponding to the coefficients in Figure 24e is

$$Q_{-4}(z) = \sum_{k=0}^{31} \langle z, \psi_{-4,k} \rangle \, \psi_{-4,k},$$

as defined in equation (3.87). By equation (3.88), the next level projection is

$$P_{-3}(z) = Q_{-4}(z) + P_{-4}(z).$$

The vector $P_{-3}(z)$ is plotted in Figure 25b; it is made up of the terms in the wavelet expansion of Z corresponding to the coefficients in Figures 24e and 24f. Thus it contains the information in the two lowest frequency levels of the wavelet expansion. This is exhibited by the picture, as the next lowest level of oscillation of the vector z begins to show up. In Figure 25c, we plot $P_{-2}(z)$, corresponding to the coefficients in Figures 24d–f, which includes the next lowest frequency level. Figure 25d shows $P_{-1}(z)$, including all but the terms coming from the highest frequency wavelet coefficients in Figure 24b. Note that the more slowly oscillating segment of z (for $128 \leq n \leq 255$) is relatively faithfully reproduced, except for the right edge (points near $n = 255$), where the cutoff requires higher frequency terms for synthesis. Somewhat unexpectedly, $P_{-1}(z)$ does a better job of approximating z near the right endpoint of the region $384 \leq n \leq 447$ than near the left endpoint. This is similar to

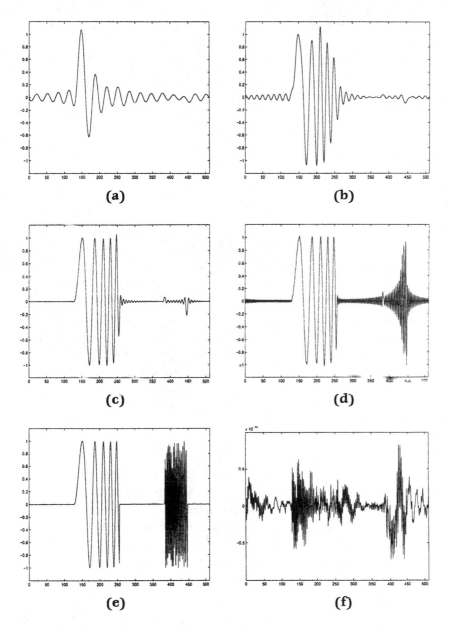

FIGURE 25 (a) $P_{-4}(z)$, (b) $P_{-3}(z)$, (c) $P_{-2}(z)$, (d) $P_{-1}(z)$, (e) $P_0(z)$, (f) $E = z - P_0(z)$

the situation in Figure 6: although formula (3.100) suggests more rapid oscillation near the right endpoint, this cannot be seen on the discrete scale $n = 0, 1, \ldots, 511$. On the discrete scale, z is actually made up of lower frequency terms near $n = 447$ than near $n = 384$. Figure 25e shows $P_0(z)$, which we define as in equation (3.88) by

$$P_0(z) = Q_{-1}(z) + P_{-1}(z).$$

Of course, $P_0(z)$ is the full reconstruction of z from all its wavelet coefficients, so it should just be z. We carried out the computation of the wavelet coefficients and the partial reconstructions on MatLab, so Figure 25e is a test of our algorithm. Visually, it seems to agree with the original in Figure 24a. To be more precise, we use MatLab to plot the difference $E = z - P_0(z)$ in Figure 25f. The most important thing to notice in this picture is the symbol $\times 10^{-14}$ above the upper left corner of the box bounding the graph. It shows that the scale on the $y-$axis is plotted in multiples of 10^{-14}. Thus the error in the reconstruction is less than 10^{-14} at every point; this error is a consequence of round-off error resulting from the computer's inability to do computations to an infinite degree of accuracy. From Figure 25 we get a sense of the information regarding z that is contributed by each level of wavelet coefficients.

Example 3.35
(Daubechies's D6 wavelets on $\mathbb{Z}_N$) The real Shannon wavelet system was designed to divide the frequency scale very sharply. Thus the real Shannon wavelets have well-localized DFTs. Ingrid Daubechies constructed families of wavelets that are very well localized in space rather than in frequency. Her construction was originally done in the contexts of $\mathbb{Z}$ (see section 4.7) and $\mathbb{R}$ (see section 5.5), but here we adapt her construction to $\mathbb{Z}_N$.

We assume that N is divisible by 2^p, for some positive integer p, and that $N/2^p > 6$. Let $M = N/2$. Our goal is to construct $u \in \ell^2(\mathbb{Z}_N)$ such that u has only six nonzero components and satisfies equation (3.15). Then we apply Lemma 3.12 to find v so that $\{R_{2k}v\}_{k=0}^{M-1} \cup \{R_{2k}u\}_{k=0}^{M-1}$ is a first-stage wavelet basis for $\ell^2(\mathbb{Z}_N)$. We begin with the trivial identity

$$\left(\cos^2\left(\frac{\pi n}{N}\right) + \sin^2\left(\frac{\pi n}{N}\right)\right)^5 = 1 \quad \text{for all} \quad n.$$

Expanding this out, we have

$$\cos^{10}\left(\frac{\pi n}{N}\right) + 5\cos^8\left(\frac{\pi n}{N}\right)\sin^2\left(\frac{\pi n}{N}\right) + 10\cos^6\left(\frac{\pi n}{N}\right)\sin^4\left(\frac{\pi n}{N}\right)$$

$$+ 10\cos^4\left(\frac{\pi n}{N}\right)\sin^6\left(\frac{\pi n}{N}\right) + 5\cos^2\left(\frac{\pi n}{N}\right)\sin^8\left(\frac{\pi n}{N}\right) + \sin^{10}\left(\frac{\pi n}{N}\right) = 1.$$
(3.101)

Define

$$b(n) = \cos^{10}\left(\frac{\pi n}{N}\right) + 5\cos^8\left(\frac{\pi n}{N}\right)\sin^2\left(\frac{\pi n}{N}\right) + 10\cos^6\left(\frac{\pi n}{N}\right)\sin^4\left(\frac{\pi n}{N}\right).$$

Note that

$$\cos\left(\frac{\pi(n+M)}{N}\right) = \cos\left(\frac{\pi n}{N} + \frac{\pi}{2}\right) = -\sin\left(\frac{\pi n}{N}\right)$$

and similarly

$$\sin\left(\frac{\pi(n+M)}{N}\right) = \cos\left(\frac{\pi n}{N}\right).$$

Hence

$$b(n+M) = 10\cos^4\left(\frac{\pi n}{N}\right)\sin^6\left(\frac{\pi n}{N}\right)$$

$$+ 5\cos^2\left(\frac{\pi n}{N}\right)\sin^8\left(\frac{\pi n}{N}\right) + \sin^{10}\left(\frac{\pi n}{N}\right).$$

Thus by equation (3.101),

$$b(n) + b(n+M) = 1 \quad \text{for all} \quad n.$$

We select $u \in \ell^2(\mathbb{Z}_N)$ so that

$$|\hat{u}(n)|^2 = 2b(n). \tag{3.102}$$

Then we have

$$|\hat{u}(n)|^2 + |\hat{u}(n+M)|^2 = 2 \quad \text{for all} \quad n = 0, 1, \ldots, M-1,$$

that is, condition (3.15).

We could obtain equation (3.102) just by setting $u = (\sqrt{2b})^{\vee}$ (where $\sqrt{2b}$ is the vector whose value at n is $\sqrt{2b(n)}$), but this would not give us a vector u that is nonzero only at six points. Instead, following Strichartz (1993), we write

$$b(n) = \cos^6\left(\frac{\pi n}{N}\right)\left[\cos^4\left(\frac{\pi n}{N}\right) + 5\cos^2\left(\frac{\pi n}{N}\right)\sin^2\left(\frac{\pi n}{N}\right)\right.$$
$$\left. +10\sin^4\left(\frac{\pi n}{N}\right)\right]$$
$$= \cos^6\left(\frac{\pi n}{N}\right)\left[\left(\cos^2\left(\frac{\pi n}{N}\right) - \sqrt{10}\sin^2\left(\frac{\pi n}{N}\right)\right)^2\right.$$
$$\left. +(5 + 2\sqrt{10})\cos^2\left(\frac{\pi n}{N}\right)\sin^2\left(\frac{\pi n}{N}\right)\right].$$

Define $\hat{u} \in \ell^2(\mathbb{Z}_N)$ by

$$\hat{u}(n) = \sqrt{2}e^{-5\pi i n/N}\cos^3\left(\frac{\pi n}{N}\right)\left[\cos^2\left(\frac{\pi n}{N}\right) - \sqrt{10}\sin^2\left(\frac{\pi n}{N}\right)\right.$$
$$\left. +i\sqrt{5 + 2\sqrt{10}}\cos\left(\frac{\pi n}{N}\right)\sin\left(\frac{\pi n}{N}\right)\right].$$

Then equation (3.102) holds. By applying Euler's formula and the double angle identities, we can write

$$\hat{u}(n) = \sqrt{2}e^{-2\pi i 4n/N}e^{3\pi i n/N}\left(\frac{e^{i\pi n/N} + e^{-i\pi n/N}}{2}\right)^3$$
$$\times\left[\frac{1}{2}\left(1 + \cos\left(\frac{2\pi n}{N}\right)\right) - \frac{\sqrt{10}}{2}\left(1 - \cos\left(\frac{2\pi n}{N}\right)\right)\right.$$
$$\left. +i\frac{\sqrt{5 + 2\sqrt{10}}}{2}\sin\left(\frac{2\pi n}{N}\right)\right].$$

To simplify the notation, let

$$a = 1 - \sqrt{10}, \quad b = 1 + \sqrt{10}, \quad \text{and} \quad c = \sqrt{5 + 2\sqrt{10}}.$$

Using Euler's formulas further, we obtain

$$\hat{u}(n) = \frac{\sqrt{2}}{8}e^{-2\pi i 4n/N}(e^{2\pi i n/N} + 1)^3$$
$$\times\left[\frac{a}{2} + \frac{b}{4}(e^{2\pi i n/N} + e^{-2\pi i n/N}) + \frac{c}{4}(e^{2\pi i n/N} - e^{-2\pi i n/N})\right].$$

At this point we can see that

$$\hat{u}(n) = \sum_{k=0}^{5} u(k)e^{-2\pi i k n/N}, \tag{3.103}$$

for some numbers $u(0), u(1), \ldots, u(5)$. Multiplying out and doing the algebra gives

$$u = (u(0), u(1), u(2), u(3), u(4), u(5), 0, 0, \ldots, 0)$$
$$= \frac{\sqrt{2}}{32}(b + c, 2a + 3b + 3c, 6a + 4b + 2c,$$
$$6a + 4b - 2c, 2a + 3b - 3c, b - c, 0, 0, \ldots, 0).$$

Hence we obtain $u \in \ell^2(\mathbb{Z})$ such that equation (3.15) holds.

Define $v \in \ell^2(\mathbb{Z}_N)$ by $v(k) = (-1)^{k-1}u(1 - k)$ for all k, which agrees with equation (3.22) since u is real. By Lemma 3.12, u and v generate a first-stage wavelet basis for $\ell^2(\mathbb{Z}_N)$.

To be explicit,

$$v(0) = -u(1), v(1) = u(0), v(N - 4) = -u(5),$$
$$v(N - 3) = u(4), v(N - 2) = -u(3), v(N - 1) = u(2),$$

and $v(n) = 0$ for $2 \le n \le N - 4$. That is,

$$v = (-u(1), u(0), 0, 0, \ldots, 0, 0, -u(5), u(4), -u(3), u(2)). \tag{3.104}$$

Hence v also has only six nonzero entries.

Notice that this works for any $N > 6$, and u and v at different levels have the same form, the only difference being the number of zeros in the vectors u and v. Thus we can define $u_1, v_1 \in \ell^2(\mathbb{Z}_N)$ of this form, then $u_2, v_2 \in \ell^2(\mathbb{Z}_{N/2})$ similarly, down to $u_p, v_p \in \ell^2(\mathbb{Z}_{N/2}^p)$ (recall we assumed that $N/2^p$ is an integer larger than 6). This is what we would obtain by the folding lemma (Lemma 3.30) also. We now follow Recipe 3.29 to obtain a pth stage wavelet basis: define $f_1 = v_1, g_1 = u_1$, then define f_ℓ and g_ℓ for $2 \le \ell \le p$ by equation (3.83), and $\psi_{-j,k}, \varphi_{-p,k}$ by equation (3.84). We call the resulting orthonormal system Daubechies's *D6 wavelet basis* for $\ell^2(\mathbb{Z}_N)$, where the "6" refers to the number of nonzero components of u and v. Daubechies constructed a similar basis with $2L$ nonzero components, for each positive integer L (see Exercise 3.3.4 for the case of D2, which are the Haar wavelets, and Exercise 3.3.5 for D4). For different values of L, these wavelets have slightly different properties, as we discuss when we come back to this subject in the context of $\mathbb{R}$ in chapter 5.

The case $N = 512$ and $p = 4$ is illustrated in Figure 26. These plots should be compared with the corresponding plots for the real Shannon wavelet basis in Figure 22. In Figure 26a we have plotted the first-generation wavelet $\psi_{-1,128}$, which is the same as $R_{256}v_1$. However, since this vector has only six nonzero components, we have restricted the plot to the small interval $245 \leq n \leq 265$ containing those nonzero components. The graphs appear continuous because we have used connect-the-dots graphs in Figures 26 a–e, for consistency, because such graphs look better on a scale of 511. However, in Figure 26a we have superimposed an "x" plot to show that $\psi_{-1,128}$ has only 6 nonzero components. The second generation wavelet plotted in Figure 26b has 16 nonzero values, and the third generation wavelet in Figure 26c has 36 nonzero components. For these we have continued to restrict the graph to a small portion of the full domain $0 \leq n \leq 511$. For Figures 26d and 26e, which depict $\psi_{-4,16}$ and $\varphi_{-4,16}$, respectively, we have plotted the full domain so that the degree of localization can be clearly seen. These vectors have 76 nonzero components. By comparing with Figure 22, we see that the D6 wavelets are much more sharply localized in space than the real Shannon wavelets. On the other hand, the D6 wavelets are not as precisely localized in frequency, as we can see by comparing Figure 26f (showing the magnitude of the DFT of a first-generation D6 wavelet) with Figure 22f (a corresponding plot for a first-generation real Shannon wavelet).

In Figure 27, we plot the D6 wavelet coefficients of vector z defined previously and studied in Figure 24. The coefficients are plotted in the same order as in Figure 24. We see that the basic features of the two sets of pictures are similar; the highest frequency wavelet coefficients (the terms $\langle z, \psi_{-1,n} \rangle$) are largest near the most rapidly oscillating portion of z, and as we go down in level (and hence in frequency) the wavelet coefficients are sensitive to the more slowly oscillating portions of z. Thus the two different wavelet systems (real Shannon and D6), although constructed on very different principles, give roughly the same information. The main

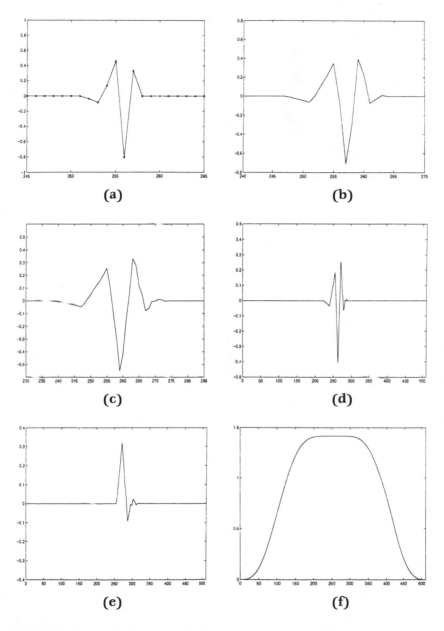

FIGURE 26 (a) $\psi_{-1,128}$, (b) $\psi_{-2,64}$, (c) $\psi_{-3,32}$, (d) $\psi_{-4,16}$, (e) $\varphi_{-4,16}$, (f) $|\hat{\psi}_{-1,k}|$

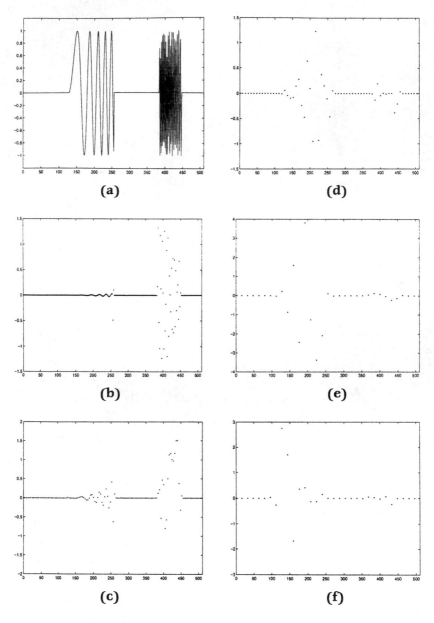

FIGURE 27   **(a)** z,   **(b)** $\langle z, \psi_{-1,k} \rangle$,   **(c)** $\langle z, \psi_{-2,k} \rangle$,   **(d)** $\langle z, \psi_{-3,k} \rangle$,
(e) $\langle z, \psi_{-4,k} \rangle$,   **(f)** $\langle z, \varphi_{-4,k} \rangle$

difference in the two sets of graphs is that many of the D6 wavelet coefficients are 0, instead of just being small as in the case of the real Shannon basis. This is because the vector z has large intervals in which it is 0. If we have a wavelet ψ such that all its nonzero coefficients occur in an interval where z is identically 0, then the wavelet coefficient

$$\langle z, \psi_{j,k} \rangle = \sum_n z(n) \overline{\psi_{j,k}(n)}$$

will be 0, because at every n, either $z(n) = 0$ or $\psi_{j,k}(n) = 0$. This is the advantage of having such a highly localized basis.

In Figure 28a, we have plotted the vector

$$z(n) = \begin{cases} 1 - \frac{n}{64} & 0 \leq n \leq 63 \\ 0, & 64 \leq n \leq 255 \\ 5 - \frac{n}{64} & 256 \leq n \leq 319 \\ 0, & 320 \leq n \leq 511. \end{cases}$$

In Figures 28b–f, we have plotted the D6 wavelet coefficients of z at different levels. Note that the highest frequency wavelet coefficients (corresponding to $\psi_{-1,k}, 0 \leq k \leq 255$) have only a few significantly large values, corresponding to the edges of the two spikes in z. As we move down to lower frequency wavelets (those with more nonzero components) we see more significant values in a slightly larger region than just where z is nonzero. There are two important observations to make regarding this example. The first is that in Figures 28c–f, we see nonzero wavelet coefficients near the right edge of the picture (near $n = 511$), despite the fact that the last nonzero component of z occurs at $n = 319$. This is because we are working on $\mathbb{Z}_N$, where every vector should be regarded as periodic with period N. Thus we should regard z as having another spike just to the right of the picture, which is the reason the wavelets located near the right edge yield nonzero coefficients. This will no longer be the case when we consider wavelets on $\mathbb{Z}$ in chapter 4. The second major observation regarding Figure 28 is that very few of the D6 wavelet coefficients are significantly large. Thus it should not take very many terms in the D6 wavelet expansion to approximate z very accurately. The essential information about z can be contained in

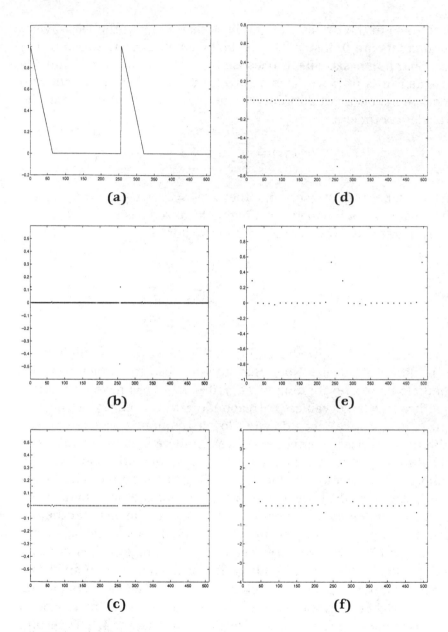

FIGURE 28 **(a)** z, **(b)** $\langle z, \psi_{-1,k} \rangle$, **(c)** $\langle z, \psi_{-2,k} \rangle$, **(d)** $\langle z, \psi_{-3,k} \rangle$, **(e)** $\langle z, \psi_{-4,k} \rangle$, **(f)** $\langle z, \varphi_{-4,k} \rangle$

only a few D6 wavelet coefficient values. This leads to one of the major applications of wavelets, namely *compression*.

Figures 29–34 show the results of some basic compression experiments. In each figure, the top left graph is a signal z, which belongs to $\ell^2(\mathbb{Z}_{512})$ and hence is represented completely by 512 numbers. For any orthonormal basis $B = \{v_j\}_{j=0}^{511}$, we have

$$z = \sum_{j=0}^{511} \langle z, v_j \rangle \, v_j. \tag{3.105}$$

We select some number K, the number of terms we will use for our compressed approximation to z; for example, in Figure 29, $K = 20$. We sort the coefficients $\langle z, v_j \rangle$ in order of magnitude, and use only the terms in the expansion corresponding to the K largest ones. In other words, if S is the set consisting of the K values of j for which $|\langle z, v_j \rangle|$ is the largest, we approximate z by

$$w = \sum_{j \in S} \langle z, v_j \rangle \, v_j. \tag{3.106}$$

(For the case of "ties," where more than one coefficient has the borderline value, we arbitrary choose to use the terms with highest index until the correct number K of terms is obtained. A different choice from among the ties would yield a different picture, but the same relative error, as described below.) This is the best approximation possible using only K of the terms in equation (3.105).

Results of this compression procedure are plotted for four different bases in Figures 29–34. The four bases used are the Fourier basis, the real Shannon basis, Daubechies's D6 basis, and the Euclidean basis. For the Euclidean basis, this compression just amounts to zeroing out all values below the K largest (in absolute value). For the Fourier basis, it is possible for the approximation w to have nonzero imaginary part, but since we have selected the original vector z to be real, the imaginary part is part of the error in w. So we have plotted only the real part of the approximation w for the Fourier basis.

We define the *error* made in approximating z by w as $\|z - w\|$. We define the *relative error* as $\|z - w\| / \|z\|$. This adjusts for the size of z, giving an absolute scale that can be used to compare the compression

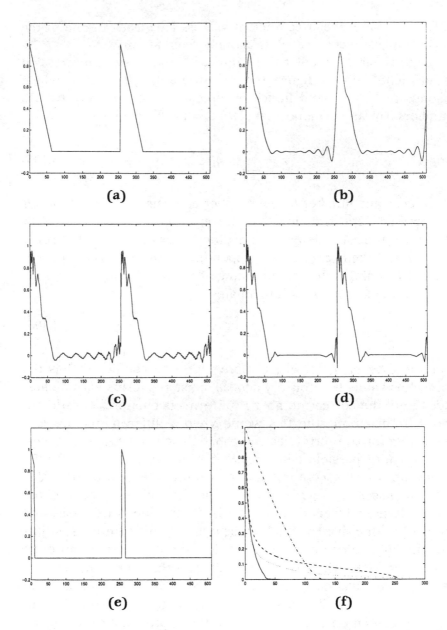

FIGURE 29 (a) Original, (b) Fourier, $K = 20$, (c) Shannon Wavelet, $K = 20$, (d) D6 Wavelet, $K = 20$, (e) Euclidean, $K = 20$, (f) Relative Error

of different signals. In Figures 29f, 31f, 32f, 33f, and 34f, we have plotted the relative error for each type of basis as a function of K, the number of terms in the approximation. The four graphs correspond to the four bases as follows:

solid line: D6 basis
dotted line: real Shannon basis
dashed line: Fourier basis
dash-dot line: Euclidean basis

In Figure 29a, the original vector z is the same "double spike" as in Figure 28a. For Figures 29c–e, we take $K = 20$. The relative errors for these cases are:

Figure 29b, Fourier basis: Relative error .2431
Figure 29c, real Shannon basis: Relative error .1694
Figure 29d, D6 basis: Relative error .1238
Figure 29e, Euclidean basis: Relative error .7767.

Figure 29f displays the graphs of the relative errors. In this case we present only the graph for $1 \leq K \leq 300$ because the graphs are too small to be visible for $K > 300$. In Figures 30b–e, we consider the same original vector z but we increase K to 75. The relative errors are:

Figure 30b, Fourier basis: Relative error .1223
Figure 30c, real Shannon basis: Relative error .0650
Figure 30d, D6 basis: Relative error 5.55×10^{-15}
Figure 30e, Euclidean basis: Relative error .2709.

These results confirm our intuition from the graphs that Daubechies's D6 basis does very well at compressing z. This is because of the sharp localization of the D6 wavelets and the fact that z is zero such a large proportion of the time. The result is that many of the D6 wavelet coefficients are zero or very small. When we compress, we omit the corresponding terms from the expansion (3.105), but since these coefficients are very small this has little effect. To put it another way, the D6 compression does not waste terms on the part of the graph where z is zero. Interestingly, the D6 expansion is virtually perfect with $K = 75$ terms, although the vector z has 128 nonzero components. This is because the

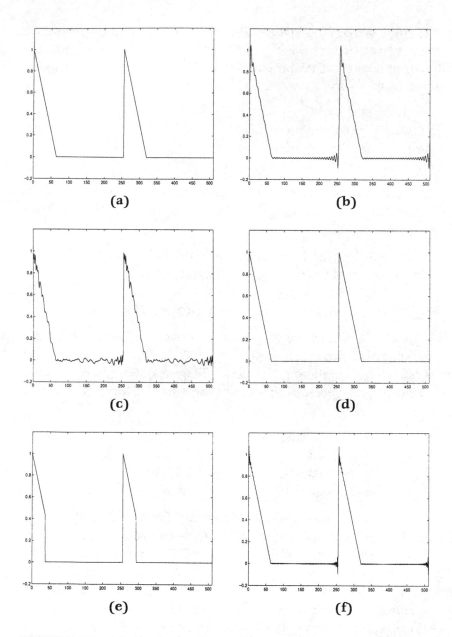

FIGURE 30   **(a)** Original, **(b)** Fourier, $K = 75$, **(c)** Shannon Wavelet, $K = 75$, **(d)** D6 Wavelet, $K = 75$, **(e)** Euclidean, $K = 75$, **(f)** Fourier, $K = 200$

multilevel structure of the wavelet basis allows one lower frequency wavelet to carry information about a number of points. The real Shannon wavelets, being somewhat localized, do fairly well, but not as well as the D6 basis. The Fourier basis has difficulty dealing with large jumps, because the basis exponentials $e^{2\pi i nm/N}$ come from very smooth functions. It takes many large high-frequency terms to synthesize a big jump, resulting in poor compression. This is further demonstrated in Figure 30f, when we consider the Fourier compression with $K = 200$ terms. Even with such a large number of terms, the approximation is not good near the large jumps of z, and the relative error in this case is .0512. Compression with the Euclidean basis does the obvious thing; it keeps the K largest values. That gives a perfect reconstruction of this particular z once $K \geq 128$, but for small values of K the approximation is not very good.

In Figure 31 we consider the signal z from Figures 23, 24 and 27. This signal is more difficult to compress, so we have taken $K = 75$, and the relative error in Figure 31f is shown over the entire interval $1 \leq K \leq 512$. For $K = 75$, the relative errors are:

Figure 31b, Fourier basis: Relative error .5899
Figure 31c, real Shannon basis: Relative error .2062
Figure 31d, D6 basis: Relative error .0763
Figure 31e, Euclidean basis: Relative error .5393.

The vector z has both spatial localization (being zero on a major part of its graph) and frequency localization (being made up primarily of two major frequency ranges), as shown in Figure 23b. Hence the two wavelet bases do a good job of compressing it. The D6 basis has the advantage of being able to more effectively ignore the regions where z is zero.

In Figure 32a, the signal

$$z(n) = \sin(n^{1.5}/64)$$

is plotted. It is a chirp of steadily increasing frequency. The relative errors for $K = 50$ are:

Figure 32b, Fourier basis: Relative error .4396
Figure 32c, real Shannon basis: Relative error .1781
Figure 32d, D6 basis: Relative error .3473
Figure 32e, Euclidean basis: Relative error .8972.

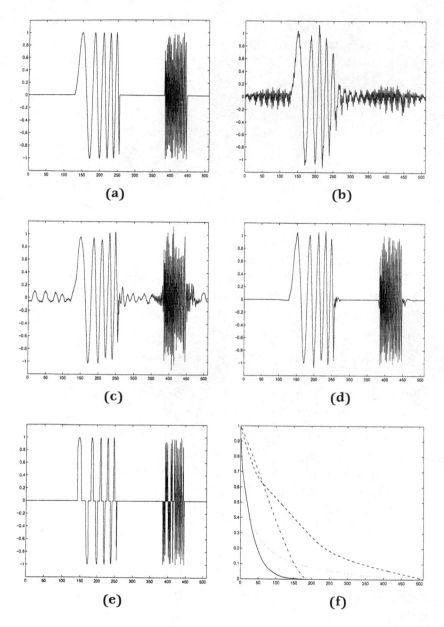

FIGURE 31 **(a)** Original, **(b)** Fourier, $K = 75$, **(c)** Shannon Wavelet, $K = 75$, **(d)** D6 Wavelet, $K = 75$, **(e)** Euclidean, $K = 75$, **(f)** Relative Error

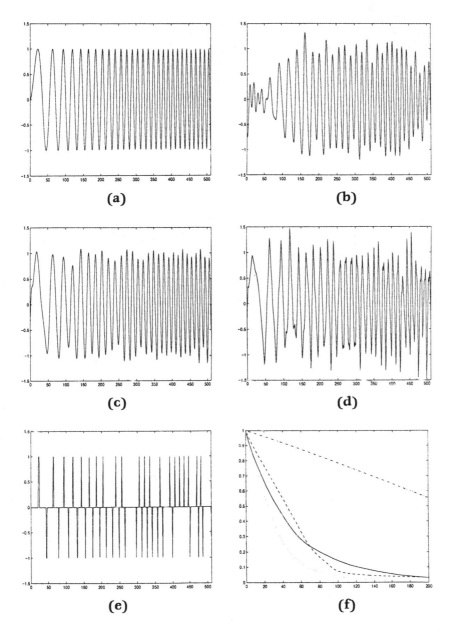

FIGURE 32 **(a)** Original, **(b)** Fourier, $K = 50$, **(c)** Shannon Wavelet, $K = 50$, **(d)** D6 Wavelet, $K = 50$, **(e)** Euclidean, $K = 50$, **(f)** Relative Error

This signal is not spatially localized, so the D6 basis does not do so well compressing it. The real Shannon basis does well because of its frequency localization. The Fourier basis does not do badly, for the same reason, and the Euclidean basis does a terrible job, as one might expect.

The advantages of wavelet bases for compressing certain signals at very high compression rates (i.e., using very few terms in the expansion) is shown in Figure 33, where

$$z(n) = (n - 256)e^{-(n-256)^2/512}.$$

This vector (a multiple of the derivative of a Gaussian) is well localized in both space and frequency. Thus even with $K = 8$, the wavelet bases do very well at compressing z. The relative errors are:

Figure 33b, Fourier basis: Relative error .6399
Figure 33c, real Shannon basis: Relative error .0168
Figure 33d, D6 basis: Relative error .0636
Figure 33e, Euclidean basis: Relative error .8915.

In Figure 34a, the vector z is

$$z(n) = \begin{cases} 1, & 32 \le n \le 95 \\ 2, & 132 \le n \le 259 \\ 4, & 416 \le n \le 511 \\ 0, & \text{other } n \text{ between 0 and 511.} \end{cases}$$

In this case, z is not so well spatially localized, and because of the sharp jumps at the edges of the steps it is not well localized in frequency either. The result is that the wavelet compressions and the Fourier compression are of similar quality. The relative errors for $K = 16$ are

Figure 34b, Fourier basis: Relative error .2389
Figure 34c, real Shannon basis: Relative error .2406
Figure 34d, D6 basis: Relative error .2566
Figure 34e, Euclidean basis: Relative error .9374.

No basis works best for all types of original signals. However, if one expects the signals with which one is working to have, to some degree, both spatial and frequency localization, then a wavelet

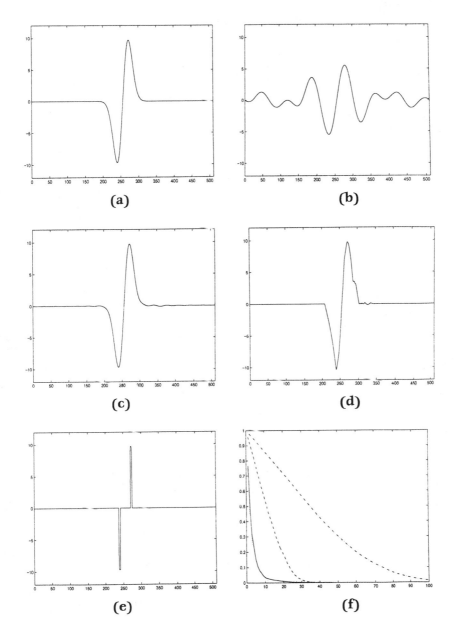

FIGURE 33 (a) Original, (b) Fourier, $K = 8$, (c) Shannon Wavelet, $K = 8$, (d) D6 Wavelet, $K = 8$, (e) Euclidean, $K = 8$, (f) Relative Error

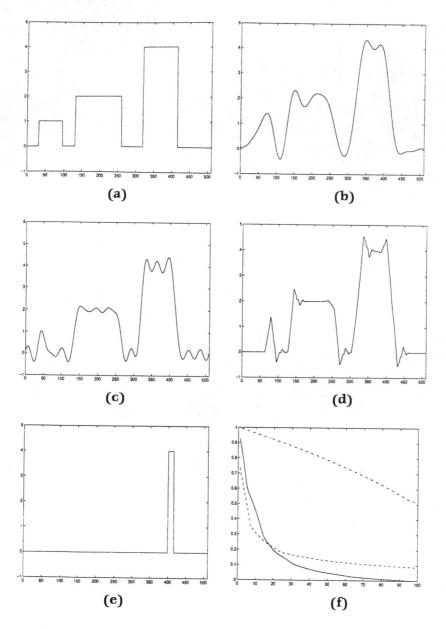

FIGURE 34 **(a)** Original, **(b)** Fourier, $K = 16$, **(c)** Shannon Wavelet, $K = 16$, **(d)** D6 Wavelet, $K = 16$, **(e)** Euclidean, $K = 16$, **(f)** Relative Error

basis is likely to do well at compressing these signals. In particular it should do better at picking up local features than the Fourier basis, and also better at recognizing frequency characteristics than the Euclidean basis.

With the advent of digital television and the transmission of digital pictures over the Internet, there is great interest in compressing images (i.e., two-dimensional signals). As described in the Prologue in reference to digital fingerprint images, an image is represented by small squares, called *pixels*, each of which is given a gray-scale value describing the darkness of that pixel (this is for a black and white picture; for color the same is done for each of the primary colors in the picture). For example, if the image is broken into a grid of 200 pixels by 250 pixels, then there are 50,000 pixel gray-scale values to assign, so the image is represented in a computer as a vector of length 50,000. This is a great deal of data, so we may need to compress it. We can do so by two-dimensional versions of the methods we have just considered. (For a description of the two-dimensional DFT, see Exercises 2.1.15–2.1.18. For the two-dimensional wavelet basis in the case of $\mathbb{Z}_{N_1} \times \mathbb{Z}_{N_2}$, see Exercises 3.1.12 and 3.2.13.)

An example of this is shown in Figure 35. In the upper left corner is the original image, a picture of Simon Zhang. In the upper right corner, we see the result of compressing with Daubechies's wavelets, in this case retaining the largest 10 percent of the terms in the wavelet expansion. The picture in the lower left corner shows the result when the largest 5 percent of the wavelet terms are retained. The result of retaining the largest 10 percent of the terms in the Fourier expansion is exhibited in the lower right corner. The odd-looking ripples are apparently a wave reflection of the sharp outline of Simon's shirt. This picture is more fuzzy than the others. To get a better sense of the difference, we focus on some local feature. A good example is the button on Simon's clothes (the one on Simon's right; the other button is not so clear in the original). Simon's button is much sharper and clearer even in the 5 percent wavelet compression picture than in the 10 percent Fourier compression picture. The basic reason is that the Fourier basis elements are not localized; they have constant magnitude at all points. Thus for every m, the inner product $\langle z, F_m \rangle$ (written for the one-dimensional case, as an example, but the

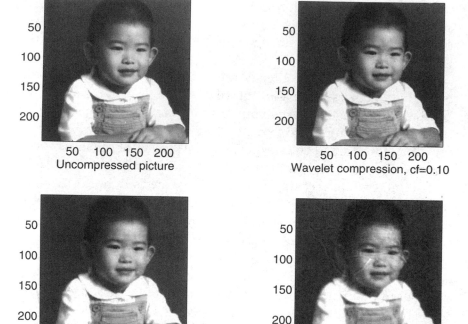

50   100   150   200
Uncompressed picture

50   100   150   200
Wavelet compression, cf=0.10

50   100   150   200
Wavelet compression, cf=0.05

50   100   150   200
Fourier compression, cf=0.10

FIGURE 35   (Courtesy of Shangqian Zhang.)

principle is the same for two dimensions) is affected by the part of z near the local feature (the button). Thus when we compress the image by deleting terms in the Fourier expansion, a large part of the information regarding the button is lost, resulting in the blurring of the button. In the wavelet basis, because of its spatial localization, relatively few terms are affected by the button, so only a few terms in the expansion are needed for the clear reconstruction of the button.

The compression examples we have given here are of a very basic nature. They are simple enough that they can be reproduced (except Figure 35) by a student familiar with a computational package such as MatLab or Mathematica (these examples were done with MatLab). They are not intended as realistic examples of signal compression. For example, with JPEG—the current industry standard for image compression—the picture is first subdivided into smaller blocks. Then a Fourier expansion (actually a variant known as the discrete cosine transform expansion) is compressed in each block. This gives a way of incorporating a certain degree of localization into the Fourier expansion. For example if an entire block is essentially constant (as in the background behind Simon), it can be represented with very few terms. However, the choice of block size is somewhat arbitrary, whereas with wavelets the different scales incorporate different degrees of localization in a natural manner.

One problem with JPEG is that since the compressions are done independently on adjacent blocks, they do not necessarily line up smoothly at the edges between blocks. This is the reason for the blocking artifacts in the JPEG fingerprint compression image in Figure 2 in the Prologue. Because the wavelet expansion carries a natural multilevel structure, the wavelet fingerprint compression image in Figure 3 does not show block lines. This is one of the main reasons that the FBI fingerprint compression contract was awarded to the group using wavelets.

A more realistic example of image compression is shown in Figure 36. Figure 36a displays the original, uncompressed "Lena" image. It is a file of size 262,159 data bytes. Figure 36b shows the same image, compressed by a factor of about 10 to 1 using JPEG (see the Prologue for a discussion of JPEG). More precisely, the file for the image in Figure 36b contains 26,264 data bytes. At this compression ratio, the image is represented with good visual accuracy. However, at higher compression ratios, there are problems. Figure 36c shows the JPEG compression at a ratio of about 40 to 1 (more precisely, using 6,646 bytes), and Figure 36d shows the JPEG compression at a ratio of about 103 to 1 (2,533 bytes). Some loss of image quality can be seen in Figure 36c, for example near the woman's shoulder. In Figure 36d the degradation of the image is severe. The JPEG program is not designed to work well at such extreme compression ratios. Figure 36e shows the Lena image at a compression ratio of about 40 to 1 (6506 bytes, comparable to Figure 36c), using a compression program based on wavelets. Figure 36f shows the wavelet compression at a ratio of about 103 to 1 (2562 bytes, comparable to Figure 36d). Especially at the higher compression ratios, the wavelet compression method gives better images. Figures

FIGURE 36 **(a)** Original, **(b)** JPEG10-1, **(c)** JPEG 40-1, **(d)** JPEG103-1,
(e) Wavelet 40-1 (Courtesy of Summus Technologies, Inc.), **(f)** Wavelet
103-1 (Courtesy of Summus Technologies, Inc.).

36b, c, and d were created using the commercially available program WinJPG v. 2.84. Figures 36e and f were done using the program Summus 4U2C 3.0, provided to the author by Björn Jawerth and Summus Technologies, Inc. Work is ongoing to incorporate wavelet compression methods into commercial digital signal processing.

Incidentally, the lady in Figure 36 is Lena Sjöblom. Her picture was digitized and used in an early paper on image compression. Since then this image has become a standard, as later papers used the same image to make comparisons. Many people in the field of image compression were unaware of the origin of this picture. In fact it is a portion of the *Playboy* magazine centerfold for November 1972. As the signal analysis field has become more gender-mixed, some have suggested that a different image be used. (Notice how the author of this text has avoided controversy by using Figure 35.) However, this is difficult to do, because one would like to compare new methods with older ones, which have used the Lena image. As a student once put it, "politically incorrect or not, the Lena image must stay, for the good of science." Recently Lena herself was located by signal analysts, and was a guest of honor at a major symposium on digital signal processing.

Exercises

3.3.1. i. Prove equations (3.89) and (3.90).

ii. Deduce that

$$
U^{\ell-1}(v_\ell)(n) = \begin{cases} \frac{1}{\sqrt{2}}, & n = 0, \\ -\frac{1}{\sqrt{2}}, & n = 2^{\ell-1} \\ 0, & n = 1, \ldots, 2^{\ell-1} - 1; \\ & 2^{\ell-1} + 1, \ldots, N - 1 \end{cases}
$$

and

$$
U^{\ell-1}(u_\ell)(n) = \begin{cases} \frac{1}{\sqrt{2}}, & n = 0, 2^{\ell-1} \\ 0, & n = 1, \ldots, 2^{\ell-1} - 1; \\ & 2^{\ell-1} + 1, \ldots, N - 1. \end{cases}
$$

iii. Prove equations (3.91) and (3.92).

iv. Prove equations (3.93) and (3.94).

v. Prove equation (3.95).

3.3.2. Prove equations (3.96) and (3.97). Hint: Proceed by induction on ℓ. By Exercise 3.2.1 (ii), and equation (3.83),

$$\hat{\psi}_{-\ell,0}(n) = \hat{f}_\ell(n) = \hat{g}_{\ell-1}(n)\hat{v}_\ell(n).$$

By induction, $\hat{g}_{\ell-1}(n)$ is only nonzero for $0 \le n \le (N/2^\ell) - 1$ and $N - N/2^\ell \le n \le N - 1$, so we need to consider only these regions. For $0 \le n \le (N/2^\ell)-1$ we consider the regions $0 \le n \le (N/2^{\ell+1})-1$ and $N/2^{\ell+1} \le n \le (N/2^\ell)-1$ separately and apply the formulas for $\hat{u}_\ell$ and $\hat{v}_\ell$ above, along with the induction assumption. Similarly, we break $N - N/2^\ell \le n \le N-1$ into the two regions $N-N/2^\ell \le n \le N-N/2^{\ell+1}-1$ and $N-N/2^{\ell+1} \le n \le N-1$. On the first of these regions, $\hat{v}_\ell$ is $\sqrt{2}$ and $\hat{u}_\ell$ is 0, and vice versa on the second of the two regions. The easiest way to see that is to note that $\hat{u}_\ell$ and $\hat{v}_\ell$ have period $2^{\ell-1}$, and, for example, the first region is equivalent, modulo $2^{\ell-1}$, to the region $N/2^\ell \le n \le (3N/2^{\ell+1})-1$, which, by the formulae in the text, is a subset of the region where $\hat{v}_\ell$ is $\sqrt{2}$ and $\hat{u}_\ell$ is 0.

3.3.3. Prove equations (3.98) and (3.99).

3.3.4. Assume N is even. Show that the Haar wavelets (except that the vector v is multiplied by -1) can be obtained by starting with the identity

$$\cos^2\left(\frac{\pi n}{N}\right) + \sin^2\left(\frac{\pi n}{N}\right) = 1$$

and proceeding as in Example 3.35.

3.3.5. Assume $N/2^p$ is an integer greater than 4, where p is some positive integer. Proceed as in Example 3.35, but starting with the identity

$$\left(\cos^2\left(\frac{\pi n}{N}\right) + \sin^2\left(\frac{\pi n}{N}\right)\right)^3 = 1,$$

to obtain a pth level wavelet basis for $\ell^2(\mathbb{Z}_N)$ for which u_1 and v_1 have only four nonzero components.

One answer:

$$u_1 = \frac{\sqrt{2}}{8}(1 + \sqrt{3}, 3 + \sqrt{3}, 3 - \sqrt{3}, 1 - \sqrt{3}, 0, 0, 0, \ldots, 0),$$

$$v_1 = \frac{\sqrt{2}}{8}(-3 - \sqrt{3}, 1 + \sqrt{3}, 0, 0, \ldots, 0, 0, -1 + \sqrt{3}, 3 - \sqrt{3}).$$

These are Daubechies's D4 wavelets for $\ell^2(\mathbb{Z}_N)$.

4

CHAPTER

Wavelets on $\mathbb{Z}$

4.1 $\ell^2(\mathbb{Z})$

So far we have considered signals (vectors) of finite length, which
we have extended periodically to be defined at all integers. In
this chapter we deal with infinite signals, which are generally
not periodic. More explicitly, we consider sequences of complex
numbers defined at the integers, denoted

$$z = (\ldots z(-2), z(-1), z(0), z(1), z(2), \ldots),$$

or, more concisely,

$$z = (z(n))_{n \in \mathbb{Z}}.$$

To do calculations in a meaningful way (e.g., so that the sums
we consider will converge), z should not be too big. Specifically,
we restrict attention to sequences that are *square-summable*, which
means that

$$\sum_{n \in \mathbb{Z}} |z(n)|^2 < +\infty.$$

We make the following remarks to clarify this because previously
we have not considered series summed over $\mathbb{Z}$.

Definition 4.1   *A series of complex numbers $\sum_{n\in\mathbb{Z}} w(n)$ converges if the sequence of symmetric partial sums*

$$s_N = \sum_{n=-N}^{+N} w(n)$$

converges as a sequence of complex numbers (Definition 1.10). We say $\sum_{n\in\mathbb{Z}} w(n)$ converges absolutely if $\sum_{n\in\mathbb{Z}} |w(n)|$ converges.

For many purposes it is more natural to require the convergence of both $\sum_{n=0}^{N} w(n)$ and $\sum_{n=-N}^{0} w(n)$ as $N \to \infty$. However, for our later work with Fourier series, Definition 4.1 is more convenient. If the terms are nonnegative real numbers, the two senses of convergence agree (Exercise 4.1.2 (ii)).

If a series $\sum_{n\in\mathbb{Z}} w(n)$ converges absolutely, then it converges (Exercise 4.1.1 (iii)). If the series consists of positive terms, as in the case of $\sum_{n\in\mathbb{Z}} |z(n)|^2$, the partial sums are increasing, hence they either diverge to $+\infty$ or converge.

We denote the space of all square-summable complex sequences on $\mathbb{Z}$ by $\ell^2(\mathbb{Z})$. Formally,

$$\ell^2(\mathbb{Z}) = \left\{ z = (z(n))_{n\in\mathbb{Z}} : z(n) \in \mathbb{C} \text{ for all } n, \text{ and } \sum_{n\in\mathbb{Z}} |z(n)|^2 < +\infty \right\}.$$

One can check (Exercise 4.1.3) that $\ell^2(\mathbb{Z})$ forms a vector space over $\mathbb{C}$, under the natural operations of componentwise addition and scalar multiplication. For $z, w \in \ell^2(\mathbb{Z})$, define

$$\langle z, w \rangle = \sum_{n\in\mathbb{Z}} z(n)\overline{w(n)}. \tag{4.1}$$

It follows (Exercise 4.1.4) that $\langle \cdot, \cdot \rangle$ is a complex inner product on $\ell^2(\mathbb{Z})$. We say that z and w are orthogonal if $\langle z, w \rangle = 0$. Thus we have a notion of perpendicularity in this infinite dimensional context that may be geometrically nonintuitive. In particular the notion of an orthonormal set of vectors (elements) in $\ell^2(\mathbb{Z})$ makes sense.

We define a norm on $\ell^2(\mathbb{Z})$ as in Definition 1.90. That is, for $z \in \ell^2(\mathbb{Z})$,

$$\|z\| = \left(\sum_{n\in\mathbb{Z}} |z(n)|^2 \right)^{1/2}. \tag{4.2}$$

By Exercise 1.6.5, this makes $\ell^2(\mathbb{Z})$ a normed space, hence a metric space with the distance $d(z, w) = \|z - w\|$. The Cauchy-Schwarz inequality $|\langle z, w \rangle| \leq \|z\|\|w\|$ (proved for any complex inner product space in Lemma 1.91) takes the following form here: for $z, w \in \ell^2(\mathbb{Z})$,

$$\left| \sum_{n \in \mathbb{Z}} z(n)\overline{w(n)} \right| \leq \left(\sum_{n \in \mathbb{Z}} |z(n)|^2 \right)^{1/2} \left(\sum_{n \in \mathbb{Z}} |w(n)|^2 \right)^{1/2}.$$

We can apply this to the sequences $|z|$ and $|w|$, whose values at n are $|z(n)|$ and $|w(n)|$, respectively. Noting that $\||z|\| = \|z\|$, and similarly for w, and recalling that $|\overline{w(n)}| = |w(n)|$, we obtain

$$\sum_{n \in \mathbb{Z}} |z(n)w(n)| \leq \left(\sum_{n \in \mathbb{Z}} |z(n)|^2 \right)^{1/2} \left(\sum_{n \in \mathbb{Z}} |w(n)|^2 \right)^{1/2}. \qquad (4.3)$$

Also, by Corollary 1.92, we have the triangle inequality $\|z + w\| \leq \|z\| + \|w\|$ in $\ell^2(\mathbb{Z})$:

$$\left(\sum_{n \in \mathbb{Z}} |z(n) + w(n)|^2 \right)^{1/2} \leq \left(\sum_{n \in \mathbb{Z}} |z(n)|^2 \right)^{1/2} + \left(\sum_{n \in \mathbb{Z}} |w(n)|^2 \right)^{1/2}.$$
$$(4.4)$$

In Definition 4.2, do not confuse z_k with $z(k)$: for each k, $z_k = (z_k(n))_{n \in \mathbb{Z}}$ is a sequence in $\ell^2(\mathbb{Z})$.

Definition 4.2 *Suppose $M \in \mathbb{Z}$, $z_k \in \ell^2(\mathbb{Z})$ for each $k \in \mathbb{Z}$ with $k \geq M$, and $z \in \ell^2(\mathbb{Z})$. The sequence $\{z_k\}_{k=M}^{\infty}$ converges to z in $\ell^2(\mathbb{Z})$ if, for all $\epsilon > 0$, there exists a positive integer N such that*

$$\|z_k - z\| < \epsilon$$

for all $k > N$. Also $\{z_k\}_{k=M}^{\infty}$ is a Cauchy sequence in $\ell^2(\mathbb{Z})$ if, for all $\epsilon > 0$, there exists a positive integer N such that

$$\|z_k - z_m\| < \epsilon$$

for all $k, m > N$.

Theorem 4.3 *(Completeness of $\ell^2(\mathbb{Z})$) Suppose $\{z_k\}_{k=M}^{\infty}$ is a Cauchy sequence in $\ell^2(\mathbb{Z})$. Then there exists $z \in \ell^2(\mathbb{Z})$ such that $\{z_k\}_{k=M}^{\infty}$ converges to z in $\ell^2(\mathbb{Z})$.*

Proof
Exercise 4.1.7. ∎

The main difference in dealing with the vector space $\ell^2(\mathbb{Z})$ instead of $\ell^2(\mathbb{Z}_N)$ is that $\ell^2(\mathbb{Z})$ is infinite dimensional. The natural analogs for $\ell^2(\mathbb{Z})$ of the standard basis vectors in $\ell^2(\mathbb{Z}_N)$ are the vectors $e_j \in \ell^2(\mathbb{Z})$ defined for $j \in \mathbb{Z}$ by

$$e_j(n) = \begin{cases} 1 & \text{if } n = j \\ 0 & \text{if } n \neq j. \end{cases} \tag{4.5}$$

However, the set $\{e_j\}_{j \in \mathbb{Z}}$ does not form a basis for $\ell^2(\mathbb{Z})$ in the sense of Definition 1.37 because we cannot write every element of $\ell^2(\mathbb{Z})$ as a *finite* linear combination of the vectors $\{e_j\}_{j \in \mathbb{Z}}$. Nevertheless, it seems clear that every element $z = (z(n))_{n \in \mathbb{Z}}$ can be written as the infinite sum $z = \sum_{j \in \mathbb{Z}} z(j) e_j$. To be precise about this, we must first discuss convergence of series in the space $\ell^2(\mathbb{Z})$. Because we will consider several cases of this type, we include a general discussion of the convergence of sequences and series in infinite dimensional inner product spaces in section 4.2.

Exercises

4.1.1. Let $\sum_{n \in \mathbb{Z}} w(n)$ be a series of complex numbers.

   i. (Cauchy criterion) Prove that $\sum_{n \in \mathbb{Z}} w(n)$ converges if and only if, for all $\epsilon > 0$, there exists an integer N such that

$$\left| \sum_{n=-m}^{-k} w(n) + \sum_{n=k}^{m} w(n) \right| < \epsilon$$

   for all $m \geq k > N$. Suggestion: Apply either Theorem 1.12 to the partial sums, or Lemma 1.14 to the series $\sum_{n=0}^{\infty} z_n$, where $z_0 = w(0)$ and, for all $n \geq 1$, $z_n = w(n) + w(-n)$.

   ii. (Comparison test) Let $\{a(n)\}_{n \in \mathbb{Z}}$ be a sequence of nonnegative real numbers such that $|w(n)| \leq a(n)$ for all n such that $|n| \geq N$, for some $N \in \mathbb{Z}$. If $\sum_{n \in \mathbb{Z}} a(n)$ converges, prove that $\sum_{n \in \mathbb{Z}} w(n)$ converges.

   iii. If $\sum_{n \in \mathbb{Z}} w(n)$ converges absolutely, prove that $\sum_{n \in \mathbb{Z}} w(n)$ converges.

4.1.2. i. Show that the series $\sum_{n \in \mathbb{Z}} n$ converges, but does not converge absolutely. In particular, it is not true that the convergence of $\sum_{n \in \mathbb{Z}} w(n)$ implies that $\lim_{n \to \infty} w(n) = 0$. This is one reason many texts require a stronger definition of convergence than the one in Definition 4.1.

ii. Prove that $\sum_{n \in \mathbb{Z}} w(n)$ converges absolutely if and only if $\sum_{n=0}^{\infty} w(n)$ and $\sum_{n=1}^{\infty} w(-n)$ converge absolutely.

iii. Show that if $\sum_{n \in \mathbb{Z}} w(n)$ converges absolutely, then $\lim_{n \to \infty} w(n) = 0$ and $\lim_{n \to \infty} w(-n) = 0$.

4.1.3. For $z, w \in \ell^2(\mathbb{Z})$, and $\alpha \in \mathbb{C}$, define $z + w$ and αz by

$$(z + w)(n) = z(n) + w(n), \quad \text{and} \quad (\alpha z)(n) = \alpha z(n),$$

for all $n \in \mathbb{Z}$. With these operations, prove that $\ell^2(\mathbb{Z})$ is a vector space over $\mathbb{C}$. Remark: The only property in Definition 1.30 that is not obvious is A1. See the hint for Exercise 1.6.3 (ii).

4.1.4. Prove that $\langle \cdot, \cdot \rangle$, defined in equation (4.1), is a complex inner product on $\ell^2(\mathbb{Z})$. (See the hint for Exercise 1.6.3. The main point is to show the absolute convergence of the sum in equation (4.1).)

4.1.5. (Monotone convergence theorem for sequences) Suppose $M \in \mathbb{Z}$, and, for each $k \in \mathbb{Z}$ with $k \geq M$, $x_k = \{x_k(n)\}_{n \in \mathbb{Z}}$ is a sequence of nonnegative real numbers (i.e., $x_k(n) \geq 0$ for each k, n). Suppose that for each n and k, $x_k(n) \leq x_{k+1}(n)$, that is, $\{x_k(n)\}_{k=M}^{\infty}$ is a nondecreasing sequence for each n. By the monotone sequence lemma, $\lim_{k \to \infty} x_k(n)$ exists (although it could be $+\infty$) for each n. Set $x(n) = \lim_{k \to \infty} x_k(n)$. Prove that

$$\sum_{n \in \mathbb{Z}} x(n) = \lim_{k \to \infty} \sum_{n \in \mathbb{Z}} x_k(n). \tag{4.6}$$

(Note that $\sum_{n \in \mathbb{Z}} x_k(n)$ is increasing in k, so the limit on the right side exists.) Here the equality is interpreted in the sense that if one side is finite, the other must be also, and the values are the same, but we also allow both sides to be $+\infty$. Hint: Since $x_k(n) \leq x(n)$, the fact that the left side of (4.6) is $\geq$ the right side is easy. For the other inequality, note that for any fixed integer $J > 0$,

$$\sum_{n=-J}^{J} x(n) = \lim_{k \to \infty} \sum_{n=-J}^{J} x_k(n) \leq \lim_{k \to \infty} \sum_{n \in \mathbb{Z}} x_k(n).$$

Now let $J \to +\infty$.

4.1.6. (Fatou's lemma for sequences) Suppose $M \in \mathbb{Z}$, and, for each $k \in \mathbb{Z}$ with $k \geq M$, $x_k = \{x_k(n)\}_{n \in \mathbb{Z}}$ is a sequence of nonnegative real numbers. Suppose $x(n) = \lim_{k \to \infty} x_k(n)$ exists for each $n \in \mathbb{Z}$. Prove that

$$\sum_{n \in \mathbb{Z}} x(n) \leq \liminf_{k \to \infty} \sum_{n \in \mathbb{Z}} x_k(n). \tag{4.7}$$

(By definition, for any sequence $\{a_k\}_{k=M}^{\infty}$ of real numbers, $\liminf_{k \to \infty} a_k = \lim_{m \to \infty} (\inf_{j \geq m} a_j)$. This limit exists (it could be infinite) because $\inf_{j \geq m} a_j$ is increasing in m.)

Hint: Let $y_m(n) = \inf_{j \geq m} x_j(n)$. Prove that $y_m(n)$ increases to $x(n)$ for each n. Therefore by Exercise 4.1.5,

$$\sum_{n \in \mathbb{Z}} x(n) = \lim_{m \to \infty} \sum_{n \in \mathbb{Z}} y_m(n).$$

Now prove that

$$\sum_{n \in \mathbb{Z}} \inf_{j \geq m} x_j(n) \leq \inf_{j \geq m} \sum_{n \in \mathbb{Z}} x_j(n).$$

4.1.7. Suppose $\{z_k\}_{k=M}^{\infty}$ is a Cauchy sequence in $\ell^2(\mathbb{Z})$ (Definition 4.2).

   i. Prove that for each $n \in \mathbb{Z}$, $\{z_k(n)\}_{k=M}^{\infty}$ is a Cauchy sequence in $\mathbb{C}$.

   ii. By part i and the completeness of $\mathbb{C}$ (Theorem 1.12), $z(n) = \lim_{k \to \infty} z_k(n)$ exists for every $n \in \mathbb{Z}$. Let $z = \{z(n)\}_{n \in \mathbb{Z}}$. Prove that $\{z_k(n)\}_{k=M}^{\infty}$ converges to z in $\ell^2(\mathbb{Z})$ (Definition 4.2). Hint: For k sufficiently large but fixed, apply Exercise 4.1.6 to obtain

$$\sum_{n \in \mathbb{Z}} |z_k(n) - z(n)|^2 \leq \liminf_{m \to \infty} \sum_{n \in \mathbb{Z}} |z_k(n) - z_m(n)|^2.$$

Remark: By using relation (4.4) to write $\|z\| \leq \|z - z_k\| + \|z_k\|$, we obtain $z \in \ell^2(\mathbb{Z})$. This proves the completeness of $\ell^2(\mathbb{Z})$.

4.1.8. i. Suppose $\{z_k\}_{k=0}^{\infty}$ is a sequence of elements of $\ell^2(\mathbb{Z})$ and z is an element of $\ell^2(\mathbb{Z})$ such that $\{z_k\}_{k\in\mathbb{Z}}$ converges to z in $\ell^2(\mathbb{Z})$ (see Definition 4.2; this is called *norm* convergence). Prove that

$$\lim_{k\to\infty} z_k(n) = z(n)$$

for each $n \in \mathbb{Z}$ (this is called *pointwise* convergence).

ii. Give an example of a sequence $\{z_k\}_{k=0}^{\infty}$ of elements of $\ell^2(\mathbb{Z})$ and an element $z \in \ell^2(\mathbb{Z})$ such that

$$\lim_{k\to\infty} z_k(n) = z(n)$$

for each $n \in \mathbb{Z}$ but $\{z_k\}_{k\in\mathbb{Z}}$ does not converge to z in $\ell^2(\mathbb{Z})$. Thus norm convergence in $\ell^2(\mathbb{Z})$ is stronger than pointwise convergence.

4.1.9. Observe that $\{e_n\}_{n=0}^{\infty}$ is a bounded sequence in $\ell^2(\mathbb{Z})$ (this is trivial: $\|e_n\| = 1$ for each n). Prove that there is no subsequence $\{e_{n_k}\}_{k=1}^{\infty}$ that converges in $\ell^2(\mathbb{Z})$ (Definition 4.2).

This is a key difference between the infinite dimensional and finite dimensional cases: by the Bolzano-Weierstrass theorem, every bounded sequence in $\mathbb{R}^n$ or $\mathbb{C}^n$ has a convergent subsequence.

4.2 Complete Orthonormal Sets in Hilbert Spaces

Let X be an infinite dimensional complex inner product space with inner product $\langle \cdot, \cdot \rangle$. (Everything we say will be true in the finite dimensional case for trivial reasons, but it simplifies our notation to consider only the infinite dimensional case here.) Define a norm $\| \cdot \|$ on X as in Definition 1.90.

Definition 4.4 *Let $M \in \mathbb{Z}$. A sequence $\{x_n\}_{n=M}^{\infty}$ of elements in X converges in X to some $x \in X$ if, for all $\epsilon > 0$, there exists $N \in \mathbb{N}$ such that $\|x_n - x\| < \epsilon$ for all $n > N$. Also $\{x_n\}_{n=M}^{\infty}$ is a Cauchy sequence* in X if, for all $\epsilon > 0$, there exists $N \in \mathbb{N}$ such that $\|x_n - x_m\| < \epsilon$ for all $n, m > N$.

If $\{x_n\}_{n=M}^{\infty}$ converges in X, then $\{x_n\}_{n=M}^{\infty}$ is a Cauchy sequence (Exercise 4.2.2).

Definition 4.5 *A complex inner product space X is* complete *if every Cauchy sequence in X converges. A complete complex inner product space is a* Hilbert space.

Note that for $H = \ell^2(\mathbb{Z})$ (which is a Hilbert space, by Exercises 4.1.3 and 4.1.4, and Theorem 4.3), Definition 4.4 is consistent with Definition 4.2.

From now on we focus on Hilbert spaces because completeness is necessary for the following theory, and our examples of interest are complete. Next we define convergence of a series in a Hilbert space. This should not be confused with Definition 4.1, in which we consider only series of complex numbers. Here the terms in the series are elements of the Hilbert space.

Definition 4.6 *Let H be a Hilbert space, and let $\{w_n\}_{n\in\mathbb{Z}}$ be a sequence of elements of H. For $N = 1, 2, 3, \ldots$, let s_N be the* symmetric partial sum

$$s_N = \sum_{n=-N}^{+N} w_n.$$

We say the series $\sum_{n\in\mathbb{Z}} w_n$ converges *in H to some $s \in H$ if the sequence $\{s_N\}_{N=1}^{\infty}$ converges to s in H (in the sense of Definition 4.4).*

Recall (Definitions 1.93 and 1.94) that the notions of orthogonality and orthonormal sets are defined in any (complex) inner product space. The next result shows that square summable sequences play a natural role in the study of any infinite dimensional inner product space.

Lemma 4.7 *Suppose H is a Hilbert space, $\{a_j\}_{j\in\mathbb{Z}}$ is an orthonormal set in H, and $z = (z(j))_{j\in\mathbb{Z}} \in \ell^2(\mathbb{Z})$. Then the series*

$$\sum_{j\in\mathbb{Z}} z(j)a_j$$

converges in H, and

$$\left\| \sum_{j \in \mathbb{Z}} z(j) a_j \right\|^2 = \sum_{j \in \mathbb{Z}} |z(j)|^2. \tag{4.8}$$

Proof

For $N = 1, 2, 3, \ldots$, let s_N be the partial sum

$$s_N = \sum_{j=-N}^{+N} z(j) a_j.$$

Then for $N > M$,

$$\| s_N - s_M \|^2 = \left\| \sum_{M < |j| \le N} z(j) a_j \right\|^2 = \sum_{M < |j| \le N} |z(j)|^2,$$

because this is a finite sum and the a_j are orthonormal (Exercise 4.2.3). Because the series $\sum_{j \in \mathbb{Z}} |z(j)|^2$ converges, its sequence of symmetric partial sums must be Cauchy (Theorem 1.12). Hence given $\epsilon > 0$, there exists K such that $\sum_{M < |j| \le N} |z(j)|^2 \le \epsilon$ for all $N > M > K$. By the above inequality, this proves that the sequence $\{s_N\}_{N \in \mathbb{N}}$ is Cauchy. Since H is complete, $\{s_N\}_{N \in \mathbb{N}}$ converges, that is, $\sum_{j \in \mathbb{Z}} z(j) a_j$ converges in H.

For the proof of (4.8), see Exercise 4.2.4. ∎

Lemma 4.7 shows part of the relation between a general infinite dimensional Hilbert space and $\ell^2(\mathbb{Z})$: given an orthonormal set $\{a_j\}_{j \in \mathbb{Z}}$ and $z \in \ell^2(\mathbb{Z})$, the series $\sum_{j \in \mathbb{Z}} z(j) a_j$ converges to an element of H. However, we can also go the other way: given an element f of H, we can obtain a sequence in $\ell^2(\mathbb{Z})$ by considering $\{\langle f, a_j \rangle\}_{j \in \mathbb{Z}}$.

Lemma 4.8 *Suppose H is a Hilbert space, $\{a_j\}_{j \in \mathbb{Z}}$ is an orthonormal set in H, and $f \in H$. Then the sequence $\{\langle f, a_j \rangle\}_{j \in \mathbb{Z}}$ belongs to $\ell^2(\mathbb{Z})$ with*

$$\sum_{j \in \mathbb{Z}} |\langle f, a_j \rangle|^2 \le \| f \|^2. \tag{4.9}$$

Proof

For $N = 1, 2, \ldots$, let $s_N = \sum_{j=-N}^{+N} \langle f, a_j \rangle a_j$. Then

$$\| f - s_N \|^2 = \langle f, f \rangle - \langle f, s_N \rangle - \langle s_N, f \rangle + \langle s_N, s_N \rangle.$$

Note (using equation (1.43)) that

$$\langle f, s_N \rangle = \left\langle f, \sum_{j=-N}^{N} \langle f, a_j \rangle a_j \right\rangle = \sum_{j=-N}^{N} \overline{\langle f, a_j \rangle} \langle f, a_j \rangle = \sum_{j=-N}^{N} |\langle f, a_j \rangle|^2.$$

By I3 in Definition 1.86,

$$\langle s_N, f \rangle = \overline{\langle f, s_N \rangle} = \overline{\sum_{j=-N}^{N} |\langle f, a_j \rangle|^2} = \sum_{j=-N}^{N} |\langle f, a_j \rangle|^2,$$

because the last quantity is real. Also, by equation (4.8),

$$\|s_N\|^2 = \sum_{j=-N}^{N} |\langle f, a_j \rangle|^2.$$

Substituting these three facts above gives

$$\|f - s_N\|^2 = \|f\|^2 - 2 \sum_{j=-N}^{+N} |\langle f, a_j \rangle|^2 + \sum_{j=-N}^{+N} |\langle f, a_j \rangle|^2$$

$$= \|f\|^2 - \sum_{j=-N}^{+N} |\langle f, a_j \rangle|^2.$$

Since $\|f - s_N\|^2 \geq 0$, this implies that

$$\sum_{j=-N}^{+N} |\langle f, a_j \rangle|^2 \leq \|f\|^2.$$

Taking the limit as $N \to \infty$ yields the conclusion. ∎

If $\{a_j\}_{j \in \mathbb{Z}}$ is an orthonormal set in H and $f \in H$, then by Lemma 4.8, the sequence $\{\langle f, a_j \rangle\}_{j \in \mathbb{Z}}$ belongs to $\ell^2(\mathbb{Z})$. So by Lemma 4.7, the series $\sum_{j \in \mathbb{Z}} \langle f, a_j \rangle a_j$ converges in H. We are interested in finding a condition on the orthonormal set that guarantees that $\sum_{j \in \mathbb{Z}} \langle f, a_j \rangle a_j = f$, for any $f \in H$, as in the case of an orthonormal basis in a finite dimensional inner product space (Lemma 1.101 i).

Definition 4.9 *Suppose H is a Hilbert space and $\{a_j\}_{j \in \mathbb{Z}}$ is a set of elements in H. We say that $\{a_j\}_{j \in \mathbb{Z}}$ is a* complete orthonormal set *or* complete orthonormal system *if $\{a_j\}_{j \in \mathbb{Z}}$ is an orthonormal set with the property that the only element $w \in H$ such that $\langle w, a_j \rangle = 0$ for all $j \in \mathbb{Z}$ is $w = 0$.*

Note that the word "complete" is used here in a sense different from the sense above relating to the convergence of Cauchy sequences.

Theorem 4.10   *Suppose H is a Hilbert space and $\{a_j\}_{j\in\mathbb{Z}}$ is an orthonormal set in H. Then $\{a_j\}_{j\in\mathbb{Z}}$ is a complete orthonormal set if and only if*

$$f = \sum_{j\in\mathbb{Z}} \langle f, a_j\rangle a_j, \quad \text{for all } f \in H.$$

Proof

First suppose $\{a_j\}_{j\in\mathbb{Z}}$ is a complete orthonormal set. Given $f \in H$, let $g = \sum_{j\in\mathbb{Z}} \langle f, a_j\rangle a_j$ (as noted previously, this series converges in H by Lemmas 4.7–4.8). Then for all $m \in \mathbb{Z}$,

$$\langle g, a_m\rangle = \langle f, a_m\rangle,$$

by Exercise 4.2.5. Thus $\langle f - g, a_m\rangle = 0$ for all $m \in \mathbb{Z}$. Since $\{a_j\}_{j\in\mathbb{Z}}$ is a complete orthonormal set, this implies that $f - g = 0$, that is, $f = g$, which is what we want to prove.

Conversely, suppose every $f \in H$ can be written in the form $\sum_{j\in\mathbb{Z}} \langle f, a_j\rangle a_j$. If $\langle f, a_j\rangle = 0$ for all j, then all the coefficients are 0, so $f = 0$. This proves the completeness of $\{a_j\}_{j\in\mathbb{Z}}$.   ∎

Going back to our example of $\ell^2(\mathbb{Z})$, it is easy to see that the set $\{e_j\}_{j\in\mathbb{Z}}$ defined in equation (4.5) is a complete orthonormal set in $\ell^2(\mathbb{Z})$. Thus, by Theorem 4.10, every $z = (z(n))_{n\in\mathbb{Z}} \in \ell^2(\mathbb{Z})$ can be represented by the series $\sum_{j\in\mathbb{Z}} z(j)e_j$, whose partial sums converge (to z) in $\ell^2(\mathbb{Z})$. (Of course, this could be seen directly, but our concern is to illustrate Theorem 4.10.) So although $\{e_j\}_{j\in\mathbb{Z}}$ does not constitute a basis in the vector space sense, it is possible to represent every element of $\ell^2(\mathbb{Z})$ as an "infinite linear combination" of the set $\{e_j\}_{j\in\mathbb{Z}}$. Many texts will therefore still refer to a complete orthonormal set as an "orthonormal basis," using the term "basis" in a sense different from Definition 1.37, but we prefer to avoid this confusion by just calling it a complete orthonormal set or system.

Another useful characterization is the following.

Lemma 4.11   *Let $\{a_j\}_{j\in\mathbb{Z}}$ be an orthonormal set in a Hilbert space H. Then the following are equivalent:*

   i. $\{a_j\}_{j\in\mathbb{Z}}$ is complete.

ii. *(Parseval's relation) For all* $f, g \in H$,

$$\langle f, g \rangle = \sum_{j \in \mathbb{Z}} \langle f, a_j \rangle \overline{\langle g, a_j \rangle}. \tag{4.10}$$

iii. *(Plancherel's formula) For all* $f \in H$,

$$\|f\|^2 = \sum_{j \in \mathbb{Z}} |\langle f, a_j \rangle|^2. \tag{4.11}$$

Proof

The implication i $\Rightarrow$ ii is Exercise 4.2.7. We have ii $\Rightarrow$ iii by letting $g = f$. For iii $\Rightarrow$ i, see Exercise 4.2.6. ∎

A Hilbert space with a finite or countably infinite complete orthonormal set is called *separable*. There are Hilbert spaces so large that any complete orthonormal set must be uncountable, but we will not encounter such nonseparable spaces in this text.

In Lemma 1.98, we considered the basic properties of the orthogonal projection operator onto a finite dimensional subspace. These results can be extended to the case of an infinite dimensional subspace.

Definition 4.12 *Suppose* $A = \{a_j\}_{j \in \mathbb{Z}}$ *is an orthonormal set in a Hilbert space* H. *Let*

$$S_A = \left\{ \sum_{j \in \mathbb{Z}} z(j) a_j : z = (z(j))_{j \in \mathbb{Z}} \in \ell^2(\mathbb{Z}) \right\}. \tag{4.12}$$

(Note that $\sum_{j \in \mathbb{Z}} z(j) a_j$ *converges in* H *for* $\{z(j)\}_{j \in \mathbb{Z}} \in \ell^2(\mathbb{Z})$ *by Lemma 4.7.)*

Any such S_A is a subspace of H (Exercise 4.2.8).

Definition 4.13 *Suppose* $\{a_j\}_{j \in \mathbb{Z}}$ *is an orthonormal set in a Hilbert space* H. *Define* $S = S_A$ *as in Definition 4.12. For* $f \in H$, *define*

$$P_S(f) = \sum_{j \in \mathbb{Z}} \langle f, a_j \rangle a_j. \tag{4.13}$$

(This series converges in H *by Lemmas 4.7 and 4.8.) We call* $P_S(f)$ *the orthogonal projection of* f *on* S. *We call the operator* P_S *itself the orthogonal projection onto* S.

Lemma 4.14 *Let* H, S, *and* P_S *be as in Definition 4.13. Then*

i. *For every $f \in H$, $P_S(f) \in S$.*
ii. *The transformation $P_S : H \rightarrow S$ is linear.*
iii. *If $s \in S$, then $P_S(s) = s$.*
iv. *(Orthogonality property) $\langle f - P_S(f), s \rangle = 0$ for any $f \in H$ and $s \in S$.*
v. *(Best approximation property) For any $f \in H$ and $s \in S$,*

$$\|f - P_S(f)\| \leq \|f - s\|,$$

with equality only for $s = P_S(f)$.

Proof
Exercise 4.2.9. ∎

Exercises

4.2.1. Let $\{x_n\}_{n=1}^{\infty}$ be a sequence in a complex inner product space X, and let $x \in X$. Prove that $\{x_n\}_{n=1}^{\infty}$ converges to x in X if and only if $\|x_n - x\|$ converges to 0 as $n \rightarrow \infty$, as a sequence of numbers.

4.2.2. Suppose $M \in \mathbb{Z}$, $\{x_n\}_{n=M}^{\infty}$ is a sequence in a complex inner product space X, and $\{x_n\}_{n=M}^{\infty}$ converges in X to some $x \in X$. Prove that $\{x_n\}_{n=M}^{\infty}$ is a Cauchy sequence (Definition 4.4).

4.2.3. Suppose X is a complex inner product space and $\{v_1, v_2, \ldots, v_n\}$ is a finite orthonormal set in X. If $z = \sum_{j=1}^{n} z(j)v_j$ and $w = \sum_{j=1}^{n} w(j)v_j$, for some scalars $z(1), z(2), \ldots, z(n)$ and $w(1), \ldots w(n)$, prove that

$$\langle z, w \rangle = \sum_{j=1}^{n} z(j)\overline{w(j)},$$

and

$$\|z\|^2 = \sum_{j=1}^{n} |z(j)|^2.$$

4.2.4. Prove equation (4.8). Hint: Let s_N be as in the proof of Lemma 4.7. By Exercise 4.2.3,

$$\| s_N \|^2 = \sum_{j=-N}^{N} |z(j)|^2.$$

Use the convergence of s_N in H to $s = \sum_{j \in \mathbb{Z}} z(j) a_j$ and the triangle inequality (4.4), to show that $\|s\| = \lim_{N \to \infty} \|s_N\|$.

4.2.5. Let H be a Hilbert space.

i. Suppose $\{f_n\}_{n=1}^{\infty}$ is a sequence of elements of H that converge in H to some element $f \in H$. Prove that for any $g \in H$,

$$\lim_{n \to \infty} \langle f_n, g \rangle = \langle f, g \rangle.$$

Hint: Apply the Cauchy-Schwarz inequality to $\langle f_n - f, g \rangle$.

ii. Suppose $\{a_j\}_{j \in \mathbb{Z}}$ is an orthonormal set in H and $z = (z(n))_{n \in \mathbb{Z}} \in \ell^2(\mathbb{Z})$. Prove that for any $m \in \mathbb{Z}$,

$$\left\langle \sum_{j \in \mathbb{Z}} z(j) a_j, a_m \right\rangle = z(m).$$

Hint: Show that this is true when the series is replaced by its partial sum s_N, for $N > m$, and apply part i.

iii. Suppose $A = \{a_j\}_{j \in \mathbb{Z}}$ is an orthonormal set in H, $z \in H$, and $\langle z, a_j \rangle = 0$ for every $j \in \mathbb{Z}$. Let S_A be defined by equation (4.12). Prove that $\langle z, s \rangle = 0$ for all $s \in S_A$.

4.2.6. Let $\{a_j\}_{j \in \mathbb{Z}}$ be an orthonormal set in a Hilbert space H. Prove that $\{a_j\}_{j \in \mathbb{Z}}$ is a complete orthonormal set if and only if equation (4.11) holds for all $f \in H$. Hint: Use the inequality obtained in the proof of Lemma 4.8 and let $N \to \infty$.

4.2.7. Let $\{a_j\}_{j \in \mathbb{Z}}$ be a complete orthonormal set in a Hilbert space H. Prove that equation (4.10) holds for all $f, g \in H$. Hint: Substitute $\sum_j \langle f, e_j \rangle e_j$ for f and use Exercise 4.2.5 to take the sum outside the inner product. Also, this follows from Exercise 4.2.6 and the polarization identity (Exercise 1.6.7 (i)) with $T = I$.

4.2.8. Define S_A as in Definition 4.12. Prove that S_A is a subspace of H.

4.2.9. Prove Lemma 4.14. Hints: i: Lemma 4.8; ii: linearity of $\langle \cdot, \cdot \rangle$ and $\sum$; iii: Exercise 4.2.5 (ii); iv: prove it for $s = a_j$ (using Exercise 4.2.5 (ii)) first, then apply Exercise 4.2.5 (i); v: see the proof of the analogous statement in Lemma 1.98.

4.2.10. Let H be a Hilbert space and E a subset of H. We say E is *closed* if whenever $\{u_j\}_{j=1}^\infty$ is a sequence of elements of E that converges in H to some $u \in H$, then $u \in E$ (i.e., E contains all of its *limit points*). Let $A = \{a_j\}_{j \in \mathbb{Z}}$ be an orthonormal set in H and let S_A be defined by equation (4.12). Prove that S_A is a *closed subspace* of H, that is, a subspace of H that is also a closed set. Hint: By the best approximation property (Lemma 4.14 v), $\|u - P_S(u)\| \leq \|u - u_j\|$ for every j.

4.3 $L^2([-\pi,\pi))$ and Fourier Series

A key example of an infinite dimensional Hilbert space other than $\ell^2(\mathbb{Z})$ is the space of complex valued functions f on the interval $[-\pi, \pi) = \{x \in \mathbb{R} : -\pi \leq x < \pi\}$ that are *square-integrable*, which means that

$$\int_{-\pi}^{\pi} |f(\theta)|^2 \, d\theta < +\infty.$$

(For those familiar with the following terms, we are using the Lebesgue integral here, not the Riemann integral. We regard two functions f and g as the same if f and g agree except on a set of "measure 0," written $f = g$ a.e., where "a.e." stands for "almost everywhere." To be more precise, the relation $f \sim g$ if $f = g$ a.e. is an equivalence relation, and the space we talk about here is a space of equivalence classes of functions modulo this equivalence relation. The condition $f = g$ a.e. is exactly the condition that $\int_{-\pi}^{\pi} |f(\theta) - g(\theta)| \, d\theta = 0$, which is the reason it arises here. For those unfamiliar with these terms, ignoring them will cause no trouble if you are willing to accept a few reasonable properties of the Lebesgue integral on faith.)

Formally, we define

$$L^2([-\pi, \pi)) = \left\{ f : [-\pi, \pi) \to \mathbb{C} : \int_{-\pi}^{\pi} |f(\theta)|^2 \, d\theta < \infty. \right\}$$

Defining pointwise addition and scalar multiplication of functions in the usual way (as in Example 1.33), one can show (Exercise 4.3.1 (i)) that $L^2([-\pi, \pi))$ is a vector space. For $f, g \in L^2([-\pi, \pi))$, define

$$\langle f, g \rangle = \frac{1}{2\pi} \int_{-\pi}^{\pi} f(\theta)\overline{g(\theta)} \, d\theta. \tag{4.14}$$

One can show (Exercise 4.3.1 (ii)) that $\langle \cdot, \cdot \rangle$ is an inner product on $L^2([-\pi, \pi))$. (This is where the identification of two functions that are equal a.e. is necessary, to obtain property I4 in Definition 1.86.) We define a norm $\| \cdot \|$ as in Definition 1.90. In this case we obtain

$$\|f\| = \left(\frac{1}{2\pi} \int_{-\pi}^{\pi} |f(\theta)|^2 \, d\theta \right)^{1/2}.$$

The Cauchy-Schwarz inequality (Lemma 1.91) gives, for $f, g \in L^2([-\pi, \pi))$,

$$\left| \frac{1}{2\pi} \int_{-\pi}^{\pi} f(\theta)\overline{g(\theta)} \, d\theta \right| \le \left(\frac{1}{2\pi} \int_{-\pi}^{\pi} |f(\theta)|^2 \, d\theta \right)^{1/2} \left(\frac{1}{2\pi} \int_{-\pi}^{\pi} |g(\theta)|^2 \, d\theta \right)^{1/2}.$$

Replacing f and g with $|f|$ and $|g|$, respectively, gives

$$\frac{1}{2\pi} \int_{-\pi}^{\pi} |f(\theta)g(\theta)| \, d\theta \le \left(\frac{1}{2\pi} \int_{-\pi}^{\pi} |f(\theta)|^2 \, d\theta \right)^{1/2} \left(\frac{1}{2\pi} \int_{-\pi}^{\pi} |g(\theta)|^2 \, d\theta \right)^{1/2}. \tag{4.15}$$

The triangle inequality (Corollary 1.92) in $L^2([-\pi, \pi))$ gives

$$\left(\frac{1}{2\pi} \int_{-\pi}^{\pi} |f(\theta) + g(\theta)|^2 \, d\theta \right)^{1/2}$$

$$\le \left(\frac{1}{2\pi} \int_{-\pi}^{\pi} |f(\theta)|^2 \, d\theta \right)^{1/2} + \left(\frac{1}{2\pi} \int_{-\pi}^{\pi} |g(\theta)|^2 \, d\theta \right)^{1/2}. \tag{4.16}$$

In addition to $L^2([-\pi, \pi))$, we consider the following class of functions.

Definition 4.15 *Let*

$$L^1([-\pi, \pi)) = \left\{ f : [-\pi, \pi) \to \mathbb{C} : \int_{-\pi}^{\pi} |f(\theta)| \, d\theta < +\infty. \right\}$$

If $f \in L^1([-\pi, \pi))$, we say f is integrable, and we define

$$\|f\|_1 = \int_{-\pi}^{\pi} |f(\theta)| \, d\theta.$$

In other words, for f to be integrable, we require that the integral $\int_{-\pi}^{\pi} f(\theta) \, d\theta$ converges absolutely. To understand the significance of this, suppose momentarily that f is real-valued (otherwise consider the real and imaginary parts of f separately). If f is not integrable, then f has either an infinite amount of positive mass, an infinite amount of negative mass, or both. Then the integral may be infinite, and in the third case, it has the ambiguous form $\infty - \infty$. To avoid this, we define only the integral of complex-valued functions that are integrable. If f is integrable, then $\int_{-\pi}^{\pi} f(\theta) \, d\theta$ is defined. Moreover, we have (Exercise 4.3.10) the inequality:

$$\left| \int_{-\pi}^{\pi} f(\theta) \, d\theta \right| \leq \int_{-\pi}^{\pi} |f(\theta)| \, d\theta. \tag{4.17}$$

If f is integrable, the integral can be broken up more or less any way one likes. That is, suppose $\{A_j\}_j$ is a finite or countably infinite collection of disjoint "reasonable" subsets of $[-\pi, \pi)$ such that $\cup_j A_j = [-\pi, \pi)$. (Here "reasonable" means *measurable*, a term coming from Lebesgue integration theory that we do not define here. We remark that nonmeasurable sets are difficult to construct, requiring the axiom of choice, so any set we encounter in this text is measurable.) Then

$$\int_{-\pi}^{\pi} f(\theta) \, d\theta = \sum_j \int_{A_j} f(\theta) \, d\theta,$$

where the integrability of f guarantees the absolute convergence of the series on the right. This will not work with a nonintegrable function because the series may diverge or we may run into the problem associated with a series of real numbers that converges conditionally: the series can be rearranged to converge to any value one likes (see Exercise 1.2.4).

Note that if $f, g \in L^2([-\pi, \pi))$, then $f \cdot g \in L^1([-\pi, \pi))$, by relation (4.15). Applying this with $g = 1$, we see that L^2 functions on $[-\pi, \pi)$ are integrable, that is, $L^2([-\pi, \pi)) \subseteq L^1([-\pi, \pi))$. We remark that $L^1([-\pi, \pi))$ is larger than $L^2([-\pi, \pi))$ (Exercise 4.3.3).

The definitions of convergence of a sequence of vectors and of a Cauchy sequence in a general complex inner product space in Definition 4.4 apply to $L^2([-\pi, \pi))$. In particular, by Exercise 4.2.1, a sequence of functions $\{f_n\}_{n=1}^{\infty}$ in $L^2([-\pi, \pi))$ converges to some $f \in L^2([-\pi, \pi))$ if and only if $\|f_n - f\|$ converges to 0 as $n \to \infty$. This definition may seem natural at this point, but this is really a new idea because we are not requiring pointwise convergence of the functions f_n to f (see Exercise 4.3.2 to clarify the distinction).

The key point that we assume is that $L^2([-\pi, \pi))$ is complete, that is, every Cauchy sequence in $L^2([-\pi, \pi))$ converges. This is a relatively deep fact, which depends on using the Lebesgue integral (it is not true for the Riemann integral). Thus $L^2([-\pi, \pi))$ is a Hilbert space. Hence the results obtained in the previous section about complete orthonormal sets apply here.

Definition 4.16 *The* trigonometric system *is the set of functions* $\{e^{in\theta}\}_{n\in\mathbb{Z}}$. *A* trigonometric polynomial *is a finite linear combination of elements of the trigonometric system, that is, a function of the form* $\sum_{n=-N}^{N} c_n e^{in\theta}$ *for some $N \in \mathbb{N}$ and some complex numbers $\{c_n\}_{n=-N}^{N}$.*

Lemma 4.17 *The trigonometric system is an orthonormal set in* $L^2([-\pi, \pi))$.

Proof
Note that

$$\langle e^{ik\theta}, e^{ij\theta} \rangle = \frac{1}{2\pi} \int_{-\pi}^{\pi} e^{ik\theta} \overline{e^{ij\theta}} \, d\theta = \frac{1}{2\pi} \int_{-\pi}^{\pi} e^{i(k-j)\theta} \, d\theta.$$

If $k = j$, this is 1. If $k \neq j$, we integrate to get

$$\frac{1}{2\pi i(k-j)} \left(e^{i(k-j)\pi} - e^{i(k-j)(-\pi)} \right) = 0,$$

because $e^{i(k-j)\theta}$ is periodic with period 2π. ∎

Our goal is to prove that the trigonometric system is complete in $L^2([-\pi, \pi))$. The following elementary lemma plays a critical role.

Lemma 4.18 *Suppose $\theta_0 \in (-\pi, \pi)$ and $\alpha > 0$ is sufficiently small that $-\pi < \theta_0 - \alpha < \theta_0 + \alpha < \pi$. Define intervals*

$$I = (\theta_0 - \alpha, \theta_0 + \alpha),$$

and

$$J = (\theta_0 - \alpha/2, \theta_0 + \alpha/2).$$

Then there exists $\delta > 0$ and a sequence of real-valued trigonometric polynomials $\{p_n(\theta)\}_{n=1}^\infty$ such that
  i. $p_n(\theta) \geq 1$ for $\theta \in I$.
  ii. $p_n(\theta) \geq (1 + \delta)^n$ for $\theta \in J$.
  iii. $|p_n(\theta)| \leq 1$ for $\theta \in [-\pi, \pi) \setminus I$.

Proof
Define

$$t(\theta) = 1 + \cos(\theta - \theta_0) - \cos\alpha.$$

Note that $\alpha < \pi$, by the requirement that $-\pi < \theta_0 - \alpha < \theta_0 + \alpha < \pi$. Observe that $\cos x$ is even on $[-\pi, \pi]$ and decreasing on $[0, \pi]$. Since $t(\theta_0 + \alpha) = 1 = t(\theta_0 - \alpha)$, we see that $t(\theta) \geq 1$ for $\theta \in I$, and, for some $\delta > 0$, $t(\theta) \geq 1 + \delta$ for $\theta \in J$. Also, looking at the graph of $t(\theta)$, the 2π-periodicity of $\cos x$ shows that $t(\theta) \leq 1$ for all points $\theta \in [-\pi, \pi) \setminus I$. However, $t(\theta) \geq -1$ at all points, because $|\cos x| \leq 1$ for all x. Therefore, $|t(\theta)| \leq 1$ for $\theta \in [-\pi, \pi) \setminus I$. We define

$$p_n(\theta) = (t(\theta))^n.$$

Then parts i, ii, and iii follow from the observations regarding t.

It is clear that $t(\theta)$ is real valued, hence so is $p_n(\theta)$. What remains is to show that p_n is a trigonometric polynomial. However, $1 - \cos\alpha$ is a constant (hence a multiple of the trigonometric system element $e^{i0\theta} = 1$) and

$$\cos(\theta - \theta_0) = \frac{1}{2}\left(e^{i(\theta - \theta_0)} + e^{i(-\theta + \theta_0)}\right) = \frac{e^{-i\theta_0}}{2}e^{i\theta} + \frac{e^{i\theta_0}}{2}e^{-i\theta}.$$

Hence $t(\theta)$ is a trigonometric polynomial. But any product of trigonometric polynomials is also a trigonometric polynomial (by multiplying out and using $e^{ik\theta}e^{im\theta} = e^{i(k+m)\theta}$ for all k, m). Therefore p_n is a trigonometric polynomial for all n. ∎

Lemma 4.18 is needed to prove the following.

Lemma 4.19 *Suppose $f : [-\pi, \pi) \to \mathbb{C}$ is continuous and bounded, say*

$$|f(\theta)| \leq M \quad \text{for all} \quad \theta.$$

If

$$\langle f, e^{in\theta} \rangle = \frac{1}{2\pi} \int_{-\pi}^{\pi} f(\theta) e^{-in\theta} \, d\theta = 0 \quad \text{for all} \quad n \in \mathbb{Z}, \qquad (4.18)$$

then $f(\theta) = 0$ for all $\theta \in [-\pi, \pi)$.

Proof

First we suppose f is real valued. We argue by contradiction. Suppose f is not identically 0. Then there exists some point $\theta_0 \in [-\pi, \pi)$ and some $\epsilon > 0$ such that $f(\theta_0) > 2\epsilon$ (or else $f(\theta_0) < -2\epsilon$, in which case we replace f by $-f$). By continuity we can assume $\theta_0 \in (-\pi, \pi)$. Again by continuity, there exists some $\alpha > 0$ such that $-\pi < \theta_0 - \alpha < \theta_0 + \alpha < \pi$ and $f(\theta) > \epsilon$ for $\theta_0 - \alpha < \theta < \theta_0 + \alpha$. Define I, J, and $\{p_n\}_{n=1}^{\infty}$ as in Lemma 4.18.

By equation (4.18) and linearity, f is orthogonal to any trigonometric polynomial. Thus

$$0 = \int_{-\pi}^{\pi} f(\theta) p_n(\theta) \, d\theta$$

$$= \int_{[-\pi,\pi)\setminus I} f(\theta) p_n(\theta) \, d\theta + \int_{I\setminus J} f(\theta) p_n(\theta) \, d\theta + \int_{J} f(\theta) p_n(\theta) \, d\theta, \qquad (4.19)$$

for all $n \geq 1$. Since $|p_n(\theta)| \leq 1$ on $[-\pi, \pi) \setminus I$ (Lemma 4.18 iii),

$$\left| \int_{[-\pi,\pi)\setminus I} f(\theta) p_n(\theta) \, d\theta \right| \leq 1 \cdot \sup_{[-\pi,\pi)} |f(\theta)| \cdot 2\pi \leq 2\pi M,$$

for any n. Since f and p_n are both positive on $I \setminus J$ (Lemma 4.18 i),

$$\int_{I\setminus J} f(\theta) p_n(\theta) \, d\theta \geq 0.$$

On J, f is bounded below by $\epsilon > 0$ and p_n is bounded below by $(1+\delta)^n$ (Lemma 4.18 ii). Hence

$$\int_{J} f(\theta) p_n(\theta) \, d\theta \geq \epsilon(1 + \delta)^n \alpha.$$

Putting these estimates together, we see that the right side of equation (4.19) goes to $+\infty$ as $n \to \infty$. But that contradicts the equality in equation (4.19). This contradiction shows that f must be identically 0.

Now suppose f is complex valued, say $f = u + iv$, with u and v real valued. Then for any $n \in \mathbb{Z}$,

$$\int_{-\pi}^{\pi} \overline{f(\theta)} e^{-in\theta} \, d\theta = \overline{\int_{-\pi}^{\pi} f(\theta) e^{in\theta} \, d\theta} = \overline{0} = 0,$$

by equation (4.18). Hence by linearity, $u = (f + \overline{f})/2$ is orthogonal to all elements of the trigonometric system. By the case of real-valued functions already considered, we obtain that u is identically 0. Similarly, $v = (f - \overline{f})/2i$ is zero. ∎

Lemma 4.19 comes close to proving the completeness of the trigonometric system, because it says that a continuous, bounded function on $[-\pi, \pi)$ that is orthogonal to all elements of the trigonometric system is identically 0. However, we would like to obtain this for all functions in $L^2([-\pi, \pi))$. In fact, we prove this for all functions in the larger class $L^1([-\pi, \pi))$. To do this we assume the following generalization of the Fundamental theorem of calculus (FTC) in the setting of Lebesgue integration.

Theorem 4.20 *Suppose $f \in L^1([-\pi, \pi))$. Define $F : [-\pi, \pi) \to \mathbb{C}$ by*

$$F(\theta) = \int_{-\pi}^{\theta} f(t) \, dt.$$

Then F is continuous on $[-\pi, \pi)$ and F is differentiable a.e., with $F'(\theta) = f(\theta)$ a.e.

Now we can state a uniqueness result for Fourier series.

Theorem 4.21 *Suppose $f \in L^1([-\pi, \pi))$ and*

$$\langle f, e^{in\theta} \rangle = \frac{1}{2\pi} \int_{-\pi}^{\pi} f(\theta) e^{-in\theta} \, d\theta = 0 \quad \text{for all} \quad n \in \mathbb{Z}.$$

Then $f(\theta) = 0$ a.e.

Proof
Define $F(\theta) = \int_{-\pi}^{\theta} f(t) \, dt$. By Theorem 4.20, F is continuous and $F' = f$ a.e. Also F is bounded: for all $\theta \in [-\pi, \pi)$, inequality (4.17) yields

$$|F(\theta)| \leq \int_{-\pi}^{\theta} |f(\theta)| \, d\theta \leq \int_{-\pi}^{\pi} |f(\theta)| \, d\theta,$$

which is finite since $f \in L^1([-\pi, \pi))$. For $n \neq 0$,

$$\int_{-\pi}^{\pi} F(\theta) e^{-in\theta} \, d\theta = \int_{-\pi}^{\pi} \int_{-\pi}^{\theta} f(t) \, dt \, e^{-in\theta} \, d\theta.$$

Interchanging the order of integration (allowed by a result regarding Lebesgue integration called *Fubini's theorem*), the last expression is equal to

$$\int_{-\pi}^{\pi} f(t) \int_{t}^{\pi} e^{-in\theta} \, d\theta \, dt = \int_{-\pi}^{\pi} f(t) \frac{1}{-in} \left(e^{-in\pi} - e^{-int} \right) dt = 0,$$

by our assumptions on f. Let

$$A = \frac{1}{2\pi} \int_{-\pi}^{\pi} F(\theta) \, d\theta.$$

For $n \neq 0$,

$$\int_{-\pi}^{\pi} (F(\theta) - A) e^{-in\theta} \, d\theta = 0,$$

by the above result for F because A is constant, and hence orthogonal to $e^{-in\theta}$ for $n \neq 0$. For $n = 0$,

$$\int_{-\pi}^{\pi} (F(\theta) - A) e^{-i0\theta} \, d\theta = \int_{-\pi}^{\pi} F(\theta) \, d\theta - 2\pi A = 0,$$

by definition of A. Thus $F(\theta) - A$ is a continuous, bounded function that is orthogonal to all elements of the trigonometric system. So by Lemma 4.19, $F(\theta) - A = 0$ for all θ, that is, $F(\theta) = A$ for all θ. Hence $f(\theta) = 0$ a.e., since $f = F'$ a.e. ∎

Corollary 4.22 *The trigonometric system is complete in $L^2([-\pi, \pi))$.*

Proof

This corollary follows from Theorem 4.21 since $L^2([-\pi, \pi)) \subseteq L^1([-\pi, \pi))$. ∎

Definition 4.23 *Suppose $f \in L^1([-\pi, \pi))$. For $n \in \mathbb{Z}$, $\langle f, e^{in\theta} \rangle$ is the nth Fourier coefficient of f. The series*

$$\sum_{n \in \mathbb{Z}} \langle f, e^{in\theta} \rangle e^{in\theta} \tag{4.20}$$

is the Fourier series of *f*. *The* N^{th} *partial sum of the Fourier series of f is*

$$s_N(f) = \sum_{n=-N}^{N} \langle f, e^{in\theta} \rangle e^{in\theta}. \qquad (4.21)$$

By applying the results in section 4.2, we obtain Corollary 4.24.

Corollary 4.24

  i. *Suppose* $z = (z(n))_{n \in \mathbb{Z}} \in \ell^2(\mathbb{Z})$. *Then the series*

$$\sum_{n \in \mathbb{Z}} z(n) e^{in\theta}$$

converges to an element of $L^2([-\pi, \pi))$.

  ii. *(Plancherel's formula) Suppose* $f \in L^2([-\pi, \pi))$. *Then the sequence* $\{\langle f, e^{in\theta} \rangle\}_{n \in \mathbb{Z}} \in \ell^2(\mathbb{Z})$, *and*

$$\sum_{n \in \mathbb{Z}} |\langle f, e^{in\theta} \rangle|^2 = \|f\|^2 = \frac{1}{2\pi} \int_{-\pi}^{\pi} |f(\theta)|^2 \, d\theta. \qquad (4.22)$$

  iii. *(Parseval's relation) Suppose* $f, g \in L^2([-\pi, \pi))$. *Then*

$$\langle f, g \rangle = \sum_{n \in \mathbb{Z}} \langle f, e^{in\theta} \rangle \overline{\langle g, e^{in\theta} \rangle}. \qquad (4.23)$$

  iv. *(Fourier inversion) For any* $f \in L^2([-\pi, \pi))$,

$$f(\theta) = \sum_{n \in \mathbb{Z}} \langle f, e^{in\theta} \rangle e^{in\theta}, \qquad (4.24)$$

in the sense that the partial sums $s_N(f)$ *(defined in equation (4.21)) of the series on the right side of equation (4.24) converge in* $L^2([-\pi, \pi))$ *to f, that is,*

$$\|s_N(f) - f\| \to 0 \quad as \quad N \to \infty. \qquad (4.25)$$

Proof

Part i follows from Lemmas 4.7 and 4.16. Parts ii and iii follow from Lemma 4.11 and Corollary 4.22. Part iv follows from Theorem 4.10. ∎

Equation (4.24) should be compared to equation (2.15) in the finite dimensional case. We regard $e^{in\theta}$ as a pure frequency because its real and imaginary parts are $\cos n\theta$ and $\sin n\theta$, respectively. Notice that $\cos n\theta$ and $\sin n\theta$ oscillate n times over the interval $[-\pi, \pi)$. Therefore, as n increases, the frequency of $e^{in\theta}$ increases without bound. This is unlike the case in chapter 2, where there were only a finite set of possible pure frequencies.

The Fourier inversion formula (4.24) states that a general function in $L^2([-\pi, \pi))$ can be written as a superposition of pure frequencies. The Fourier coefficient $\langle f, e^{in\theta} \rangle$ in equation (4.24) is the strength of the pure frequency $e^{in\theta}$ in f. If we think of a function f on $[-\pi, \pi)$ as an audio signal, and if the sound is high pitched, then there must be at least one large value of $|n|$ such that $\langle f, e^{in\theta} \rangle$ is large. If the sound is low pitched, its Fourier coefficients must be large for some small value of $|n|$.

Part iv of Corollary 4.24 shows that any $f \in L^2([-\pi, \pi))$ is represented by its Fourier series in the sense of convergence in the norm on $L^2([-\pi, \pi))$. The question of pointwise convergence of $s_N(f)(\theta)$ to $f(\theta)$ is much more delicate and has been extensively studied. On the negative side, du Bois Reymond showed in 1876 that the Fourier series of a continuous function can diverge at a point. In 1926 Kolmogoroff gave an example of $f \in L^1([-\pi, \pi))$, for which the Fourier series diverges everywhere. On the positive side, if f has a reasonable amount of smoothness, an elementary argument shows that the partial sums $s_N(f)$ converge to f at every point (Exercise 4.3.16). There are many more refined results than this, with weaker assumptions. Finally in 1966 Carleson proved that the Fourier series of any $f \in L^2([-\pi, \pi))$ converges to f a.e. This last result is extremely deep and difficult. For our purposes, we need only the basic result in Corollary 4.24 iv.

When working with linear transformations on an infinite dimensional space, one has to be careful. For example, the derivative operator is a linear transformation on the class of differentiable functions. In particular,

$$\frac{d}{d\theta}e^{in\theta} = ine^{in\theta}. \tag{4.26}$$

By analogy with the finite dimensional case, one might think that one could use linearity and the Fourier inversion formula (4.24) to deduce that for any $f \in L^2([-\pi, \pi))$ and any θ,

$$f'(\theta) = \sum_{n \in \mathbb{Z}} \langle f, e^{in\theta} \rangle \frac{d}{d\theta} e^{in\theta} = \sum_{n \in \mathbb{Z}} in \langle f, e^{in\theta} \rangle e^{in\theta}. \tag{4.27}$$

However, this last series may not converge (in L^2 or pointwise at θ), and f may not be differentiable at all points (for example an L^2 function can be discontinuous at many points). (We remark that there is a certain generalized definition of the derivative, arising in a subject known as *distribution theory*, for which equation (4.27) is true.) The source of this problem is that the derivative is not a bounded operator on $L^2([-\pi, \pi))$.

Definition 4.25 *Suppose H_1 and H_2 are Hilbert spaces (possibly the same) with norms $\| \cdot \|_1$ and $\| \cdot \|_2$, respectively, as in Definition 1.90. Suppose $T : H_1 \to H_2$ is a linear transformation. We say T is* bounded *if there exists $C > 0$ such that*

$$\|T(x)\|_2 \leq C\|x\|_1 \tag{4.28}$$

for all $x \in H_1$. The infimum of all C such that equation (4.28) holds is called the operator norm *of T; it is denoted $\|T\|$.*

Lemma 4.2.6 shows how the boundedness property plays a role.

Lemma 4.26 *Suppose H is a Hilbert space and $T : H \to H$ is a bounded linear transformation. Suppose the series $\sum_{n \in \mathbb{Z}} x_n$ converges in H (Definition 4.6). Then*

$$T\left(\sum_{n \in \mathbb{Z}} x_n\right) = \sum_{n \in \mathbb{Z}} T(x_n),$$

where the series on the right converges in H.

Proof
Exercise 4.3.8. ∎

Like the Fourier basis for $\ell^2(\mathbb{Z}_N)$, the trigonometric system diagonalizes translation-invariant (bounded) linear transformations, as defined below. For this and other definitions, we extend $f \in L^2([-\pi, \pi))$ periodically with period 2π to all of $\mathbb{R}$, that is, so that

$$f(\theta + 2\pi) = f(\theta) \tag{4.29}$$

for all $\theta \in \mathbb{R}$.

Definition 4.27   *For $\varphi \in \mathbb{R}$, define the translation operator τ_φ :* $L^2([-\pi, \pi)) \to L^2([-\pi, \pi))$ *by*

$$(\tau_\varphi f)(\theta) = f(\theta - \varphi).$$

A linear transformation $T : L^2([-\pi, \pi)) \to L^2([-\pi, \pi))$ is translation-invariant if it commutes with τ_φ for every $\varphi \in \mathbb{R}$:

$$T(\tau_\varphi f)(\theta) = \tau_\varphi(T(f))(\theta),$$

where $\tau_\varphi(T(f))(\theta) = T(f)(\theta - \varphi)$.

Theorem 4.28   *Suppose $T : L^2([-\pi, \pi)) \to L^2([-\pi, \pi))$ is a bounded, translation-invariant linear transformation. Then for each $m \in \mathbb{Z}$, there exists $\lambda_m \in \mathbb{C}$ such that*

$$T(e^{im\theta}) = \lambda_m e^{im\theta}. \tag{4.30}$$

Proof

Fix $m \in \mathbb{Z}$. By Corollary 4.24 iv, we can write

$$T(e^{im\theta}) = \sum_{n \in \mathbb{Z}} c_n e^{in\theta}, \tag{4.31}$$

where $c_n = \langle T(e^{im\theta}), e^{in\theta} \rangle$. Let $\varphi \in \mathbb{R}$ be arbitrary. Then

$$\tau_\varphi(e^{im\theta}) = e^{im(\theta - \varphi)} = e^{-im\varphi} e^{im\theta}.$$

Hence by linearity of T and equation (4.31),

$$T(\tau_\varphi(e^{im\theta})) = e^{-im\varphi} T(e^{im\theta}) = \sum_{n \in \mathbb{Z}} c_n e^{-im\varphi} e^{in\theta}.$$

On the other hand, equation (4.31) also implies

$$\tau_\varphi T(e^{im\theta}) = \sum_{n \in \mathbb{Z}} c_n e^{in(\theta - \varphi)} = \sum_{n \in \mathbb{Z}} c_n e^{-in\varphi} e^{in\theta}.$$

However, $T(\tau_\varphi(e^{im\theta})) = \tau_\varphi T(e^{im\theta})$ by the assumption of translation invariance of T. Hence by the uniqueness of Fourier coefficients (Exercise 4.3.9),

$$c_n e^{-im\varphi} = c_n e^{-in\varphi},$$

for every n and φ. If $n \neq m$, this implies that $c_n = 0$. Returning to equation (4.31), this means that

$$T(e^{im\theta}) = c_m e^{im\theta}.$$

Since $m \in \mathbb{Z}$ is arbitrary, this completes the proof. ∎

You may have seen the analog of Fourier series (i.e., equation (4.24)) written in real notation (see Exercise 4.3.6). This has the advantage that if f is real valued, all the expansion coefficients are real valued also. However, this expansion lacks the key property described in Theorem 4.28 (see Exercise 4.3.13): the functions $\sin n\theta$ and $\cos n\theta$ do not diagonalize translation-invariant linear transformations.

Observe that the proof of Theorem 4.28 is analogous to the proof of Theorem 2.18. Also note that Theorem 4.28 shows that a linear transformation on an infinite dimensional space can have infinitely many eigenvectors. Because these vectors are functions in this case, they are called *eigenfunctions*.

Suppose $T : L^2([-\pi, \pi)) \to L^2([-\pi, \pi))$ is a bounded, translation-invariant linear transformation. Writing any $f \in L^2([-\pi, \pi))$ in its Fourier series expansion

$$f(\theta) = \sum_{n \in \mathbb{Z}} c(n) e^{in\theta},$$

where $c(n) = \langle f, e^{in\theta} \rangle$, we get (by Lemma 4.26)

$$T(f)(\theta) = \sum_{n \in \mathbb{Z}} c(n) T(e^{in\theta}) = \sum_{n \in \mathbb{Z}} c(n) \lambda_n e^{in\theta}.$$

Thus from the point of view of the Fourier coefficients (the components of f in the trigonometric system), the effect of T is just to multiply the nth component $c(n)$ by λ_n. This is just like multiplying a vector by a diagonal matrix in the finite dimensional case (and we can do that similarly here if we develop the theory of infinite matrices). So we say that T is diagonalized by the trigonometric system.

We remark that the derivative operator is translation invariant:

$$\frac{d}{d\theta}(f(\theta - \varphi)) = \frac{df}{d\theta}(\theta - \varphi),$$

by the chain rule. However, the derivative is not defined on all of $L^2([-\pi, \pi))$, just on the subspace of differentiable functions, and it is not a bounded operator even on that subspace. So Theorem 4.28 does not apply, but nevertheless heuristically we understand that $d/d\theta$ is diagonalized by the trigonometric system because

$$\frac{d}{d\theta} e^{in\theta} = in e^{in\theta}.$$

This shows that a result may be true in a more broad sense than we know how to prove or even to formulate. Finding a version of Theorem 4.28 that includes the derivative operator requires a substantial amount of sophisticated material, so we stop here with Theorem 4.28 and a hint that the result is more general. The diagonalization of translation-invariant linear transformations is the real reason for the effectiveness of Fourier series in the study of differential equations: the trigonometric system diagonalizes $d/d\theta$ and, more generally, all constant coefficient differential operators because they are translation invariant.

Exercises

4.3.1. For $f, g \in L^2([-\pi, \pi))$, and $\alpha \in \mathbb{C}$, define $f + g$ and αf by

$$(f + g)(\theta) = f(\theta) + g(\theta) \quad \text{and} \quad (\alpha f)(\theta) = \alpha f(\theta),$$

for $-\pi \le \theta < \pi$.

i. Prove that $L^2([-\pi, \pi))$, with these operations, is a vector space. Remark: All properties in Definition 1.30 are obvious except A1. Use Exercise 1.6.3 (i).

ii. Prove that $\langle \cdot, \cdot \rangle$ defined in equation (4.14) is an inner product on $L^2([-\pi, \pi))$. Remark: The main point is to see that the integral in equation (4.14) converges absolutely, so that $\langle f, g \rangle$ is defined. We cannot use inequality (4.15) because its proof depends on knowing that $\langle \cdot, \cdot \rangle$ is an inner product on $L^2([-\pi, \pi))$. Use the inequality in Exercise 1.6.3 (i).

4.3.2. i. For $n \in \mathbb{N}$, define a function f_n on $[-\pi, \pi)$ by setting $f_n(\theta) = \sqrt{n}$ if $0 < \theta < 1/n$ and $f_n(\theta) = 0$ otherwise.

Observe that $f_n \in L^2([-\pi, \pi))$ for each n. Prove that the sequence $\{f_n\}_{n \in \mathbb{N}}$ converges *pointwise* to 0, which means that for each $\theta \in [-\pi, \pi)$, $\lim_{n \to \infty} f_n(\theta) = 0$. However, prove that $\{f_n\}_{n \in \mathbb{N}}$ does not converge to 0 in *norm*, that is, in the space $L^2([-\pi, \pi))$ (recall Definition 4.4).

ii. Define a sequence $\{g_n\}_{n \in \mathbb{N}}$ of functions on $[-\pi, \pi)$ as follows. For $n \in \mathbb{N}$, we can write $n = 2^j + k$ for some $k \in \{0, 1, \ldots, 2^j - 1\}$ in a unique way, with $j \in \mathbb{Z}$ and $j \geq 0$. Define $g_{2^j+k}(\theta) = 1$ for $2^{-j}k\pi \leq |\theta| \leq 2^{-j}(k+1)\pi$, and $g_{2^j+k}(\theta) = 0$ otherwise. Prove that the sequence $\{g_n\}_{n \in \mathbb{N}}$ converges to 0 in $L^2([-\pi, \pi))$ (i.e., in norm), but at every point $\theta \in [-\pi, \pi)$, the sequence of numbers $g_n(\theta)$ does not converge as $n \to \infty$ (in particular, we do not have pointwise convergence).

Remark: This shows that in $L^2([-\pi, \pi))$, there is no implication between norm convergence and pointwise convergence, unlike the case of $\ell^2(\mathbb{Z})$ discussed in Exercise 4.1.8.

4.3.3. Define $f(\theta) = 1/\sqrt{|\theta|}$ for $\theta \neq 0$, and $f(0) = 0$. Show that $f \in L^1([-\pi, \pi))$ but $f \notin L^2([-\pi, \pi))$.

4.3.4. i. Define $f : [-\pi, \pi) \to \mathbb{R}$ by

$$f(\theta) = \begin{cases} 1 & \text{if } 0 \leq \theta < \pi \\ 0 & \text{if } -\pi \leq \theta < 0. \end{cases}$$

Compute the Fourier coefficients of f. Note that their decay is only of order $1/|n|$; this reflects the discontinuity of f at the origin. Intuitively, large high-frequency components are required to synthesize a discontinuity.

ii. Define $g : [-\pi, \pi) \to \mathbb{R}$ by $g(\theta) = |\theta|$. Compute the Fourier coefficients of g. Note that they decay on the order of $1/n^2$. This decay is faster than that in part i because of the continuity of g (or, more precisely, the continuity of the 2π-periodic extension of g).

4.3.5. For $a > 0$, define

$$L^2([-a, a)) = \left\{ f : [-a, a) \to \mathbb{C} : \int_{-a}^{a} |f(t)|^2 \, dt < \infty \right\}.$$

As in Exercises 4.3.1 and 4.3.2, $L^2([-a, a))$ is a complex inner product space with inner product

$$\langle f, g \rangle = \frac{1}{2a} \int_{-a}^{a} f(t)\overline{g(t)}\, dt.$$

Assume this fact and also the fact that $L^2([-a, a))$ is a Hilbert space. Prove that the system $\{e^{in\pi t/a}\}_{n \in \mathbb{Z}}$ is a complete orthonormal set in $L^2([-a, a))$. Hint: Change variables to reduce to the case of $[-\pi, \pi)$.

4.3.6. (Fourier series in real notation)
  i. Show that the set of functions

$$\{1\} \cup \{\sqrt{2} \cos n\theta\}_{n=1}^{\infty} \cup \{\sqrt{2} \sin n\theta\}_{n=1}^{\infty}$$

  is a complete orthonormal set in $L^2([-\pi, \pi))$. Hint: For completeness, use Corollary 4.22 and Euler's formulas.
  ii. Deduce that each $f \in L^2([-\pi, \pi))$ has an expansion

$$\frac{a_0}{2} + \sum_{n=1}^{\infty} a_n \cos n\theta + \sum_{n=1}^{\infty} b_n \sin n\theta,$$

where

$$a_n = \frac{1}{\pi} \int_{-\pi}^{\pi} f(\theta) \cos n\theta\, d\theta$$

for all $n \geq 0$, and

$$b_n = \frac{1}{\pi} \int_{-\pi}^{\pi} f(\theta) \sin n\theta\, d\theta$$

for all $n \geq 1$. Describe the sense in which this series converges to f.
  iii. Let f, $\{a_n\}_{n=0}^{\infty}$, and $\{b_n\}_{n=1}^{\infty}$ be as in part ii. For $n \in \mathbb{Z}$, let $c_n = \langle f, e^{in\theta} \rangle$. Deduce the relations

$$a_0 = 2c_0, \quad a_n = c_n + c_{-n}\, (n \geq 1), \quad b_n = i(c_n - c_{-n})\, (n \geq 1)$$

and, going the other way,

$$c_0 = a_0/2,$$
$$c_n = (a_n - ib_n)/2\ (n \geq 1),$$
$$c_n = (a_{-n} - ib_{-n})/2\ (n \leq -1).$$

Hence it is straightforward to pass back and forth between the expansion in part ii and equation (4.24).

4.3.7. Define

$$L^2([0, \pi)) = \left\{ f : [0, \pi) \to \mathbb{C} : \int_0^\pi |f(\theta)|^2 \, d\theta < \infty \right\}.$$

As in Exercises 4.3.1 and 4.3.2, $L^2([0, \pi))$ is a complex inner product space with inner product

$$\langle f, g \rangle = \frac{1}{\pi} \int_0^\pi f(\theta)\overline{g(\theta)} \, d\theta.$$

Assume this fact and also the fact that $L^2([0, \pi))$ is a Hilbert space.

  i. Prove that the set $\{\sqrt{2} \sin(n\theta)\}_{n=1}^\infty$ is a complete orthonormal set in $L^2([0, \pi))$. Hint: To prove the completeness, define $g : [-\pi, \pi) \to \mathbb{C}$ by setting $g(\theta) = f(\theta)$ for $0 \le \theta < \pi$, and $g(\theta) = -f(-\theta)$ for $-\pi \le \theta < 0$ (g is the *odd extension* of f). Show that g is orthogonal to the trigonometric system in $L^2([-\pi, \pi))$.

  ii. Prove that the set $\{1, \sqrt{2} \cos \theta, \sqrt{2} \cos 2\theta, \sqrt{2} \cos 3\theta, \dots \}$ is a complete orthonormal set in $L^2([0, \pi))$. Hint: Consider the *even extension* of f, appropriately defined.

4.3.8. Prove Lemma 4.26. Hint: First prove that $\sum_{n \in \mathbb{Z}} T(x_n)$ converges by showing that its partial sums are Cauchy. Let $s_N = \sum_{n=-N}^N x_n$ and $s = \sum_{n \in \mathbb{Z}} x_n$. Since s_N is defined by a finite sum, linearity gives $T(s_N) = \sum_{n=-N}^N T(x_n)$. Now use

$$\|T(s_N) - T(s)\| \le C\|s_N - s\|.$$

4.3.9. (Uniqueness for Fourier series) Suppose

$$\{c(n)\}_{n \in \mathbb{Z}}, \{d(n)\}_{n \in \mathbb{Z}} \in \ell^2(\mathbb{Z}),$$

and

$$\sum_{n \in \mathbb{Z}} c(n)e^{in\theta} = \sum_{n \in \mathbb{Z}} d(n)e^{in\theta}$$

(where both series converge in $L^2([-\pi, \pi))$, by Corollary 4.24 i). Prove that $c(n) = d(n)$ for all n.

4.3.10. Suppose $f \in L^1([-\pi, \pi))$. Prove relation (4.17). Hint: This is relatively easy for f real valued. If f is complex valued, we can write $\int_{-\pi}^{\pi} f(\theta)\, d\theta = re^{i\alpha}$ for some $r \geq 0$ and $\alpha \in [0, 2\pi)$. Show that

$$r = \int_{-\pi}^{\pi} \operatorname{Re}\left(e^{-i\alpha} f(\theta)\right)\, d\theta.$$

4.3.11. Suppose $f, g \in L^1([-\pi, \pi))$, and extend f and g periodically to all of $\mathbb{R}$ with period 2π, as in equation (4.29). Define $f * g$, the *convolution* of f and g, by

$$f * g(\theta) = \int_{-\pi}^{\pi} f(\theta - \varphi) g(\varphi)\, d\varphi. \tag{4.32}$$

Prove that $f * g \in L^1([-\pi, \pi))$. Hint:

$$\int_{-\pi}^{\pi} \left| \int_{-\pi}^{\pi} f(\theta - \varphi) g(\varphi)\, d\varphi \right| d\theta \leq \int_{-\pi}^{\pi} \int_{-\pi}^{\pi} |f(\theta - \varphi) g(\varphi)|\, d\varphi\, d\theta,$$

by inequality (4.17). Because the last integrand is nonnegative, you can switch the order of integration (assume this; it follows from a result called *Tonelli's theorem*). Then make a change of variable.

4.3.12. Let f, g and $f * g$ be as in Exercise 4.3.11.

i. Prove that

$$\langle f * g, e^{in\theta} \rangle = 2\pi \langle f, e^{in\theta} \rangle \langle g, e^{in\theta} \rangle. \tag{4.33}$$

In other words, up to a factor of 2π, taking Fourier coefficients turns convolution into multiplication, much like the case in Lemma 2.30.

ii. Prove that $f * g = g * f$.

iii. Let τ_φ be the translation operator from Definition 4.27. Set $T_g(f) = f * g$. Show that T_g is translation invariant, that is, that $T_g(\tau_\varphi f) = \tau_\varphi(T_g(f))$.

4.3.13. Let $j \in \mathbb{Z}$. For $f \in L^2([-\pi, \pi))$, define

$$T_j(f) = f * e^{ij\theta}.$$

See Exercise 4.3.11 for the definition of convolution.

i. Prove that $T_j(f) = \langle f, e^{ij\theta} \rangle e^{ij\theta}$, for all $f \in L^2([-\pi, \pi))$.

ii. Deduce that $T_j(f) \in L^2([-\pi, \pi))$, if $f \in L^2([-\pi, \pi))$. Then show that $T_j : L^2([-\pi, \pi)) \to L^2([-\pi, \pi))$ is bounded (Definition 4.25).

iii. By Exercise 4.3.12 (iii), T_j is translation invariant, so by the previous part, T_j satisfies the conditions of Theorem 4.28. Determine the values λ_m, for $m \in \mathbb{Z}$, in Theorem 4.28, for $T = T_j$.

iv. We have just noted that T_j is diagonalized by the trigonometric system. Observe that for $j \neq 0$, $\cos j\theta$ and $\sin j\theta$ are not eigenfunctions of T_j (i.e., $T_j(\cos j\theta)$ is not a multiple of $\cos j\theta$, and similarly for $\sin j\theta$). Thus the system associated with Fourier series in real notation (Exercise 4.3.6) does not diagonalize bounded translation-invariant linear transformations.

4.3.14. Suppose $f \in L^1([-\pi, \pi))$. Define $s_N(f)$ by equation (4.21). Prove that

$$s_N(f) = \frac{1}{2\pi} D_N * f,$$

where convolution $*$ is defined in Exercise 4.3.11, and

$$D_N(\theta) = \sum_{n=-N}^{N} e^{in\theta}.$$

The function D_N is the *Dirichlet kernel*. Hint: Writing out the partial sum gives

$$s_N(f)(\theta) = \sum_{n=-N}^{N} \frac{1}{2\pi} \int_{-\pi}^{\pi} f(\varphi) e^{-in\varphi} \, d\varphi e^{in\theta}.$$

Bringing the sum inside the integral gives $D_N * f(\theta)$.

4.3.15. Define D_N as in Exercise 4.3.14. Prove that

$$D_N(\theta) = \frac{\sin(N + 1/2)\theta}{\sin(\theta/2)},$$

for $\theta \neq 0$, and $D_N(0) = 2N + 1$. Hint: After factoring out $e^{-iN\theta}$, what is left is the partial sum of a geometric series.

4.3.16. Suppose $f : \mathbb{R} \to \mathbb{C}$ is periodic with period 2π (i.e., f satisfies equation (4.29)). Suppose also that at every point $t \in \mathbb{R}$, $f''(t)$ exists, and f'' is continuous. Prove that at every point t, $s_N(f)(t)$ converges to $f(t)$ as $N \to \infty$. Hints: Note from the definition of D_N in Exercise 4.3.14 that

$$\frac{1}{2\pi} \int_{-\pi}^{\pi} D_N(\theta) \, d\theta = 1.$$

Therefore, using Exercises 4.3.14, 4.3.12 (ii), and 4.3.15, we can write

$$s_N(f)(t) - f(t) = \frac{1}{2\pi} \int_{-\pi}^{\pi} f(t - \theta) D_N(\theta) \, d\theta$$

$$- \frac{1}{2\pi} \int_{-\pi}^{\pi} f(t) D_N(\theta) \, d\theta$$

$$= \frac{1}{2\pi} \int_{-\pi}^{\pi} (f(t - \theta) - f(t)) D_N(\theta) \, d\theta$$

$$= \frac{1}{2\pi} \int_{-\pi}^{\pi} \frac{f(t - \theta) - f(t)}{\sin(\theta/2)} \sin((N + 1/2)\theta) \, d\theta.$$

Fixing t, define

$$g(\theta) = \frac{f(t - \theta) - f(t)}{\sin(\theta/2)}$$

for $\theta \neq 0$ and $g(0) = -2f'(t)$. Prove that g is differentiable with a continuous derivative on $\mathbb{R}$ (this uses the assumptions on f''). Then integrate by parts to obtain

$$s_N(f)(t) - f(t) = \frac{1}{\pi(2N + 1)} \int_{-\pi}^{\pi} g'(\theta) \cos((N + 1/2)\theta) \, d\theta.$$

4.4 The Fourier Transform and Convolution on $\ell^2(\mathbb{Z})$

Our goal in this section is to develop an analog for $\ell^2(\mathbb{Z})$ of the DFT. We begin by considering the critical properties of the vectors $g_m(n) = e^{-2\pi i m n/N}$, $m = 0, 1, \ldots, N - 1$, used in the context of $\ell^2(\mathbb{Z}_N)$ in chapters 2 and 3. A look at some of the proofs (e.g., Theorem 2.18) suggests that the key property is that each g_m is multiplicative, defined as follows. A function $\chi : \mathbb{Z}_N \to \mathbb{C}$ is *multiplicative* if

$$\chi(j + k) = \chi(j)\chi(k), \tag{4.34}$$

for all j, k. Note that if χ is not identically 0, this forces $\chi(0) = 1$ (let $j = k = 0$ in equation (4.34)) and

$$\chi(n) = \big(\chi(1)\big)^n, \tag{4.35}$$

by induction (let $j = n - 1$ and $k = 1$ for the inductive step). However, for χ defined on $\mathbb{Z}_N$, χ must be periodic with period N. Thus, equation (4.35) gives

$$\big(\chi(1)\big)^N = \chi(N) = \chi(0) = 1.$$

Hence $\chi(1)$ must be an Nth root of unity $e^{-2\pi i m/N}$ for some m. Then equation (4.35) gives $\chi(n) = e^{-2\pi i m n/N}$. Because only N of these exponentials are different, we obtain the system of functions $\{g_m\}_{0 \le m \le N-1}$ in the DFT.

Now consider the case of infinite sequences $z \in \ell^2(\mathbb{Z})$. A function $\chi : \mathbb{Z} \to \mathbb{C}$ is multiplicative if equation (4.34) holds for all $j, k \in \mathbb{Z}$. A multiplicative function χ that is not identically 0 will, by the same reasoning as above, satisfy $\chi(0) = 1$ and equation (4.35). However, because χ is not necessarily periodic, we do not obtain the restriction that $\chi(1)$ is a root of unity. If $|\chi(1)| > 1$, then $|\chi(n)| = |\chi(1)|^n$ grows so rapidly as $n \to +\infty$ that such a function is not useful for us. Similarly, if $|\chi(1)| < 1$, then $|\chi(n)| = |\chi(1)|^n$ grows too rapidly as $n \to -\infty$. So we restrict our attention to the case where $|\chi(1)| = 1$. Then $\chi(1) = e^{i\theta}$ for some $\theta \in [-\pi, \pi)$, and we end up with $\chi(n) = e^{in\theta}$ for some such θ. For each different θ, we obtain a multiplicative function, so we consider all of these.

This leads to the following analog of the DFT.

Definition 4.29   *The Fourier transform on $\ell^2(\mathbb{Z})$ is the map $\hat{\ }$: $\ell^2(\mathbb{Z}) \to L^2([-\pi, \pi))$ defined for $z \in \ell^2(\mathbb{Z})$ by*

$$\hat{z}(\theta) = \sum_{n \in \mathbb{Z}} z(n) e^{in\theta},$$

where the series is interpreted as its limit in $L^2([-\pi, \pi))$.

The existence of this limit is guaranteed by Lemmas 4.7 and 4.17. We sometimes regard $\hat{z}$ as defined on all of $\mathbb{R}$ by extending it to have period 2π. Note that the formula defining $\hat{z}$ has period 2π, so it can be used as the definition of $\hat{z}(\theta)$ for all $\theta \in \mathbb{R}$.

Given a function $f \in L^2([-\pi, \pi))$, Lemmas 4.8 and 4.17 show that its sequence of Fourier coefficients belong to $\ell^2(\mathbb{Z})$. This allows the following definition.

Definition 4.30   *The* inverse Fourier transform on $L^2([-\pi, \pi))$ *is the map* $^{\vee} : L^2([-\pi, \pi)) \to \ell^2(\mathbb{Z})$ *defined for* $f \in L^2([-\pi, \pi))$ *by*

$$\check{f}(n) = \langle f, e^{in\theta} \rangle = \frac{1}{2\pi} \int_{-\pi}^{\pi} f(\theta) e^{-in\theta} \, d\theta.$$

As the notation suggests, $^{\wedge}$ and $^{\vee}$ are inverse maps.

Lemma 4.31   *The map* $^{\wedge}$ *in Definition 4.29 is one-to-one and onto, with inverse* $^{\vee}$. *For* $z \in \ell^2(\mathbb{Z})$,

$$z(n) = (\hat{z})^{\vee}(n) = \frac{1}{2\pi} \int_{-\pi}^{\pi} \hat{z}(\theta) e^{-in\theta} \, d\theta. \tag{4.36}$$

For all $z, w \in \ell^2(\mathbb{Z})$ *we have Parseval's relation:*

$$\langle z, w \rangle = \sum_{n \in \mathbb{Z}} z(n)\overline{w(n)} = \frac{1}{2\pi} \int_{-\pi}^{\pi} \hat{z}(\theta)\overline{\hat{w}(\theta)} \, d\theta = \langle \hat{z}, \hat{w} \rangle \tag{4.37}$$

(where the inner product on the right side is as in equation (4.14)), and Plancherel's formula

$$\|z\|^2 = \sum_{n \in \mathbb{Z}} |z(n)|^2 = \frac{1}{2\pi} \int_{-\pi}^{\pi} |\hat{z}(\theta)|^2 \, d\theta = \|\hat{z}\|^2. \tag{4.38}$$

Proof
Suppose that $z \in \ell^2(\mathbb{Z})$. By Exercise 4.2.5 (ii),

$$\langle \hat{z}, e^{in\theta} \rangle = \left\langle \sum_{m \in \mathbb{Z}} z(m) e^{im\theta}, e^{in\theta} \right\rangle = z(n), \tag{4.39}$$

that is, $(\hat{z})^{\vee} = z$, or equation (4.36). This shows that $^{\wedge}$ is one-to-one. Suppose $f \in L^2([-\pi, \pi))$. Corollary 4.24 iv shows that $f = (\check{f})^{\wedge}$. Hence $^{\wedge}$ is onto, with inverse $^{\vee}$.

Applying Corollary 4.24 iii with $f = \hat{z}$ and $g = \hat{w}$ gives equation (4.37), by equation (4.39). Letting $z = w$ in equation (4.37) implies equation (4.38). ∎

Equation (4.36) is the Fourier inversion formula for $\ell^2(\mathbb{Z})$. It expresses $z(n)$ as an integral average over $\theta \in [-\pi, \pi)$ of the pure frequencies $e^{-in\theta}$. It is curious, however, that for every θ, the sequence $e^{-in\theta}$ (as a sequence in n) does not belong to $\ell^2(\mathbb{Z})$. Thus in equation (4.36), elements of $\ell^2(\mathbb{Z})$ are written as averages of elements not in this space. This suggests the need for an integral average: no single term $e^{-in\theta}$ should be given more than infinitesimal weight. With this understanding, equation (4.36) is analogous to a basis representation. Then $\hat{z}(\theta)$ is the weight of the pure frequency $e^{-in\theta}$ in the integral average in equation (4.36) that represents z.

Next we define convolution. Observe that for $z, w \in \ell^2(\mathbb{Z})$, and $m \in \mathbb{Z}$, we have, by relation (4.3),

$$\sum_{n\in\mathbb{Z}} |z(m-n)w(n)| \leq \left(\sum_{n\in\mathbb{Z}} |z(m-n)|^2\right)^{1/2} \left(\sum_{n\in\mathbb{Z}} |w(n)|^2\right)^{1/2} = \|z\|\|w\|,$$

(4.40)

where we changed summation index (let $k = m - n$) to obtain the last equality. This shows that the sum in Definition 4.32 converges absolutely.

Definition 4.32   *Suppose $z, w \in \ell^2(\mathbb{Z})$. For $m \in \mathbb{Z}$, define*

$$z * w(m) = \sum_{n\in\mathbb{Z}} z(m - n)w(n).$$

(4.41)

The sequence $z * w$ is called the convolution *of z and w.*

We note that

$$|z * w(m)| \leq \sum_{n\in\mathbb{Z}} |z(m - n)w(n)| \leq \|z\|\|w\|,$$

by inequality (4.40). Thus for $z, w \in \ell^2(\mathbb{Z})$, $z * w$ is a bounded sequence. However, it is not necessarily the case that $z * w \in \ell^2(\mathbb{Z})$ (Exercise 4.4.6 (ii)). To obtain $z * w \in \ell^2(\mathbb{Z})$, we make a stronger assumption on one of the two vectors.

Definition 4.33   *Let $z = (z(n))_{n\in\mathbb{Z}}$ be a sequence of complex numbers. We say z is* summable *if the series*

$$\sum_{n\in\mathbb{Z}} |z(n)| < +\infty.$$

Let

$$\ell^1(\mathbb{Z}) = \left\{z = (z(n))_{n\in\mathbb{Z}} : z(n) \in \mathbb{C} \text{ for all } n, \text{and } z \text{ is summable}\right\}.$$

For $z \in \ell^1(\mathbb{Z})$, *define*

$$\|z\|_1 = \sum_{n \in \mathbb{Z}} |z(n)|.$$

By Exercise 4.4.2, $\ell^1(\mathbb{Z})$ is a vector space (but not an inner product space), and $\| \cdot \|_1$ is a norm (called the ℓ^1-*norm*) on $\ell^1(\mathbb{Z})$, in the sense of Exercise 1.6.5. Moreover, $\ell^1(\mathbb{Z})$ is a proper subspace of $\ell^2(\mathbb{Z})$ (meaning that it is a subspace that is not the whole space). We continue to use $\| \cdot \|$ to denote the norm in $\ell^2(\mathbb{Z})$ defined in equation (4.2), but we denote the ℓ^1 norm by $\| \cdot \|_1$.

Lemma 4.34 *Suppose* $z \in \ell^2(\mathbb{Z})$ *and* $w \in \ell^1(\mathbb{Z})$. *Then* $z * w \in \ell^2(\mathbb{Z})$, *and*

$$\|z * w\| \le \|w\|_1 \|z\|. \tag{4.42}$$

Proof
For any $m \in \mathbb{Z}$,

$$\left| \sum_{n \in \mathbb{Z}} z(m-n)w(n) \right| \le \sum_{n \in \mathbb{Z}} |z(m-n)||w(n)|^{1/2}|w(n)|^{1/2}$$

$$\le \left(\sum_{n \in \mathbb{Z}} |z(m-n)|^2 |w(n)| \right)^{1/2} \left(\sum_{n \in \mathbb{Z}} |w(n)| \right)^{1/2}$$

$$= \|w\|_1^{1/2} \left(\sum_{n \in \mathbb{Z}} |z(m-n)|^2 |w(n)| \right)^{1/2},$$

by inequality (4.3). Therefore,

$$\|z * w\|^2 = \sum_{m \in \mathbb{Z}} \left| \sum_{n \in \mathbb{Z}} z(m-n)w(n) \right|^2$$

$$\le \|w\|_1 \sum_{m \in \mathbb{Z}} \sum_{n \in \mathbb{Z}} |z(m-n)|^2 |w(n)|$$

$$= \|w\|_1 \sum_{n \in \mathbb{Z}} |w(n)| \sum_{m \in \mathbb{Z}} |z(m-n)|^2.$$

(The interchange of order of summation is justified by a theorem in analysis, because all terms are nonnegative.) Changing summation index (say, letting $k = m - n$) gives that $\sum_{m \in \mathbb{Z}} |z(m-n)|^2 = \|z\|^2$, for any n. Substituting this gives

$$\|z * w\|^2 \le \|w\|_1^2 \|z\|^2.$$

The result follows by taking the square root. ∎

Convolution has the following basic properties.

Lemma 4.35 *Suppose $v, w \in \ell^1(\mathbb{Z})$ and $z \in \ell^2(\mathbb{Z})$. Then*
  i. $(z * w)\hat{}(\theta) = \hat{z}(\theta)\hat{w}(\theta)$ a.e.
  ii. $z * w = w * z$.
  iii. $v * (w * z) = (v * w) * z$.

Proof
To prove part i, we first suppose $z \in \ell^1(\mathbb{Z})$. Then for each $\theta \in [-\pi, \pi)$,

$$(z * w)\hat{}(\theta) = \sum_{n \in \mathbb{Z}} z * w(n)e^{in\theta} = \sum_{n \in \mathbb{Z}}\sum_{k \in \mathbb{Z}} z(n-k)w(k)e^{i(n-k)\theta}e^{ik\theta}$$

$$= \sum_{k \in \mathbb{Z}} w(k)e^{ik\theta}\sum_{n \in \mathbb{Z}} z(n-k)e^{i(n-k)\theta}$$

$$= \sum_{k \in \mathbb{Z}} w(k)e^{ik\theta}\sum_{m \in \mathbb{Z}} z(m)e^{im\theta} = \hat{z}(\theta)\hat{w}(\theta),$$

by the change of index $m = n - k$. The interchange of the order of summation is justified because $z, w \in \ell^1(\mathbb{Z})$, so the double sum converges absolutely.

The extension to the case $z \in \ell^2(\mathbb{Z})$ requires care because $\hat{z}$ and $(z * w)\hat{}$ are interpreted in the $L^2([-\pi, \pi))$ sense as in Definition 4.29. For each positive integer N, define a sequence z_N by setting $z_N(n) = z(n)$ if $|n| \le N$ and $z_N(n) = 0$ if $|n| > N$. Then $z_N \in \ell^1(\mathbb{Z})$, so by the $\ell^1(\mathbb{Z})$ case,

$$(z_N * w)\hat{} = \hat{z}_N\hat{w}. \tag{4.43}$$

We would like to take the limit in norm as $N \to \infty$ of both sides of (4.43).

For the left side, note that $\{z_N\}_{N=1}^{\infty}$ converges to z in norm; that is, $\|z_N - z\| \to 0$ as $N \to \infty$. Hence, by Lemma 4.34,

$$\|z_N * w - z * w\| = \|(z_N - z) * w\| \le \|z_N - z\|\|w\|_1 \to 0$$

as $N \to \infty$. By Plancherel's formula (4.38), then,

$$\|(z_N * w)\hat{} - (z * w)\hat{}\| \to 0,$$

as $N \to \infty$.

For the right side of equation (4.43), note that for all θ,

$$|\hat{w}(\theta)| = \left| \sum_{n\in\mathbb{Z}} w(n)e^{in\theta} \right| \le \sum_{n\in\mathbb{Z}} |w(n)| = \|w\|_1.$$

By using this inequality and equation (4.38), we get

$$\|\hat{z}_N\hat{w} - \hat{z}\hat{w}\|^2 = \frac{1}{2\pi} \int_{-\pi}^{\pi} |\hat{z}_N(\theta)\hat{w}(\theta) - \hat{z}(\theta)\hat{w}(\theta)|^2 \, d\theta$$

$$\le \|w\|_1^2 \frac{1}{2\pi} \int_{-\pi}^{\pi} |\hat{z}_N(\theta) - \hat{z}(\theta)|^2 \, d\theta = \|w\|_1^2 \|\hat{z}_N - \hat{z}\|^2$$

$$= \|w\|_1^2 \|z_N - z\|^2 \to 0$$

as $N \to \infty$.

Hence, as $N \to \infty$, the left side of equation (4.43) converges in $L^2([-\pi,\pi))$ to $(z * w)\hat{}$ whereas the right side converges to $\hat{z}\hat{w}$. Therefore $(z*w)\hat{}$ and $\hat{z}\hat{w}$ agree in $L^2([-\pi,\pi))$, hence a.e. This proves part i.

We leave the proof of parts ii and iii as Exercise 4.4.4. ∎

The technique in the last proof of verifying the result first under weaker conditions, which justify the formal calculation, and then using this result and a limiting argument to justify the general result, is common in analysis.

By analogy to Theorems 2.18 and 4.28, we expect the trigonometric system $\{e^{-in\theta}\}_{\theta\in[-\pi,\pi)}$ to diagonalize any bounded translation-invariant linear transformation $T : \ell^2(\mathbb{Z}) \to \ell^2(\mathbb{Z})$. In some sense this is true, but care must be taken in the interpretation of this statement. We cannot say that $e^{-in\theta}$ is an eigenvector of T, because $e^{-in\theta}$ is not in the domain $\ell^2(\mathbb{Z})$ of T. In particular, the analog of the first step of the proofs of Theorems 2.18 and 4.28, namely, to apply the Fourier inversion formula to $T(e^{-in\theta})$, breaks down immediately because $T(e^{-in\theta})$ is not defined. However, the approach of the alternate proof in Theorem 2.19 can be carried out, with the proper interpretation. We begin with the definition of translation invariance in this context.

Definition 4.36 *For $k \in \mathbb{Z}$, the* translation operator $R_k : \ell^2(\mathbb{Z}) \rightarrow \ell^2(\mathbb{Z})$ *is defined by*

$$R_k z(n) = z(n - k),$$

for all $n \in \mathbb{Z}$. A linear transformation $T : \ell^2(\mathbb{Z}) \rightarrow \ell^2(\mathbb{Z})$ is translation invariant *if, for all $z \in \ell^2(\mathbb{Z})$ and $k \in \mathbb{Z}$,*

$$T(R_k z) = R_k T(z),$$

that is, if T commutes with each R_k.

Example 4.37
Suppose $b \in \ell^1(\mathbb{Z})$. For $z \in \ell^2(\mathbb{Z})$, define

$$T_b(z) = b * z.$$

By Lemma 4.34, $T_b(z)$ is defined and belongs to $\ell^2(\mathbb{Z})$, that is, $T : \ell^2(\mathbb{Z}) \rightarrow \ell^2(\mathbb{Z})$. Moreover, Lemma 4.34 shows that T_b is bounded (Definition 4.25) on $\ell^2(\mathbb{Z})$. One can check that T_b is translation invariant (Exercise 4.4.5).

Definition 4.38 *Define the* delta function δ *by*

$$\delta(n) = \begin{cases} 1, & if\ n = 0 \\ 0, & if\ n \neq 0. \end{cases} \tag{4.44}$$

This terminology is redundant because $\delta = e_0$, but convenient and standard, as in the case of $\ell^2(\mathbb{Z}_N)$ (Definition 2.28).

Lemma 4.39 *Suppose $T : \ell^2(\mathbb{Z}) \rightarrow \ell^2(\mathbb{Z})$ is a bounded, translation-invariant linear transformation. Define $b \in \ell^2(\mathbb{Z})$ by*

$$b = T(\delta).$$

Then for all $z \in \ell^2(\mathbb{Z})$,

$$T(z) = b * z.$$

Proof
Because $\{e_j\}_{j \in \mathbb{Z}}$ is a complete orthonormal set in $\ell^2(\mathbb{Z})$, we can write

$$T(e_j) = \sum_{k \in \mathbb{Z}} a_{j,k} e_k,$$

for some scalars $\{a_{j,k}\}_{k\in\mathbb{Z}}$. Taking the inner product of both sides with a standard basis vector shows that

$$a_{j,k} = \langle T(e_j), e_k\rangle = T(e_j)(k).$$

By the translation invariance of T,

$$a_{j+1,k+1} = T(e_{j+1})(k+1) = T(R_1 e_j)(k+1)$$
$$= R_1(T(e_j))(k+1) = T(e_j)(k) = a_{j,k}.$$

(This says that the infinite matrix $\{a_{j,k}\}_{j,k\in\mathbb{Z}}$ representing T in the standard basis is circulant.) Repeating this ℓ times gives $a_{j+\ell,k+\ell} = a_{j,k}$ for all j, k, ℓ.

Suppose $z \in \ell^2(\mathbb{Z})$. We can write $z = \sum_{j\in\mathbb{Z}} z(j)e_j$. By Lemma 4.26 and the boundedness of T,

$$T(z) = \sum_{j\in\mathbb{Z}} z(j)T(e_j).$$

These remarks give

$$T(e_j)(n) = a_{j,n} = a_{0,n-j} = T(e_0)(n-j) = T(\delta)(n-j) = b(n-j).$$

Hence

$$T(z)(n) = \sum_{j\in\mathbb{Z}} z(j)b(n-j) = b * z(n),$$

for all n. ∎

Hence for T translation invariant and $b = T(\delta)$, we can write

$$T(z)(n) = b * z(n) = ((b * z)\hat{\,})\check{\,}(n) = (\hat{b}\hat{z})\check{\,}(n)$$
$$= \frac{1}{2\pi}\int_{-\pi}^{\pi} \hat{b}(\theta)\hat{z}(\theta)e^{-in\theta}\,d\theta,$$

by Lemma 4.35. By comparing this with equation (4.36), we see that the effect of T on z is to replace the "coefficient" $\hat{z}(\theta)$ of $e^{-in\theta}$ in equation (4.36) by $\hat{b}(\theta)\hat{z}(\theta)$. In this sense the system $\{e^{-in\theta}\}_{\theta\in[-\pi,\pi)}$ diagonalizes T, like the cases of the Fourier basis for $\ell^2(\mathbb{Z}_N)$ in Theorem 2.18 and the trigonometric system for $L^2([-\pi,\pi))$ in Theorem 4.28.

Now we introduce some definitions in the context of $\ell^2(\mathbb{Z})$ that are similar to those for $\ell^2(\mathbb{Z}_N)$ in chapter 3.

Definition 4.40 *Suppose* $z \in \ell^2(\mathbb{Z})$. *For* $n, k \in \mathbb{Z}$, *define the* conjugate reflection *of* z:

$$\tilde{z}(n) = \overline{z(-n)}. \tag{4.45}$$

Also define

$$z^*(n) = (-1)^n z(n). \tag{4.46}$$

This leads to the following results, analogous to those in chapter 3.

Lemma 4.41 *Suppose* $z, w \in \ell^2(\mathbb{Z})$. *Then*
   i. $\tilde{z}, z^* \in \ell^2(\mathbb{Z})$, *and* $R_k z \in \ell^2(\mathbb{Z})$, *for all* $k \in \mathbb{Z}$.
   ii. $(\tilde{z})\hat{}(\theta) = \overline{\hat{z}(\theta)}$.
   iii. $(z^*)\hat{}(\theta) = \hat{z}(\theta + \pi)$.
   iv. $(R_k z)\hat{}(\theta) = e^{ik\theta}\hat{z}(\theta)$.
   v. $\langle R_j z, R_k w \rangle = \langle z, R_{k-j} w \rangle$, *for all* $j, k \in \mathbb{Z}$.
   vi. $\langle z, R_k w \rangle = z * \tilde{w}(k)$ *for all* $k \in \mathbb{Z}$.
   vii. $\hat{\delta}(\theta) = 1$, *for all* θ.

Proof
We give the proof of part iii, because its statement looks a little different from the corresponding statement for $\ell^2(\mathbb{Z}_N)$. By definition,

$$(z^*)\hat{}(\theta) = \sum_{n \in \mathbb{Z}} z^*(n)e^{in\theta} = \sum_{n \in \mathbb{Z}}(-1)^n z(n)e^{in\theta}$$

$$= \sum_{n \in \mathbb{Z}}(e^{i\pi})^n z(n)e^{in\theta} = \sum_{n \in \mathbb{Z}} z(n)e^{in(\theta+\pi)} = \hat{z}(\theta + \pi).$$

We leave parts i, ii, iv, v, vi, and vii as Exercise 4.4.8. ∎

Note that the basic machinery of Fourier analysis that we have just constructed in the context of $\ell^2(\mathbb{Z})$ is closely analogous to the machinery we developed in chapter 2 for $\ell^2(\mathbb{Z}_N)$. Because of this, we are able to develop wavelets on $\mathbb{Z}$ in a manner closely corresponding to the construction on $\mathbb{Z}_N$ in chapter 3.

Exercises

4.4.1. Suppose $\chi : \mathbb{R} \to \mathbb{C}$ satisfies

    i. χ is 2π periodic on $\mathbb{R}$: $\chi(\theta + 2\pi) = \chi(\theta)$ for all $\theta \in \mathbb{R}$.

    ii. χ is multiplicative: $\chi(\theta + \varphi) = \chi(\theta)\chi(\varphi)$ for all $\theta, \varphi \in \mathbb{R}$.

    iii. χ is not identically 0 on $\mathbb{R}$.

    iv. χ is differentiable at 0.

    Prove that there exists $n \in \mathbb{R}$ such that $\chi(\theta) = e^{in\theta}$. Hint: Prove that χ is differentiable on $\mathbb{R}$ and $\chi'(\theta) = \chi'(0)\chi(\theta)$.

    Remark: These functions correspond to the group characters on the circle $\{e^{i\theta} : -\pi \leq \theta < \pi\}$ defined by $\chi(e^{i\theta}) = e^{in\theta}$ or just $\chi(z) = z^n$. Their appearance in equation (4.24) is analogous to the appearance of the characters $\chi(n) = e^{2\pi i mn/N}$ (for $m \in \mathbb{Z}_N$) on $\mathbb{Z}_N$ in equation (2.10) and the characters $\chi(n) = e^{in\theta}$ (for $\theta \in [-\pi, \pi)$) on $\mathbb{Z}$ in equation (4.36). In every case, a general ℓ^2 or L^2 function is written as some sort of superposition (sum or integral) of the associated characters. There is a general theory (Fourier analysis on *locally compact abelian groups*) behind this.

4.4.2. i. Prove that $\ell^1(\mathbb{Z})$ is a vector space, with the usual componentwise addition and scalar multiplication.

    ii. Prove that $\| \cdot \|_1$ is a norm on $\ell^1(\mathbb{Z})$ (see Exercise 1.6.5 for the definition of a norm on a vector space).

    iii. Prove that $\ell^1(\mathbb{Z})$ is not an inner product space. (Hint: See Exercise 1.6.6.)

    iv. Prove that $\ell^1(\mathbb{Z}) \subseteq \ell^2(\mathbb{Z})$. Hint: If $z \in \ell^1(\mathbb{Z})$, prove that $\sup_{n \in \mathbb{Z}} |z(n)| < +\infty$ and then that

$$\sum_{n \in \mathbb{Z}} |z(n)|^2 \leq \sup_{n \in \mathbb{Z}} |z(n)| \sum_{n \in \mathbb{Z}} |z(n)|.$$

    Remark: Observe that this containment goes the opposite way to the containment $L^2([-\pi, \pi)) \subseteq L^1([-\pi, \pi))$.

    v. Give an example of $z \in \ell^2(\mathbb{Z})$ such that $z \notin \ell^1(\mathbb{Z})$.

4.4.3. Suppose $z, w \in \ell^1(\mathbb{Z})$.

    i. Prove that the series $\sum_{n \in \mathbb{Z}} z(m - n)w(n)$ converges absolutely for each $m \in \mathbb{Z}$.

    ii. Prove that $z * w \in \ell^1(\mathbb{Z})$.

4.4.4. Prove Lemma 4.35 ii and iii. Remark: For part ii, note that $w \in \ell^2(\mathbb{Z})$ by Exercise 4.4.2 (iv), so $\hat{w}$ is defined. For part iii, note that $v * w \in \ell^1(\mathbb{Z})$ by Exercise 4.4.3, so $(v * w) * z$ is defined. Hint: This is very similar to the case of $\ell^2(\mathbb{Z}_N)$.

4.4.5. Define T_b as in Example 4.37. Prove that T_b is translation invariant.

4.4.6. i. Let

$$f(\theta) = \frac{1}{\sqrt[4]{|\theta|}}$$

for $\theta \neq 0$, and $f(0) = 0$. Prove that $f \in L^2([-\pi, \pi))$ but $f^2 \notin L^2([-\pi, \pi))$.

ii. Prove that there exist $z, w \in \ell^2(\mathbb{Z})$ such that $z * w \notin \ell^2(\mathbb{Z})$. Hint: Let $z = w = \check{f}$, for f as in part i, and apply Lemma 4.35 i.

4.4.7. Suppose $z \in \ell^2(\mathbb{Z})$. Prove that $z * \delta = z$, for δ as in Definition 4.38.

4.4.8. Prove Lemma 4.41, parts i, ii, iv, v, vi, and vii.

4.4.9. Suppose $z \in \ell^1(\mathbb{Z})$. Prove that the series $\sum_{n \in \mathbb{Z}} z(n) e^{in\theta}$ converges absolutely and uniformly. Deduce that $\hat{z}(\theta)$ is continuous. Hint: Apply a theorem in analysis that says that the limit of a uniformly convergent sequence of continuous functions is continuous.

4.4.10. Convolution arises in multiplication of polynomials as follows. Let $p(x) = \sum_{k=0}^{M} a_k x^k$ and $q(x) = \sum_{k=0}^{N} b_k x^k$ be polynomials (for some $M, N \in \mathbb{N}$). Define $a, b \in \ell^2(\mathbb{Z})$ by setting $a(k) = a_k$ for $0 \leq k \leq M$ and $a(k) = 0$ otherwise, and similarly $b(k) = b_k$ for $0 \leq k \leq N$ and 0 otherwise. Prove that

$$p(x)q(x) = \sum_{k=0}^{M+N} a * b(k) x^k.$$

With appropriate convergence considerations, this result can be extended to power series $\sum_{k=0}^{\infty} a_k x^k$ or even to Laurent series $\sum_{k \in \mathbb{Z}} a_k z^k$.

4.5 First-Stage Wavelets on $\mathbb{Z}$

In this section, we characterize first stage wavelet bases on $\mathbb{Z}$. For the most part, this is done by adapting the techniques in chapter

3 to $\ell^2(\mathbb{Z})$. The most substantive differences come from the infinite dimensionality of $\ell^2(\mathbb{Z})$. In the finite dimensional case, we could say that an orthonormal set with the right number of elements must span the space, and hence is a basis. Here an infinite orthonormal set is not necessarily complete (e.g., take one element away from a complete orthonormal set). So some extra work will be required to show completeness.

The first step is to prove the analog of Lemma 3.6.

Lemma 4.42 *Suppose $w, z \in \ell^1(\mathbb{Z})$.*

  i. *The set $\{R_{2k}w\}_{k \in \mathbb{Z}}$ is orthonormal if and only if*

$$|\hat{w}(\theta)|^2 + |\hat{w}(\theta + \pi)|^2 = 2 \quad \text{for all } \theta \in [0, \pi). \tag{4.47}$$

 ii. *We have*

$$\langle R_{2k}z, R_{2j}w \rangle = 0 \quad \text{for all} \quad k, j \in \mathbb{Z} \tag{4.48}$$

  if and only if

$$\hat{z}(\theta)\overline{\hat{w}(\theta)} + \hat{z}(\theta + \pi)\overline{\hat{w}(\theta + \pi)} = 0 \text{ for all } \theta \in [0, \pi). \tag{4.49}$$

Proof

For part i, we first observe that the elements $R_{2k}w$ must be distinct for $k \in \mathbb{Z}$, that is, $R_{2k}w = R_{2j}w$ implies $k = j$ (Exercise 4.5.1 (ii)). Next, note that the quantity $|\hat{w}(\theta)|^2 + |\hat{w}(\theta + \pi)|^2$ is periodic with period π (because $\hat{w}$ has period 2π), so the identity in equation (4.47) holds for all θ if and only if it holds for all $\theta \in [0, \pi)$. By Exercise 4.5.1 (iii), $\{R_{2k}w\}_{k \in \mathbb{Z}}$ is orthonormal if and only if

$$\langle w, R_{2k}w \rangle = \begin{cases} 1 & \text{if } k = 0 \\ 0 & \text{if } k \neq 0. \end{cases} \tag{4.50}$$

By Lemma 4.41 vi, equation (4.50) is equivalent to

$$w * \tilde{w}(2k) = \begin{cases} 1 & \text{if } k = 0 \\ 0 & \text{if } k \neq 0. \end{cases}$$

For any $y \in \ell^2(\mathbb{Z})$,

$$(y + y^*)(n) = (1 + (-1)^n)y(n) = \begin{cases} 2y(n) & \text{if } n \text{ is even,} \\ 0 & \text{if } n \text{ is odd.} \end{cases}$$

Using this fact with $y = u * \tilde{u}$, we see that equation (4.50) is equivalent to

$$w * \tilde{w} + (w * \tilde{w})^* = 2\delta.$$

By Fourier inversion (Lemma 4.31) and Lemma 4.41 vii, this is equivalent to

$$(w * \tilde{w})\hat{\ }(\theta) + ((w * \tilde{w})^*)\hat{\ }(\theta) = 2 \text{ for all } \theta. \tag{4.51}$$

By Lemma 4.35 i and Lemma 4.41 ii,

$$(w * \tilde{w})\hat{\ }(\theta) = \hat{w}(\theta)(\tilde{w})\hat{\ }(\theta) = \hat{w}(\theta)\overline{\hat{w}(\theta)} = |\hat{w}(\theta)|^2.$$

By using this fact and Lemma 4.41 iii, we get

$$((w * \tilde{w})^*)\hat{\ }(\theta) = (w * \tilde{w})\hat{\ }(\theta + \pi) = |\hat{w}(\theta + \pi)|^2.$$

By substituting these last two facts in equation (4.51), we obtain the equivalence of equations (4.50) and (4.47). This completes the proof of part i. Part ii is similar, and is left as Exercise 4.5.2. ∎

There is an alternate proof of Lemma 4.42 based on Parseval's formula (Exercise 4.5.1 (iv)).

The following definitions correspond to definitions in chapter 3.

Definition 4.43 *For a sequence $z = (z(n))_{n \in \mathbb{Z}}$, define sequences $D(z)$ and $U(z)$ on $\mathbb{Z}$ by*

$$D(z)(n) = z(2n)$$

and

$$U(z)(n) = \begin{cases} z(n/2) & \text{if } n \text{ is even} \\ 0 & \text{if } n \text{ is odd}. \end{cases}$$

Clearly $D : \ell^2(\mathbb{Z}) \to \ell^2(\mathbb{Z})$ and $U : \ell^2(\mathbb{Z}) \to \ell^2(\mathbb{Z})$. The m-fold composition of D with itself is denoted D^m, and similarly for U^m. Then

$$D^m(z)(n) = z(2^m n),$$

and

$$U^m(z)(n) = \begin{cases} z(n/2^m) & \text{if } n = 2^m j \text{ for some } j \in \mathbb{Z} \\ 0 & \text{if } n \text{ is not divisible by } 2^j . \end{cases}$$

We call D the downsampling *operator and U the* upsampling *operator.*

Definition 4.44 *Suppose $u, v \in \ell^1(\mathbb{Z})$. Let*

$$B = \{R_{2k}v\}_{k\in\mathbb{Z}} \cup \{R_{2k}u\}_{k\in\mathbb{Z}}. \tag{4.52}$$

If B is a complete orthonormal set in $\ell^2(\mathbb{Z})$, we call B a first-stage wavelet system *for $\ell^2(\mathbb{Z})$.*

Recall that $\ell^1(\mathbb{Z}) \subseteq \ell^2(\mathbb{Z})$ (Exercise 4.4.2 (iv)). It might seem more natural in Definition 4.44 to assume only that $u, v \in \ell^2(\mathbb{Z})$. However, we will be considering $z * u$ and $z * v$, for $z \in \ell^2(\mathbb{Z})$, and we need these convolutions to belong to $\ell^2(\mathbb{Z})$. By Lemma 4.34, that condition is guaranteed under the assumption $u, v \in \ell^1(\mathbb{Z})$. It seems that there is no harm in making this assumption because it is satisfied by most interesting examples. This assumption guarantees that $\hat{u}$ and $\hat{v}$ are continuous functions (see Exercise 4.4.9).

Definition 4.45 *Suppose $u, v \in \ell^2(\mathbb{Z})$. The* system matrix *of u and v is*

$$A(\theta) = \frac{1}{\sqrt{2}} \begin{bmatrix} \hat{u}(\theta) & \hat{v}(\theta) \\ \hat{u}(\theta + \pi) & \hat{v}(\theta + \pi) \end{bmatrix}. \tag{4.53}$$

The characterization of first-stage wavelet systems for $\ell^2(\mathbb{Z})$ is analogous to the characterization in $\ell^2(\mathbb{Z}_N)$. The proof is similar as well except that the additional step of proving completeness is required.

Theorem 4.46 *Suppose that $u, v \in \ell^1(\mathbb{Z})$. Then*

$$B = \{R_{2k}v\}_{k\in\mathbb{Z}} \cup \{R_{2k}u\}_{k\in\mathbb{Z}}$$

is a complete orthonormal set in $\ell^2(\mathbb{Z})$ if and only if the system matrix $A(\theta)$ is unitary for all $\theta \in [0, \pi)$.

Proof
Applying Lemma 4.42 i with $w = u$, the set $\{R_{2k}u\}_{k\in\mathbb{Z}}$ is orthonormal if and only if the first column of $A(\theta)$ has norm one for all θ. Similarly, letting $w = v$, $\{R_{2k}v\}_{k\in\mathbb{Z}}$ is orthonormal if and only if the second column of $A(\theta)$ has norm one for all θ. Applying Lemma 4.42 ii, the elements of $\{R_{2k}u\}_{k\in\mathbb{Z}}$ are orthogonal to those in $\{R_{2k}v\}_{k\in\mathbb{Z}}$ if and only if the two columns of $A(\theta)$ are orthogonal for all θ. Thus, B is an orthonormal set if and only if $A(\theta)$ is unitary for all θ.

To complete the proof, we must show that if $A(\theta)$ is unitary for all $\theta \in [0, \pi)$, then the orthonormal set B is complete in $\ell^2(\mathbb{Z})$. The proof of completeness is inspired by the reconstruction procedure for the filter bank in chapter 3, Figure 14. Observe that the unitarity of $A(\theta)$ for all $\theta \in [0, \pi)$ implies the unitarity of $A(\theta)$ for all $\theta \in \mathbb{R}$ because $\hat{u}$ and $\hat{v}$ are 2π-periodic. We claim that

$$v * U(D(z * \tilde{v})) + u * U(D(z * \tilde{u})) = z, \qquad (4.54)$$

for all $z \in \ell^2(\mathbb{Z})$. To see this, note that $U \circ D(z) = (z + z^*)/2$ (Exercise 4.5.4 (ii)) and proceed as in the proof of Lemma 4.42 i, to obtain

$$\left(v * U(D(z * \tilde{v}))\right)\hat{}(\theta) + \left(u * U(D(z * \tilde{u}))\right)\hat{}(\theta)$$

$$= \hat{v}(\theta)\frac{1}{2}\left[\hat{z}(\theta)\overline{\hat{v}(\theta)} + \hat{z}(\theta + \pi)\overline{\hat{v}(\theta + \pi)}\right]$$

$$+ \hat{u}(\theta)\frac{1}{2}\left[\hat{z}(\theta)\overline{\hat{u}(\theta)} + \hat{z}(\theta + \pi)\overline{\hat{u}(\theta + \pi)}\right]$$

$$= \hat{z}(\theta)\frac{1}{2}\left[|\hat{u}(\theta)|^2 + |\hat{v}(\theta)|^2\right]$$

$$+ \hat{z}(\theta + \pi)\frac{1}{2}\left[\hat{u}(\theta)\overline{\hat{u}(\theta + \pi)} + \hat{v}(\theta)\overline{\hat{v}(\theta + \pi)}\right].$$

However, the unitarity of $A(\theta)$ implies that the rows of $A(\theta)$ are orthonormal (Lemma 1.105), so the last expression reduces to

$$\hat{z}(\theta) \cdot 1 + \hat{z}(\theta + \pi) \cdot 0 = \hat{z}(\theta).$$

By Fourier inversion, this implies equation (4.54).

Note that

$$D(z * \tilde{v})(k) = z * \tilde{v}(2k) = \langle z, R_{2k}v \rangle,$$

and similarly $D(z * \tilde{u})(k) = \langle z, R_{2k}u \rangle$. Hence, if $\langle z, R_{2k}v \rangle = 0$ and $\langle z, R_{2k}u \rangle = 0$ for all $k \in \mathbb{Z}$, then $D(z * \tilde{v}) = 0$ and $D(z * \tilde{u}) = 0$. Hence, $z = 0$, by equation (4.54), which proves the completeness of B. ∎

The last part of the proof can also be carried out by using equation (4.54) and noting (Exercise 4.5.2 (ii)) that

$$v * U(D(z * \tilde{v})) + u * U(D(z * \tilde{u})) = \sum_{k \in \mathbb{Z}}\langle z, R_{2k}v \rangle R_{2k}v + \sum_{k \in \mathbb{Z}}\langle z, R_{2k}u \rangle R_{2k}u.$$

$$(4.55)$$

The proof of Theorem 4.46 is remarkable because the first part shows that the orthonormality of B is equivalent to the unitarity of $A(\theta)$ for all θ, yet by the second part, the unitarity of $A(\theta)$ for all θ implies the completeness of B. Hence the orthonormality of

B implies its completeness; we get the completeness free. This is similar to the finite dimensional case, in which a linearly independent set of maximal size automatically spans the space. In the infinite dimensional case in general, there is no reason that an infinite orthonormal set should be complete. Theorem 4.46 shows that such an unusual implication does hold for sets of the form of B. Curiously the reason behind this is the elementary fact that if the columns of a square matrix are orthonormal, then so are the rows.

The identity (4.55) shows that the same filter bank arrangement (Figure 14 with $\tilde{t} = v$ and $\tilde{s} = u$) as for wavelets on $\mathbb{Z}_N$ can be used to carry out the analysis and synthesis phases associated with the identity

$$z = \sum_{k \in \mathbb{Z}} \langle z, R_{2k}v \rangle R_{2k}v + \sum_{k \in \mathbb{Z}} \langle z, R_{2k}u \rangle R_{2k}u.$$

We saw in Lemma 3.12 that a vector $u \in \ell^2(\mathbb{Z}_N)$ with the property that its even integer translates are orthonormal always has a companion v such that u and v generate a first-stage wavelet basis for $\ell^2(\mathbb{Z}_N)$. A similar result is true for $\ell^2(\mathbb{Z})$, as stated in Lemma 4.47. This result is more surprising in the infinite dimensional context because, for example, the subspace orthogonal to an infinite dimensional subspace does not have to be infinite dimensional. However, because of the structure of the closed subspace generated by the even integer translates of u, its orthogonal subspace is not only infinite dimensional, it is generated by the even integer translates of a vector v which is easily determined by u.

Lemma 4.47 *Suppose $u \in \ell^1(\mathbb{Z})$ and $\{R_{2k}u\}_{k \in \mathbb{Z}}$ is orthonormal in $\ell^2(\mathbb{Z})$. Define a sequence $v \in \ell^1(\mathbb{Z})$ by*

$$v(k) = (-1)^{k-1}\overline{u(1-k)}. \tag{4.56}$$

Then

$$\{R_{2k}v\}_{k \in \mathbb{Z}} \cup \{R_{2k}u\}_{k \in \mathbb{Z}}$$

is a complete orthonormal system (hence a first-stage wavelet system) in $\ell^2(\mathbb{Z})$.

Proof

Since $u \in \ell^1(\mathbb{Z})$, it is clear that $v \in \ell^1(\mathbb{Z})$. By Lemma 4.42 i, the orthonormality of $\{R_{2k}u\}_{k\in\mathbb{Z}}$ is equivalent to

$$|\hat{u}(\theta)|^2 + |\hat{u}(\theta + \pi)|^2 = 2 \quad \text{for all } \theta.$$

By changing the summation index, we observe that

$$\hat{v}(\theta) = \sum_{k\in\mathbb{Z}}(-1)^{k-1}\overline{u(1-k)}e^{ik\theta} = \sum_{j\in\mathbb{Z}}(-1)^{-j}\overline{u(j)}e^{i(1-j)\theta}.$$

By writing $(-1)^{-j} = (e^{i\pi})^{-j} = e^{-i\pi j}$, we obtain

$$\hat{v}(\theta) = \sum_{j\in\mathbb{Z}}e^{-i\theta}\overline{u(j)}e^{ij(\theta+\pi)} = e^{i\theta}\overline{\hat{u}(\theta + \pi)}. \tag{4.57}$$

Hence

$$\hat{v}(\theta + \pi) = e^{i(\theta+\pi)}\overline{\hat{u}(\theta + 2\pi)} = -e^{i\theta}\overline{\hat{u}(\theta)}.$$

It follows (Exercise 4.5.9) that the matrix $A(\theta)$ defined in equation (4.53) is unitary for all θ. The result follows by Theorem 4.46. ∎

Lemma 4.47 reduces the construction of a first-stage wavelet basis to the construction of a vector $u \in \ell^1(\mathbb{Z})$ such that the set $\{R_{2k}u\}_{k\in\mathbb{Z}}$ is orthonormal.

A first-stage wavelet system $\{R_{2k}v\}_{k\in\mathbb{Z}} \cup \{R_{2k}u\}_{k\in\mathbb{Z}}$ for $\ell^2(\mathbb{Z})$ automatically yields a first-stage wavelet basis for $\ell^2(\mathbb{Z}_N)$ (Definition 3.4) by a process known as *periodization*.

Lemma 4.48 (Periodized wavelets for $\ell^2(\mathbb{Z}_N)$) *Suppose $M \in \mathbb{N}$ and $N = 2M$. Suppose $u, v \in \ell^1(\mathbb{Z})$ are such that $\{R_{2k}v\}_{k\in\mathbb{Z}} \cup \{R_{2k}u\}_{k\in\mathbb{Z}}$ is a first-stage wavelet system for $\ell^2(\mathbb{Z})$. Define $u_{(N)}, v_{(N)} \in \ell^2(\mathbb{Z}_N)$ by*

$$u_{(N)}(n) = \sum_{k\in\mathbb{Z}}u(n + kN) \quad \text{and} \quad v_{(N)}(n) = \sum_{k\in\mathbb{Z}}v(n + kN). \tag{4.58}$$

(Note that these sums converge absolutely by the assumption that $u, v \in \ell^1(\mathbb{Z})$.) Then

$$\{R_{2k}v_{(N)}\}_{k=0}^{M-1} \cup \{R_{2k}u_{(N)}\}_{k=0}^{M-1}$$

is a first-stage wavelet basis for $\ell^2(\mathbb{Z}_N)$.

Proof

By computing the *DFT* of $u_{(N)}$, we obtain

$$\hat{u}_{(N)}(m) = \sum_{n=0}^{N-1} u_{(N)}(n)e^{-2\pi inm/N} = \sum_{n=0}^{N-1} \sum_{k\in\mathbb{Z}} u(n+kN)e^{-2\pi inm/N}$$

$$= \sum_{n=0}^{N-1} \sum_{k\in\mathbb{Z}} u(n+kN)e^{-2\pi i(n+kN)m/N}$$

$$= \sum_{\ell\in\mathbb{Z}} u(\ell)e^{-2\pi i\ell m/N} = \hat{u}(-2\pi m/N),$$

where on the right we mean the Fourier transform of u in the sense of Definition 4.29. As a consequence,

$$\hat{u}_{(N)}(m+M) = \hat{u}(-2\pi(m+M)/N)$$
$$= \hat{u}(-2\pi m/N - \pi) = \hat{u}(-2\pi m/N + \pi),$$

because $\hat{u}$ is 2π periodic. By applying the same identities with u replaced by v, we see that the system matrix $A(n)$ for $u_{(N)}$ and $v_{(N)}$ in Definition 3.7 is equal to the system matrix $A(\theta)$ for u, v, as in Definition 4.45, when $\theta = -2\pi n/N$. By Theorem 4.46, $A(\theta)$ is unitary for all θ, and hence $A(n)$ is unitary for all n. By Theorem 3.8, $\{R_{2k}v_{(N)}\}_{k=0}^{M-1} \cup \{R_{2k}u_{(N)}\}_{k=0}^{M-1}$ is a first-stage wavelet basis for $\ell^2(\mathbb{Z}_N)$. ∎

Lemma 4.48 can be proved directly, without using the DFT (Exercise 4.5.10).

Corollary 4.49 *Suppose $u, v \in \ell^1(\mathbb{Z})$ are such that $\{R_{2k}v\}_{k\in\mathbb{Z}} \cup \{R_{2k}u\}_{k\in\mathbb{Z}}$ is a first-stage wavelet system for $\ell^2(\mathbb{Z})$. Suppose also that*

$$u(n) = v(n) = 0 \quad \text{for all} \quad n < 0 \quad \text{and} \quad n > N - 1.$$

Define $u_{(N)}, v_{(N)} \in \ell^2(\mathbb{Z}_N)$ by

$$u_{(N)}(n) = u(n) \quad \text{and} \quad v_{(N)}(n) = v(n) \quad \text{for} \quad n = 0, 1, \ldots, N - 1.$$

Then $\{R_{2k}v_{(N)}\}_{k=0}^{M-1} \cup \{R_{2k}u_{(N)}\}_{k=0}^{M-1}$ is a first-stage wavelet basis for $\ell^2(\mathbb{Z}_N)$.

Proof

Because of the finiteness assumptions on u and v, $u_{(N)}$ and $v_{(N)}$ agree with the definitions in equation (4.58). Hence the result follows from Lemma 4.48. ∎

For m a positive integer, Daubechies's D_{2m} wavelets for $\ell^2(\mathbb{Z})$ (see Example 4.57 for the case $m = 3$) are generated by vectors u and v such that $u(n) = 0$ for $n < 0$ or $n > 2m - 1$, whereas $v(n) = 0$ for $n < -2m + 2$ or $n > 1$. If we replace v with $R_{2m-2}v$, which does not change the set $\{R_{2k}v\}_{k \in \mathbb{Z}}$, the assumptions of Corollary 4.49 apply to these wavelets if $N > 2m$. Sometimes people implementing Daubechies's wavelets numerically are unsure whether they should use circular convolution (i.e., convolution on $\ell^2(\mathbb{Z}_N)$, as in Definition 2.23) or linear convolution (convolution on $\ell^2(\mathbb{Z})$, as in Definition 4.32). The surprising answer is that either will work if done correctly. This is explained by Corollary 4.49. If circular convolution is used, one is computing wavelets on $\mathbb{Z}_N$, as in chapter 3, whereas linear convolution is used when computing wavelets on $\mathbb{Z}$, as in this chapter.

Exercises

4.5.1. Suppose $w \in \ell^2(\mathbb{Z})$.
   i. If there exists $k \in \mathbb{Z}$ such that $R_k w = w$, prove that $w = 0$ (i.e., $w(n) = 0$ for all n). Hint: If $w(m) \neq 0$, prove that there are infinitely many $n \in \mathbb{Z}$ such that $w(n) = w(m)$, which contradicts the assumption that $w \in \ell^2(\mathbb{Z})$.
   ii. If there exist $k, j \in \mathbb{Z}$ with $k \neq j$ such that $R_k w = R_j w$, prove that $w = 0$.
   iii. Prove that $\{R_{2k}w\}_{k \in \mathbb{Z}}$ is orthonormal if and only if equation (4.50) holds.
   iv. Complete the following proof of Lemma 4.42. For part i, by Parseval's formula (4.37) and Lemma 4.41 iv,

$$\langle w, R_{2k}w \rangle = \frac{1}{2\pi} \int_{-\pi}^{\pi} |\hat{w}(\theta)|^2 e^{-i2k\theta} \, d\theta.$$

Write $\int_{-\pi}^{\pi}$ as $\int_0^{\pi} + \int_{-\pi}^0$ and replace θ by $\theta + \pi$ in the second integral. By noting that $\hat{w}$ is 2π-periodic, obtain

$$\langle w, R_{2k}w \rangle = \frac{1}{2\pi} \int_0^{\pi} \left(|\hat{w}(\theta)|^2 + |\hat{w}(\theta + \pi)|^2 \right) e^{-i2k\theta} \, d\theta.$$

Let $\varphi = 2\theta$ and use Fourier inversion to deduce equation (4.47). Part ii can be proved similarly.

4.5.2.   i. Prove Lemma 4.42 ii by methods analogous to those used in the text for part i.

ii. Prove equation (4.55).

4.5.3.   i. Prove that $D : \ell^2(\mathbb{Z}) \to \ell^2(\mathbb{Z})$ is a linear transformation that is onto but not one-to-one.

ii. Prove that $U : \ell^2(\mathbb{Z}) \to \ell^2(\mathbb{Z})$ is a linear transformation that is one-to-one but not onto.

Recall from Exercise 1.4.8 (v) that examples like this do not exist for a linear transformation from a finite dimensional vector space to itself.

4.5.4. Let $z = (z(n))_{n \in \mathbb{Z}}$ be a sequence.

i. Prove that $D \circ U(z) = z$.

ii. Prove that $U \circ D(z) = (z + z^*)/2$.

4.5.5. Suppose $w \in \ell^1(\mathbb{Z})$.

i. Prove that $\{R_k w\}_{k \in \mathbb{Z}}$ is a complete orthonormal set for $\ell^2(\mathbb{Z})$ if and only if $|\hat{w}(\theta)| = 1$ for all $\theta \in [-\pi, \pi)$.

Remark: As for Lemma 3.3 in the context of $\ell^2(\mathbb{Z}_N)$, this proof shows that frequency localization cannot be obtained for an orthonormal basis of the form $\{R_k w\}_{k \in \mathbb{Z}}$.

ii. Prove that $\{R_{2k} w\}_{k \in \mathbb{Z}}$ cannot be a complete orthonormal set in $\ell^2(\mathbb{Z})$. Hint: See Exercises 3.1.10 and 3.1.13.

4.5.6. (Perfect reconstruction in a first-stage filter bank on $\mathbb{Z}$) Suppose $u, v, s, t \in \ell^1(\mathbb{Z})$. Prove that

$$z = \tilde{t} * U(D(z * \tilde{v})) + \tilde{s} * U(D(z * \tilde{u}))$$

for all $z \in \ell^2(\mathbb{Z})$ if and only if

$$A(\theta) \begin{bmatrix} \hat{s}(\theta) \\ \hat{t}(\theta) \end{bmatrix} = \begin{bmatrix} \sqrt{2} \\ 0 \end{bmatrix}$$

a.e. (Hint: Compare with Lemma 3.15. Let $z = \delta$ and $z = R_1 \delta$.)

4.5.7. Suppose $z, w \in \ell^2(\mathbb{Z})$. Prove that

i. $U(z * w) = U(z) * U(w)$.

ii. $(U(z))\tilde{} = U(\tilde{z})$.

iii. $(z * w)\tilde{} = \tilde{z} * \tilde{w}$.

4.5.8. (Generalization of Theorem 4.46 to ℓ functions) Suppose ℓ is a positive integer.

  i. Let $u, v, w \in \ell^1(\mathbb{Z})$. Prove that $\{R_{\ell k}w\}_{k \in \mathbb{Z}}$ is orthonormal if and only if

$$\sum_{k=0}^{\ell-1} |\hat{w}(\theta + 2\pi k/\ell)|^2 = \ell \quad \text{for all } \theta.$$

Also prove that

$$\langle R_{\ell k}u, R_{\ell j}v \rangle = 0$$

for all $j, k \in \mathbb{Z}$ if and only if

$$\sum_{k=0}^{\ell-1} \hat{u}(\theta + 2\pi k/\ell)\overline{\hat{v}(\theta + 2\pi k/\ell)} = 0 \quad \text{for all } \theta.$$

Hint: Method 1: Prove that

$$\sum_{m=0}^{\ell-1} e^{2\pi imn/\ell} = \begin{cases} \ell & \text{if } \ell | n \\ 0 & \text{if } \ell \nmid n. \end{cases}$$

Deduce that

$$\langle u, R_{\ell j}v \rangle = u * \tilde{v}(\ell j) = \begin{cases} \delta(j) & \text{if } u = v \\ 0 & \text{if } u \neq v \end{cases}$$

if and only if

$$\sum_{k=0}^{\ell-1} e^{2\pi ikn/\ell} u * \tilde{v}(n) = \begin{cases} \ell\delta(n) & \text{if } u = v \\ 0 & \text{if } u \neq v. \end{cases}$$

Take the Fourier transform in $\ell^2(\mathbb{Z})$ on both sides of this equation, noting that for $z \in \ell^1(\mathbb{Z})$ and $\varphi \in \mathbb{R}$,

$$(z(n)e^{in\varphi})\hat{}(\theta) = \hat{z}(\theta + \varphi).$$

Method 2 (compare with Exercise 4.5.1 (iv)): By Parseval's relation (4.37) and Lemma 4.41 iv,

$$\langle u, R_{\ell j}v \rangle = \frac{1}{2\pi} \int_{-\pi}^{\pi} \hat{u}(\theta)\overline{\hat{v}(\theta)}e^{-i\ell j\theta} \, d\theta$$

$$= \frac{1}{2\pi} \sum_{k=0}^{\ell-1} \int_{-\pi+2\pi k/\ell}^{-\pi+2\pi(k+1)/\ell} \hat{u}(\theta)\overline{\hat{v}(\theta)}e^{-i\ell j\theta} \, d\theta.$$

Make the substitution $\varphi = \ell\theta - 2\pi k + \pi(\ell - 1)$, which brings every integral back to $[-\pi, \pi]$. Taking the sum on k inside the integral, obtain

$$\langle u, R_{\ell j} v\rangle = \frac{1}{2\pi}\int_{-\pi}^{\pi} g(\varphi)e^{-ij\varphi}\, d\varphi,$$

for

$$g(\varphi)=e^{ij\pi(\ell-1)}\frac{1}{\ell}\sum_{k=0}^{\ell-1}\hat{u}\left(\frac{\varphi}{\ell}-\pi+\frac{\pi}{\ell}+2\pi\frac{k}{\ell}\right)\times\overline{\hat{v}\left(\frac{\varphi}{\ell}-\pi+\frac{\pi}{\ell}+\frac{2\pi k}{\ell}\right)}.$$

Note that g is 2π-periodic and apply Fourier inversion.

ii. Let $u_0, u_1, \dots\ u_{\ell-1} \in \ell^1(\mathbb{Z})$. Prove that

$$B = \{R_{\ell k}u_0\}_{k\in\mathbb{Z}} \cup \{R_{\ell k}u_1\}_{k\in\mathbb{Z}} \cup \cdots \cup \{R_{\ell k}u_{\ell-1}\}_{k\in\mathbb{Z}}$$

is a complete orthonormal set in $\ell^2(\mathbb{Z})$ if and only if the matrix

$$A(\theta) =$$

$$\frac{1}{\sqrt{\ell}}\begin{bmatrix} \hat{u}_0(\theta) & \hat{u}_1(\theta) & \cdots & \hat{u}_{\ell-1}(\theta) \\ \hat{u}_0\left(\theta+\frac{2\pi}{\ell}\right) & \hat{u}_1\left(\theta+\frac{2\pi}{\ell}\right) & \cdots & \hat{u}_{\ell-1}\left(\theta+\frac{2\pi}{\ell}\right) \\ \hat{u}_0\left(\theta+\frac{4\pi}{\ell}\right) & \hat{u}_1\left(\theta+\frac{4\pi}{\ell}\right) & \cdots & \hat{u}_{\ell-1}\left(\theta+\frac{4\pi}{\ell}\right) \\ \vdots & \vdots & \cdots & \vdots \\ \hat{u}_0\left(\theta+\frac{(\ell-1)2\pi}{\ell}\right) & \hat{u}_1\left(\theta+\frac{(\ell-1)2\pi}{\ell}\right) & \cdots & \hat{u}_{\ell-1}\left(\theta+\frac{(\ell-1)2\pi}{\ell}\right) \end{bmatrix}$$

is unitary for all $\theta \in [-\pi, \pi]$. Hint: Part i shows that the unitarity of all $A(\theta)$ is equivalent to the orthonormality of B. As in Theorem 4.46, the completeness of B is left to prove. Prove that

$$\frac{1}{\ell}\sum_{j=0}^{\ell-1}\left(u_j * \sum_{m=0}^{\ell-1}e^{2\pi imn/\ell}z * \tilde{u}_j(n)\right)^{\hat{}}(\theta)$$

$$= \frac{1}{\ell}\sum_{j=0}^{\ell-1}\hat{u}_j(\theta)\sum_{k=0}^{\ell-1}\hat{z}(\theta+2\pi k/\ell)\overline{\hat{u}_j(\theta+2\pi k/\ell)}.$$

Change the order of summation and apply to unitarity of A to deduce that the last expression equals $\hat{z}(\theta)$. As in the hint for part i, Method 1 above, observe that

$$\frac{1}{\ell} \sum_{m=0}^{\ell-1} e^{2\pi i m n/\ell} z * \tilde{u}_j(n) = \begin{cases} \langle z, R_n u_j \rangle & \text{if } \ell | n \\ 0 & \text{if } \ell \nmid n. \end{cases}$$

4.5.9. Prove that the matrix $A(\theta)$ in the proof of Lemma 4.47 is unitary for all θ. Hint: The proof is similar to the corresponding part of the proof of Lemma 3.12.

4.5.10. Prove Lemma 4.48 directly (without using the DFT). Hint: For example, to show that $\{R_{2k} u_{(N)}\}_{k=0}^{M-1}$ is orthonormal, write, for $0 \le j \le M - 1$,

$$\langle u_{(N)}, R_{2j} u_{(N)} \rangle = \sum_{n=0}^{N-1} \sum_{k \in \mathbb{Z}} u(n + kN) \sum_{\ell \in \mathbb{Z}} \overline{u(n - 2j + \ell N)}.$$

In the inside sum, k is fixed, so we can let $p = \ell - k$, to obtain

$$\sum_{n=0}^{N-1} \sum_{k \in \mathbb{Z}} u(n + kN) \sum_{p \in \mathbb{Z}} \overline{u(n + kN - 2j + pN)}$$

$$= \sum_{p \in \mathbb{Z}, \, m \in \mathbb{Z}} u(m) \overline{u(m - 2j + pN)} = \sum_{p \in \mathbb{Z}} \langle u, R_{2j - 2pM} u \rangle,$$

where the last inner product is in $\ell^2(\mathbb{Z})$.

4.6 The Iteration Step for Wavelets on $\mathbb{Z}$

Theorem 4.46 states that if the system matrix $A(\theta)$ of u and v (belonging to $\ell^1(\mathbb{Z})$) is unitary for all θ, then $\ell^2(\mathbb{Z})$ splits into two pieces, generated by the even integer translates of u and v, respectively. As in section 3.3, the next step is to iterate this splitting. As in section 4.5, the proofs correspond to those in chapter 3, except that extra steps are required to prove completeness because of the infinite dimensionality of the spaces considered. The next result is analogous to Lemma 3.24.

Lemma 4.50 *Suppose ℓ is a positive integer, $g_{\ell-1} \in \ell^2(\mathbb{Z})$, and*

$$\{R_{2^{\ell-1}k} g_{\ell-1}\}_{k \in \mathbb{Z}}$$

is orthonormal in $\ell^2(\mathbb{Z})$. Suppose also that $u, v \in \ell^1(\mathbb{Z})$ and the system matrix $A(\theta)$ of u and v (Definition 4.45) is unitary for all θ. Define

$$f_\ell = g_{\ell-1} * U^{\ell-1}(v) \quad \text{and} \quad g_\ell = g_{\ell-1} * U^{\ell-1}(u). \tag{4.59}$$

Then

$$\{R_{2^\ell k} f_\ell\}_{k \in \mathbb{Z}} \cup \{R_{2^\ell k} g_\ell\}_{k \in \mathbb{Z}} \tag{4.60}$$

is orthonormal.

Proof

We first observe that $U^{\ell-1}(u)$ and $U^{\ell-1}(v)$ belong to $\ell^1(\mathbb{Z})$, hence $f_\ell, g_\ell \in \ell^2(\mathbb{Z})$, by their definition and Lemma 4.34.

Next we note that by Exercise 4.5.7 (ii, iii),

$$\tilde{f}_\ell = \left(g_{\ell-1} * U^{\ell-1}(v)\right)^{\tilde{}} = \tilde{g}_{\ell-1} * U^{\ell-1}(\tilde{v}).$$

Hence by Lemma 4.41 vi, Lemma 4.35 ii and iii, and Exercise 4.5.7 (i),

$$\begin{aligned}
\langle f_\ell, R_{2^\ell k} f_\ell \rangle &= f_\ell * \tilde{f}_\ell(2^\ell k) \\
&= g_{\ell-1} * U^{\ell-1}(v) * \tilde{g}_{\ell-1} * U^{\ell-1}(\tilde{v})(2^\ell k) \\
&= (g_{\ell-1} * \tilde{g}_{\ell-1}) * \left(U^{\ell-1}(v * \tilde{v})\right)(2^\ell k) \\
&= \sum_{n \in \mathbb{Z}} (g_{\ell-1} * \tilde{g}_{\ell-1})(2^\ell k - n) U^{\ell-1}(v * \tilde{v})(n).
\end{aligned}$$

Note that $U^{\ell-1}(v * \tilde{v})(n) = 0$ unless n is of the form $2^{\ell-1} m$, for some $m \in \mathbb{Z}$, in which case it is $v * \tilde{v}(m)$, so we can rewrite the last term and obtain

$$\langle f_\ell, R_{2^\ell k} f_\ell \rangle = \sum_{m \in \mathbb{Z}} (g_{\ell-1} * \tilde{g}_{\ell-1})(2^{\ell-1}(2k - m)) v * \tilde{v}(m).$$

By Lemma 4.41 vi,

$$g_{\ell-1} * \tilde{g}_{\ell-1}(2^{\ell-1}(2k - m)) = \langle g_{\ell-1}, R_{2^{\ell-1}(2k-m)} g_{\ell-1} \rangle,$$

which is 1 when $m = 2k$ and 0 otherwise, by assumption. Hence

$$\langle f_\ell, R_{2^\ell k} f_\ell \rangle = v * \tilde{v}(2k) = \langle v, R_{2k} v \rangle.$$

Since $A(\theta)$ is unitary for all θ by assumption, Theorem 4.46 shows that

$$\langle f_\ell, R_{2^\ell k} f_\ell \rangle = \begin{cases} 1 & \text{if } k = 0 \\ 0 & \text{if } k \neq 0. \end{cases} \tag{4.61}$$

By nearly the same argument with v in place of u, we obtain

$$\langle g_\ell, R_{2^\ell k} g_\ell \rangle = \begin{cases} 1 & \text{if } k = 0 \\ 0 & \text{if } k \neq 0. \end{cases} \tag{4.62}$$

The unitarity of $A(\theta)$ also implies that $\langle v, R_{2k} u \rangle = 0$ for all $k \in \mathbb{Z}$. By using this implication and Lemma 4.42 ii, we obtain (Exercise 4.6.1) that

$$\langle f_\ell, R_{2^\ell k} g_\ell \rangle = 0 \text{ for all } k \in \mathbb{Z}. \tag{4.63}$$

Equations (4.61), (4.62), and (4.63) yield the result. ∎

As in the corresponding case in chapter 3, this result also has a proof based on Fourier transform methods (Exercise 4.6.3). Next we see that Lemma 4.50 gives an orthogonal splitting of subspaces.

Lemma 4.51 *Let $u, v, g_{\ell-1}, f_\ell$, and g_ℓ be as in Lemma 4.50. Define*

$$V_{-\ell+1} = \left\{ \sum_{k \in \mathbb{Z}} z(k) R_{2^{\ell-1}k} g_{\ell-1} : z = (z(k))_{k \in \mathbb{Z}} \in \ell^2(\mathbb{Z}) \right\},$$

$$V_{-\ell} = \left\{ \sum_{k \in \mathbb{Z}} z(k) R_{2^\ell k} g_\ell : z = (z(k))_{k \in \mathbb{Z}} \in \ell^2(\mathbb{Z}) \right\}, \tag{4.64}$$

and

$$W_{-\ell} = \left\{ \sum_{k \in \mathbb{Z}} z(k) R_{2^\ell k} f_\ell : z = (z(k))_{k \in \mathbb{Z}} \in \ell^2(\mathbb{Z}) \right\}. \tag{4.65}$$

Then

$$V_{-\ell} \oplus W_{-\ell} = V_{-\ell+1}.$$

Proof
By Exercise 4.2.8, $V_{-\ell+1}$, $V_{-\ell}$, and $W_{-\ell}$ are subspaces of $\ell^2(\mathbb{Z})$. By Exercise 4.6.4 and equation (4.63), $V_{-\ell}$ and $W_{-\ell}$ are orthogonal.

Now we prove that $V_{-\ell}$ and $W_{-\ell}$ are subsets of $V_{-\ell+1}$. First, note that

$$R_{2^\ell k} f_\ell(n) = f_\ell(n - 2^\ell k) = g_{\ell-1} * U^{\ell-1}(v)(n - 2^\ell k)$$
$$= \sum_{m \in \mathbb{Z}} g_{\ell-1}(n - 2^\ell k - m) U^{\ell-1}(v)(m)$$

$$= \sum_{j \in \mathbb{Z}} g_{\ell-1}(n - 2^\ell k - 2^{\ell-1}j)v(j)$$

$$= \sum_{j \in \mathbb{Z}} v(j) R_{2^{\ell-1}(j+2k)} g_{\ell-1}(n),$$

where, in the next to last line, we used the fact that $U^{\ell-1}(v)(m) = 0$ unless m is of the form $2^{\ell-1}j$ for some $j \in \mathbb{Z}$, in which case it is just $v(j)$. Changing summation index (let $m = j + 2k$) in the last expression gives

$$R_{2^\ell k} f_\ell = \sum_{m \in \mathbb{Z}} v(m - 2k) R_{2^{\ell-1}m} g_{\ell-1}, \tag{4.66}$$

which shows that the elements of the complete orthonormal set $\{R_{2^\ell k} f_\ell\}_{k \in \mathbb{Z}}$ for $W_{-\ell}$ belong to $V_{-\ell+1}$. By Exercise 4.6.5, this implies that $W_{-\ell}$ is a subspace of $V_{-\ell+1}$.

A similar argument applied to g_ℓ leads to

$$R_{2^\ell k} g_\ell = \sum_{m \in \mathbb{Z}} u(m - 2k) R_{2^{\ell-1}m} g_{\ell-1}, \tag{4.67}$$

and hence that $V_{-\ell}$ is a subspace of $V_{-\ell+1}$.

What remains to prove is that $V_{-\ell+1} \subseteq V_{-\ell} \oplus W_{-\ell}$. We begin by observing a certain identity. By Theorem 4.46, $\{R_{2k}v\}_{k \in \mathbb{Z}} \cup \{R_{2k}u\}_{k \in \mathbb{Z}}$ is a complete orthonormal set in $\ell^2(\mathbb{Z})$. Let $j \in \mathbb{Z}$ and let e_j be as in equation (4.5). Then for all $m \in \mathbb{Z}$,

$$e_j(m) = \sum_{k \in \mathbb{Z}} \langle e_j, R_{2k}v \rangle R_{2k}v(m) + \sum_{k \in \mathbb{Z}} \langle e_j, R_{2k}u \rangle R_{2k}u(m).$$

Note that

$$\langle e_j, R_{2k}v \rangle = \sum_{n \in \mathbb{Z}} e_j(n)\overline{v(n - 2k)} = \overline{v(j - 2k)} = \tilde{v}(2k - j),$$

and similarly for u. Substituting these identities into the previous equation yields

$$e_j(m) = \sum_{k \in \mathbb{Z}} \tilde{v}(2k - j)v(m - 2k) + \sum_{k \in \mathbb{Z}} \tilde{u}(2k - j)u(m - 2k). \tag{4.68}$$

Now we move on to our main goal, which is to prove that, for $j \in \mathbb{Z}$,

$$R_{2^{\ell-1}j} g_{\ell-1} = \sum_{k \in \mathbb{Z}} \tilde{v}(2k - j) R_{2^\ell k} f_\ell + \sum_{k \in \mathbb{Z}} \tilde{u}(2k - j) R_{2^\ell k} g_\ell. \tag{4.69}$$

(See Exercise 4.6.6 for the reason one would guess this result.) To prove equation (4.69), substitute equations (4.66) and (4.67) to obtain

$$
\sum_{k \in \mathbb{Z}} \tilde{v}(2k - j) R_{2^\ell k} f_\ell + \sum_{k \in \mathbb{Z}} \tilde{u}(2k - j) R_{2^\ell k} g_\ell
$$

$$
= \sum_{k \in \mathbb{Z}} \tilde{v}(2k - j) \sum_{m \in \mathbb{Z}} v(m - 2k) R_{2^{\ell-1} m} g_{\ell - 1}
$$

$$
+ \sum_{k \in \mathbb{Z}} \tilde{u}(2k - j) \sum_{m \in \mathbb{Z}} u(m - 2k) R_{2^{\ell-1} m} g_{\ell - 1}
$$

$$
= \sum_{m \in \mathbb{Z}} \left(\sum_{k \in \mathbb{Z}} \big(\tilde{v}(2k - j) v(m - 2k) + \tilde{u}(2k - j) u(m - 2k) \big) \right) R_{2^{\ell-1} m} g_{\ell - 1}
$$

$$
= \sum_{m \in \mathbb{Z}} e_j(m) R_{2^{\ell-1} m} g_{\ell - 1} = R_{2^{\ell-1} j} g_{\ell - 1},
$$

by equation (4.68). Thus $R_{2^{\ell-1} j} g_{\ell - 1} \in V_{-\ell} \oplus W_{-\ell}$ for each $j \in \mathbb{Z}$. As a consequence, Exercise 4.6.5 shows that $V_{-\ell+1} \subseteq V_{-\ell} \oplus W_{-\ell}$. This completes the proof. ∎

Definition 4.52 *Suppose p is a positive integer and $f_1, f_2, \ldots$* *$f_{p-1}, f_p, g_p \in \ell^2(\mathbb{Z})$. Let*

$$
B = \{R_{2k} f_1\} \cup \{R_{4k} f_2\} \cup \cdots \cup \{R_{2^p k} f_p\} \cup \{R_{2^p k} g_p\}, \tag{4.70}
$$

where in each case k runs over $\mathbb{Z}$, or equivalently

$$
B = \{R_{2^\ell k} f_\ell : k \in \mathbb{Z}, \ell = 1, 2, \ldots, p\} \cup \{R_{2^p k} g_p : k \in \mathbb{Z}\}. \tag{4.71}
$$

If B is a complete orthonormal set in $\ell^2(\mathbb{Z})$, then B is called a pth stage wavelet system for $\ell^2(\mathbb{Z})$. We say that $f_1, f_2, \ldots, f_p, g_p$ generate B.

Theorem 4.53 *Let $p \in \mathbb{N}$. For $\ell = 1, 2, \ldots, p$, suppose that $u_\ell, v_\ell \in \ell^1(\mathbb{Z})$ and the system matrix*

$$
A_\ell(\theta) = \frac{1}{\sqrt{2}} \begin{bmatrix} \hat{u}_\ell(\theta) & \hat{v}_\ell(\theta) \\ \hat{u}_\ell(\theta + \pi) & \hat{v}_\ell(\theta + \pi) \end{bmatrix} \tag{4.72}
$$

is unitary for all $\theta \in [0, \pi)$. Define $f_1 = v_1, g_1 = u_1$, and, inductively, for $\ell = 2, 3, \ldots, p$,

$$
f_\ell = g_{\ell-1} * U^{\ell-1}(v_\ell), \quad g_\ell = g_{\ell-1} * U^{\ell-1}(u_\ell). \tag{4.73}
$$

Define B as in equation (4.71). Then B is a complete orthonormal set (hence a pth-stage wavelet system) for $\ell^2(\mathbb{Z})$.

Proof

For $\ell = 1, 2, \ldots, p$, define $V_{-\ell}$ and $W_{-\ell}$ as in equations (4.64) and (4.65). To prove that B is orthonormal, we see from Lemma 4.50 and induction that $\{R_{2^\ell k}f_\ell\}_{k \in \mathbb{Z}}$ is orthonormal for each $\ell = 1, 2, \ldots, p$ and that $\{R_{2^p k}g_p\}_{k \in \mathbb{Z}}$ is orthonormal. Hence we need to check orthogonality only between these different sets. Note that for $m < \ell \le p$, and $k \in \mathbb{Z}$,

$$R_{2^\ell k}f_\ell \in V_{-\ell} \subseteq V_{-\ell+1} \subseteq \cdots \subseteq V_{-m},$$

and similarly with $R_{2^p k}g_p$ in place of $R_{2^\ell k}f_\ell$. However, for $j \in \mathbb{Z}$, $R_{2^m j}f_m \in W_{-m}$, which is orthogonal to V_{-m}, by Lemma 4.51.

We now prove that the orthonormal system B is complete. Suppose $z \in \ell^2(\mathbb{Z})$ is orthogonal to every element of B. Applying Exercise 4.2.5 (i), z is orthogonal to every element in $W_{-1} \cup W_{-2} \cup \cdots \cup W_{-p} \cup V_{-p}$. By Lemma 4.51, $W_{-p} \oplus V_{-p} = V_{-p+1}$. This implies z is orthogonal to every element in V_{-p+1}. Similarly, $W_{-p+1} \oplus V_{-p+1} = V_{-p+2}$, so this implies that z is orthogonal to every element of V_{-p+2}, etc., until we obtain that z is orthogonal to every element of $W_{-1} \oplus V_{-1} = \ell^2(\mathbb{Z})$. Hence $z = 0$. ∎

Writing out f_ℓ and g_ℓ, we have

$$f_\ell = u_1 * U(u_2) * \cdots * U^{\ell-2}(u_{\ell-1}) * U^{\ell-1}(v_\ell) \tag{4.74}$$

and

$$g_\ell = u_1 * U(u_2) * \cdots * U^{\ell-2}(u_{\ell-1}) * U^{\ell-1}(u_\ell), \tag{4.75}$$

exactly as in the finite dimensional case (equations (3.43) and (3.44)).

Because we are in the infinite dimensional case, we can carry out the iteration step infinitely many times. This leads to the following variation.

Definition 4.54 *Let $f_\ell \in \ell^2(\mathbb{Z})$ for all $\ell \in \mathbb{N}$. Let*

$$B = \{R_{2^\ell k}f_\ell : k \in \mathbb{Z}, \ell \in \mathbb{N}\}. \tag{4.76}$$

If B is a complete orthonormal system for $\ell^2(\mathbb{Z})$, we say B is a homogeneous wavelet system for $\ell^2(\mathbb{Z})$.

Theorem 4.55 *Suppose $u_\ell, v_\ell \in \ell^1(\mathbb{Z})$ for each $\ell \in \mathbb{N}$, and the system matrix $A_\ell(\theta)$ defined in equation (4.72) is unitary for all $\theta \in [0, \pi)$. Define*

$f_1 = u_1, g_1 = v_1$, and, inductively, for $\ell \in \mathbb{N}, \ell \geq 2$, define f_ℓ and g_ℓ by equation (4.73). For each $\ell \in \mathbb{N}$, define $V_{-\ell}$ as in equation (4.64). Suppose

$$\bigcap_{\ell \in \mathbb{N}} V_{-\ell} = \{0\}. \tag{4.77}$$

Define B as in equation (4.76). Then B is a complete orthonormal set (hence a homogeneous wavelet system) in $\ell^2(\mathbb{Z})$.

Proof

Define $W_{-\ell}$ as in equation (4.65), for all $\ell \in \mathbb{N}$. The orthonormality of B follows by the same argument as in the proof of Theorem 4.53, or directly from Theorem 4.53 by noting that for any two elements of B, there exists p sufficiently large so that the pth-stage wavelet system obtained as in Theorem 4.53 contains these two elements.

To prove the completeness of B, suppose $\langle z, R_{2^\ell k} f_\ell \rangle = 0$ for all $k \in \mathbb{Z}$ and all $\ell \in \mathbb{N}$. Then (e.g., using Exercise 4.2.5 (i)), z is orthogonal to every element of each space W_{-j}. We claim that this implies that $z \in V_{-j}$ for all $j \in \mathbb{N}$. We proceed inductively. First, to prove $z \in V_{-1}$, note that since $\ell^2(\mathbb{Z}) = V_{-1} \oplus W_{-1}$ (by Theorem 4.46), there exist $v_{-1} \in V_{-1}$ and $w_{-1} \in W_{-1}$ such that $z = v_{-1} + w_{-1}$. But $V_{-1} \perp W_{-1}$, so $\langle v_{-1}, w_{-1} \rangle = 0$. Hence

$$\|w_{-1}\|^2 = \langle w_{-1}, w_{-1} \rangle - \langle v_{-1} + w_{-1}, w_{-1} \rangle = \langle z, w_{-1} \rangle = 0,$$

since $w_{-1} \in W_{-1}$. Therefore $w_{-1} = 0$. Hence $z = v_{-1} \in V_{-1}$. Now suppose $z \in V_{-\ell+1}$. By Lemma 4.51, $V_{-\ell+1} = V_{-\ell} \oplus W_{-\ell}$. Thus we can write $z = v_{-\ell} + w_{-\ell}$ for some $v_{-\ell} \in V_{-\ell}$ and $w_\ell \in W_{-\ell}$. By the same argument as in the case $\ell = 1$, the orthogonality of z to $W_{-\ell}$ implies that $w_{-\ell} = 0$, so $z = v_{-\ell} \in V_{-\ell}$. This completes the induction, which shows that $z \in \bigcap_{\ell \in \mathbb{N}} V_{-\ell}$. But our assumption is that $\bigcap_{\ell \in \mathbb{N}} V_{-\ell} = \{0\}$. Hence $z = 0$, proving the completeness of B. ∎

Notice that it is possible to pick $u_\ell = u_1$ and $v_\ell = v_1$ for all ℓ in either Theorem 4.53 or Theorem 4.55. We call this *repeated filters*.

The more standard notation for wavelets is

$$\psi_{-j,k} = R_{2^j k} f_j \tag{4.78}$$

and

$$\varphi_{-j,k} = R_{2^j k} g_j. \tag{4.79}$$

In this notation, the pth-stage wavelet system in Theorem 4.53 is

$$B = \{\psi_{-j,k} : k \in \mathbb{Z}, 1 \le j \le p\} \cup \{\varphi_{-p,k} : k \in \mathbb{Z}\}. \tag{4.80}$$

The homogeneous wavelet system in Theorem 4.55 is

$$\{\psi_{-j,k} : j \in \mathbb{N}, k \in \mathbb{Z}\}. \tag{4.81}$$

Thus in the homogeneous case, the *father wavelet* φ disappears from the record, leaving only the *mother wavelet* ψ and her daughters. In the next chapter, when we consider wavelets on $\mathbb{R}$, the chain of spaces V_j will not only be infinite, it will extend infinitely in both directions (i.e., j will run throughout $\mathbb{Z}$).

Exercises

4.6.1. Prove equation (4.63).

4.6.2. i. Suppose $z \in \ell^2(\mathbb{Z})$. Prove that

$$(U(z))\hat{\ }(\theta) = \hat{z}(2\theta)$$

for all θ.

ii. Define f_ℓ and g_ℓ as in Theorem 4.53 (i.e., by equations (4.74) and (4.75)). Prove that

$$\hat{f}_\ell(\theta) = \hat{u}_1(\theta)\hat{u}_2(2\theta)\hat{u}_3(4\theta)\cdots\hat{u}_{\ell-1}(2^{\ell-2}\theta)\hat{v}_\ell(2^{\ell-1}\theta)$$

and

$$\hat{g}_\ell(\theta) = \hat{u}_1(\theta)\hat{u}_2(2\theta)\hat{u}_3(4\theta)\cdots\hat{u}_{\ell-1}(2^{\ell-2}\theta)\hat{u}_\ell(2^{\ell-1}\theta).$$

If the filters are repeated, that is, $u_\ell = u$ and $v_\ell = v$ for all ℓ, we obtain

$$\hat{\varphi}_{-\ell,0}(\theta) = \prod_{j=0}^{\ell-1} \hat{u}(2^j\theta)$$

and

$$\hat{\psi}_{-\ell,0}(\theta) = \hat{v}(2^{\ell-1}\theta) \prod_{j=0}^{\ell-2} \hat{u}(2^j\theta).$$

4.6.3. Prove Lemma 4.50 using Exercises 4.6.2(i) and Exercise 4.5.8(i). Hint: For example, to show $\{R_{2^\ell k}f_\ell\}_{k\in\mathbb{Z}}$ is orthonormal, write

$$\sum_{k=0}^{2^\ell-1}\left|\hat{f}_\ell\left(\theta+\frac{2\pi k}{2^\ell}\right)\right|^2$$

$$=\sum_{k=0}^{2^\ell-1}\left|\hat{g}_{\ell-1}\left(\theta+\frac{2\pi k}{2^\ell}\right)\right|^2\left|\hat{v}(2^{\ell-1}\theta+\pi k)\right|^2$$

$$=|\hat{v}(2^{\ell-1}\theta)|^2\sum_{m=0}^{2^{\ell-1}-1}\left|\hat{g}_{\ell-1}\left(\theta+\frac{2\pi m}{2^{\ell-1}}\right)\right|^2$$

$$+|\hat{v}(2^{\ell-1}\theta+\pi)|^2\sum_{m=0}^{2^{\ell-1}-1}\left|\hat{g}_{\ell-1}\left(\theta+\frac{\pi}{2^{\ell-1}}+\frac{2\pi m}{2^{\ell-1}}\right)\right|^2,$$

by breaking the sum on k into its even and odd parts. Now apply the assumptions.

4.6.4. Suppose H is a Hilbert space, and $\{a_k\}_{k\in\mathbb{Z}}$ and $\{b_k\}_{k\in\mathbb{Z}}$ are orthonormal sets in H with $\langle a_j, b_k\rangle = 0$ for all $j, k \in \mathbb{Z}$. Let

$$V = \left\{\sum_{k\in\mathbb{Z}}z(k)a_k : z = (z(k))_{k\in\mathbb{Z}} \in \ell^2(\mathbb{Z})\right\}$$

and

$$W = \left\{\sum_{k\in\mathbb{Z}}z(k)b_k : z = (z(k))_{k\in\mathbb{Z}} \in \ell^2(\mathbb{Z})\right\}.$$

Prove that $V \perp W$ (i.e., for all $v \in V$ and $w \in W$, $\langle v, w\rangle = 0$). Hint: First show that $\langle v, b_k\rangle = 0$ for all $k \in \mathbb{Z}$, by Exercise 4.2.5 (iii).

4.6.5. Suppose H is a Hilbert space, and $\{a_k\}_{k\in\mathbb{Z}}$ and $\{b_k\}_{k\in\mathbb{Z}}$ are orthonormal sets in H. Let

$$V = \left\{\sum_{k\in\mathbb{Z}}z(k)a_k : z = (z(k))_{k\in\mathbb{Z}} \in \ell^2(\mathbb{Z})\right\}$$

and

$$W = \left\{\sum_{k\in\mathbb{Z}}z(k)b_k : z = (z(k))_{k\in\mathbb{Z}} \in \ell^2(\mathbb{Z})\right\}.$$

Suppose $a_k \in W$ for each k. Prove that V is a subspace of W. Hint: Let $v = \sum_{k \in \mathbb{Z}} z(k)a_k \in V$. If you try to write each a_k as a sum of the form $\sum_{j \in \mathbb{Z}} c_k(j)b_j$ and work with these sums, you will run into a problem similar to that in Exercise 4.4.6 (ii). Instead, note that the partial sums of $\sum_{k \in \mathbb{Z}} z(k)a_k$ belong to W and use Exercise 4.2.10.

4.6.6. For $u, v, g_{\ell-1}, f_\ell, g_\ell$ as in Lemma 4.51, prove that

$$\langle R_{2^{\ell-1}j}g_{\ell-1}, R_{2^\ell k}f_\ell \rangle = \tilde{v}(2k - j), \text{ and}$$
$$\langle R_{2^{\ell-1}j}g_{\ell-1}, R_{2^\ell k}g_\ell \rangle = \tilde{u}(2k - j).$$

If $R_{2^{\ell-1}j}g_{\ell-1} \in V_{-\ell} \oplus W_{-\ell}$, it must be true that

$$R_{2^{\ell-1}j}g_{\ell-1} = \sum_{k \in \mathbb{Z}} \langle R_{2^{\ell-1}j}g_{\ell-1}, R_{2^\ell k}f_\ell \rangle R_{2^\ell k}f_\ell$$
$$+ \sum_{k \in \mathbb{Z}} \langle R_{2^{\ell-1}j}g_{\ell-1}, R_{2^\ell k}g_\ell \rangle R_{2^\ell k}g_\ell,$$

so we are led to equation (4.69). Hint: Apply Lemma 4.41 v and vi, and follow the type of argument leading to equation (4.61).

4.7   Implementation and Examples

We discuss the computation of wavelets on $\mathbb{Z}$, which is analogous to the theory on $\mathbb{Z}_N$ in chapter 3. We can construct examples of wavelet systems in $\mathbb{Z}$ along the same lines as some of the examples for $\mathbb{Z}_N$. In particular, we discuss Haar wavelets and Daubechies's D6 wavelets on $\mathbb{Z}$.

Recall the procedure in Theorem 4.53. We start with $u_\ell, v_\ell \in \ell^1(\mathbb{Z})$, for $1 \le \ell \le p$, such that the system matrix $A_\ell(\theta)$ in equation (4.72) is unitary for all θ. We set $f_1 = u_1, f_2 = u_2$, and define f_ℓ and g_ℓ inductively for $\ell = 2, 3, \ldots, p$ by equation (4.73). Define $\psi_{-j,k}$ and $\varphi_{-p,k}$ for $k \in \mathbb{Z}$ and $1 \le j \le p$ by equations (4.78) and (4.79). Then B defined in equation (4.80) is a pth-stage wavelet system in $\ell^2(\mathbb{Z})$. We have already remarked (after the proof of Theorem 4.46) that the first-stage wavelet coefficients and the reconstruction of a vector from its first-stage wavelet coefficients can be computed by the same filter bank arrangement as in Figure 14 (with $\tilde{s} = u$ and $\tilde{t} = v$). The only difference is that the vectors and convolutions are defined in $\ell^2(\mathbb{Z})$. In the same way, the pth-stage transform and its inverse can be computed as in the nonrecursive filter-bank structure in Figure 20. To see this, note that

$$\langle z, \psi_{-\ell,k} \rangle = \langle z, R_{2^\ell k} f_\ell \rangle = z * \tilde{f}_\ell(2^\ell k) = D^\ell(z * \tilde{f}_\ell)(k),$$

by Lemma 4.41 vi and the definition of D^ℓ. Similarly,

$$\langle z, \varphi_{-p,k} \rangle = D^p(z * \tilde{g}_p)(k).$$

These relations show that the pth-stage wavelet coefficients of z can be computed by the analysis phase in the left portion of Figure 20. To see that z can be reconstructed using the synthesis phase in the right half of Figure 20, note (Exercise 4.7.1) that

$$f_\ell * U^\ell(D^\ell(z * \tilde{f}_\ell)) = \sum_{k \in \mathbb{Z}} \langle z, \psi_{-\ell,k} \rangle \psi_{-\ell,k} \tag{4.82}$$

and similarly

$$g_p * U^p(D^p(z * \tilde{g}_p)) = \sum_{k \in \mathbb{Z}} \langle z, \varphi_{-p,k} \rangle \varphi_{-p,k}. \tag{4.83}$$

The wavelet coefficients can also be computed recursively as in Figures 16–19. Define $x_1 = D(z * \tilde{v}_1) \in \ell^2(\mathbb{Z})$ and $y_1 = D(z * \tilde{u}_1) \in \ell^2(\mathbb{Z})$. Then define $x_2, y_2, \dots, x_p, y_p \in \ell^2(\mathbb{Z})$ inductively by

$$x_\ell = D(y_{\ell-1} * \tilde{v}_\ell)$$

and

$$y_\ell = D(y_{\ell-1} * \tilde{u}_\ell),$$

analogous to Definition 3.16. Then (Exercise 4.7.2)

$$x_\ell = D^\ell(z * \tilde{f}_\ell) \tag{4.84}$$

and

$$y_\ell = D^\ell(z * \tilde{g}_\ell,) \tag{4.85}$$

as in Lemma 3.21. With these facts, one can see that the analysis phase of the nonrecursive structure in Figure 20 (now regarded as taking place in $\mathbb{Z}$ rather than $\mathbb{Z}_N$) is equivalent to the recursive structure indicated in Figure 17 for $p = 3$ and at the general step in Figure 18. Similarly, the synthesis phase can be computed by a recursive structure whose general step is represented in Figure 19. The proof of this is the analog for $\mathbb{Z}$ of the proof of Lemma 3.22.

In principle, these convolutions in $\mathbb{Z}$ involve infinite sums and hence do not appear computable in practice. However, as we see by example later, if the vector z has only finitely many nonzero components and if the elements of the wavelet basis are each zero except at finitely many points, the convolution can be computed in finitely many steps. Hence we focus attention on examples of wavelet bases on $\mathbb{Z}$ that have this property that each basis element has only finitely many nonzero components. The simplest example is the Haar system.

Example 4.56

(Haar wavelets on $\mathbb{Z}$) Define $u, v \in \ell^1(\mathbb{Z})$ by

$$u(n) = \begin{cases} 1/\sqrt{2} & \text{if } n = 0 \text{ or } n = 1 \\ 0 & \text{otherwise} \end{cases} \tag{4.86}$$

and

$$v(n) = \begin{cases} 1/\sqrt{2} & \text{if } n = 0 \\ -1/\sqrt{2} & \text{if } n = 1 \\ 0 & \text{otherwise.} \end{cases} \tag{4.87}$$

One can check (Exercise 4.7.3) that the pair u, v generates a first-stage wavelet system for $\ell^2(\mathbb{Z})$. For each $\ell \in \mathbb{N}$, let $u_\ell = u$ and $v_\ell = v$. Define f_ℓ and g_ℓ inductively by equation (4.73). One can show (Exercise 4.7.4 (i)) that

$$g_\ell(n) = \begin{cases} 2^{-\ell/2} & 0 \le n \le 2^\ell - 1 \\ 0 & \text{otherwise} \end{cases} \tag{4.88}$$

and

$$f_\ell(n) = \begin{cases} 2^{-\ell/2} & 0 \le n \le 2^{\ell-1} - 1 \\ -2^{-\ell/2} & 2^{\ell-1} \le n \le 2^\ell - 1 \\ 0 & \text{otherwise.} \end{cases} \tag{4.89}$$

(Compare with Exercise 3.3.1.) Then by Theorem 4.53, B defined by equation (4.71) is a pth-stage wavelet system for $\ell^2(\mathbb{Z})$.

For $j \in \mathbb{N}$, define V_{-j} as in equation (4.64). This space has a simple characterization, described in Exercise 4.7.4 (ii), which shows that

$\cap_{j\in\mathbb{Z}}V_{-j} = \{0\}$. By Theorem 4.55, B defined by equation (4.76) is a homogeneous wavelet system on $\ell^2(\mathbb{Z})$, called the *Haar system* for $\ell^2(\mathbb{Z})$.

For $j \geq 1$, define the orthogonal projection operator P_{-j} onto V_{-j} as in Definition 4.13, by

$$P_{-j}(z) = \sum_{k\in\mathbb{Z}}\langle z, \varphi_{-j,k}\rangle\varphi_{-j,k}. \tag{4.90}$$

For the Haar system, the projection $P_{-j}(z)$ of z onto V_{-j} has a natural interpretation. For each $n \in \mathbb{Z}$, we can find $k \in \mathbb{Z}$ such that $2^j k \leq n < 2^j(k+1)$ (because for fixed j, these intervals in n are disjoint and their union covers $\mathbb{Z}$). Then (Exercise 4.7.4 (iii))

$$P_{-j}z(n) = 2^{-j}\left(z(2^j k) + z(2^j k + 1) + \cdots + z(2^j k + 2^j - 1)\right). \tag{4.91}$$

In other words, P_{-j} replaces z on each block $\{n \in \mathbb{Z} : 2^j k \leq n < 2^j(k+1)\}$ by its average on that block.

We also define the partial sums $Q_{-j}(z)$ of the wavelet expansion of z by

$$Q_{-j}(z) = \sum_{k\in\mathbb{Z}}\langle z, \psi_{-j,k}\rangle\psi_{-j,k}. \tag{4.92}$$

It follows (Exercise 4.7.5) that

$$P_{-j+1}(z) = P_{-j}(z) + Q_{-j}(z), \tag{4.93}$$

for all $j \geq 1$, where for $j = 0$ we define $P_0(z) = z$. Thus, as in the case of $\mathbb{Z}_N$, we interpret $Q_{-j}(z)$ as the detail regarding z needed to pass from the $-j^{\text{th}}$ level approximation $P_{-j}(z)$ to the $-j + 1$ level approximation $P_{-j+1}(z)$.

An example of the Haar wavelet transform is pictured in Figures 37 and 38. Figure 37a shows the graph of

$$z(n) = \begin{cases} \sin(\pi n/19) & \text{if } 0 \leq n \leq 38 \\ 0 & \text{otherwise.} \end{cases}$$

Of course, the graph window is finite, but keep in mind that z is defined on $\mathbb{Z}$ and equals 0 for all integers less than 0 or greater than 38. The Haar wavelet coefficients $\langle z, \psi_{-j,k}\rangle$ for $j = 1, 2, 3, 4$ are plotted in Figures 37b, c, d, and e, respectively. The coefficients $\langle z, \varphi_{-4,k}\rangle$ are plotted in Figure 37f. As usual, we plot the value $\langle z, \psi_{-j,k}\rangle$ at the point

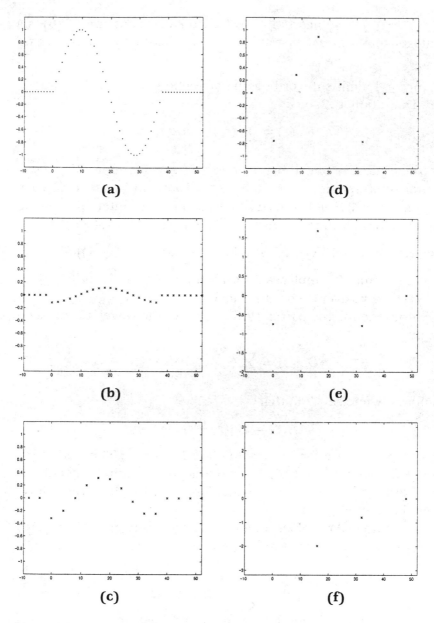

FIGURE 37 **(a)** z, **(b)** $\langle z, \psi_{-1,k} \rangle$, **(c)** $\langle z, \psi_{-2,k} \rangle$, **(d)** $\langle z, \psi_{-3,k} \rangle$,
(e) $\langle z, \psi_{-4,k} \rangle$, **(f)** $\langle z, \varphi_{-4,k} \rangle$

$2^j k$ near which $\psi_{-j,k}$ is concentrated. Similarly $\langle z, \varphi_{-4,k} \rangle$ is plotted at $2^4 k$.

In this case it appears (and turns out to be true) that the wavelet coefficients $\langle z, \psi_{-j,k} \rangle$ are nonzero only at values $2^j k$ such that $0 \le 2^j k \le 36$ (similarly for $\langle z, \varphi_{-4,k} \rangle$). However, as will be more evident with Daubechies's D6 wavelets later, it is not necessarily true that the nonzero wavelet coefficients are confined to the region where z is nonzero (strictly speaking this is not true here because $z(0) = 0$, but the wavelet coefficients corresponding to the point 0 are not zero). How do we know in Figure 37 that there are no nonzero wavelet coefficients corresponding to points outside the region graphed? Exercise 4.7.6 gives a very useful general result characterizing the maximal nonzero region for the convolution of two vectors, in terms of their nonzero regions. Note that

$$\langle z, \psi_{-j,k} \rangle = \sum_{n \in \mathbb{Z}} z(n) \overline{R_{2^j k} f_j(n)} = \sum_{n \in \mathbb{Z}} z(n) \overline{f_j(n - 2^j k)}$$
$$= \sum_{n \in \mathbb{Z}} z(n) \tilde{f}_j(2^j k - n) = z * \tilde{f}_j(2^j k).$$

By equation (4.89), $\tilde{f}_j(n)$ is nonzero only for $-2^j + 1 \le n \le 0$. By definition, z is nonzero only for $1 \le n \le 37$. Hence by Exercise 4.7.6, $z * \tilde{f}_j(n)$ can be nonzero only for $-2^j + 2 \le n \le 37$. Therefore, $\langle z, \psi_{-j,k} \rangle$ could be nonzero only if $-2^j + 2 \le 2^j k \le 37$. The smallest k for which this can happen is 0, and the largest is at most 36 (since $j \ge 1$). (The analysis for $\langle z, \varphi_{-4,k} \rangle$ is the same as for $\langle z, \psi_{-4,k} \rangle$.) Thus we can determine in advance the maximal region on which we need to plot the wavelet coefficients. If we count, we get at most 41 wavelet coefficients that could be nonzero (19, 10, 5, 3, and 3, respectively, for Figures 37b, c, d, e, and f, respectively), although some in-between values actually turn out to be 0. Note that this is not exactly the same as the number of nonzero values of z, but it is close.

To understand the graphs in Figure 37, recall that the wavelet coefficients $\langle z, \psi_{-j,k} \rangle$ measure the activity in z going on at a scale of 2^j near the point $2^j k$. Thus the fact that the wavelet coefficients at the smallest scale (for $j = 1$) in Figure 37b are relatively small shows that the small-scale variation of z is mild. This reflects the smoothness and slow variation of z. As we increase j, the

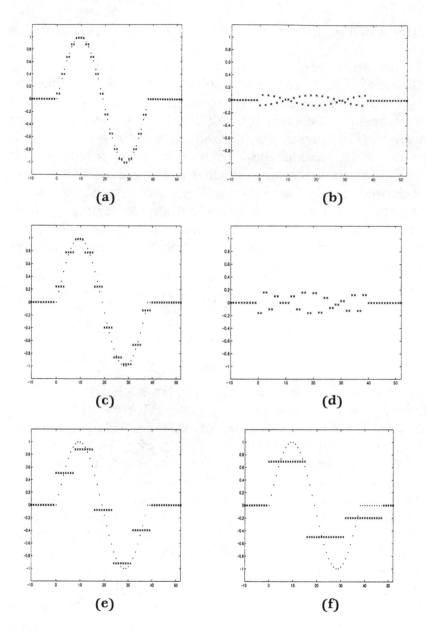

FIGURE 38   **(a)** $P_{-1}(z)$, **(b)** $Q_{-1}(z)$, **(c)** $P_{-2}(z)$, **(d)** $Q_{-2}(z)$, **(e)** $P_{-3}(z)$, **(f)** $P_{-4}(z)$

typical wavelet coefficients become larger, indicating the substantial large-scale variation of z.

Figures 38a, c, e, and f show the projections $P_{-j}(z)$, for $j = 1$, 2, 3, and 4, respectively. These projection vectors are plotted with x's, and the original vector z is plotted with dots. These pictures exhibit the property that $P_{-j}(z)$ is obtained by replacing z on a block $\{n : 2^j k \leq n < 2^j(k + 1)\}$ of length 2^j by its average on this block. Thus we can think of $P_{-j}(z)$ as z averaged out at a scale of 2^j, or seen at a resolution of 2^j. Since $z = P_{-1}(z) + Q_{-1}(z)$ (by equation (4.93)), $Q_{-1}(z)$ is the difference between z and its scale-of-2 average $P_{-1}(z)$. Thus $Q_{-1}(z)$, plotted in Figure 38b, contains the fine-scale information needed to pass from z seen at a resolution of 2 ($P_{-1}(z)$ in Figure 38a) to z at resolution 1 (z itself). Similarly, since $P_{-1}(z) = P_{-2}(z) + Q_{-2}(z)$, $Q_{-2}(z)$ (plotted in Figure 38d) contains the information required to improve from z at a resolution of 4 ($P_{-2}(z)$ in Figure 38c) to z at a resolution of 2 ($P_{-1}(z)$).

Example 4.57
(Daubechies's D6 wavelets on $\mathbb{Z}$) In Example 3.35, we constructed vectors with only six nonzero components, which generate a first-stage wavelet system for $\ell^2(\mathbb{Z}_N)$, for certain N. By virtually the same construction, we can obtain vectors $u, v \in \ell^2(\mathbb{Z})$ with only six nonzero components, which generate a first-stage wavelet system for $\ell^2(\mathbb{Z})$. Using these as repeated filters, we can obtain a pth-stage wavelet system for $\ell^2(\mathbb{Z})$ for any $p \in \mathbb{N}$. We call these *Daubechies's D6 wavelets for $\ell^2(\mathbb{Z})$*.

By Theorems 4.46 and 4.53, and Lemma 4.47, our main concern is to find $u \in \ell^2(\mathbb{Z})$ with only six nonzero components such that $\{R_{2k}u\}_{k \in \mathbb{Z}}$ is an orthonormal set in $\ell^2(\mathbb{Z})$. This is done by the same method as in Example 3.35, so we leave some computations that are virtually the same to the reader (Exercise 4.7.7). The main difference is that we work with all $\theta \in [0, \pi)$ instead of the discrete set of values $\pi n/N$, for $0 \leq n < N/2$. There are also a few minor differences coming from the fact that the DFT is defined with a minus sign in the complex exponential, whereas the Fourier transform on $\mathbb{Z}$ is defined with a positive sign (both definitions are arbitrary but traditional).

We start with the identity

$$\left(\cos^2\left(\frac{\theta}{2}\right) + \sin^2\left(\frac{\theta}{2}\right)\right)^5 = 1 \quad \text{for all} \quad \theta. \tag{4.94}$$

Define

$$b(\theta) = \cos^{10}\left(\frac{\theta}{2}\right) + 5\cos^8\left(\frac{\theta}{2}\right)\sin^2\left(\frac{\theta}{2}\right) + 10\cos^6\left(\frac{\theta}{2}\right)\sin^4\left(\frac{\theta}{2}\right).$$

By elementary trigonometric identities,

$$b(\theta+\pi) = 10\cos^4\left(\frac{\theta}{2}\right)\sin^6\left(\frac{\theta}{2}\right) + 5\cos^2\left(\frac{\theta}{2}\right)\sin^8\left(\frac{\theta}{2}\right) + \sin^{10}\left(\frac{\theta}{2}\right).$$

By expanding equation (4.94), we see that

$$b(\theta) + b(\theta+\pi) = 1 \quad \text{for all} \quad \theta. \tag{4.95}$$

We select $u \in \ell^2(\mathbb{Z})$ with only six nonzero components such that

$$|\hat{u}(\theta)|^2 = 2b(\theta) \quad \text{for all} \quad \theta. \tag{4.96}$$

To do this, we first write

$$b(\theta) = \cos^6\left(\frac{\theta}{2}\right)\left[\left(\cos^2\left(\frac{\theta}{2}\right) - \sqrt{10}\sin^2\left(\frac{\theta}{2}\right)\right)^2\right.$$
$$\left. + \left(5 + 2\sqrt{10}\right)\cos^2\left(\frac{\theta}{2}\right)\sin^2\left(\frac{\theta}{2}\right)\right].$$

Define

$$\hat{u}(\theta) = \sqrt{2}e^{5i\theta/2}\cos^3\left(\frac{\theta}{2}\right)\left[\cos^2\left(\frac{\theta}{2}\right) - \sqrt{10}\sin^2\left(\frac{\theta}{2}\right)\right.$$
$$\left. -i\sqrt{5 + 2\sqrt{10}}\cos\left(\frac{\theta}{2}\right)\sin\left(\frac{\theta}{2}\right)\right].$$

Hence equation (4.96) holds. As in Example 3.35, it is convenient to set

$$a = 1 - \sqrt{10}, \ b = 1 + \sqrt{10}, \ \text{and} \ c = \sqrt{5 + 2\sqrt{10}}.$$

Then the double-angle formulae, Euler's formula, and some computations yield

$$\hat{u}(\theta) = \sqrt{2}e^{i\theta}e^{3i\theta/2}\left(\frac{e^{i\theta/2}+e^{-i\theta/2}}{2}\right)^3\left[\frac{a}{2}+\frac{b}{4}\left(e^{i\theta}+e^{-i\theta}\right)+\frac{c}{4}\left(e^{-i\theta}-e^{i\theta}\right)\right]$$

$$=\frac{\sqrt{2}}{32}\left(e^{i\theta+1}\right)^3\left[b+c+2ae^{i\theta}+(b-c)e^{2i\theta}\right]=\sum_{k=0}^{5}u(k)e^{ik\theta},$$

for

$$(u(0),u(1),u(2),u(3),u(4),u(5)) \tag{4.97}$$

$$=\frac{\sqrt{2}}{32}(b+c,2a+3b+3c,6a+4b+2c, \tag{4.98}$$

$$6a+4b-2c,2a+3b-3c,b-c). \tag{4.99}$$

Define $u \in \ell^2(\mathbb{Z})$ by letting $u(n)$ be as in equation (4.97) for $0 \le n \le 5$ and $u(n) = 0$ for all other $n \in \mathbb{Z}$. By equations (4.95) and (4.96),

$$|\hat{u}(\theta)|^2 + |\hat{u}(\theta+\pi)|^2 = 2 \quad \text{for all} \quad \theta.$$

By Lemma 4.42 i, $\{R_{2k}u\}_{k\in\mathbb{Z}}$ is an orthonormal set in $\ell^2(\mathbb{Z})$.

Define $v \in \ell^2(\mathbb{Z})$ by setting $v(n) = (-1)^{n-1}u(1-n)$ for all $n \in \mathbb{Z}$. (This agrees with equation (4.56) because u is real-valued in this case.) Specifically,

$$v(-4) = -u(5), v(-3) = u(4), v(-2) = -u(3),$$
$$v(-1) = u(2), v(0) = -u(1), v(1) = u(0),$$

and $v(n) = 0$ for all other $n \in \mathbb{Z}$. By Lemma 4.47, u and v generate a first-stage wavelet system for $\ell^2(\mathbb{Z})$.

Let $p \in \mathbb{N}$ be given. Let $u_\ell = u$ and $v_\ell = v$ for $1 \le \ell \le p$. Define $f_1 = v_1$ and $g_1 = u_1$. Define $f_2, g_2, \ldots, f_p, g_p$ inductively by equation (4.73). Define $\psi_{-j,k} = R_{2^jk}f_j$ and $\varphi_{-j,k} = R_{2^jk}g_j$ for $1 \le j \le p$ and $k \in \mathbb{Z}$. Define B by equation (4.80). By Theorem 4.53, B is a pth stage wavelet system for $\ell^2(\mathbb{Z})$. We call B the pth-stage D6 wavelet system for $\ell^2(\mathbb{Z})$.

Note that all components of u and v are real-valued, which simplifies computations.

We remark on the main step of this construction. By using the half-angle trigonometric formulae ($\cos^2(\theta/2) = (1 + \cos\theta)/2$ and $\sin^2(\theta/2) = (1 - \cos\theta)/2$), we can rewrite $b(\theta)$ as a polynomial in the variable $\cos\theta$. By using Euler's formulas, we can then write b as a trigonometric polynomial. It is not obvious that the trigonometric polynomial $2b(\theta)$ has a "modular square root," that is, a function u satisfying $|\hat{u}(\theta)|^2 = 2b(\theta)$, which is also a trigonometric polynomial. Following Strichartz (1993), we have obtained this example by hand. However, there is a general result of Fejér and Riesz (sometimes called the *Riesz lemma*) from the early 1900s that states that any nonnegative trigonometric polynomial has a modular square root that is also a trigonometric polynomial.

In addition to the D6 wavelet system, Daubechies constructed similar wavelet systems with u and v having $2N$ nonzero components, for each $N \in \mathbb{Z}$. The case $N = 2$ gives the Haar system (Exercise 4.7.8).

Figure 39 shows the 4th-level D6 wavelet coefficients of the vector

$$z(n) = \begin{cases} \sin(\frac{n^{1.5}}{64}) & \text{if } 0 \le n \le 511 \\ 0 & \text{otherwise.} \end{cases}$$

This is the extension by 0 of the vector in $\ell^2(\mathbb{Z}_N)$ considered in Figure 32. Figure 39a shows z, restricted to the interval $0 \le n \le 511$. Figures 39b, c, d, and e show the wavelet coefficients $\langle z, \psi_{-j,k} \rangle$, for $j = 1, 2, 3$ and 4, respectively, where $\langle z, \psi_{-j,k} \rangle$ is plotted at the point $2^j k$, as in Figure 37, and similarly for $\langle z, \varphi_{-4,k} \rangle$ plotted in Figure 39f. Notice that in Figure 39b, the larger coefficients occur in the right half of the graph, where z is oscillating more rapidly. This reflects the fact that the wavelet coefficients $\langle z, \psi_{-1,k} \rangle$ measure the finest scale (or highest frequency, or most rapid oscillation) behavior of z. As j increases, the wavelet coefficients pick up the larger scale (more slowly oscillating) behavior of z, so that in Figure 39f, the largest values are near zero and in the region just to the right of zero. These graphs also show the phenomenon that a few wavelet coefficients corresponding to points outside the original range $0 \le n \le 511$ can be nonzero; this is most clear in Figure 39f.

The partial reconstructions $P_{-4}(z)$, $P_{-3}(z)$, $P_{-2}(z)$, and $P_{-1}(z)$ are plotted in Figures 40a, b, c, and d, respectively. These exhibit the same phenomenon as in Figure 39: the more slowly oscillating

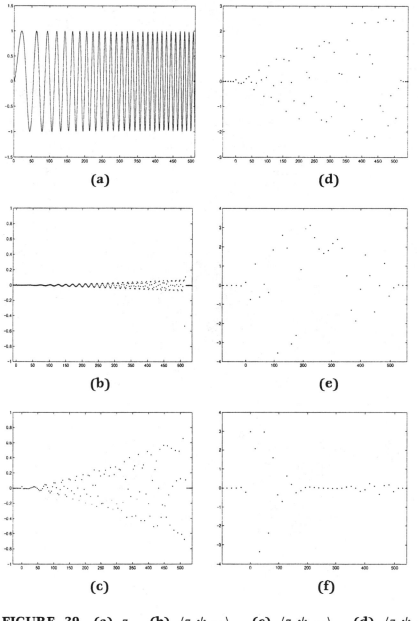

FIGURE 39 **(a)** z, **(b)** $\langle z, \psi_{-1,k} \rangle$, **(c)** $\langle z, \psi_{-2,k} \rangle$, **(d)** $\langle z, \psi_{-3,k} \rangle$, **(e)** $\langle z, \psi_{-4,k} \rangle$, **(f)** $\langle z, \varphi_{-4,k} \rangle$

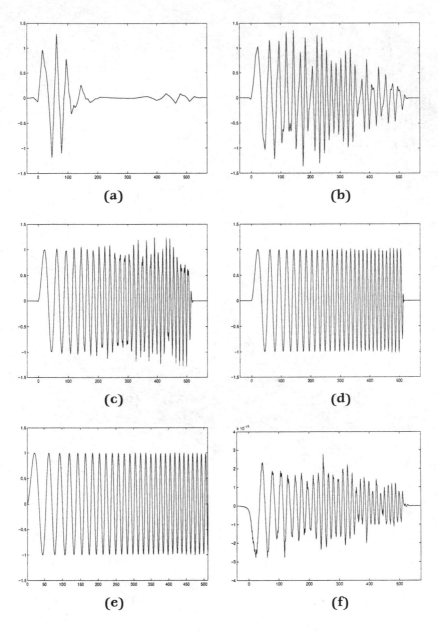

FIGURE 40   **(a)** $P_{-4}(z)$,   **(b)** $P_{-3}(z)$,   **(c)** $P_{-2}(z)$,   **(d)** $P_{-1}(z)$,   **(e)** $P_0(z)$,   **(f)** $E = z - P_0(z)$

portions of z are picked up by $P_{-4}(z)$, the part of the wavelet expansion corresponding to the terms $\{\varphi_{-4,k}\}_{k\in\mathbb{Z}}$. As we include terms corresponding to $\psi_{-4,k}$ (in $P_{-3}(z)$), then $\psi_{-3,k}$ (in $P_{-2}(z)$), $\psi_{-2,k}$ (in $P_{-1}(z)$), and finally $\psi_{-1,k}$ (in $P_0(z)$), we fill in the more rapid oscillations and fine detail of z. Of course $P_0(z)$ should agree with z. This appears to be the case in Figure 40e. To test this, we have plotted $z - P_0(z)$ in Figure 40f. Note that the scale here is in multiples of 10^{-14}. This small error comes from round-off error in MatLab, the program used to do these computations and graphs.

In Figure 41a we have plotted the vector

$$z(n) = \begin{cases} 0 & \text{if } n < 0 \\ 1 - n/64 & \text{if } 0 \le n \le 63 \\ 0 & \text{if } 64 \le n \le 255 \\ 5 - n/64 & \text{if } 256 \le n \le 319 \\ 0 & \text{if } n > 320. \end{cases}$$

This is the extension by 0 of the vector in $\ell^2(\mathbb{Z}_N)$ whose D6 wavelet coefficients in the sense of $\mathbb{Z}_N$ were displayed in Figure 28. The only difference is that values that appeared near the right edge in Figure 28 appear slightly to the left of 0 in Figure 41 (this is easiest to see in Figures 28f and 41f). Note that all wavelet coefficients in Figure 41b–f corresponding to points near the right edge of the graph are 0, because these wavelets do not have any nonzero values at points where z is nonzero. This was not the case for Figure 28 because of the periodicity of vectors in $\ell^2(\mathbb{Z}_N)$.

The vector

$$z(n) = \begin{cases} \sin(\pi n/128) & \text{if } 0 \le n \le 511 \\ 0 & \text{otherwise} \end{cases}$$

is plotted in Figure 42a. Its 4th-level D6 wavelet coefficients are plotted in Figures 42b–f. For a vector z with such a slow oscillation, the only terms of significant size are those in Figure 42f, corresponding to larger scale behavior (note the widely varying scales used in these different graphs). One curiosity in Figure 42b–e is the relatively large size of the values near 0 and 511 (compared to the values in the middle). This comes from the fact that the underlying function $f(x)$ on $\mathbb{R}$ defined by $f(x) = \sin(\pi x/128)$ for $0 \le x \le 512$ and $f(x) = 0$ otherwise, is not differentiable at $x = 0$

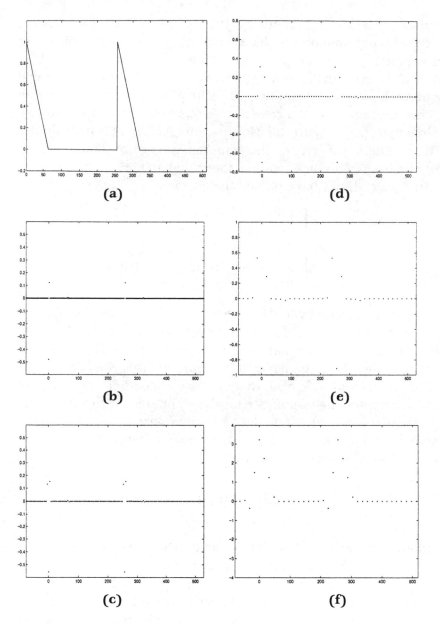

FIGURE 41   **(a)** z,   **(b)** $\langle z, \psi_{-1,k} \rangle$,   **(c)** $\langle z, \psi_{-2,k} \rangle$,   **(d)** $\langle z, \psi_{-3,k} \rangle$,
(e) $\langle z, \psi_{-4,k} \rangle$,   **(f)** $\langle z, \varphi_{-4,k} \rangle$

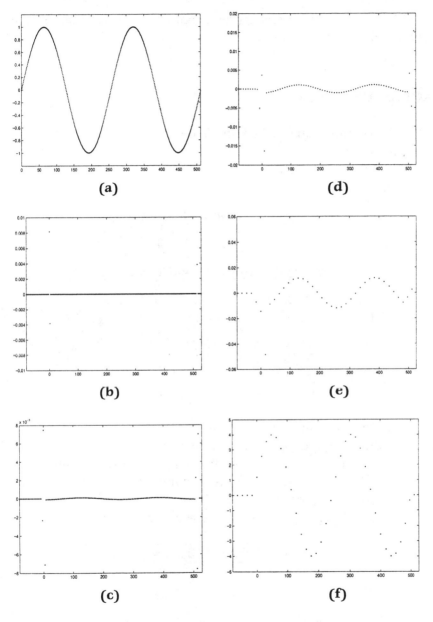

FIGURE 42 **(a)** z, **(b)** $\langle z, \psi_{-1,k} \rangle$, **(c)** $\langle z, \psi_{-2,k} \rangle$, **(d)** $\langle z, \psi_{-3,k} \rangle$, **(e)** $\langle z, \psi_{-4,k} \rangle$, **(f)** $\langle z, \varphi_{-4,k} \rangle$

or $x = 512$. This lack of smoothness requires relatively large high frequency terms to synthesize.

Exercises

4.7.1. Prove equation (4.82). Hint: See equations (4.54) and (4.55) for the case $\ell = 1$.

4.7.2. Suppose $z, w \in \ell^2(\mathbb{Z})$.

  i. Prove that

$$D(z) * w = D(z * U(w))$$

and

$$U(z) * U(w) = U(z * w).$$

Hint: See Lemma 3.18 for the case of $\mathbb{Z}_N$.

  ii. Let ℓ be a positive integer. Prove that

$$D^\ell(z) * w = D^\ell(z * U^\ell(w))$$

and

$$U^\ell(z * w) = U^\ell(z) * U^\ell(z) * U^\ell(w).$$

  iii. Prove equations (4.84) and (4.85).

4.7.3. Define $u, v \in \ell^1(\mathbb{Z})$ by equaitons (4.86) and (4.87).

  i. Prove directly (without using Theorem 4.46) that $\{R_{2k}v\}_{k \in \mathbb{Z}} \cup \{R_{2k}u\}_{k \in \mathbb{Z}}$ is a complete orthonormal set (hence a first-stage wavelet system) in $\ell^2(\mathbb{Z})$. Hint: To prove completeness, prove that for $z \in \ell^2(\mathbb{Z})$,

$$z(2m) = \frac{1}{\sqrt{2}} \left(\langle z, R_{2m}u \rangle + \langle z, R_{2m}v \rangle \right),$$

and

$$z(2m + 1) = \frac{1}{\sqrt{2}} \left(\langle z, R_{2m}u \rangle - \langle z, R_{2m}v \rangle \right).$$

  ii. Show directly that the sytem matrix $A(\theta)$ (Definition 4.45) is unitary for all θ. This gives another proof of the result in part i.

iii. For $z \in \ell^2(\mathbb{Z})$, let

$$P_{-1}(z) = \sum_{k \in \mathbb{Z}} \langle z, R_{2k}u \rangle R_{2k}u.$$

Prove that for $m \in \mathbb{Z}$,

$$P_{-1}(z)(2m) = P(z)(2m+1) = \frac{z(2m) + z(2m+1)}{2}.$$

Thus $P_{-1}(z)$ replaces $z(2m)$ and $z(2m+1)$ by their average.

4.7.4. Define $u, v \in \ell^1(\mathbb{Z})$ by equations (4.86) and (4.87). For each $\ell \in \mathbb{N}$, let $u_\ell = u$ and $v_\ell = v$. Define f_ℓ and g_ℓ inductively by equation (4.73).

  i. Prove equations (4.88) and (4.89).

  ii. Define $V_{-\ell}$ for $\ell \in \mathbb{N}$ as in equation (4.64). Show that $z \in V_{-\ell}$ if and only if $z \in \ell^2(\mathbb{Z})$ and z is constant on each block of values

$$k2^\ell, k2^\ell + 1, k2^\ell + 2, \ldots, k2^\ell + 2^\ell - 1$$

  for $k \in \mathbb{Z}$. Deduce that $\bigcap_{\ell \in \mathbb{N}} V_{-\ell} = \{0\}$.

  iii. Let P_{-j} be the orthogonal projection onto V_{-j}, as in equation (4.90). For $z \in \ell^2(\mathbb{Z})$ and $n \in \mathbb{Z}$ such that $2^j k \le n < 2^j(k+1)$ for some $k \in \mathbb{Z}$, prove equation (4.91).

4.7.5. Prove equation (4.93). Hint: See the proof of equation (3.88).

4.7.6. Suppose $a, b, c, d \in \mathbb{Z}$ with $a < b$ and $c < d$. Suppose $z, w \in \ell^2(\mathbb{Z})$ satisfy

$$z(n) = 0 \quad \text{if} \quad n < a \quad \text{or} \quad n > b$$

and

$$w(n) = 0 \quad \text{if} \quad n < c \quad \text{or} \quad n > d.$$

  i. Prove that $z * w(n) = 0$ if $n < a + c$ or $n > b + d$.

  ii. Let $z(n) = 1$ for $a \le n \le b$ and 0 otherwise, and let $w(n) = 1$ for $c \le n \le d$ and 0 otherwise. Prove that $z * w(n) \ne 0$ for $a + c \le n \le b + d$. Thus the result in part i cannot be improved.

4.7.7. Verify the computational assertions in Example 4.57, including equations (4.95) and (4.97–4.99). Hint: See Example 3.35.

4.7.8. Starting with $\cos^2(\theta/2) + \sin^2(\theta/2) = 1$ and following the procedure in Example 4.57, derive the Haar wavelets.

4.7.9. Starting with

$$\left(\cos^2(\theta/2) + \sin^2(\theta/2)\right)^3 = 1$$

and proceeding as in Example 4.57, construct $u, v \in \ell^1(\mathbb{Z})$, each having only four nonzero components, which generate a first-stage wavelet basis for $\ell^2(\mathbb{Z})$. One answer:

$$(u(0), u(1), u(2), u(3))$$
$$= \frac{\sqrt{2}}{8}\left(1 + \sqrt{3}, 3 + \sqrt{3}, 3 - \sqrt{3}, 1 - \sqrt{3}\right),$$

$u(n) = 0$ for all other $n \in \mathbb{Z}$,

$$(v(-2), v(-1), v(0), v(1))$$
$$= \frac{\sqrt{2}}{8}\left(-1 + \sqrt{3}, 3 - \sqrt{3}, -3 - \sqrt{3}, 1 + \sqrt{3}\right),$$

and $v(n) = 0$ for all other $n \in \mathbb{Z}$. These generate *Daubechies's D4 wavelets* on $\ell^2(\mathbb{Z})$.

5

CHAPTER

Wavelets on ℝ

5.1 $L^2(\mathbb{R})$ and Approximate Identities

Despite the previous few chapters, the term "wavelets" usually refers to wavelets on $\mathbb{R}$, examples of which we construct in this chapter. The first two sections present the basics of Fourier analysis on $\mathbb{R}$.

We consider complex-valued functions defined on $\mathbb{R}$. As one might suspect from chapter 4, to obtain a suitable notion of orthogonality we must restrict ourselves to functions are not too large. Specifically, we consider functions f that are *square-integrable*, that is, such that

$$\int_{\mathbb{R}} |f(x)|^2 \, dx < +\infty.$$

As in the case of $L^2([-\pi, \pi))$ in chapter 4, we are using the Lebesgue integral and identifying two functions that agree a.e. (almost everywhere). (Also as in chapter 4, the reader unfamiliar with these terms can just ignore them, if he or she is willing to accept a few consequences of this theory.) Formally,

$$L^2(\mathbb{R}) = \left\{ f : \mathbb{R} \to \mathbb{C} : \int_{\mathbb{R}} |f(x)|^2 \, dx < +\infty \right\}.$$

349

$L^2(\mathbb{R})$ is a vector space with the operations of pointwise addition and scalar multiplication of functions (Exercise 5.1.1 (i)).

For $f, g \in L^2(\mathbb{R})$, define

$$\langle f, g \rangle = \int_{\mathbb{R}} f(x)\overline{g(x)}\, dx. \tag{5.1}$$

By Exercise 5.1.1(ii), $\langle \cdot, \cdot \rangle$ is an inner product on $L^2(\mathbb{R})$. Applying Definition 1.90 and Exercise 1.6.5, $L^2(\mathbb{R})$ is a normed space with the norm

$$\|f\| = \left(\int_{\mathbb{R}} |f(x)|^2\, dx \right)^{1/2}, \tag{5.2}$$

called the L^2 norm. The Cauchy-Schwarz inequality (Lemma 1.91) gives us

$$\left| \int_{\mathbb{R}} f(x)\overline{g(x)}\, dx \right| \leq \left(\int_{\mathbb{R}} |f(x)|^2\, dx \right)^{1/2} \left(\int_{\mathbb{R}} |g(x)|^2\, dx \right)^{1/2},$$

for $f, g \in L^2(\mathbb{R})$. By applying this inequality with f and g replaced by $|f|$ and $|g|$, respectively, we obtain

$$\int_{\mathbb{R}} |f(x)g(x)|\, dx \leq \left(\int_{\mathbb{R}} |f(x)|^2\, dx \right)^{1/2} \left(\int_{\mathbb{R}} |g(x)|^2\, dx \right)^{1/2}. \tag{5.3}$$

Also, by Corollary 1.92, we have the triangle inequality

$$\left(\int_{\mathbb{R}} |f(x) + g(x)|^2\, dx \right)^{1/2} \leq \left(\int_{\mathbb{R}} |f(x)|^2\, dx \right)^{1/2} + \left(\int_{\mathbb{R}} |g(x)|^2\, dx \right)^{1/2}. \tag{5.4}$$

We define convergence in $L^2(\mathbb{R})$ in accordance with the definitions in section 4.2 for a general inner product space. Namely, suppose $\{f_n\}_{n \in \mathbb{N}}$ is a sequence of functions in $L^2(\mathbb{R})$ and $f \in L^2(\mathbb{R})$. We say $\{f_n\}_{n \in \mathbb{N}}$ converges to f in $L^2(\mathbb{R})$ if, for all $\epsilon > 0$, there exist $N \in \mathbb{N}$ such that $\|f_n - f\| < \epsilon$ for all $n > N$. By Exercise 4.2.1, this is equivalent to $\|f_n - f\| \to 0$ as $n \to +\infty$. We say $\{f_n\}_{n \in \mathbb{N}}$ is Cauchy if, for all $\epsilon > 0$, there exists N such that $\|f_n - f_m\| < \epsilon$ for all $n, m > N$. We assume the somewhat deep fact (which depends on properties of the Lebesgue integral) that $L^2(\mathbb{R})$ is complete, meaning that every Cauchy sequence in $L^2(\mathbb{R})$ converges in $L^2(\mathbb{R})$. Thus, in the terminology of section 4.2, $L^2(\mathbb{R})$ is a Hilbert space. In particular, all of the results about complete orthonormal sets from section 4.2 apply here.

One example of a complete orthonormal set in $L^2(\mathbb{R})$ is given in Exercise 5.1.2. In this chapter, we construct wavelet systems, which are complete orthonormal sets for $L^2(\mathbb{R})$ of a particular form.

As in chapter 4, we consider the class of integrable functions, that is, those functions whose integral converges absolutely.

Definition 5.1 *Let*

$$L^1(\mathbb{R}) = \left\{ f : \mathbb{R} \to \mathbb{C} : \int_{\mathbb{R}} |f(x)| \, dx < +\infty \right\}.$$

For $f \in L^1(\mathbb{R})$, let

$$\|f\|_1 = \int_{\mathbb{R}} |f(x)| \, dx.$$

We call $\| \cdot \|_1$ the L^1 norm. If $f \in L^1(\mathbb{R})$, we say f is integrable.

Remarks corresponding to those made after Definition 4.15 apply in the context of $\mathbb{R}$, with the important exception that there is no containment between $L^1(\mathbb{R})$ and $L^2(\mathbb{R})$ (Exercise 5.1.3).

Lemma 5.2 *If $f \in L^1(\mathbb{R})$, then*

$$\left| \int_{\mathbb{R}} f(x) \, dx \right| \le \int_{\mathbb{R}} |f(x)| \, dx = \|f\|_1. \tag{5.5}$$

Proof
Exercise 5.1.4. ∎

With the norm $\|\cdot\|_1$, $L^1(\mathbb{R})$ is a normed vector space (as in Exercise 1.6.5) but not an inner product space (Exercise 5.1.5). Note that we still use the notation $\| \cdot \|$ for the norm defined in equation (5.2).

We now consider convolution on $\mathbb{R}$.

Definition 5.3 *Suppose $f, g : \mathbb{R} \to \mathbb{C}$ and*

$$\int_{\mathbb{R}} |f(x - y)g(y)| \, dy < \infty \tag{5.6}$$

for a.e. $x \in \mathbb{R}$. For x such that equation (5.6) holds, define

$$f * g(x) = \int_{\mathbb{R}} f(x - y)g(y) \, dy. \tag{5.7}$$

Define $f * g(x) = 0$ *for* x *at which equation (5.6) is false. We call* $f * g$
the convolution of f *and* g.

Under the assumptions of Definition 5.3, inequality (5.6) fails
only on a set of measure 0, so we could define $f * g$ arbitrarily there.
We selected the value 0 just to be definite.

It is easy to see (by making the change of variables $t = x - y$ in
equation (5.7)) that convolution is commutative:

$$f * g = g * f, \tag{5.8}$$

whenever either convolution (and hence the other) is defined.

Several circumstances under which the convolution of f and g is
defined follow.

Lemma 5.4

    i. *Suppose* $f, g \in L^2(\mathbb{R})$. *Then inequality (5.6) holds for all* $x \in \mathbb{R}$.
    In this case $f * g$ *is bounded and*

$$|f * g(x)| \leq \|f\| \|g\| \quad \text{for all} \quad x \in \mathbb{R}.$$

    ii. *Suppose* $f, g \in L^1(\mathbb{R})$. *Then inequality (5.6) holds for a.e.* $x \in \mathbb{R}$.
    In this case, $f * g \in L^1(\mathbb{R})$ *with*

$$\|f * g\|_1 \leq \|f\|_1 \|g\|_1.$$

    iii. *Suppose* $f \in L^2(\mathbb{R})$ *and* $g \in L^1(\mathbb{R})$. *Then (5.6) holds for*
    a.e. $x \in \mathbb{R}$. *In this case,* $f * g \in L^2(\mathbb{R})$ *with*

$$\|f * g\| \leq \|f\| \|g\|_1.$$

Proof
For part i, note that by relation (5.3),

$$\int_{\mathbb{R}} |f(x-y)g(y)| \, dy$$

$$\leq \left(\int_{\mathbb{R}} |f(x-y)|^2 \, dy \right)^{1/2} \left(\int_{\mathbb{R}} |g(y)|^2 \, dy \right)^{1/2} = \|f\| \|g\|,$$

for all $x \in \mathbb{R}$, by changing variables in the integral involving $f(x-y)$.
Thus inequality (5.6) holds for all x. Since

$$|f * g(x)| = \left| \int_{\mathbb{R}} f(x-y)g(y) \, dy \right| \leq \int_{\mathbb{R}} |f(x-y)| |g(y)| \, dy,$$

by relation(5.5), we obtain $|f*g(x)| \leq \|f\|\,\|g\|$ from the same estimate.

To prove part ii, note that by inequality (5.5),

$$\int_{\mathbb{R}} |f*g(x)|\,dx \leq \int_{\mathbb{R}} \int_{\mathbb{R}} |f(x-y)||g(y)|\,dy\,dx$$

$$= \int_{\mathbb{R}} |g(y)| \int_{\mathbb{R}} |f(x-y)|\,dx\,dy$$

$$= \|f\|_1 \int_{\mathbb{R}} |g(y)|\,dy = \|f\|_1 \|g\|_1,$$

where the next-to-last equality follows by the change of variable $t = x - y$ in the interior integral, and the interchange of the order of integration is justified by a theorem in analysis because the integrands are all positive. This shows that inequality (5.6) holds, since $\int_{\mathbb{R}} |f(x-y)g(y)|\,dy$, as a function of x, is in $L^1(\mathbb{R})$ and so must be finite a.e.

We leave the proof of part iii to the reader (Exercise 5.1.6). ∎

We remark that in case i, when $f, g \in L^2(\mathbb{R})$, it is not necessarily true (see Exercise 5.2.15) that $f*g \in L^2(\mathbb{R})$ (as in the case of $\ell^2(\mathbb{Z})$ in Exercise 4.4.6 (ii)).

The definitions of conjugate reflection and translation are similar to their analogs in chapters 3 and 4.

Definition 5.5   *Suppose* $f : \mathbb{R} \to \mathbb{C}$, *and* $y \in \mathbb{R}$. *Define the translation* $R_y f : \mathbb{R} \to \mathbb{C}$ *by*

$$R_y f(x) = f(x-y).$$

Also define the conjugate reflection $\tilde{f} : \mathbb{R} \to \mathbb{C}$ *by*

$$\tilde{f}(x) = \overline{f(-x)}.$$

We have the following properties.

Lemma 5.6   *Suppose* $f, g \in L^2(\mathbb{R})$ *and* $x, y \in \mathbb{R}$. *Then*
  i. $\langle R_x f, R_y g \rangle = \langle f, R_{y-x} g \rangle$.
  ii. $\langle f, R_y g \rangle = f * \tilde{g}(y)$.

Proof
Exercise 5.1.7. ∎

We can consider the operation of dilation for functions in $L^2(\mathbb{R})$ (unlike the cases of $\ell^2(\mathbb{Z}_N)$, $\ell^2(\mathbb{Z})$, or $L^2([-\pi, \pi))$).

Definition 5.7   *For $g : \mathbb{R} \to \mathbb{C}$ and $t \in \mathbb{R}$ with $t > 0$, define $g_t : \mathbb{R} \to \mathbb{C}$, the t-dilation of g, by*

$$g_t(x) = \frac{1}{t} g\left(\frac{x}{t}\right). \tag{5.9}$$

The factor t in equation (5.9) guarantees that $\int_{\mathbb{R}} g_t(x) dx = \int_{\mathbb{R}} g(x) dx$, by a change of variable.

In the cases of $\ell^2(\mathbb{Z}_N)$ and $\ell^2(\mathbb{Z})$, there is an identity element for convolution. That is, there exists an element $\delta \in \ell^2(\mathbb{Z}_N)$ (respectively, $\delta \in \ell^2(\mathbb{Z})$) such that $\delta * z = z$ for all $z \in \ell^2(\mathbb{Z})$ (respectively, $\ell^2(\mathbb{Z})$) (see Lemma 2.29 and Exercise 4.4.7.) However, there is no $\delta \in L^2(\mathbb{R})$ such that $\delta * f = f$ for all $f \in L^2(\mathbb{R})$ (see Exercise 5.2.16). (There is a measure δ with this property, popularly known as the *delta function* but it is not an element of $L^2(\mathbb{R})$.) Instead, there is an analog known as an approximate identity.

Definition 5.8   *Suppose $g : \mathbb{R} \to \mathbb{C}$ satisfies*

$$|g(x)| \leq \frac{c_1}{(1 + |x|)^2} \quad \text{for all} \quad x \in \mathbb{R}, \tag{5.10}$$

for some constant $c_1 > 0$, and

$$\int_{\mathbb{R}} g(x) \, dx = 1. \tag{5.11}$$

For each $t > 0$, define g_t by equation (5.9). The family $\{g_t\}_{t>0}$ is called an approximate identity.

Inequality (5.10) is a concise way of stating that g is bounded and decays at least as fast as a constant multiple of $1/x^2$ for large x. Note that these two estimates imply that $g \in L^1(\mathbb{R})$: By applying the boundedness estimate for $|x| \leq 1$ and the decay estimate for $|x| > 1$, we have

$$\int_{\mathbb{R}} |g(x)| \, dx \leq \int_{\{x : |x| \leq 1\}} c_1 \, dx + \int_{\{x : |x| > 1\}} \frac{c_1}{x^2} \, dx < +\infty.$$

Thus the integral in equation (5.11) is absolutely convergent. In many texts, decay conditions weaker than (5.10) are allowed (see Exercise 5.1.14), but this definition is sufficient for our purposes. The reason for calling $\{g_t\}_{t>0}$ an approximate identity is made clear in Theorem 5.11. To state this result, we require some preliminaries.

We assume the fundamental theorem of calculus for Lebesgue integration: if $f \in L^1(\mathbb{R})$ and

$$F(x) = \int_a^x f(t)\,dt$$

for some fixed $a \in \mathbb{R}$, then F is a.e. differentiable on $\mathbb{R}$ and $F' = f$ a.e. (as in Theorem 4.20). A consequence of this is that if $f \in L^1(\mathbb{R})$, then

$$\lim_{h \to 0^+} \frac{1}{2h} \int_{x-h}^{x+h} f(t)\,dt = f(x) \quad \text{for} \quad \text{a.e.} x \in \mathbb{R}. \tag{5.12}$$

To see this, note that

$$\frac{1}{2h} \int_{x-h}^{x+h} f(t)\,dt = \frac{F(x+h) - F(x-h)}{2h}$$

$$= \frac{1}{2} \frac{F(x+h) - F(x)}{h} + \frac{1}{2} \frac{F(x) - F(x-h)}{h}$$

and observe that each difference quotient converges to $F'(x)$ as $h \to 0$, when $F'(x)$ exists, hence a.e. Because $f(x)$ is constant with respect to the integration variable t,

$$\frac{1}{2h} \int_{x-h}^{x+h} f(x)\,dt = f(x),$$

so equation (5.12) can be rewritten as

$$\lim_{h \to 0^+} \frac{1}{2h} \int_{x-h}^{x+h} f(t) - f(x)\,dt = 0 \quad \text{a.e.,}$$

or, by changing variables (let $y = x - t$) as

$$\lim_{h \to 0^+} \frac{1}{2h} \int_{-h}^{h} f(x-y) - f(x)\,dy = 0 \quad \text{a.e.} \tag{5.13}$$

However, it is possible for these integrals to converge to 0 because of cancellation rather than because the nearby values $f(x - y)$ for $-h < y < h$ are getting close, on the average, to $f(x)$ (see Exercise 5.1.8). This suggests the following stronger notion.

Definition 5.9   *Suppose $f \in L^1(\mathbb{R})$. A point $x \in \mathbb{R}$ is a* Lebesgue point *of f if*

$$\lim_{h \to 0^+} \frac{1}{2h} \int_{-h}^{h} |f(x-y) - f(x)| \, dy = 0. \qquad (5.14)$$

A point of continuity of f is a Lebesgue point of f, but a Lebesgue point is not necessarily a point of continuity (Exercise 5.1.9). An integrable function f may not have any points of continuity (for example, the function that is 1 at every rational value of x and 0 at every irrational value), but the next result guarantees that nearly all points are Lebesgue points of f. This result appears stronger than equation (5.13), but actually follows from equation (5.13) by a measure theory argument.

Lemma 5.10   *Suppose $f \in L^1(\mathbb{R})$. Then almost every point of $\mathbb{R}$ is a Lebesgue point of f.*

Proof
Exercise 5.1.11. ∎

The notion of a Lebesgue point of a function extends easily to the class of locally integrable functions (see Exercises 5.1.10 and 5.1.12), a class larger than $L^1(\mathbb{R})$. However, we have restricted our attention to integrable functions because that restriction is sufficient for our next result, which is our main purpose.

Theorem 5.11   *Suppose $f \in L^1(\mathbb{R})$ and $\{g_t\}_{t>0}$ is an approximate identity (i.e., g satisfies relation (5.10) for some $c_1 > 0$, equation (5.11), and g_t is defined by equation (5.9)). Then for every Lebesgue point x of f (hence, by Lemma 5.10, for a.e. $x \in \mathbb{R}$),*

$$\lim_{t \to 0^+} g_t * f(x) = f(x). \qquad (5.15)$$

Proof
Suppose x is a Lebesgue point of f. Let $\epsilon > 0$. By the definition of a Lebesgue point, there exists $H > 0$ such that if $0 < h \le H$,

$$\frac{1}{2h} \int_{-h}^{h} |f(x-y) - f(x)| \, dy < \frac{\epsilon}{24c_1}. \qquad (5.16)$$

Since $g \in L^1(\mathbb{R})$ (as noted earlier), by Exercise 5.1.13 there exists a sufficiently small $t_0 > 0$ such that

$$\frac{c_1 t_0 \|f\|_1}{H^2} + |f(x)| \int_{\{y:|y| \geq H/t_0\}} |g(y)| \, dy < \epsilon/2. \tag{5.17}$$

We claim that $|g_t * f(x) - f(x)| < \epsilon$ if $0 < t < t_0$, which would complete the proof of equation (5.15). To obtain this estimate, first note that

$$\int_{\mathbb{R}} g_t(y) \, dy = \int_{\mathbb{R}} \frac{1}{t} g\left(\frac{y}{t}\right) dy = \int_{\mathbb{R}} g(u) \, du = 1,$$

by the change of variables $u = y/t$ and equation (5.11). Hence

$$f(x) = \int_{\mathbb{R}} f(x) g_t(y) \, dy.$$

Therefore, by using (5.5), we get

$$|g_t * f(x) - f(x)| = \left| \int_{\mathbb{R}} \big(f(x-y) - f(x)\big) g_t(y) \, dy \right|$$

$$\leq \int_{\mathbb{R}} \big|f(x-y) - f(x)\big| \, |g_t(y)| \, dy = I_t + II_t,$$

where

$$I_t = \int_{-H}^{H} \big|f(x-y) - f(x)\big| \, |g_t(y)| \, dy$$

and

$$II_t = \int_{\{y:|y| \geq H\}} \big|f(x-y) - f(x)\big| \, |g_t(y)| \, dy.$$

We estimate II_t first. By relation (5.10),

$$|g_t(y)| = \left| \frac{1}{t} g\left(\frac{y}{t}\right) \right| \leq \frac{c_1}{t} \left(\frac{y}{t}\right)^{-2} = \frac{c_1 t}{y^2}, \tag{5.18}$$

so for $|y| \geq H$, $|g_t(y)| \leq c_1 t/H^2$. Hence,

$$\int_{\{y:|y| \geq H\}} |f(x-y)| |g_t(y)| \, dy \leq \frac{c_1 t}{H^2} \int_{\mathbb{R}} |f(x-y)| \, dy$$

$$\leq \frac{c_1 t}{H^2} \int_{\mathbb{R}} |f(y)| \, dy = \frac{c_1 t}{H^2} \|f\|_1.$$

Therefore, by the triangle inequality,

$$II_t \leq \int_{\{y:|y|\geq H\}} |f(x-y)||g_t(y)|\, dy + |f(x)| \int_{\{y:|y|\geq H\}} |g_t(y)|\, dy$$

$$\leq \frac{c_1 t}{H^2} \|f\|_1 + |f(x)| \int_{\{u:|u|\geq H/t\}} |g(u)|\, du < \frac{\epsilon}{2}$$

if $t < t_0$, using a change of variables $u = y/t$, relation (5.17), and the fact that the last expression involving t decreases when t is reduced.

Hence, the proof will be complete if we show that $I_t < \epsilon/2$. We use the simple estimate

$$|g_t(y)| = \left| \frac{1}{t} g \left(\frac{y}{t} \right) \right| \leq \frac{c_1}{t}, \tag{5.19}$$

which follows from inequality (5.10). If $t \geq H$, we use relations (5.19) and (5.16) to conclude that

$$I_t \leq \frac{c_1}{t} \int_{-H}^{H} |f(x-y) - f(x)|\, dy \leq \frac{c_1}{t} \frac{2H\epsilon}{24c_1} \leq \frac{\epsilon}{12} < \frac{\epsilon}{2},$$

by using the assumption that $t \geq H$.

If $t < H$, there exists a unique nonnegative integer K such that $2^K \leq H/t < 2^{K+1}$. We break up the region of integration in I_t as follows:

$$I_t = \sum_{k=1}^{K} \int_{\{y:2^{-k}H \leq y < 2^{-k+1}H\}} |f(x-y) - f(x)||g_t(y)|\, dy$$

$$+ \int_{\{y:|y| < 2^{-K}H\}} |f(x-y) - f(x)||g_t(y)|\, dy.$$

In the region $\{y : 2^{-k}H \leq y < 2^{-k+1}H\}$, relation (5.18) implies

$$|g_t(y)| \leq \frac{c_1 t}{y^2} \leq \frac{c_1 t 2^{2k}}{H^2}.$$

For the last integral we use relation (5.19). Substituting these estimates, we obtain

$$I_t \leq \frac{c_1 t}{H^2} \sum_{k=1}^{K} 2^{2k} \int_{\{y:|y|\leq 2^{-k+1}H\}} |f(x-y) - f(x)|\, dy$$

$$+ \frac{c_1}{t} \int_{\{y:|y|\leq 2^{-K}H\}} |f(x-y) - f(x)|\, dy,$$

where we have replaced the annular regions $\{y : 2^{-k}H \leq y \leq 2^{-k+1}H\}$ by the larger regions $\{y : |y| \leq 2^{-k+1}H\}$, which cannot decrease the corresponding integrals. We apply equation (5.16) to obtain

$$I_t \leq \frac{c_1 t}{H^2} \frac{4H\epsilon}{24c_1} \sum_{k=1}^{K} 2^{2k} 2^{-k} + \frac{c_1}{t} \frac{2H\epsilon}{24c_1} 2^{-K}$$

$$= \frac{\epsilon t}{6H} \sum_{k=1}^{K} 2^k + \frac{\epsilon}{12} \frac{2^{-K}H}{t} < \frac{\epsilon}{6} \frac{2t2^K}{H} + \frac{\epsilon}{6} \leq \frac{\epsilon}{3} + \frac{\epsilon}{6} = \frac{\epsilon}{2},$$

by using $\sum_{k=1}^{K} 2^k < 2^{K+1}$, and the facts, following from our definition of K, that $2^{-K}H/t < 2$ and $t2^K/H \leq 1$. This estimate completes the proof. ∎

Thus $g_t * f(x)$ is close to $f(x)$ for sufficiently small t, which justifies calling $\{g_t\}_{t>0}$ an approximate identity for convolution. We remark that one cannot obtain more than the a.e. convergence of $g_t * f(x)$ to $f(x)$, because one can change f arbitrarily on a set of measure 0 without changing $g_t * f(x)$. Theorem 5.11 plays a key role in the inversion of the Fourier transform on $\mathbb{R}$ in the next section.

Exercises

5.1.1. i. Define addition on $L^2(\mathbb{R})$ by $(f + g)(x) = f(x) + g(x)$. Define multiplication of a function $f \in L^2(\mathbb{R})$ by a scalar $\alpha \in \mathbb{C}$ by $(\alpha f)(x) = \alpha f(x)$. With these operations, prove that $L^2(\mathbb{R})$ is a vector space. (As in Exercise 4.3.1, the only property in Definition 1.30 that is not immediately clear is A1. Use Exercise 1.6.3 (i) to check A1.)

   ii. Check that $\langle \cdot, \cdot \rangle$, defined by equation (5.1), is a complex inner product on $L^2(\mathbb{R})$. (As in Exercise 4.3.1 (ii), the only difficulty is to see that the integral defining $\langle f, g \rangle$ converges absolutely for $f, g \in L^2(\mathbb{R})$. See Exercise 1.6.3 (i).)

5.1.2. For $j, k \in \mathbb{Z}$, let

$$g_{j,k}(x) = \begin{cases} (2\pi)^{-1/2} e^{ijx} & \text{if } 2\pi k \leq x < 2\pi(k+1) \\ 0 & \text{otherwise.} \end{cases}$$

Prove that $\{g_{j,k}\}_{j,k\in\mathbb{Z}}$ is a complete orthonormal set for $L^2(\mathbb{R})$. Hint: For completeness, suppose $\langle f, g_{j,k}\rangle = 0$ for all $j, k \in \mathbb{Z}$, for $f \in L^2(\mathbb{R})$. Use the completeness of the trigonometric system in $L^2([-\pi, \pi))$ (Corollary 4.22) to show that the restriction of f to $[2\pi k, 2\pi(k+1))$ is a.e. 0 for each $k \in \mathbb{Z}$.

5.1.3. i. Let $f(x) = 1/x$ for $x > 1$, and 0 otherwise. Show that $f \in L^2(\mathbb{R}) \setminus L^1(\mathbb{R})$.

ii. Give an example of a function $f \in L^1(\mathbb{R}) \setminus L^2(\mathbb{R})$. Hint: See Exercise 4.3.3.

5.1.4. Prove Lemma 5.2. Hint: See Exercise 4.3.10.

5.1.5. Prove that $(L^1(\mathbb{R}), \|\cdot\|_1)$ is a normed vector space (see the definition in Exercise 1.6.5), but not an inner product space. Hint: See Exercise 1.6.6.

5.1.6. Prove Lemma 5.4 iii. Hint: Follow the argument in the proof of Lemma 4.34, but use integrals instead of sums.

5.1.7. Prove Lemma 5.6.

5.1.8. Define $f \in L^1(\mathbb{R})$ by setting $f(x) = -1$ for $-1 < x < 0, f(0) = 0, f(x) = 1$ for $0 < x < 1$, and $f(x) = 0$ for all other $x \in \mathbb{R}$. Prove that equation (5.13) holds for $x = 0$, but 0 is not a Lebesgue point of f.

5.1.9. i. Suppose $f \in L^1(\mathbb{R})$ and f is continuous at x. Prove that x is a Lebesgue point of f.

ii. Define $f \in L^1(\mathbb{R})$ by setting

$$f(x) = 1 \quad \text{for} \quad \frac{1}{n} < x < \frac{1}{n} + \frac{1}{2^n}, \quad n = 1, 2, 3, \ldots$$

and $f(x) = 0$ for all other x. Prove that f is not continuous at 0 but 0 is a Lebesgue point of f.

5.1.10. (This exercise requires some background in Lebesgue integration theory.) A function $f : \mathbb{R} \to \mathbb{C}$ is *locally integrable* if $\int_{-T}^{T} |f(x)| \, dx < \infty$ for every $T > 0$. The class of all locally integrable functions on $\mathbb{R}$ is denoted $L^1_{\text{loc}}(\mathbb{R})$. For example, $f(x) = x$ is locally integrable but not integrable. Prove equation (5.13) for $f \in L^1_{\text{loc}}$. (Note that the local integrability of f guarantees the existence and finiteness of the integral in equation (5.13).) Hint: By the fact that a countable union of sets of measure 0 has measure 0, it is enough to show that the set of points x in the interval $[n, n+1]$ for which equation (5.13) fails has measure 0 for each integer n. To see this, apply equation (5.13) to an integrable function that agrees with f on a neighborhood of $[n, n+1]$.

5.1.11. (This exercise requires some background in Lebesgue integration theory.) Prove Lemma 5.10, assuming equation (5.13). Hint: Select some enumeration $\{r_i\}_{i=1}^{\infty}$ of the rational numbers. For each $i \in \mathbb{N}$, apply Exercise 5.1.10 to $|f(x) - r_i|$, which is locally integrable, to deduce that there is a set E_i of measure 0 such that for all $x \in \mathbb{R} \setminus E_i$,

$$\lim_{h \to 0} \frac{1}{2h} \int_{-h}^{h} \left| f(x-y) - r_i \right| - \left| f(x) - r_i \right| \, dy = 0.$$

Let $E = \cup_{i=1}^{\infty} E_i$. The objective is to show that every $x \in \mathbb{R} \setminus E$ is a Lebesgue point of f. To see this, let $\epsilon > 0$ be given and pick r_i sufficiently close to $f(x)$. Then write

$$\int_{-h}^{h} |f(x-y) - f(x)| \, dy \leq \int_{-h}^{h} |f(x-y) - r_i| \, dy + \int_{-h}^{h} |r_i - f(x)| \, dy.$$

5.1.12. (This exercise requires some background in Lebesgue integration theory.) Suppose $f \in L^1_{\text{loc}}(\mathbb{R})$ (see Exercise 5.1.10). We say that $x \in \mathbb{R}$ is a Lebesgue point of f if equation (5.14) holds. Prove that almost every $x \in \mathbb{R}$ is a Lebesgue point of f.

5.1.13. Suppose $g \in L^1(\mathbb{R})$. Prove that

$$\lim_{T \to \infty} \int_{\{x : |x| > T\}} |g(x)| \, dx = 0.$$

Hint: Write $\int_{\mathbb{R}} |g(x)| \, dx = \lim_{N \to \infty} \sum_{k=-N}^{N} \int_{k}^{k+1} |g(x)| \, dx$ and use facts about series.

5.1.14. In the definition of an approximate identity, replace relation (5.10) with the weaker assumption that there exists some $\epsilon > 0$ and some $c_1 > 0$ such that

$$|g(x)| \leq c_1 (1 + |x|)^{-1-\epsilon} \quad \text{for all} \quad x \in \mathbb{R}.$$

Obtain the result of Theorem 5.11 under this assumption.

5.2   The Fourier Transform on $\mathbb{R}$

In this section we develop the Fourier transform for $L^2(\mathbb{R})$; it is analogous to the DFT on $\ell^2(\mathbb{Z}_N)$ in chapter 2 and the Fourier transform on $\ell^2(\mathbb{Z})$ in chapter 4. As suggested by the discussion at the beginning of section 4.4, we consider functions χ defined on $\mathbb{R}$ that are *multiplicative*, that is, such that

$$\chi(x+y) = \chi(x)\chi(y) \quad \text{for all} \quad x, y \in \mathbb{R}. \tag{5.20}$$

If χ is not identically 0, this implies that $\chi(0) = 1$. If we assume that χ is reasonably smooth, say differentiable at the origin, it follows (Exercise 5.2.1) that

$$\chi(x) = e^{cx},$$

for some constant $c = \eta + i\xi \in \mathbb{C}$, with $\eta, \xi \in \mathbb{R}$. Then

$$|e^{cx}| = |e^{x\eta}e^{ix\xi}| = e^{x\eta}.$$

If $\eta > 0$, $e^{x\eta}$ is too big as $x \to +\infty$, whereas if $\eta < 0$, $e^{x\eta}$ is too big as $x \to -\infty$. So we restrict to the case $\eta = 0$, that is,

$$\chi(x) = e^{ix\xi},$$

for some $\xi \in \mathbb{R}$. For $\ell^2(\mathbb{Z})$, we noted that for $n \in \mathbb{Z}$, $e^{in\theta}$ is periodic with period 2π in the variable θ, for all $n \in \mathbb{Z}$, so we restricted θ to $[-\pi, \pi)$. Here, the functions $\{e^{ix\xi}\}_{\xi \in \mathbb{R}}$ are all distinct, so we consider all real ξ. Based on our previous experience, we expect to obtain some analog of the DFT using expressions of the form

$$\langle f, e^{ix\xi} \rangle = \int_{\mathbb{R}} f(x)\overline{e^{ix\xi}}\, dx = \int_{\mathbb{R}} f(x)e^{-ix\xi}\, dx.$$

Unfortunately, this integral may not converge absolutely for $f \in L^2(\mathbb{R})$ (this is related to the fact that the inner product $\langle f, e^{ix\xi} \rangle$ is not defined in the usual sense, since $e^{ix\xi} \notin L^2(\mathbb{R})$), so this requires some interpretation. However, if $f \in L^1(\mathbb{R})$, then by relation (5.5),

$$\left| \int_{\mathbb{R}} f(x)e^{-ix\xi}\, dx \right| \le \int_{\mathbb{R}} |f(x)|\, dx = \|f\|_1, \tag{5.21}$$

for any $\xi \in \mathbb{R}$. Thus the following definition is possible.

Definition 5.12   *For* $f \in L^1(\mathbb{R})$ *and* $\xi \in \mathbb{R}$, *define*

$$\hat{f}(\xi) = \int_{\mathbb{R}} f(x) e^{-ix\xi} \, dx. \tag{5.22}$$

We call $\hat{f}$ *the* Fourier transform *of* f; *the mapping* $\hat{\ }$ *is the* Fourier transform.

For $g \in L^1(\mathbb{R})$ *and* $x \in \mathbb{R}$, *define* $\check{g}$, *the* inverse Fourier transform *of* g, *by*

$$\check{g}(x) = \frac{1}{2\pi} \int_{\mathbb{R}} g(\xi) e^{ix\xi} \, d\xi. \tag{5.23}$$

The mapping $\check{\ }$ *is the* inverse Fourier transform.

We are hoping for an inversion formula

$$f(x) = (\hat{f})^{\check{}}(x) = \frac{1}{2\pi} \int_{\mathbb{R}} \hat{f}(\xi) e^{ix\xi} \, d\xi, \tag{5.24}$$

which would be analogous to the inversion formulas (2.10) for the DFT and (4.36) for the Fourier transform on $\ell^2(\mathbb{Z})$. However there are several difficulties. We expect equation (5.24) to hold for $f \in L^2(\mathbb{R})$, but (Exercise 5.1.3) functions in $L^2(\mathbb{R})$ need not belong to $L^1(\mathbb{R})$, and the integral in equation (5.22) converges absolutely only for $f \in L^1(\mathbb{R})$. Also, we cannot prove equation (5.24) by orthogonality methods, as we did for formulas (2.10) and (4.36) because the functions $e^{-ix\xi}$ do not belong to $L^2(\mathbb{R})$, as functions of x or ξ, and there are uncountably many of these functions. Even if $f \in L^1(\mathbb{R})$, so that $\hat{f}(\xi)$ is defined for each ξ, we can interpret the integral in equation (5.24) as an absolutely convergent integral only if $\hat{f} \in L^1(\mathbb{R})$, which is not generally the case. However, we will see (Theorem 5.15) that when $f \in L^1(\mathbb{R})$ is such that $\hat{f} \in L^1(\mathbb{R})$, the Fourier inversion formula (5.24) does hold a.e. Moreover, this allows us (Definition 5.21) to define $\hat{f}$ more abstractly for $f \in L^2(\mathbb{R})$ in such a way that Fourier inversion holds in an appropriate sense for all $f \in L^2(\mathbb{R})$ (Theorem 5.24).

First we require a lemma regarding the Gaussian function G, which is incidentally the density function of a standard normal random variable, whose graph is the famous bell-shaped curve.

Lemma 5.13   *Define* $G : \mathbb{R} \to \mathbb{R}$ *by*

$$G(x) = \frac{1}{\sqrt{2\pi}} e^{-x^2/2}.$$

Then

     i. $\int_{\mathbb{R}} G(x)\, dx = 1$
     ii. *There exists* $c_1 > 0$ *such that* $G(x) \le \frac{c_1}{(1+|x|)^2}$.
     iii. $\hat{G}(\xi) = e^{-\xi^2/2}$, *or* $\hat{G} = \sqrt{2\pi} G$.

Proof

The proof of part i is a famous trick involving conversion to polar coordinates, which is outlined in Exercise 5.2.2. Property ii is trivial: G is bounded above by $1/\sqrt{2\pi}$, and decays exponentially at $\pm\infty$, which is much faster than c_1/x^2. To prove part iii, first note that

$$G'(x) + xG(x) = \frac{-x}{\sqrt{2\pi}} e^{-x^2/2} + x\frac{1}{\sqrt{2\pi}} e^{-x^2/2} = 0, \qquad (5.25)$$

for all x. Exercise 5.2.3 (i) implies that

$$(G')\hat{}\,(\xi) = i\xi\hat{G}(\xi).$$

Also (Exercise 5.2.3 (ii)) $\hat{G}$ is differentiable and

$$(xG(x))\hat{}\,(\xi) = i(\hat{G})'(\xi). \qquad (5.26)$$

Thus, taking the Fourier transform of both sides of equation (5.25) yields

$$i\xi\hat{G}(\xi) + i(\hat{G})'(\xi) = 0.$$

This is a first-order ordinary differential equation that can be solved by multiplying by the integrating factor $e^{\xi^2/2}$ to obtain

$$\left(e^{\xi^2/2}\hat{G}(\xi) \right)' = \xi e^{\xi^2/2}\hat{G}(\xi) + e^{\xi^2/2}(\hat{G})'(\xi) = 0.$$

Hence,

$$\hat{G}(\xi) = Ce^{-\xi^2/2}$$

for some constant C. Letting $\xi = 0$ shows that

$$C = \hat{G}(0) = \int_{\mathbb{R}} G(x)\, dx = 1,$$

by property i.   ■

Observe that properties i and ii imply that the family $\{G_t\}_{t>0}$ is an approximate identity (Definition 5.8). We also require the following lemma. Here $\hat{g}_t(\xi)$ denotes the Fourier transform of g_t, that is $(g_t)\hat{\ }(\xi)$, not the dilate of $\hat{g}$ (which we would write as $(\hat{g})_t$).

Lemma 5.14 *Suppose $g \in L^1(\mathbb{R})$ and $t > 0$. Then for all $\xi \in \mathbb{R}$,*

$$\hat{g}_t(\xi) = \hat{g}(t\xi). \tag{5.27}$$

Proof
Exercise 5.2.4. ∎

We now prove our first version of the inversion formula for the Fourier transform.

Theorem 5.15 (Fourier inversion on $L^1(\mathbb{R})$) *Suppose $f \in L^1(\mathbb{R})$ and $\hat{f} \in L^1(\mathbb{R})$. Then*

$$\frac{1}{2\pi} \int_{\mathbb{R}} \hat{f}(\xi) e^{ix\xi} \, d\xi = f(x)$$

at every Lebesgue point x of f (hence a.e., by Lemma 5.10).

Proof
Let

$$I_t(x) = \frac{1}{2\pi} \int_{\mathbb{R}} \hat{f}(\xi) e^{-t^2 \xi^2 / 2} e^{ix\xi} \, d\xi = \frac{1}{2\pi} \int_{\mathbb{R}} \hat{f}(\xi) \hat{G}_t(\xi) e^{ix\xi} \, d\xi,$$

by Lemmas 5.13 iii and 5.14. An elementary argument (Exercise 5.2.5), or an application of the dominated convergence theorem (if this is familiar to you; we discuss it in section 5.4) shows that

$$\lim_{t \to 0^+} I_t(x) = \frac{1}{2\pi} \int_{\mathbb{R}} \hat{f}(\xi) e^{ix\xi} \, d\xi \tag{5.28}$$

for every x. We evaluate this limit by another route. By writing out $\hat{f}(\xi)$ in the definition of I_t, we obtain

$$I_t(x) = \frac{1}{2\pi} \int_{\mathbb{R}} \left(\int_{\mathbb{R}} f(y) e^{-iy\xi} \, dy \right) \hat{G}_t(\xi) e^{ix\xi} \, d\xi$$

$$= \int_{\mathbb{R}} f(y) \frac{1}{2\pi} \int_{\mathbb{R}} \hat{G}_t(\xi) e^{i(x-y)\xi} \, d\xi \, dy,$$

where the interchange in the order of integration is justified by Fubini's theorem (a result in measure theory) because f and $\hat{G}_t$ are integrable. However, by using Lemma 5.14, the change of variables $\gamma = t\xi$, and Lemma 5.13 iii twice, we get

$$
\frac{1}{2\pi} \int_{\mathbb{R}} \hat{G}_t(\xi) e^{i(x-y)\xi} \, d\xi = \frac{1}{2\pi} \int_{\mathbb{R}} \hat{G}(t\xi) e^{i(x-y)\xi} \, d\xi
$$

$$
= \frac{1}{2\pi t} \int_{\mathbb{R}} \hat{G}(\gamma) e^{i(x-y)\gamma/t} \, d\gamma
$$

$$
= \frac{1}{t\sqrt{2\pi}} \int_{\mathbb{R}} G(\gamma) e^{-i\gamma(y-x)/t} \, d\gamma
$$

$$
= \frac{1}{t\sqrt{2\pi}} \hat{G}((y-x)/t) = \frac{1}{t} G((y-x)/t)
$$

$$
= \frac{1}{t} G((x-y)/t) = G_t(x-y),
$$

since G is an even function. Substituting this equation into the one above it yields

$$
I_t(x) = \int_{\mathbb{R}} f(y) G_t(x-y) \, dy = G_t * f(x).
$$

We have already noted that $\{G_t\}_{t>0}$ is an approximate identity, so by Theorem 5.11,

$$
\lim_{t \to 0^+} I_t(x) = \lim_{t \to 0^+} G_t * f(x) = f(x) \tag{5.29}
$$

at every Lebesgue point x of f. Thus equations (5.28) and (5.29) complete the proof. ∎

This gives an important uniqueness property.

Corollary 5.16 (Uniqueness of the Fourier transform in $L^1(\mathbb{R})$) *Suppose $f, g \in L^1(\mathbb{R})$ and $\hat{f} = \hat{g}$ a.e. Then $f = g$ a.e.*

Proof
Apply Theorem 5.15 to $f - g$ to deduce that $f - g = (\hat{f} - \hat{g})^{\vee} = 0^{\vee} = 0$ a.e. ∎

Theorem 5.15 states that for $f, \hat{f} \in L^1(\mathbb{R})$ we have

$$
f = (\hat{f})^{\vee} \quad \text{a.e.}
$$

Under the same assumptions, we also have

$$
f = (\check{f})^{\wedge}. \tag{5.30}
$$

To see this, first observe that from the definition of the inverse Fourier transform,

$$\check{f}(x) = \frac{1}{2\pi}\hat{f}(-x). \tag{5.31}$$

Hence the assumption that $\hat{f} \in L^1(\mathbb{R})$ implies $\check{f} \in L^1(\mathbb{R})$. Then

$$(\check{f})\hat{\ }(\xi) = \int_{\mathbb{R}} \check{f}(x)e^{-ix\xi}\, dx = \frac{1}{2\pi}\int_{\mathbb{R}}\hat{f}(-x)e^{-ix\xi}\, dx$$

$$= \frac{1}{2\pi}\int_{\mathbb{R}}\hat{f}(x)e^{ix\xi}\, dx = (\hat{f})^{\vee}(\xi) = f(\xi) \text{ a.e.}$$

The conditions $f, \hat{f} \in L^1(\mathbb{R})$ in Theorem 5.15 are quite restrictive. First note (Exercise 5.2.6) that $f \in L^1(\mathbb{R})$ implies that $\hat{f}$ is a bounded, continuous function. By equation (5.31), corresponding results hold for the inverse Fourier transform. In particular, if $\hat{f} \in L^1(\mathbb{R})$, then $(\hat{f})^{\vee}$ is bounded and continuous. Thus by Theorem 5.15, if $f, \hat{f} \in L^1(\mathbb{R})$, f must be a.e. equal to the bounded, continuous function $(\hat{f})^{\vee}$. That is, the assumptions of Theorem 5.15 imply that f can by modified on a set of measure 0 (yielding an equivalent function from the standpoint of Lebesgue integration theory) so that the result is bounded and continuous.

On the other hand, if a function f is C^2 (meaning that at every point, f has two derivatives, and the second derivative is continuous) and f has compact support, defined as follows, then f and $\hat{f}$ belong to $L^1(\mathbb{R})$ (Exercise 5.2.7).

Definition 5.17 *Let* $f : \mathbb{R} \to \mathbb{C}$ *be a function. The* support *of* f, *denoted* supp f, *is the closure of the set*

$$\{x \in \mathbb{R} : f(x) \neq 0\}.$$

We say f *has* compact support *if* supp f *is a compact set.*

Because supp f is closed by definition, f has compact support if and only if supp f is bounded (by the Heine-Borel theorem). In other words, f has compact support if there exists $r < \infty$ such that supp $f \subseteq [-r, r]$, that is, such that $f(x) = 0$ for all x satisfying $|x| > r$.

Although the class of C^2 functions with compact support may not seem large, it is large enough to be *dense* in $L^2(\mathbb{R})$, in the following sense.

Lemma 5.18 *Suppose $f \in L^2(\mathbb{R})$ and let $\epsilon > 0$. Then there exists a C^2 function g with compact support (hence, by Exercise 5.2.7, a function $g \in L^1(\mathbb{R})$ such that $\hat{g} \in L^1(\mathbb{R})$), satisfying*

$$\|f - g\| < \epsilon.$$

Proof

We assume the fact from Lebesgue integration theory that there is a step function h such that

$$\|f - h\| < \epsilon/2.$$

(By definition, a *step function* is any function of the form

$$h = \sum_{k=1}^{n} c_k \chi_{[a_k, b_k]},$$

where $\chi_{[a_k, b_k]}$ is the function that is 1 on the interval $[a_k, b_k]$ and 0 elsewhere; $c_1, c_2, \ldots, c_n$ are constants; and n can be any positive integer.) By Exercise 5.2.8, there is a C^2 function g of compact support such that $\|h - g\| < \epsilon/2$. Then by the triangle inequality (5.4),

$$\|f - g\| \le \|f - h\| + \|h - g\| < \epsilon/2 + \epsilon/2 = \epsilon.$$

This completes the proof. ∎

In fact one can find an infinitely differentiable function of compact support that is within ϵ of f in L^2 norm, but this is more difficult to prove and we will not need this result.

Lemma 5.18 says that we can approximate an L^2 function by functions that satisfy the conditions of Theorem 5.15. We use this lemma to define the Fourier transform on $L^2(\mathbb{R})$. The following preliminary forms of Parseval's relation and Plancherel's theorem play a key role.

Lemma 5.19 *Suppose $f, g \in L^1(\mathbb{R})$ and $\hat{f}, \hat{g} \in L^1(\mathbb{R})$. Then*

i. $f, g, \hat{f}, \hat{g} \in L^2(\mathbb{R})$.

ii. (Parseval's relation) $\langle \hat{f}, \hat{g} \rangle = 2\pi \langle f, g \rangle$.

iii. (Plancherel's formula) $\|\hat{f}\| = \sqrt{2\pi}\|f\|$.

Proof

We leave the proof of part i as Exercise 5.2.9. For part ii,

$$\langle \hat{f}, \hat{g} \rangle = \int_{\mathbb{R}} \hat{f}(\xi)\overline{\hat{g}(\xi)}\, d\xi = \int_{\mathbb{R}} \hat{f}(\xi) \overline{\int_{\mathbb{R}} g(x)e^{-ix\xi}\, dx}\, d\xi$$

$$= \int_{\mathbb{R}} \hat{f}(\xi) \int_{\mathbb{R}} \overline{g(x)}e^{ix\xi}\, dx\, d\xi = \int_{\mathbb{R}} \int_{\mathbb{R}} \hat{f}(\xi)e^{ix\xi}\, d\xi\, \overline{g(x)}\, dx$$

$$= 2\pi \int_{\mathbb{R}} f(x)\overline{g(x)}\, dx = 2\pi\langle f, g \rangle,$$

by Theorem 5.15 (and where the interchange in the order of integration is justified by Fubini's theorem since $g, \hat{f} \in L^1(\mathbb{R})$). Then part iii follows from part ii by taking $g = f$. ∎

For $f \in L^2(\mathbb{R})$, we cannot define $\hat{f}$ pointwise by equation (5.22) because the integral may not be absolutely convergent. However, we can find a sequence of functions $\{f_n\}_{n=1}^{\infty}$ such that $f_n, \hat{f}_n \in L^1(\mathbb{R})$ for each n, and $f_n \to f$ in $L^2(\mathbb{R})$ as $n \to \infty$ (recall this means that $\lim_{n\to\infty} \|f_n - f\| = 0$). For example, we can apply Lemma 5.18 with $\epsilon = 1/n$ to obtain f_n as desired such that $\|f_n - f\| \le 1/n$, for each n. Since $f_n \in L^1(\mathbb{R})$, $\hat{f}_n$ is defined for each n. We would like to define $\hat{f} = \lim_{n\to\infty} \hat{f}_n$. Lemma 5.20 shows that this makes sense.

Lemma 5.20 *Suppose $f \in L^2(\mathbb{R})$. Let $\{f_n\}_{n=1}^{\infty}$ be a sequence of functions such that $f_n, \hat{f}_n \in L^1(\mathbb{R})$ for each n, and $f_n \to f$ in $L^2(\mathbb{R})$ as $n \to \infty$.*

- *i. The sequence $\{\hat{f}_n\}_{n=1}^{\infty}$ converges to some $F \in L^2(\mathbb{R})$ (i.e., $\|\hat{f}_n - F\| \to 0$ as $n \to \infty$).*
- *ii. Let $\{g_n\}_{n=1}^{\infty}$ be another sequence of functions such that $g_n, \hat{g}_n \in L^1(\mathbb{R})$ for each n, and $g_n \to f$ in $L^2(\mathbb{R})$ as $n \to \infty$. By part i, $\hat{g}_n$ converges in L^2 to some $G \in L^2(\mathbb{R})$. Then $G = F$ a.e., for F as in part i.*
- *iii. Suppose $f \in L^1(\mathbb{R}) \cap L^2(\mathbb{R})$. Let F be as in part i. Then $F = \hat{f}$.*

Proof

To prove part i, note that by Lemma 5.19 i, $\hat{f}_n \in L^2(\mathbb{R})$ for each n. Since $\{f_n\}_{n=1}^{\infty}$ is convergent in $L^2(\mathbb{R})$, it is a Cauchy sequence in $L^2(\mathbb{R})$. By Lemma 5.19 iii,

$$\|\hat{f}_n - \hat{f}_m\| = \sqrt{2\pi}\|f_n - f_m\|$$

for each n and m. Hence $\{\hat{f}_n\}$ is also a Cauchy sequence in $L^2(\mathbb{R})$. Since $L^2(\mathbb{R})$ is complete, there exists some $F \in L^2(\mathbb{R})$ such that $\{\hat{f}_n\}$ converges to F in L^2.

For part ii, the triangle inequality (5.4) implies that

$$\|F - G\| \le \|F - \hat{f}_n\| + \|\hat{f}_n - \hat{g}_n\| + \|\hat{g}_n - G\| \tag{5.32}$$

for each n. By Lemma 5.19 iii and triangle inequality (5.4),

$$\|\hat{f}_n - \hat{g}_n\| = \sqrt{2\pi}\|f_n - g_n\| \le \sqrt{2\pi}(\|f_n - f\| + \|f - g_n\|).$$

Hence, as $n \to \infty$, the right side of relation (5.32) goes to 0. Therefore, $\|F - G\| = 0$, so $F = G$ a.e.

To establish part iii, by Exercise 5.2.10 we can find a sequence $\{g_n\}_{n=1}^{\infty}$ of L^1 functions with $\hat{g}_n \in L^1(\mathbb{R})$ for each n such that $\{g_n\}$ converges to f in L^2, and also

$$\|g_n - f\|_1 \to 0 \quad \text{as} \quad n \to \infty. \tag{5.33}$$

By relation (5.5), for each $\xi \in \mathbb{R}$,

$$\left| \hat{g}_n(\xi) - \hat{f}(\xi) \right| = \left| \int_{\mathbb{R}} (g_n(x) - f(x))e^{-ix\xi}\, dx \right| \le \int_{\mathbb{R}} |g_n(x) - f(x)|\, dx.$$

Hence by relation (5.33), $\hat{g}_n$ converges uniformly to $\hat{f}$ as $n \to \infty$. By part ii, however, $\hat{g}_n$ converges to F in $L^2(\mathbb{R})$. This guarantees that $F = \hat{f}$ a.e., as follows. For any positive integer N, by inequality (5.4) we have

$$\left(\int_{-N}^{N} |F(\xi) - \hat{f}(\xi)|^2\, d\xi \right)^{1/2} \le \left(\int_{-N}^{N} |F(\xi) - \hat{g}_n(\xi)|^2\, d\xi \right)^{1/2}$$

$$+ \left(\int_{-N}^{N} |\hat{g}_n(\xi) - \hat{f}(\xi)|^2\, d\xi \right)^{1/2} \le \|F - \hat{g}_n\| + \sqrt{2N}\sup_{\xi \in \mathbb{R}} |\hat{g}_n(\xi) - \hat{f}(\xi)|$$

for every n. Letting $n \to \infty$, the right side goes to 0. Hence $\int_{-N}^{N} |F(\xi) - \hat{f}(\xi)|^2\, d\xi = 0$. Because this is true for all $N \in \mathbb{N}$, we obtain that $F = \hat{f}$ a.e. ∎

This allows the following definition.

Definition 5.21   *Suppose $f \in L^2(\mathbb{R})$. Let $\{f_n\}_{n=1}^{\infty}$ be a sequence such that $f_n, \hat{f}_n \in L^1(\mathbb{R})$ for all n, and such that f_n converges to f in L^2 as*

$n \to \infty$. Define $\hat{f}$, *the* Fourier transform *of* f, *to be the limit in* L^2 *of the sequence* $\{\hat{f}_n\}_{n=1}^{\infty}$.

We define $\check{f}$, *the* inverse Fourier transform *of* f, *to be the* L^2 *limit of the sequence* $\check{f}_n\}_{n=1}^{\infty}$.

Note that for $f \in L^2(\mathbb{R})$, $\hat{f}$ is defined as an L^2 function, not pointwise as in Definition 5.12 when $f \in L^1(\mathbb{R})$. Thus $\hat{f}$ is only determined a.e., that is, up to a set of measure 0, for $f \in L^2(\mathbb{R})$. Lemma 5.20 i guarantees that the L^2 limit of $\{\hat{f}_n\}_{n=1}^{\infty}$ exists in $L^2(\mathbb{R})$, and Lemma 5.20 ii shows that this limit is independent of the choice of the sequence $\{f_n\}_{n=1}^{\infty}$. Hence $\hat{f}$ is well defined. In the case $f \in L^1(\mathbb{R}) \cap L^2(\mathbb{R})$, Lemma 5.20 iii guarantees that $\hat{f}$ in Definition 5.21 is the same L^2 function as $\hat{f}$ in Definition 5.12.

The relation (5.31) between $\hat{\ }$ and $\check{\ }$ shows that the analog for the inverse Fourier transform of Lemma 5.20 holds also, which justifies the definition of $\check{\ }$ in Definition 5.21 in the same way as for $\hat{\ }$. Moreover, for $f \in L^2(\mathbb{R})$, by equation (5.31) we have

$$\check{f}(x) = \lim_{n \to \infty} \check{f}_n(x) = \lim_{n \to \infty} \frac{1}{2\pi} \hat{f}_n(-x) = \frac{1}{2\pi} \hat{f}(-x),$$

for $\{f_n\}_{n=1}^{\infty}$, as in Definition 5.21, where the limits are L^2 limits. Thus equation (5.31) is true for $f \in L^2(\mathbb{R})$ also.

For $f \in L^2(\mathbb{R})$, it is part of the definition of $\hat{f}$ that $\hat{f} \in L^2(\mathbb{R})$. That is, we have $\hat{\ }: L^2(\mathbb{R}) \to L^2(\mathbb{R})$. By equation (5.31), interpreted in the L^2 sense, we also have $\check{\ }: L^2(\mathbb{R}) \to L^2(\mathbb{R})$.

We can now extend Lemma 5.19 to $L^2(\mathbb{R})$.

Theorem 5.22   *Suppose* $f, g \in L^2(\mathbb{R})$. *Then we have*
  i. *(Parseval's relation)* $\langle \hat{f}, \hat{g} \rangle = 2\pi \langle f, g \rangle$.
  ii. *(Plancherel's formula)* $\|\hat{f}\| = \sqrt{2\pi} \|f\|$.
     We also have
  iii. $\langle \check{f}, \check{g} \rangle = \frac{1}{2\pi} \langle f, g \rangle$.
     and
  iv. $\|\check{f}\| = \frac{1}{\sqrt{2\pi}} \|f\|$.

Proof

Let $\{f_n\}_{n=1}^{\infty}$ and $\{g_n\}_{n=1}^{\infty}$ be sequences of functions with $f_n, g_n, \hat{f}_n, \hat{g}_n \in L^1(\mathbb{R})$ for each n, such that $f_n \to f$ and $g_n \to g$ in $L^2(\mathbb{R})$ as $n \to \infty$. By definition, $\hat{f} = \lim_{n \to \infty} \hat{f}_n$ and $\hat{g} = \lim_{n \to \infty} \hat{g}_n$, where these are L^2 limits. This implies

$$\langle \hat{f}, \hat{g} \rangle = \lim_{n \to \infty} \langle \hat{f}_n, \hat{g}_n \rangle. \tag{5.34}$$

(Proof: by relations (5.4) and (5.3),

$$\left| \langle \hat{f}, \hat{g} \rangle - \langle \hat{f}_n, \hat{g}_n \rangle \right| \le \left| \langle \hat{f} - \hat{f}_n, \hat{g} \rangle \right| + \left| \langle \hat{f}_n, \hat{g} - \hat{g}_n \rangle \right|$$

$$\le \| \hat{f} - \hat{f}_n \| \| \hat{g} \| + \| \hat{f}_n \| \| \hat{g} - \hat{g}_n \|,$$

which goes to 0 as $n \to \infty$, because the convergence of $\hat{f}_n$ in $L^2(\mathbb{R})$ implies the boundedness of $\| \hat{f}_n \|$.) We can apply Lemma 5.19 ii to f_n, g_n to obtain

$$\langle \hat{f}_n, \hat{g}_n \rangle = 2\pi \langle f_n, g_n \rangle$$

for each n. Hence by equation (5.34),

$$\langle \hat{f}, \hat{g} \rangle = \lim_{n \to \infty} 2\pi \langle f_n, g_n \rangle = 2\pi \langle f, g \rangle,$$

by the same argument as in the proof of equation (5.34). This proves part i. Then part ii follows from part i by taking $g = f$. Part iii follows either by a similar argument or from equation (5.31) (for $f \in L^2(\mathbb{R})$) and part i because

$$\langle \check{f}, \check{g} \rangle = \frac{1}{4\pi^2} \int_{\mathbb{R}} \hat{f}(-x) \overline{\hat{g}(-x)} \, dx = \frac{1}{4\pi^2} \langle \hat{f}, \hat{g} \rangle = \frac{1}{2\pi} \langle f, g \rangle.$$

Then part iv follows from part iii by taking $g = f$.   ∎

For $f \in L^2(\mathbb{R})$, we defined $\hat{f}$ by taking a sequence $\{f_n\}_{n=1}^{\infty}$ of sufficiently nice functions (i.e., such that $f_n, \hat{f}_n \in L^1(\mathbb{R})$ for all n) that converge to f in $L^2(\mathbb{R})$, and letting $\hat{f}$ be the L^2 limit of $\hat{f}_n$ as $n \to \infty$. However, at this point we can see that for any sequence of functions, nice or otherwise, converging in L^2 to f, their Fourier transforms converge to $\hat{f}$.

Corollary 5.23   *Suppose $f \in L^2(\mathbb{R})$, and $\{f_n\}_{n=1}^{\infty}$ is a sequence of L^2 functions such that $f_n \to f$ in $L^2(\mathbb{R})$. Then*

$$\hat{f}_n \to \hat{f} \quad \text{in } L^2(\mathbb{R}) \text{ as } n \to \infty \tag{5.35}$$

and

$$\check{f}_n \to \check{f} \quad \text{in } L^2(\mathbb{R}) \text{ as } n \to \infty. \tag{5.36}$$

Proof

By Theorem 5.22 ii,

$$\|\hat{f}_n - \hat{f}\| = \sqrt{2\pi}\|f_n - f\| \to 0 \quad \text{as} \quad n \to \infty,$$

which proves relation (5.35). Theorem 5.22 iv implies relation (5.36) similarly. ∎

Hence, if we take any sequence $\{f_n\}_{n=1}^{\infty}$ of functions in $L^1(\mathbb{R}) \cap L^2(\mathbb{R})$ such that $f_n \to f$ in $L^2(\mathbb{R})$, then $\hat{f}_n$ is defined pointwise by equation (5.22) and $\hat{f}$ is the L^2 limit of $\hat{f}_n$ as $n \to \infty$. By Exercise 5.2.11, an example of such a sequence $\{f_n\}_{n=1}^{\infty}$ is

$$f_n(x) = \begin{cases} f(x), & \text{if } |x| \leq n \\ 0, & \text{if } |x| > n. \end{cases} \tag{5.37}$$

Note that

$$\hat{f}_n(x) = \int_{\mathbb{R}} f_n(x) e^{-ix\xi}\, dx = \int_{-n}^{n} f(x) e^{-ix\xi}\, dx.$$

Thus for $f \in L^2(\mathbb{R})$,

$$\hat{f}(\xi) = \lim_{n \to \infty} \int_{-n}^{n} f(x) e^{-ix\xi}\, dx, \tag{5.38}$$

where the limit is in the L^2 sense. This "principal value" interpretation of $\hat{f}$ for $f \in L^2(\mathbb{R})$ is as natural and explicit as possible, so many texts take it as the definition of $\hat{f}$.

After all this, the proof of our main objective is anticlimactic.

Theorem 5.24 (*Fourier inversion on $L^2(\mathbb{R})$*)

   i. *Suppose $f \in L^2(\mathbb{R})$. Then*

$$f = (\hat{f})^{\vee} \tag{5.39}$$

   and

$$f = (\check{f})^{\wedge}. \tag{5.40}$$

  ii. *The Fourier transform* $\hat{}$: $L^2(\mathbb{R}) \to L^2(\mathbb{R})$ *is one-to-one and onto with inverse* $\check{}$: $L^2(\mathbb{R}) \to L^2(\mathbb{R})$.

Proof

Select a sequence $\{f_n\}_{n=1}^{\infty}$ with $f_n, \hat{f}_n \in L^1(\mathbb{R})$ for all n, such that $f_n \to f$ in $L^2(\mathbb{R})$ as $n \to \infty$. Then $\hat{f}_n \to \hat{f}$ in $L^2(\mathbb{R})$ as $n \to \infty$, by definition of $\hat{f}$. By relation (5.36),

$$(\hat{f_n})^{\vee} \to (\hat{f})^{\vee} \quad \text{in } L^2(\mathbb{R}) \text{ as } n \to \infty.$$

By Theorem 5.15, however,

$$(\hat{f_n})^{\vee} = f_n \to f \quad \text{in } L^2(\mathbb{R}) \text{ as } n \to \infty.$$

This proves equation (5.39). In the same way, using relations (5.35) and (5.30),

$$(\check{f})^{\wedge} = \lim_{n \to \infty} (\check{f_n})^{\wedge} = \lim_{n \to \infty} f_n = f,$$

where these are L^2 limits, proving equation (5.40). Then part ii follows easily from part i. ∎

The Fourier inversion formula (5.24) has essentially the same interpretation here as in the DFT context (see formula (2.10)), the context of $L^2([-\pi, \pi))$ (see equation (4.24)), or in $\ell^2(\mathbb{Z})$ (see equation (4.36)). As ξ increases in magnitude, $e^{ix\xi}$ is a more and more rapidly oscillating function of x. We think of each such function as representing a pure frequency, which is higher as $|\xi|$ gets larger. These different frequencies can be "added" together with weights $\hat{f}(\xi)$ via the integral in equation (5.24) to form any L^2 function f. Thus $\hat{f}(\xi)$ measures the weight or strength of the pure frequency $e^{ix\xi}$ used in making up f. We think of equation (5.24) as analogous to a basis representation in the case of a finite dimensional vector space; every element of $L^2(\mathbb{R})$ is written in terms of the fixed system $\{e^{ix\xi}\}_{\xi \in \mathbb{R}}$ via a weighted superposition (in this case an integral instead of a linear combination). As in the case of equation (4.36), the representing elements $e^{ix\xi}$, as functions of x, are not in the space $L^2(\mathbb{R})$, so none of these elements can be given more than infinitesimal weight in the representation (5.24).

We now consider how some operations we have introduced, such as convolution or translation, interact with the Fourier transform.

Lemma 5.25   *Suppose $g, h \in L^1(\mathbb{R})$, and either $f \in L^1(\mathbb{R})$ or $f \in L^2(\mathbb{R})$. Then*

  i. $(f * g)^{\wedge} = \hat{f}\hat{g}$.
  ii. $f * (g * h) = (f * g) * h$.

Proof

Exercise 5.2.13. ∎

Lemma 5.26 *Suppose $f \in L^1(\mathbb{R})$ or $f \in L^2(\mathbb{R})$, and $y, \xi \in \mathbb{R}$. Then*

i. $(\tilde{f})\hat{}(\xi) = \overline{\hat{f}(\xi)}$ *a.e.*

ii. $(R_y f)\hat{}(\xi) = e^{-iy\xi}\hat{f}(\xi)$ *a.e.*

Proof

Exercise 5.2.14. ∎

As noted above, the Fourier inversion formula (5.24) can be thought of as a representation of $f(x)$ as a superposition of the pure frequencies $e^{ix\xi}$. This representation has the same key property for $L^2(\mathbb{R})$ as the DFT expansion for $\ell^2(\mathbb{Z}_N)$ in Theorem 2.18, the Fourier series expansion for $L^2([-\pi, \pi))$ in Theorem 4.28, and the Fourier system expansion for $\ell^2(\mathbb{Z})$ in Lemma 4.39: it diagonalizes (in a sense to be described) all bounded translation-invariant linear transformations. We say that a bounded (Definition 4.25) linear transformation $T : L^2(\mathbb{R}) \rightarrow L^2(\mathbb{R})$ is *translation invariant* if, for all $y \in \mathbb{R}$ and all $f \in L^2(\mathbb{R})$,

$$T(R_y f) = R_y T(f),$$

where $R_y T(f)(x) = T(f)(x - y)$, by definition. It turns out, like the cases noted above, that every translation-invariant bounded linear transformation T on $L^2(\mathbb{R})$ is a convolution operator, that is, there exists b such that $T(f) = b * f$. The class b belongs to is not easy to specify at this point, so we leave this point vague. By Lemma 5.25 i, we have (in an appropriate sense)

$$(T(f))\hat{}(\xi) = \hat{b}(\xi)\hat{f}(\xi)$$

for all $\xi \in \mathbb{R}$. If we apply Fourier inversion (5.24) to $T(f)$, we obtain

$$T(f)(x) = \frac{1}{2\pi} \int_{\mathbb{R}} \hat{b}(\xi)\hat{f}(\xi)e^{ix\xi} \, d\xi.$$

Thus the result of applying T is to replace the "coefficients" $\hat{f}(\xi)$ in the Fourier representation formula (5.24) by $\hat{b}(\xi)\hat{f}(\xi)$. Thus T acts as a diagonal operator with respect to the system $\{e^{ix\xi}\}_{\xi \in \mathbb{R}}$. In this sense, the Fourier transform diagonalizes bounded translation-invariant operators on $L^2(\mathbb{R})$.

A key example of a translation-invariant operator is the derivative, since

$$\frac{d}{dx}\left(R_y f\right)(x) = \frac{d}{dx}(f(x-y)) = \frac{df}{dx}(x-y) = R_y\left(\frac{df}{dx}\right)(x).$$

Although the derivative is not defined on all of $L^2(\mathbb{R})$, and is not a bounded operator on its domain, the above argument is correct in some generalized sense. In particular, if we formally bring the derivative with respect to x inside the integral sign in equation (5.24), we obtain

$$f'(x) = \frac{1}{2\pi}\int_{\mathbb{R}} i\xi \hat{f}(\xi) e^{ix\xi}\, d\xi.$$

This argument can be justified for nice enough functions f. Thus the derivative operator corresponds to the Fourier multiplier $i\xi$, and the derivative is diagonalized by the Fourier system. This is the reason for the prevalence of Fourier techniques in the study of differential equations: the Fourier system diagonalizes the derivative operator.

Exercises

5.2.1. Suppose $\chi : \mathbb{R} \to \mathbb{C}$ is multiplicative (i.e., equation (5.20) holds), χ is not identically 0, and χ is differentiable at 0. Prove that $\chi(x) = e^{cx}$ for some $c \in \mathbb{C}$. Hint: Use the definition of the derivative to show that χ is differentiable at every point of $\mathbb{R}$ with $\chi'(x) = \chi(x)\chi'(0)$.

5.2.2. Prove Lemma 5.13 i. Hint: Let $I = \int_{\mathbb{R}} G(x)\, dx$. Then

$$I^2 = \int_{\mathbb{R}} \frac{1}{\sqrt{2\pi}} e^{-x^2/2}\, dx \int_{\mathbb{R}} \frac{1}{\sqrt{2\pi}} e^{-y^2/2}\, dy$$

$$= \frac{1}{2\pi} \int_{\mathbb{R}}\int_{\mathbb{R}} e^{-(x^2+y^2)/2}\, dx\, dy.$$

Then conversion to polar coordinates yields an integral that can be explicitly evaluated.

5.2.3. i. Suppose $g \in L^1(\mathbb{R})$ is a differentiable function such that $g' \in L^1(\mathbb{R})$ and $\lim_{x\to\pm\infty} g(x) = 0$. Prove

$$(g')^{\hat{}}(\xi) = i\xi \hat{g}(\xi).$$

Hint: Integrate by parts.

ii. Suppose $g \in L^1(\mathbb{R})$ is continuous and $xg(x) \in L^1(\mathbb{R})$. Prove that $\hat{g}$ is differentiable at every point in $\mathbb{R}$ and

$$\hat{g}'(\xi) = -i(xg(x))\hat{}(\xi).$$

Hint: The assumptions justify differentiating under the integral sign.

5.2.4. Prove Lemma 5.14.

5.2.5. Prove equation (5.28). Hint: Show that

$$\left| I_t - \frac{1}{2\pi} \int_{\mathbb{R}} \hat{f}(\xi) e^{ix\xi}\, d\xi \right| \leq \frac{1}{2\pi} \int_{\mathbb{R}} |\hat{f}(\xi)| \left| e^{-t^2\xi^2/2} - 1 \right|\, d\xi.$$

Break this last integral into two parts, the integral over $\{\xi : |\xi| > T\}$ and the integral over $\{\xi : |\xi| \leq T\}$. Given $\epsilon > 0$, pick T using Exercise 5.1.13 to estimate the integral over the infinite region. Show that $e^{-t^2\xi^2/2}$ converges uniformly to 1 on the finite region as $t \to 0$.

5.2.6. Suppose $f \in L^1(\mathbb{R})$.

i. Prove that $\hat{f}$ is a bounded function with $|\hat{f}(\xi)| \leq \|f\|$ for all ξ.

ii. Prove that $\hat{f}$ is a continuous function on $\mathbb{R}$. Hint: Show that

$$\left| \hat{f}(\xi) - \hat{f}(\xi_0) \right| \leq \int_{\mathbb{R}} |f(x)| \left| e^{-ix(\xi-\xi_0)} - 1 \right|\, dx,$$

and see the hint in Exercise 5.2.5.

5.2.7. Suppose $f : \mathbb{R} \to \mathbb{C}$ is C^2 and has compact support. Prove that $f, \hat{f} \in L^1(\mathbb{R})$. Hint: Recall the theorem that a continuous function on a compact set is bounded. With this it is easy to show $f \in L^1(\mathbb{R})$. By Exercise 5.2.6(i), $\hat{f}$ is bounded. Integrate by parts twice in the definition of $\hat{f}(\xi)$ and estimate to show that $|\hat{f}(\xi)| \leq c/\xi^2$ for some constant c.

5.2.8. Let $\epsilon > 0$. For each of the functions h defined in parts i, ii, and iii of this exercise prove that there exists a C^2 function g of compact support such that $\|h - g\| < \epsilon$.

i. Let $h(x) = 1$ for $0 \leq x \leq 1$ and $h(x) = 0$ for all other $x \in \mathbb{R}$. Suggestion: Round off the corners of h. For example, for

δ chosen sufficiently small, let

$$g(x) = \begin{cases} 0 & \text{if } x < -\delta \text{ or } x > 1 + \delta \\ 1 + \frac{x}{\delta} - \frac{1}{2\pi} \sin\left(\frac{2\pi(x+\delta)}{\delta}\right) & \text{if } -\delta \le x \le 0 \\ 1 & \text{if } 0 < x < 1 \\ 1 + \frac{1-x}{\delta} - \frac{1}{2\pi} \sin\left(\frac{2\pi(1+\delta-x)}{\delta}\right) & \text{if } 1 \le x \le 1 + \delta. \end{cases}$$

   ii. For $a, b \in \mathbb{R}$ with $a < b$ and $c \in \mathbb{C}$, let $h(x) = c$ for $a \le x \le b$ and $h(x) = 0$ for all other real numbers x.

  iii. Let $h : \mathbb{R} \to \mathbb{C}$ be any step function (defined in the proof of Lemma 5.18).

5.2.9. Prove Lemma 5.19 i. Hint: See Exercise 5.2.6(i).

5.2.10. Suppose $f \in L^1(\mathbb{R}) \cap L^2(\mathbb{R})$. Let $\epsilon > 0$.

   i. Prove that there exists a step function h (defined in the proof of Lemma 5.18) such that

$$\|f - h\| < \epsilon \quad \text{and} \quad \|f - h\|_1 < \epsilon.$$

Suggestion: We can find N (which we can assume is greater than 1) such that

$$\int_{\{x:|x|>N\}} |f(x)| \, dx < \frac{\epsilon}{2} \quad \text{and} \quad \int_{\{x:|x|>N\}} |f(x)|^2 \, dx < \frac{\epsilon^2}{4}.$$

Assume the result from Lebesgue integration theory that there exists a step function H such that

$$\int_{\mathbb{R}} |f(x) - H(x)|^2 \, dx < \frac{\epsilon^2}{8N}.$$

Apply inequality (5.3) to the functions $f(x) - H(x)$ and 1 to obtain

$$\int_{\{x:|x|\le N\}} |f(x) - H(x)| \, dx \le \frac{\epsilon}{\sqrt{8N}} \sqrt{2N} = \frac{\epsilon}{2}.$$

Let $h(x) = H(x)$ for $|x| \le N$ and $h(x) = 0$ for $|x| > N$. Observe that h is a step function. Use the triangle inequalities for $L^1(\mathbb{R})$ and $L^2(\mathbb{R})$ to complete the proof.

   ii. Prove that there is a C^2 function g of compact support (hence, by Exercise 5.2.7, a function $g \in L^1(\mathbb{R})$ such that $\hat{g} \in L^1(\mathbb{R})$) satisfying

$$\|f - g\| < \epsilon \quad \text{and} \quad \|f - g\|_1 < \epsilon.$$

Hint: See Exercise 5.2.8.

5.2.11. Suppose $f \in L^2(\mathbb{R})$. For $n \in \mathbb{N}$, define f_n by equation (5.37).
  i. Prove that $f_n \in L^1(\mathbb{R}) \cap L^2(\mathbb{R})$ and

$$\|f_n\|_1 \le \sqrt{2n}\|f\|.$$

  Hint: Let $g(x) = 1$ if $|x| \le n$, and $g(x) = 0$ if $|x| > N$, and apply inequality (5.3).

  ii. Prove that $f_n \to f$ in $L^2(\mathbb{R})$ as $n \to \infty$ (i.e., $\|f_n - f\| \to 0$ as $n \to \infty$). Hint: See Exercise 5.1.13.

5.2.12. Suppose $f, g \in L^2(\mathbb{R})$. Prove that

$$\int_{\mathbb{R}} \hat{f}(y)g(y)\,dy = \int_{\mathbb{R}} f(y)\hat{g}(y)\,dy.$$

5.2.13. Prove Lemma 5.25. Hint: When $f \in L^1(\mathbb{R})$, part i can be proved by interchanging the order of integration. For $f \in L^2(\mathbb{R})$, apply the L^1 result and a limiting argument, using Exercise 5.2.6(i) and Lemma 5.4 iii. The easy way to prove part ii is to use part i.

5.2.14. Prove Lemma 5.26. Remark: For $f \in L^1(\mathbb{R})$, the conclusions hold for every ξ.

5.2.15. Show that there exist $f, g \in L^2(\mathbb{R})$ such that $f * g \notin L^2(\mathbb{R})$. Hint: Compare to Exercise 4.4.b(ii).

5.2.16. Prove that there is a no $\delta \in L^2(\mathbb{R})$ such that $\delta * f = f$ for all $f \in L^2(\mathbb{R})$. Hint: If such a δ exists, prove that $\hat{\delta}(\xi) = 1$ for all ξ.

5.2.17. Let $L^1 \oplus L^2(\mathbb{R})$ denote the class of all functions $f : \mathbb{R} \to \mathbb{C}$ such that $f = f_1 + f_2$ for some $f_1 \in L^1(\mathbb{R})$ and $f_2 \in L^2(\mathbb{R})$. (Warning: This representation is not unique.)
  i. Give an example of $f \in L^1 \oplus L^2(\mathbb{R})$ that is not an element of $L^1(\mathbb{R}) \cup L^2(\mathbb{R})$. Hint: See, Exercise 5.1.3.
  ii. If $f_1 + f_2 = g_1 + g_2$ with $f_1, g_1 \in L^1(\mathbb{R})$ and $f_2, g_2 \in L^2(\mathbb{R})$, prove that

$$\hat{f}_1 + \hat{f}_2 = \hat{g}_1 + \hat{g}_2 \quad \text{a.e.,}$$

  where $\hat{f}_1$ and $\hat{g}_1$ are defined pointwise by Definition 5.12, and $\hat{f}_2$ and $\hat{g}_2$ are defined in $L^2(\mathbb{R})$ by Definition 5.21.

  Remark: This allows us to extend the definition of the Fourier transform to $L^1 \oplus L^2(\mathbb{R})$ by setting $\hat{f} = \hat{f}_1 + \hat{f}_2$ for any representation $f = f_1 + f_2$ with $f_1 \in L^1(\mathbb{R})$ and $f_2 \in L^2(\mathbb{R})$. Part ii guarantees that this is well defined.

5.3   Multiresolution Analysis and Wavelets

Our goal in this chapter is to construct a wavelet system, which is a complete orthonormal set in $L^2(\mathbb{R})$ consisting of a certain set of translates and dilates of a single function ψ (see Definition 5.28). In this section, we reduce the construction of a wavelet system to the construction of a multiresolution analysis. First we set some notation.

Definition 5.27   *For* $\varphi, \psi \in L^2(\mathbb{R})$ *and* $j, k \in \mathbb{Z}$, *define* $\varphi_{j,k}, \psi_{j,k} \in L^2(\mathbb{R})$ *by*

$$\varphi_{j,k}(x) = 2^{j/2}\varphi(2^jx - k) \quad and \quad \psi_{j,k}(x) = 2^{j/2}\psi(2^jx - k). \tag{5.41}$$

The factor of $2^{j/2}$ in the definitions of $\varphi_{j,k}$ and $\psi_{j,k}$ is included so that the L^2 norms will be the same for all j, k:

$$\|\psi_{j,k}\|^2 = \int_{\mathbb{R}} |2^{j/2}\psi(2^jx - k)|^2 \, dx$$

$$= \int_{\mathbb{R}} 2^j |\psi(2^jx - k)|^2 \, dx = \int_{\mathbb{R}} |\psi(y)|^2 \, dy = \|\psi\|^2,$$

and similarly for $\varphi_{j,k}$, by changing variables (let $y = 2^jx - k$) in the integral.

We can write

$$\psi_{j,k}(x) = 2^{j/2}\psi(2^j(x - 2^{-j}k)). \tag{5.42}$$

Thus the definition of $\psi_{j,k}$ involves a normalization, as just noted, a dilation, and a translation. To understand the dilation, note that for $j > 0$, the graph of $\psi(2^jx)$ is obtained by contracting the graph of ψ along the x-axis by a factor of 2^j (for $j < 0$ the graph is expanded in the x direction). For example, suppose ψ has compact support (Definition 5.17), and let $r > 0$ be the smallest number such that $\psi(x) = 0$ for all x such that $|x| > r$. Then $\psi(2^jx)$ has compact support inside the interval $[-r/2^j, r/2^j]$ since $\psi(2^j x) = 0$ whenever $|2^j x| > r$, that is, when $|x| > r/2^j$.

The graph of $\psi(2^j x - k) = \psi(2^j (x - 2^{-j} k))$ is obtained by translating the graph of $\psi(2^j x)$ by $2^{-j} k$ along the x axis (to the right if $k > 0$, to the left if $k < 0$). Hence, if ψ has compact support in the interval $[-r, r]$, then $\psi(2^j x - k)$ has support inside $[2^{-j} k - 2^{-j} r, 2^{-j} k + 2^{-j} r]$. Finally, by equation (5.42), the graph of $\psi_{j,k}$ is obtained from the graph of $\psi(2^j x - k)$ by multiplying by $2^{j/2}$, which stretches the graph in the y direction by this factor. Similar remarks hold for $\varphi_{j,k}$.

Roughly speaking, the functions φ and ψ that we consider are centered near 0 and concentrated on a scale comparable to 1 (which means that most of the mass of the function is located within an interval around the origin of length about n, where n is a reasonably small positive integer). Then $\varphi_{j,k}$ and $\psi_{j,k}$ are centered near the point $2^{-j} k$ on a scale comparable to 2^{-j}.

Definition 5.28 *A wavelet system for $L^2(\mathbb{R})$ is a complete orthonormal set in $L^2(\mathbb{R})$ of the form*

$$\{\psi_{j,k}\}_{j,k \in \mathbb{Z}},$$

for some $\psi \in L^2(\mathbb{R})$. The functions $\psi_{j,k}$ are called wavelets. *The function ψ is called the* mother wavelet.

At the moment, it is not clear that any wavelet system exists. We will eventually construct one. If $\{\psi_{j,k}\}_{j,k \in \mathbb{Z}}$ is a wavelet system, then (by Theorem 4.10) every $f \in L^2(\mathbb{R})$ can be written in the form

$$f = \sum_{j \in \mathbb{Z}} \sum_{k \in \mathbb{Z}} \langle f, \psi_{j,k} \rangle \psi_{j,k}. \tag{5.43}$$

This is called the *wavelet identity*, and the map taking f to the sequence of coefficients $\{\langle f, \psi_{j,k} \rangle\}_{j,k \in \mathbb{Z}}$ is called the (discrete) *wavelet transform*. The wavelet identity should be interpreted as follows. By the above discussion, $\psi_{j,k}$ is centered near the point $2^{-j} k$ and has a scale of about 2^{-j}. The wavelet transform coefficient $\langle f, \psi_{j,k} \rangle$ is the weight or strength of the term $\psi_{j,k}$ in the expansion (5.43). So we think of the value $\langle f, \psi_{j,k} \rangle$ as measuring the part of f near the point $2^{-j} k$ at the scale 2^{-j}. We think of the wavelet identity (5.43) as breaking down f into its components at different scales 2^{-j}, centered at different locations $2^{-j} k$, for $j, k \in \mathbb{Z}$.

Before discussing the construction of discrete wavelet systems, we digress for a moment and discuss a formula similar to identity (5.43), except that it involves an integral instead of a sum and includes all possible translations and positive dilations. It has the advantage that it is relatively easy to derive.

Lemma 5.29   (Calderón formula) *Suppose* $\psi \in L^1(\mathbb{R}) \cap L^2(\mathbb{R})$ *is such that*

$$\int_0^{+\infty} |\hat{\psi}(s)|^2 \frac{ds}{s} = 1 \tag{5.44}$$

and

$$\int_0^{+\infty} |\hat{\psi}(-s)|^2 \frac{ds}{s} = 1. \tag{5.45}$$

For $t > 0$ *and* $y \in \mathbb{R}$, *define* $\psi_t(x) = (1/t)\psi(x/t)$ *as in equation (5.9), and*

$$\psi_t^y(x) = \frac{1}{\sqrt{t}} \psi\left(\frac{x-y}{t}\right).$$

Then for all $f \in L^2(\mathbb{R})$,

$$f(x) = \int_0^{+\infty} \psi_t * \tilde{\psi}_t * f(x) \frac{dt}{t} \tag{5.46}$$

or, equivalently,

$$f(x) = \int_0^{+\infty} \int_{\mathbb{R}} \langle f, \psi_t^y \rangle \psi_t^y(x) \, dy \, \frac{dt}{t^2}. \tag{5.47}$$

Proof

We take the Fourier transform of the right side of equation (5.46) and change the order of integration:

$$\left(\int_0^{+\infty} \psi_t * \tilde{\psi}_t * f(x) \frac{dt}{t}\right)\hat{}(\xi) = \int_{\mathbb{R}} \int_0^{+\infty} \psi_t * \tilde{\psi}_t * f(x) \frac{dt}{t} e^{-ix\xi} \, dx$$

$$= \int_0^{+\infty} \int_{\mathbb{R}} \psi_t * \tilde{\psi}_t * f(x) e^{-ix\xi} \, dx \, \frac{dt}{t}$$

$$= \int_0^{+\infty} \left(\psi_t * \tilde{\psi}_t * f\right)\hat{}(\xi) \frac{dt}{t}$$

$$= \int_0^{+\infty} (\psi_t)\hat{}(\xi)(\tilde{\psi}_t)\hat{}(\xi)\hat{f}(\xi) \frac{dt}{t}$$

$$= \hat{f}(\xi) \int_0^{+\infty} |(\psi_t)\hat{\ }(\xi)|^2 \, \frac{dt}{t}$$

$$= \hat{f}(\xi) \int_0^{+\infty} |\hat{\psi}(t\xi)|^2 \, \frac{dt}{t},$$

where we have used Lemmas 5.25 i, 5.26 i, and 5.14. If $\xi > 0$, we make the change of variables $s = t\xi$ in the last integral, obtaining

$$\hat{f}(\xi) \int_0^{+\infty} |\hat{\psi}(s)|^2 \, \frac{ds}{s} = \hat{f}(\xi),$$

by equation (5.44). If $\xi < 0$, we set $s = -t\xi$, and obtain

$$\hat{f}(\xi) \int_0^{+\infty} |\hat{\psi}(-s)|^2 \, \frac{ds}{s} = \hat{f}(\xi),$$

by equation (5.45). We now apply Fourier inversion (Theorem 5.24) to obtain the identity (5.46). To obtain identity (5.47), write

$$\psi_t * \tilde{\psi}_t * f(x) = \int_{\mathbb{R}} \psi_t(x - y)\tilde{\psi}_t * f(y) \, dy = \frac{1}{t} \int_{\mathbb{R}} \langle f, \psi_t^y \rangle \psi_t^y \, dy,$$

using Lemma 5.6 ii and the definitions. Substituting this equation into the identity (5.46) gives identity (5.47). ∎

The right-hand side of identity (5.47) should be interpreted as the limit in $L^2(\mathbb{R})$ as $n \to \infty$ of

$$\int_{1/n}^n \int_{\mathbb{R}} \langle f, \psi_t^y \rangle \psi_t^y(x) \, dy \, \frac{dt}{t^2}.$$

The identity (5.46) is named for A. P. Calderón. In the form (5.47), it is a continuous analog (i.e., an integral version) of the discrete wavelet identity (5.43) for which we are searching. The identity (5.47) shows that every $f \in L^2(\mathbb{R})$ can be written as an integral superposition of the basic functions $\{\psi_t^y\}_{t>0, y \in \mathbb{R}}$. Reasoning as in our discussion of $\psi_{j,k}$, we regard the function ψ_t^y as having scale t and being centered at y. The map taking f to the set of coefficient values $\{\langle f, \psi_t^y \rangle\}_{t>0, y \in \mathbb{R}}$ is called the *continuous wavelet transform*. The size of $\langle f, \psi_t^y \rangle$ is a measure the part of f near y on a scale of t. The identity (5.47) is sometimes called the *continuous wavelet identity*. It is useful for certain applications, particularly those in which wavelets are used as diagnostic tools to understand the behavior of f at different scales.

The main disadvantage of identities (5.46) and (5.47) is that they are difficult to use computationally. Because each formula involves a continuum of values, it is difficult to approximate by a finite set. One can think of identity (5.46) as an "unfolding" of f, replacing one function of $x \in \mathbb{R}$ by the family $\{\psi_t * \tilde{\psi}_t * f(x)\}_{t>0}$, each element of which is a function of $x \in \mathbb{R}$. This is in some sense the opposite of compression. The advantage of the discrete wavelet identity (5.43) is its concision: no information in equation (5.43) is redundant. It is remarkable how much more difficult it is to obtain the tight representation (5.43) than relation (5.47).

Some approaches to discrete wavelet theory start with identity (5.4.7) and then show that one can replace the full set $\{\psi_t^y\}_{t>0,y\in\mathbb{R}}$ by a discrete sampled subset of these functions and still recover f. This approach reflects some aspects of the history of the subject more accurately than the linear algebra approach of this text. Exercises 5.3.1 and 5.3.3 show one way the Calderón formula can be discretized, although this does not yield an orthonormal wavelet system.

The reader might expect that the derivation of equation (5.43) will proceed by arguments analogous to those in sections 3.1 and 3.2 for $\ell^2(\mathbb{Z}_N)$ and in sections 4.5 and 4.6 for $\ell^2(\mathbb{Z})$, by using the Fourier transform on $L^2(\mathbb{R})$ in place of the previous versions. Unfortunately, this does not work. It is instructive to see why. In $\ell^2(\mathbb{Z}_N)$ and $\ell^2(\mathbb{Z})$, there is an element δ such that the set of all translates $\{R_k\delta\}_k$ ($k \in \mathbb{Z}_N$ or $k \in \mathbb{Z}$, respectively) is a complete orthonormal set for the space. Then we could split the space to obtain a complete orthonormal set of the form $\{R_{2k}v\}_k \cup \{R_{2k}u\}_k$. In other words, the first step was to replace the minimal scale translations $\{R_k\delta\}_k$ with two sets of translations at twice the minimal scale. For $L^2(\mathbb{R})$, there is no minimal scale (and no $\delta \in L^2(\mathbb{R})$— see Exercise 5.2.16). Although we can find necessary and sufficient conditions for a set $\{\varphi_{0,k}\}_{k\in\mathbb{Z}}$ (recall that $\varphi_{0,k}(x) = \varphi(x - k)$) to be orthonormal (see Lemma 5.42), we cannot expect such a set to be complete in $L^2(\mathbb{R})$. For example, if φ has compact support, then only finitely many functions $\varphi_{0,k}$ have support intersecting a given interval, say $[-\pi, \pi)$. If $\{\varphi_{0,k}\}_{k\in\mathbb{Z}}$ is complete, then any $f \in L^2([-\pi, \pi))$ could be written as a linear combination of those $\varphi_{0,k}$ with support intersecting $[-\pi, \pi)$. This would imply that $L^2([-\pi, \pi))$ is finite dimensional, which is false. The same consideration would apply if we used smaller scale translations, say $\varphi(x - \epsilon k)$, $k \in \mathbb{Z}$ for some $\epsilon > 0$.

For $L^2(\mathbb{R})$, we start at a scale of 1 by considering φ such that $\{\varphi_{0,k}\}_{k \in \mathbb{Z}}$ is orthonormal. Then we consider dilations to both larger and smaller scales. We still construct an increasing sequence of subspaces, each of which is then split into two parts, as in the cases of $\ell^2(\mathbb{Z}_N)$ and $\ell^2(\mathbb{Z})$. However, for $L^2(\mathbb{R})$, the sequence is infinite in both directions, to capture smaller scales as well as larger scales. Due to the lack of completeness at any single scale and the nonexistence of $\delta \in L^2(\mathbb{R})$, the techniques for subspace splitting in $\ell^2(\mathbb{Z}_N)$ and $\ell^2(\mathbb{Z})$ do not carry over directly to $L^2(\mathbb{R})$. Conveniently, however, we are able to apply the results for $\ell^2(\mathbb{Z})$ to carry out the subspace splitting in $L^2(\mathbb{R})$. Stéphane Mallat determined the conditions that the sequence of subspaces should satisfy so that it leads to a wavelet system, as follows.

Definition 5.30 *A multiresolution analysis (or MRA) with scaling function or father wavelet φ is a sequence $\{V_j\}_{j \in \mathbb{Z}}$ of subspaces of $L^2(\mathbb{R})$ having the following properties:*

  i. *(Monotonicity) The sequence is increasing, that is, $V_j \subseteq V_{j+1}$ for all $j \in \mathbb{Z}$.*
  ii. *(Existence of the scaling function) There exists a function $\varphi \in V_0$ such that the set $\{\varphi_{0,k}\}_{k \in \mathbb{Z}}$ is orthonormal and*

$$V_0 = \left\{ \sum_{k \in \mathbb{Z}} z(k)\varphi_{0,k} : z = (z(k))_{k \in \mathbb{Z}} \in \ell^2(\mathbb{Z}) \right\}. \qquad (5.48)$$

  iii. *(Dilation property) For each j, $f(x) \in V_0$ if and only if $f(2^j x) \in V_j$.*
  iv. *(Trivial intersection property) $\bigcap_{j \in \mathbb{Z}} V_j = \{0\}$.*
  v. *(Density) $\bigcup_{j \in \mathbb{Z}} V_j$ is dense in $L^2(\mathbb{R})$.*

By definition, part v means that for any $f \in L^2(\mathbb{R})$, there exists a sequence $\{f_n\}_{n=1}^{+\infty}$ such that each $f_n \in \bigcup_{j \in \mathbb{Z}} V_j$ and $\{f_n\}_{n=1}^{+\infty}$ converges to f in $L^2(\mathbb{R})$, that is, $\|f_n - f\| \to 0$ as $n \to +\infty$.

This definition is difficult to understand at first. Example 5.31 may help.

Example 5.31
(Haar MRA) For each $j, k \in \mathbb{Z}$, let $I_{j,k}$ be the interval $[2^{-j}k, 2^{-j}(k+1))$. An interval of the form $I_{j,k}$ is called a *dyadic* interval. For each $j \in \mathbb{Z}$,

let

$$V_j = \{f \in L^2(\mathbb{R}) : \text{ for all } k \in \mathbb{Z}, f \text{ is constant on } I_{j,k} \}.$$

Note that any dyadic interval of length 2^{-j-1} (i.e., an $I_{j+1,k}$) is contained in a dyadic interval of length 2^{-j} (specifically, if k is even, then $I_{j+1,k} \subseteq I_{j,k/2}$, whereas if k is odd, $I_{j+1,k} \subseteq I_{j,(k-1)/2}$). If $f \in V_j$, in other words, if f is constant on dyadic intervals of length 2^{-j}, then f is constant on dyadic intervals of length 2^{-j-1}, and hence $f \in V_{j+1}$. Therefore $\{V_j\}_{j\in\mathbb{Z}}$ is an increasing sequence of subspaces. If we set

$$\varphi(x) = \begin{cases} 1 & \text{if } 0 \leq x < 1 \\ 0 & \text{if } x < 0 \text{ or } x \geq 1, \end{cases} \tag{5.49}$$

then the set $\{\varphi_{0,k}\}_{k\in\mathbb{Z}}$ is orthonormal because the supports of different $\varphi_{0,k}$ do not overlap. Moreover, every $f \in V_0$ can be written as

$$f = \sum_{k\in\mathbb{Z}} c_k \, \varphi_{0,k},$$

where c_k is the value of f on $[k, k+1)$. Note that $\sum_{k\in\mathbb{Z}} |c_k|^2 = \|f\|^2 < +\infty$, so equation (5.48) holds. The dilation property (Definition 5.30 iii) for a multiresolution analysis follows directly from the definition of $\{V_j\}_{j\in\mathbb{Z}}$. If $f \in \cap_{j\in\mathbb{Z}} V_j$, then f is constant on the intervals $[0, 2^{-j})$ and $[-2^{-j}, 0)$ for all $j \in \mathbb{Z}$. This implies that f is constant on $[0, +\infty)$ and on $(-\infty, 0)$ (take any two points in either region and let $j \to -\infty$). Since $f \in L^2(\mathbb{R})$, these two constants must be 0, so $f = 0$. Thus $\cap_{j\in\mathbb{Z}} V_j = \{0\}$. The final property, Definition 5.30 v, that $\cup_{j\in\mathbb{Z}} V_j$ is dense in $L^2(\mathbb{R})$, is also true, but a little more difficult to prove. We will not prove it here because it will follow from a general result below (Lemma 5.48). Assuming this, $\{V_j\}_{j\in\mathbb{Z}}$ is a multiresolution analysis with scaling function φ.

Another example is given in Exercise 5.3.4. Eventually we will give more examples. Our main goal in this section is to prove that a multiresolution analysis gives rise to a wavelet system.

Note that the orthonormality of the set $\{\varphi_{0,k}\}_{k\in\mathbb{Z}}$ implies that for each $j \in \mathbb{Z}$, $\{\varphi_{j,k}\}_{k\in\mathbb{Z}}$ is an orthonormal set, because changing variables shows that for $j, k, k' \in \mathbb{Z}$,

$$\langle \varphi_{j,k}, \varphi_{j,k'} \rangle = \langle \varphi_{0,k}, \varphi_{0,k'} \rangle = \begin{cases} 1 & \text{if } k = k' \\ 0 & \text{if } k \neq k'. \end{cases}$$

The dilation condition iii states that V_j consists of the dilates by 2^j of the elements of V_0. From this condition and equation (5.48), it follows (Exercise 5.3.5) that for each $j \in \mathbb{Z}$,

$$V_j = \left\{ \sum_{k \in \mathbb{Z}} z(k)\varphi_{j,k} : z = (z(k))_{k \in \mathbb{Z}} \in \ell^2(\mathbb{Z}) \right\}. \tag{5.50}$$

Thus $\{\varphi_{j,k}\}_{k \in \mathbb{Z}}$ is a complete orthonormal system for the subspace V_j.

One might think that the union of these systems would be a complete orthonormal system for $L^2(\mathbb{R})$. However, that is not true because the functions $\varphi_{j,k}$ are not necessarily orthogonal at different levels j. For example, since $\varphi \in V_0 \subseteq V_1$, equation (5.50) implies that

$$\varphi(x) = \sum_{k \in \mathbb{Z}} u(k)\varphi_{1,k}(x) = \sum_{k \in \mathbb{Z}} u(k)\sqrt{2}\varphi(2x - k), \tag{5.51}$$

for some coefficient sequence $u = (u(k))_{k \in \mathbb{Z}} \in \ell^2(\mathbb{Z})$. Note that $u(k)$ must equal $\langle \varphi, \varphi_{1,k} \rangle$ because $\{\varphi_{1,k}\}_{k \in \mathbb{Z}}$ is a complete orthonormal system for V_1. In particular, $\langle \varphi_{0,0}, \varphi_{1,k} \rangle = \langle \varphi, \varphi_{1,k} \rangle$ is not always 0. Equation 5.51 explains why φ is called the scaling function.

Definition 5.32 *Suppose $\{V_j\}_{j \in \mathbb{Z}}$ is a multiresolution analysis with scaling function φ. Equation (5.51) is known as the* scaling equation, *the* scaling relation, *or the* refinement equation. *The sequence $u = (u(k))_{k \in \mathbb{Z}}$ in equation (5.51) is called the* scaling sequence.

The following observation is a crucial clue suggesting that we can apply what we know about $\ell^2(\mathbb{Z})$ directly to the construction of wavelet systems in $L^2(\mathbb{R})$.

Lemma 5.33 *Suppose $\{V_j\}_{j \in \mathbb{Z}}$ is a multiresolution analysis with scaling function φ and scaling sequence u. Then $\{R_{2k}u\}_{k \in \mathbb{Z}}$ is an orthonormal set in $\ell^2(\mathbb{Z})$.*

Proof
If we replace x by $x - k$ in the scaling identity (5.51) (and change summation index on the right because k is now fixed), we obtain

$$\varphi(x - k) = \sum_{\ell \in \mathbb{Z}} u(\ell)\sqrt{2}\varphi(2x - 2k - \ell) = \sum_{m \in \mathbb{Z}} u(m - 2k)\sqrt{2}\varphi(2x - m),$$

that is,

$$\varphi_{0,k} = \sum_{m \in \mathbb{Z}} u(m - 2k)\varphi_{1,m}. \tag{5.52}$$

Substituting this equation and identity (5.51) gives

$$\langle \varphi, \varphi_{0,k} \rangle = \left\langle \sum_{j \in \mathbb{Z}} u(j)\varphi_{1,j}, \sum_{m \in \mathbb{Z}} u(m - 2k)\varphi_{1,m} \right\rangle$$

$$= \sum_{j \in \mathbb{Z}} u(j) \sum_{m \in \mathbb{Z}} \overline{u(m - 2k)} \langle \varphi_{1,j}, \varphi_{1,m} \rangle.$$

But the set $\{\varphi_{1,\ell}\}_{\ell \in \mathbb{Z}}$ is orthonormal, so $\langle \varphi_{1,j}, \varphi_{1,m} \rangle$ is 0 except when $m = j$, in which case $\langle \varphi_{1,j}, \varphi_{1,m} \rangle = 1$. Hence we obtain

$$\langle \varphi, \varphi_{0,k} \rangle = \sum_{j \in \mathbb{Z}} u(j)\overline{u(j - 2k)} = \langle u, R_{2k}u \rangle. \tag{5.53}$$

From our assumption, $\{\varphi(x - k)\}_{k \in \mathbb{Z}}$ is an orthonormal set in $L^2(\mathbb{R})$, so

$$\langle u, R_{2k}u \rangle = \begin{cases} 1 & \text{if } k = 0 \\ 0 & \text{if } k \neq 0. \end{cases} \tag{5.54}$$

By Exercise 4.5.1 (iii), the result follows. ∎

   Lemma 5.33 states that given a scaling function φ for an MRA, there corresponds a scaling sequence $u \in \ell^2(\mathbb{Z})$ with the property that the even integer translates of u are orthonormal in $\ell^2(\mathbb{Z})$. So we have mapped our problem in $L^2(\mathbb{R})$, which we do not know how to handle directly, into $\ell^2(\mathbb{Z})$, which we understand pretty well. We use the results in $\ell^2(\mathbb{Z})$ and map back into $L^2(\mathbb{R})$ to obtain wavelets for $\mathbb{R}$. Specifically, by Lemma 4.47 the sequence u has a companion v such that u and v generate a first-stage wavelet system for $\ell^2(\mathbb{Z})$. In the same way that φ corresponds to u, we define ψ corresponding to v. This yields the orthogonal splitting of $V_1 = V_0 \oplus W_0$, which corresponds to the orthogonal splitting of $\ell^2(\mathbb{Z})$ derived from a first-stage wavelet system for $\mathbb{Z}$. By dilation, a similar splitting $V_{j+1} = V_j \oplus W_j$ holds at every level. Taking the union of the orthonormal systems $\{\psi_{j,k}\}_{k \in \mathbb{Z}}$ for the orthogonal complement spaces W_j yields a wavelet system for $L^2(\mathbb{R})$. The plan is summarized by Figure 43.

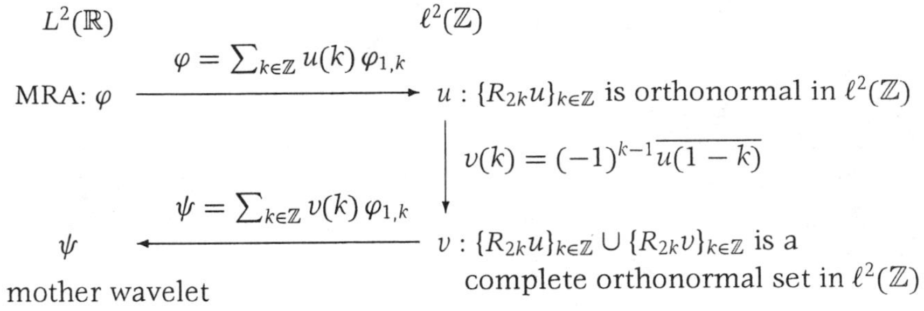

$L^2(\mathbb{R})$ $\qquad\qquad\qquad\qquad$ $\ell^2(\mathbb{Z})$

$$\varphi = \sum_{k \in \mathbb{Z}} u(k)\,\varphi_{1,k}$$

MRA: φ $\xrightarrow{\hspace{3cm}}$ $u : \{R_{2k}u\}_{k \in \mathbb{Z}}$ is orthonormal in $\ell^2(\mathbb{Z})$

$$\Big\downarrow \quad v(k) = (-1)^{k-1}\overline{u(1-k)}$$

$$\psi = \sum_{k \in \mathbb{Z}} v(k)\,\varphi_{1,k}$$

ψ $\xleftarrow{\hspace{3cm}}$ $v : \{R_{2k}u\}_{k \in \mathbb{Z}} \cup \{R_{2k}v\}_{k \in \mathbb{Z}}$ is a

mother wavelet $\qquad\qquad\qquad$ complete orthonormal set in $\ell^2(\mathbb{Z})$

FIGURE 43

Lemma 5.34 *Suppose $\{V_j\}_{j \in \mathbb{Z}}$ is a multiresolution analysis with scaling function φ and scaling sequence $u \in \ell^1(\mathbb{Z})$. Define $v \in \ell^1(\mathbb{Z})$ by*

$$v(k) = (-1)^{k-1}\overline{u(1-k)}, \quad \text{for all } k \in \mathbb{Z}. \tag{5.55}$$

Define

$$\psi(x) = \sum_{k \in \mathbb{Z}} v(k)\varphi_{1,k}(x) = \sum_{k \in \mathbb{Z}} v(k)\sqrt{2}\varphi(2x - k). \tag{5.56}$$

Then $\{\psi_{0,k}\}_{k \in \mathbb{Z}}$ is an orthonormal set in $L^2(\mathbb{R})$. Define

$$W_0 = \left\{ \sum_{k \in \mathbb{Z}} z(k)\psi_{0,k} : z = (z(k))_{k \in \mathbb{Z}} \in \ell^2(\mathbb{Z}) \right\}. \tag{5.57}$$

Then

$$V_1 = V_0 \oplus W_0. \tag{5.58}$$

Proof

By steps like these used to prove equation (5.52), identity (5.56) leads to

$$\psi_{0,k} = \sum_{m \in \mathbb{Z}} v(m - 2k)\varphi_{1,m}. \tag{5.59}$$

Then, as for the proof of equation (5.53), we obtain (Exercise 5.3.6)

$$\langle \psi, \psi_{0,k} \rangle = \langle v, R_{2k}v \rangle = \begin{cases} 1 & \text{if } k = 0 \\ 0 & \text{if } k \neq 0. \end{cases} \tag{5.60}$$

By a change of variables, for $k, \ell \in \mathbb{Z}$,

$$\langle \psi_{0,k}, \psi_{0,\ell} \rangle = \langle \psi, \psi_{0,\ell-k} \rangle.$$

Hence equation (5.60) implies that $\{\psi_{0,k}\}_{k \in \mathbb{Z}}$ is an orthonormal set in $L^2(\mathbb{R})$. By Exercise 4.2.8, W_0 is a subspace of $L^2(\mathbb{R})$. The orthonormality of $\{\varphi_{0,k}\}_{k \in \mathbb{Z}}$ is part of the definition of an MRA.

By arguments like those leading to equations (5.53) and (5.60) (Exercise 5.3.6),

$$\langle \varphi, \psi_{0,k} \rangle = \langle u, R_{2k}v \rangle = 0 \tag{5.61}$$

for all k. Hence, by changing variables,

$$\langle \varphi_{0,k}, \psi_{0,\ell} \rangle = \langle \varphi, \psi_{0,\ell-k} \rangle = 0 \quad \text{for all} \quad k, \ell \in \mathbb{Z}.$$

By Exercise 4.6.4, these results imply that the subspaces V_0 and W_0 are orthogonal.

By equation (5.59), $\psi_{0,k} \in V_1$ for each $k \in \mathbb{Z}$. By Exercise 4.6.5, it follows that W_0 is a subspace of V_1. Recall that V_0 is a subspace of V_1 by assumption.

To prove that $V_0 \oplus W_0 = V_1$, all that is left to show is that $V_1 \subseteq V_0 \oplus W_0$. The proof of this is similar to the proof of equation (4.69). Our goal is to show that

$$\varphi_{1,j} = \sum_{k \in \mathbb{Z}} \tilde{u}(2k - j)\varphi_{0,k} + \sum_{k \in \mathbb{Z}} \tilde{v}(2k - j)\psi_{0,k} \tag{5.62}$$

for each $j \in \mathbb{Z}$. To prove this, substituting equations (5.52) and (5.59) shows that the right side of equation (5.62) is

$$\sum_{k \in \mathbb{Z}} \tilde{u}(2k - j) \sum_{m \in \mathbb{Z}} u(m - 2k)\varphi_{1,m} + \sum_{k \in \mathbb{Z}} \tilde{v}(2k - j) \sum_{m \in \mathbb{Z}} v(m - 2k)\varphi_{1,m}$$

$$= \sum_{m \in \mathbb{Z}} \left(\sum_{k \in \mathbb{Z}} \tilde{u}(2k - j)u(m - 2k) + \sum_{k \in \mathbb{Z}} \tilde{v}(2k - j)v(m - 2k) \right) \varphi_{1,m}.$$

The only way that equation (5.62) can be true is if the term inside the large parentheses in the last expression is 1 when $m = j$ and 0 otherwise. Fortunately, this is exactly what equation (4.68) states. Therefore equation (5.62) holds, and shows that for each $j \in \mathbb{Z}$, $\varphi_{1j} \in V_0 \oplus W_0$. By Exercise 4.6.5, we obtain $V_1 \subseteq V_0 \oplus W_0$. ∎

Thus we have obtained a splitting of V_1. By dilation, every V_j is split similarly. This leads to a wavelet system for $L^2(\mathbb{R})$.

Theorem 5.35   *(Mallat's theorem) Suppose $\{V_j\}_{j\in\mathbb{Z}}$ is a multiresolution analysis with a scaling function φ and scaling sequence $u = (u(k))_{k\in\mathbb{Z}} \in \ell^1(\mathbb{Z})$. Define $v = (v(k))_{k\in\mathbb{Z}}$ by equation (5.55) and ψ by equation (5.56). Then $\{\psi_{j,k}\}_{j,k\in\mathbb{Z}}$ is a wavelet system in $L^2(\mathbb{R})$.*

Proof

For $j \in \mathbb{Z}$, the orthonormality of $\{\psi_{0,k}\}_{k\in\mathbb{Z}}$ implies, by a change of variable, just as in the case of φ above, that $\{\psi_{j,k}\}_{k\in\mathbb{Z}}$ is an orthonormal set. Define

$$W_j = \left\{ \sum_{k\in\mathbb{Z}} z(k)\psi_{j,k} : z = (z(k))_{k\in\mathbb{Z}} \in \ell^2(\mathbb{Z}) \right\}. \qquad (5.63)$$

It follows from this definition (Exercise 5.3.7) that the spaces $\{W_j\}_{j\in\mathbb{Z}}$ have the same dilation property that the $\{V_j\}_{j\in\mathbb{Z}}$ have, according to Definition 5.30 iii:

$$f \in W_0 \text{ if and only if } f(2^j x) \in W_j. \qquad (5.64)$$

From Lemma 5.34, $V_1 = V_0 \oplus W_0$. By Definition 5.30 iii and relation (5.64), a dilation argument (Exercise 5.3.8) shows that this splitting persists at every stage:

$$V_{j+1} = V_j \oplus W_j \quad \text{for all} \quad j \in \mathbb{Z}. \qquad (5.65)$$

We claim that

$$B = \{\psi_{j,k}\}_{j,k\in\mathbb{Z}}$$

is an orthonormal set in $L^2(\mathbb{R})$. For each fixed j, we already know that $\{\psi_{j,k}\}_{k\in\mathbb{Z}}$ is orthonormal. We must show that $\psi_{j,k}$ is orthogonal to $\psi_{\ell,m}$, when $j \neq \ell$; we can assume by symmetry that $j > \ell$. Then $\psi_{\ell,m} \in W_\ell \subseteq V_{\ell+1} \subseteq \cdots \subseteq V_j$, by Definition 5.30 i. But $\psi_{j,k} \in W_j$, and W_j is orthogonal to V_j. So $\psi_{j,k}$ is orthogonal to $\psi_{\ell,m}$. Hence B is orthonormal.

What remains is to show the completeness of the orthonormal set B. To prove this, we use properties iv and v in Definition 5.30, which have not been used so far. We first make the following claim. Suppose $g \in V_j$ for some $j \in \mathbb{Z}$, and $g \perp W_\ell$ (which means that $\langle g, w \rangle = 0$ for all $w \in W_\ell$) for all $\ell \leq j - 1$. Then $g = 0$. The proof of this claim is almost the same as in the proof of completeness in Theorem 4.55, so we leave this as Exercise 5.3.9.

Now suppose $f \in L^2(\mathbb{R})$ is orthogonal to every element of B; that is, $\langle f, \psi_{j,k} \rangle = 0$ for all $j, k \in \mathbb{Z}$. It follows that $f \perp W_j$ for each $j \in \mathbb{Z}$ (e.g., using Exercise 4.2.5 (i)). We need to show $f = 0$. For each j, let $P_j(f)$ be the projection of f on V_j, defined (as in Definition 4.13) by

$$P_j(f) = \sum_{k \in \mathbb{Z}} \langle f, \varphi_{j,k} \rangle \varphi_{j,k}. \tag{5.66}$$

By definition (see Lemma 4.14 i), $P_j(f) \in V_j$. By Lemma 4.14 iii, $f - P_j(f)$ is orthogonal to every element of V_j. For $\ell \leq j - 1$, $W_\ell \subseteq V_{\ell+1} \subseteq V_j$, so $f - P_j(f)$ is orthogonal to W_ℓ for every $\ell \leq j - 1$. Because f is orthogonal to every W_ℓ, linearity shows that $P_j(f) \perp W_\ell$ for all $\ell \leq j - 1$. By the result stated in the previous paragraph,

$$P_j(f) = 0 \quad \text{for all} \quad j \in \mathbb{Z}.$$

However, by Lemma 4.14 v, for all $j \in \mathbb{Z}$, $P_j(f) = 0$ is the best approximation to f in V_j; that is, for all $h \in V_j$,

$$\|f\| = \|f - P_j(f)\| \leq \|f - h\|. \tag{5.67}$$

By Definition 5.30 v, there exists a sequence $\{f_n\}_{n \in \mathbb{Z}}$ such that $f_n \in \cup_{j \in \mathbb{Z}} V_j$ for all $n \in \mathbb{Z}$ and $\|f - f_n\| \to 0$ as $n \to \infty$. By relation (5.67), this implies $\|f\| = 0$, that is, $f = 0$. This proves the completeness of B.   ■

Because the subspaces V_j are increasing and their union is dense (Definition 5.30 i and v), we think of these spaces as approximations to the entire space $L^2(\mathbb{R})$. We think of the projection $P_j(f)$ of f onto V_j as the approximation of f at level j. For $f \in L^2(\mathbb{R})$, the approximations $P_j(f)$ improve and converge in norm to f (Exercise 5.3.10).

Example 5.36
We return to the Haar MRA (Example 5.31) and apply Theorem 5.35 to find the corresponding wavelet system for $L^2(\mathbb{R})$). The key is to find the coefficients $u(k), k \in \mathbb{Z}$ in the scaling relation (5.51). The orthonormality of the set $\{\varphi_{1,k}\}_{k \in \mathbb{Z}}$ shows that $u(k) = \langle \varphi, \varphi_{1,k} \rangle$ for each $k \in \mathbb{Z}$. Note that $\varphi_{1,k}(x) = \sqrt{2}\varphi(2x - k)$ is $\sqrt{2}$ for $k/2 \leq x < (k+1)/2$, and 0 for all other x. Therefore computing $\langle \varphi, \varphi_{1,k} \rangle$ is easy, and we obtain $u(0) = 1/\sqrt{2}$, $u(1) = 1/\sqrt{2}$, and $u(j) = 0$ if $j \notin \{0, 1\}$. We can check that this works: the scaling relation (5.51) with these values

is equivalent to

$$\varphi(x) = \varphi(2x) + \varphi(2x - 1),$$

which is easy to verify. Formula (5.55) for v gives $v(0) = -1/\sqrt{2}$, $v(1) = 1/\sqrt{2}$, and $v(j) = 0$ for all other j. Thus by equation (5.56), the mother wavelet is

$$\psi(x) = -\varphi(2x) + \varphi(2x - 1) = \begin{cases} -1 & \text{if } 0 \le x < \frac{1}{2} \\ 1 & \text{if } \frac{1}{2} \le x < 1 \\ 0 & \text{if } x < 0 \text{ or } x \ge 1. \end{cases}$$

Then

$$\psi_{j,k}(x) = \begin{cases} -2^{j/2} & \text{if } \frac{k}{2^j} \le x < \frac{k}{2^j} + \frac{1}{2^{j+1}} \\ 2^{j/2} & \text{if } \frac{k}{2^j} + \frac{1}{2^{j+1}} \le x < \frac{k+1}{2^j} \\ 0 & \text{if } x < \frac{k}{2^j} \text{ or } x \ge \frac{k+1}{2^j}. \end{cases} \tag{5.68}$$

It is not difficult to check directly that the set $\{\psi_{j,k}\}_{j,k\in\mathbb{Z}}$ is orthonormal (Exercise 5.3.11). The completeness of $\{\psi_{j,k}\}_{j,k\in\mathbb{Z}}$ follows from Theorem 5.35. Thus we have an example of a wavelet system. The functions $\psi_{j,k}$ are called *Haar functions*, after A. Haar, who considered them in a paper published in 1910. This was the first example of a wavelet system. The projections $P_j(f)$ of $f \in L^2(\mathbb{R})$ onto V_j defined by identity (5.66) have a natural interpretation: $P_j(f)$ is obtained (Exercise 5.3.12) from f by replacing f on each interval $I_{j,k}$, for $k \in \mathbb{Z}$, by its average (as in the discrete case—see equations (3.95) and (4.91)).

Note that the coefficient sequences u and v for the Haar system are the generators of the discrete Haar system in $\ell^2(\mathbb{Z})$ (Example 4.56) (except for the sign of v, which has been reversed).

For many purposes, the Haar system is not satisfactory, due to the lack of smoothness of the Haar functions: they are not even continuous. One purpose of wavelet theory is to construct smooth versions of the Haar functions.

The procedure in Theorem 5.35 for obtaining a wavelet system from a MRA is explicit and constructive, as we have seen in Example 5.36. Thus one can obtain examples of wavelet systems by the recipe in Theorem 5.35 if one can construct MRAs.

We remark that for a while it was an open question whether every wavelet system arises from a multiresolution analysis. The answer is "no" in general, but all nice enough wavelet systems do (see Auscher (1995) for the precise statement).

In the next section we discuss the construction of multi-resolution analyses.

Exercises

5.3.1.   i. Suppose $\varphi, \psi \in L^2(\mathbb{R})$ are such that

$$\sum_{j \in \mathbb{Z}} \overline{\hat{\varphi}(2^{-j}\xi)} \hat{\psi}(2^{-j}\xi) = 1 \quad \text{for all} \quad \xi \neq 0, \; \xi \in \mathbb{R}. \quad (5.69)$$

Show, at least formally (i.e., without discussing convergence) that, for $f \in L^2(\mathbb{R})$,

$$f = \sum_{j \in \mathbb{Z}} \tilde{\varphi}_{2^{-j}} * \psi_{2^{-j}} * f.$$

This is analogous to equation (5.46), with the integral replaced by a sum. Hint: Take the Fourier transform of both sides, assuming you can interchange the order of the resulting double integral on the right side. Then apply Lemma 5.14.

ii. It is not difficult to obtain equation (5.69). Suppose
 a. $\hat{\varphi}(\xi) = 0$ unless $1/2 \leq |\xi| \leq 2$, and
 b. $\hat{\varphi}(\xi) \neq 0$ for all $\xi \in [-5/3, -3/5] \cup [3/5, 5/3]$.
 Let

$$B(\xi) = \sum_{k \in \mathbb{Z}} |\hat{\varphi}(2^k \xi)|^2.$$

Show that by supposition a, this sum is finite at each ξ (at most two terms are nonzero), and by supposition b, $B(\xi) \neq 0$ for $\xi \neq 0$ (the dilates of the interval $[3/5, 5/3]$ overlap). Prove that

$$B(2^j \xi) = B(\xi),$$

for all $j \in \mathbb{Z}$. Let

$$\hat{\psi}(\xi) = \hat{\varphi}(\xi)/B(\xi).$$

Prove that equation (5.69) holds.

5.3.2. (Shannon sampling theorem) Suppose $h \in L^2(\mathbb{R})$ and

$$\operatorname{supp} \hat{h} \subseteq [-b, b].$$

(Such a function is said to be *of exponential type b*. If f is of some exponential type, we say f is *bandlimited*.) Suppose also that Fourier inversion holds at every point for h, that is, $(\hat{h})^{\vee}(x) = h(x)$ for all $x \in \mathbb{R}$.

  i. Prove that for $|\xi| \leq b$,

$$\hat{h}(\xi) = \sum_{n \in \mathbb{Z}} \frac{\pi}{b} h\left(\frac{n\pi}{b}\right) e^{-in\pi\xi/b}. \tag{5.70}$$

Hint: By Exercise 4.3.5, the set $\{e^{-in\pi t/b}\}_{n \in \mathbb{Z}}$ is a complete orthonormal set in $L^2([-b, b))$ (see that exercise for the definitions). Expand $\hat{h}$ in terms of this set. Note that in the equation

$$\langle \hat{h}, e^{-in\pi t/b} \rangle = \frac{1}{2b} \int_{-b}^{b} \hat{h}(t) e^{in\pi t/b} \, dt,$$

the interval of integration can be replaced by $\mathbb{R}$ because of the exponential type assumption. Use Theorem 5.24.

  ii. Prove that

$$h(x) = \sum_{n \in b\mathbb{Z}} h\left(\frac{n\pi}{b}\right) \frac{\sin(bx - n\pi)}{bx - n\pi}. \tag{5.71}$$

Hint: Apply Fourier inversion

$$h(x) = \frac{1}{2\pi} \int_{\mathbb{R}} \hat{h}(\xi) e^{ix\xi} \, d\xi.$$

By the exponential type assumption, the integral on $\mathbb{R}$ can be replaced by $\int_{-b}^{b}$. Substitute equation (5.70).

Remark: This exercise shows that h is determined by the discrete set of values $\{h(n\pi/b)\}_{n \in \mathbb{Z}}$. This statement is false without the exponential type assumption. This means that a bandlimited signal can be recovered from its sample values if the samples are sufficiently dense (with required density inversely related to the exponential type, also called the *bandwidth*). This fact underlies digital processing of audio signals, which can be assumed bandlimited because our ears hear only a finite bandwidth.

5.3.3. (Phi transform identity) Let φ, ψ be as in Exercise 5.3.1 (i), with supp $\hat{\varphi}$, supp $\hat{\psi} \subseteq [-2, -1/2] \cup [1/2, 2]$. Define $\varphi_{j,k}$ and $\psi_{j,k}$ as in Definition 5.27.

   i. Prove that $\hat{\psi}_{j,k} = 2^{-j/2}e^{-ik2^{-j}\xi}\hat{\psi}_{2^{-j}}$, where $\hat{\psi}_{2^{-j}} = (\psi_{2^{-j}})^{\hat{}}(\xi)$.

   ii. For $f \in L^2(\mathbb{R})$, prove that $\langle f, \varphi_{j,k} \rangle = 2^{-j/2}f * \tilde{\varphi}_{2^{-j}}(2^{-j}k)$.

   iii. For $f \in L^2(\mathbb{R})$, prove (formally, without discussing convergence) that

$$f = \sum_{j \in \mathbb{Z}} \sum_{k \in \mathbb{Z}} \langle f, \varphi_{j,k} \rangle \psi_{j,k}.$$

Hint: Apply Exercise 5.3.1 (i) and Fourier inversion to write

$$f = \sum_{j \in \mathbb{Z}} \left((f * \tilde{\varphi}_{2^{-j}})^{\hat{}} \hat{\psi}_{2^{-j}} \right)^{\vee}.$$

Note that

$$\text{supp}\,(f * \tilde{\varphi}_{2^{-j}})^{\hat{}} \subseteq \{\xi : 2^{j-1} \leq |\xi| \leq 2^{j+1}\}$$
$$\subseteq \{\xi : -2^{j}\pi \leq \xi \leq 2^{j}\pi\},$$

and apply equation (5.70) with $b = 2^{j}\pi$. Then use parts i and ii.

Remark: This formula is similar to the wavelet identity (5.43), although it does not come from an orthonormal set.

5.3.4. For $j \in \mathbb{Z}$, let

$$V_j = \{f \in L^2(\mathbb{R}) : \text{supp}\,\hat{f} \subseteq [-2^{j}\pi, 2^{j}\pi]\}.$$

Prove that $\{V_j\}_{j \in \mathbb{Z}}$ is a multiresolution analysis. Hint: It is easy to see that each V_j is a subspace of $L^2(\mathbb{R})$. Property i in Definition 5.30 is clear. For property ii, define

$$\chi_{[-\pi,\pi]}(\xi) = \begin{cases} 1 & \text{if } -\pi \leq \xi \leq \pi \\ 0 & \text{if } |\xi| > \pi. \end{cases}$$

Let

$$\varphi(x) = (\chi_{[-\pi,\pi]})^{\vee}(x) = \frac{\sin(\pi x)}{\pi x}.$$

If $f \in V_0$, then we can expand $\hat{f}$ in a Fourier series on $[-\pi, \pi)$:

$$\hat{f}(\xi) = \sum_{k \in \mathbb{Z}} a(k)e^{-ik\xi}\chi_{[-\pi,\pi]}(\xi),$$

for some sequence $(a(k))_{k \in \mathbb{Z}} \in \ell^2(\mathbb{Z})$. Note that

$$(\varphi_{0,k})\hat{}(\xi) = e^{-ik\xi} \chi_{[-\pi,\pi]}(\xi),$$

by Lemma 5.26 ii. By Fourier inversion, deduce that

$$f(x) = \sum_{k \in \mathbb{Z}} a(k)\varphi(x - k).$$

The dilation property (iii in Definition 5.30) follows from Lemma 5.14. Property iv follows from the definition of the spaces V_j. For property v, suppose $f \in L^2(\mathbb{R})$ and let f_n be such that $\hat{f}_n(\xi) = \hat{f}(\xi)$ if $|\xi| \leq 2^n \pi$, and 0 otherwise. Show that $\|\hat{f}_n - \hat{f}\| \to 0$ as $n \to \infty$ (compare to Exercise 5.1.13) and apply Plancherel's formula.

5.3.5. Suppose $\{V_j\}_{j \in \mathbb{Z}}$ is a multiresolution analysis with scaling function φ. Prove equation (5.50).

5.3.6. Prove equations (5.60) and (5.61).

5.3.7. Prove equation (5.64).

5.3.8. Prove equation (5.65), assuming equations (5.58) and (5.64).

5.3.9. Let $\{V_j\}_{j \in \mathbb{Z}}$ be a multiresolution analysis with scaling fuction φ. Define ψ by equation (5.56) and W_j by equation (5.63) for $j \in \mathbb{Z}$.

　　i. Prove the claim in the proof of Theorem 5.35: if $g \in V_j$ and $g \perp W_\ell$ for all $\ell \leq j - 1$, then $g = 0$. Hint: Follow the reasoning in the proof of Theorem 4.55, but start at j rather than at 0. Obtain $g \in \cap_{\ell=-\infty}^{j} V_\ell = \cap_{\ell \in \mathbb{Z}} V_\ell$, by Definition 5.30 i. Then apply Definition 5.30 iv.

　　ii. Deduce that $\{\psi_{\ell,k}\}_{\ell,k \in \mathbb{Z}: \ell < j}$ is a complete orthonormal system for V_j.

　　Remark: With the correct interpretation, this implies that $V_j = \oplus_{\ell=-\infty}^{j-1} W_\ell$ for each $j \in \mathbb{Z}$.

5.3.10. Suppose $\{V_j\}_{j \in \mathbb{Z}}$ is a multiresolution analysis. Let $f \in L^2(\mathbb{R})$. For each j, let $P_j(f)$ be the orthogonal projection of f onto the subspace V_j, as in equation (5.66). Show that the sequence $\{P_j(f)\}_{j \in \mathbb{Z}}$ converges to f in $L^2(\mathbb{R})$, that is, that

$$\|P_j(f) - f\| \to 0 \quad \text{as} \quad j \to \infty.$$

Hint: By using Definition 5.30 i, show that $\|P_j(f) - f\|$ is a decreasing sequence, and hence converges. Use the best approximation property and Definition 5.30 v to see that the limit must be 0.

5.3.11. For $j, k \in \mathbb{Z}$, define $\psi_{j,k}$ by equation (5.68). Prove that $\{\psi_{j,k}\}_{j,k \in \mathbb{Z}}$ is an orthonormal set in $L^2(\mathbb{R})$. Hint: Use the property that if two dyadic intervals $I_{j,k}$ and $I_{j',k'}$ intersect, then one is a subset of the other. This fact follows from the comments on dyadic intervals at the beginning of Example 5.31.

5.3.12. Define φ by equation (5.49) and $\varphi_{j,k}$ as in Definition 5.27, for $j, k \in \mathbb{Z}$.

    i. Prove that $\varphi_{j,k}(x) = 2^{j/2}$ if $x \in I_{j,k}$ (defined in Example 5.31) and $\varphi_{j,k}(x) = 0$ if $x \notin I_{j,k}$.

    ii. For $j \in \mathbb{Z}$ and $f \in L^2(\mathbb{R})$, define $P_j(f)$ by equation (5.66). For $x \in I_{j,k}$, show that

$$P_j(f)(x) = 2^j \int_{I_{j,k}} f(t)\, dt,$$

the average of f over $I_{j,k}$.

5.3.13. (Inhomogeneous wavelet system for $L^2(\mathbb{R})$) Let $\{V_j\}_{j \in \mathbb{Z}}$ be a multiresolution analysis with scaling fuction φ. Define ψ as in Theorem 5.34. Let $\ell \in \mathbb{Z}$. Set

$$B = \{\psi_{j,k}\}_{j,k \in \mathbb{Z}: j \geq \ell} \cup \{\varphi_{\ell,k}\}_{k \in \mathbb{Z}}.$$

Prove that B is a complete orthonormal system in $L^2(\mathbb{R})$. Hint: To prove completeness, show that if f is orthogonal to every element of B, then f is orthogonal to the wavelet system in Theorem 5.35.

5.4   Construction of Multiresolution Analyses

In section 5.3 we saw that a multiresolution analysis with scaling sequence $u \in \ell^1(\mathbb{Z})$ yields a wavelet system for $L^2(\mathbb{R})$. However, we have only two examples of MRAs so far (the Haar MRA in Example 5.31 and the MRA in Exercise 5.3.4). The Haar MRA leads to the Haar system, which has been known since 1910, so it may appear that not much progress has been made. However, we do have interesting examples of vectors $u \in \ell^1(\mathbb{Z})$ such that $\{R_{2k}u\}_{k\in\mathbb{Z}}$ is orthonormal in $\ell^2(\mathbb{Z})$, such as Daubechies's D6 wavelets for $\ell^2(\mathbb{Z})$ in Example 4.57. In this section we see that it is possible to use such a u to construct a wavelet system for $L^2(\mathbb{R})$.

From Figure 43, it may appear that we can skip the first step, start with u, define v and then obtain the mother wavelet ψ. However, this is not the case because the definition of ψ involves φ. Instead, we attempt to reverse the first arrow. That is, given u such that $\{R_{2k}u\}_{k\in\mathbb{Z}}$ is orthonormal in $\ell^2(\mathbb{Z})$, we attempt to find φ that solves the scaling relation

$$\varphi(x) = \sum_{k\in\mathbb{Z}} u(k)\varphi_{1,k}(x) = \sum_{k\in\mathbb{Z}} u(k)\sqrt{2}\varphi(2x - k), \qquad (5.72)$$

and obtain a corresponding MRA from which we can derive a wavelet system by the recipe in Theorem 5.35. Because φ occurs on both sides of the scaling equation, it is not apparent whether equation (5.72) has a non-trivial solution φ (the trivial solution being $\varphi(x) = 0$ for all x, which obviously does not yield an MRA). If equation (5.72) does have a nontrivial solution, the solution is not unique because any constant multiple of a solution is also a solution.

However, under reasonable conditions, there is a nontrivial solution to the scaling equation that is unique except for multiplication by a constant. To see why, we sketch a heuristic argument, which leads to a formula for a nontrivial solution. Precise assumptions will be given later when we state results as lemmas and theorems. We begin by taking the Fourier transform of both sides of the scaling identity. Note (Exercise 5.4.1) that

$$\hat{\varphi}_{1,k}(\xi) = \frac{1}{\sqrt{2}}e^{-ik\xi/2}\hat{\varphi}\left(\frac{\xi}{2}\right). \qquad (5.73)$$

Assuming we can interchange the sum and the integral in the definition of the Fourier transform of the right side of equation (5.72), we obtain

$$\hat{\varphi}(\xi) = \frac{1}{\sqrt{2}}\sum_{k\in\mathbb{Z}} u(k)e^{-ik\xi/2}\hat{\varphi}\left(\frac{\xi}{2}\right). \qquad (5.74)$$

This suggests that we define

$$m_0(\xi) = \frac{1}{\sqrt{2}} \sum_{k \in \mathbb{Z}} u(k) e^{-ik\xi}. \tag{5.75}$$

Then equation (5.74) gives

$$\hat{\varphi}(\xi) = m_0 \left(\frac{\xi}{2} \right) \hat{\varphi} \left(\frac{\xi}{2} \right). \tag{5.76}$$

The great thing about equation (5.76) is that it can be iterated. That is, if we apply equation (5.76) with ξ replaced by $\xi/2$, we obtain

$$\hat{\varphi} \left(\frac{\xi}{2} \right) = m_0 \left(\frac{\xi}{4} \right) \hat{\varphi} \left(\frac{\xi}{4} \right).$$

If we substitute this in equation (5.76), we obtain

$$\hat{\varphi}(\xi) = m_0 \left(\frac{\xi}{2} \right) m_0 \left(\frac{\xi}{4} \right) \hat{\varphi} \left(\frac{\xi}{4} \right).$$

We can continue this (replace ξ by $\xi/4$ in equation (5.76), etc.) and obtain, for any $n \in \mathbb{N}$,

$$\hat{\varphi}(\xi) = m_0 \left(\frac{\xi}{2} \right) m_0 \left(\frac{\xi}{4} \right) \cdots m_0 \left(\frac{\xi}{2^n} \right) \hat{\varphi} \left(\frac{\xi}{2^n} \right). \tag{5.77}$$

This suggests that we should be able to let $n \to +\infty$ (this requires $\hat{\varphi}$ to be continuous at 0) to obtain

$$\hat{\varphi}(\xi) = \hat{\varphi}(0) \prod_{j=1}^{\infty} m_0 \left(\frac{\xi}{2^j} \right). \tag{5.78}$$

If $\hat{\varphi}(0) = 0$, equation (5.78) implies that $\hat{\varphi}(\xi) = 0$ for all ξ, and hence φ is the zero function. Therefore, a nontrivial solution φ of equation (5.72) must satisfy $\hat{\varphi}(0) \neq 0$. Then equation (5.78) shows that, under reasonable conditions, the solution is unique up to scalar multiplication (namely, the choice of $\hat{\varphi}(0)$), and is explicitly determined by m_0, hence by u.

We see later that to obtain the orthonormality of the set $\{\varphi_{0,k}\}_{k \in \mathbb{Z}}$, we must have $|\hat{\varphi}(0)| = 1$. Given this, we can multiply by a unimodular constant to obtain $\hat{\varphi}(0) = 1$. This normalization will turn out to be useful (see Lemma 5.54) in the numerical implementation of the wavelet transform. Note that $\hat{\varphi}(0) = \int_{\mathbb{R}} \varphi(x) \, dx$, by definition of the Fourier transform, hence

$$\int_{\mathbb{R}} \varphi(x) \, dx = 1. \tag{5.79}$$

We obtained equation (5.78) assuming that a nontrivial solution of the scaling equation exists. This argument suggests that to show the existence of a solution, it is natural to define

$$\hat{\varphi}(\xi) = \prod_{j=1}^{\infty} m_0 \left(\frac{\xi}{2^j} \right) \tag{5.80}$$

and

$$\varphi = (\hat{\varphi})^{\vee}. \tag{5.81}$$

Then we can hope to prove that φ really satisfies the scaling equation with scaling sequence u.

We prove here that under a few additional hypotheses, this approach works. First we make some observations.

We noted previously that a nontrivial solution φ of equation (5.72) must satisfy $\hat{\varphi}(0) \neq 0$. If we substitute $\xi = 0$ in equation (5.76), we obtain $\hat{\varphi}(0) = m_0(0)\hat{\varphi}(0)$, and hence that

$$m_0(0) = 1. \tag{5.82}$$

By the definition of m_0 in equation (5.75), equation (5.82) is equivalent to

$$\sum_{k \in \mathbb{Z}} u(k) = \sqrt{2}, \tag{5.83}$$

which puts a new restriction on u that did not arise in the context of $\ell^2(\mathbb{Z})$. This assumption, in the form of equation (5.82), plays another key role: it guarantees that the terms $m_0(\xi/2^j)$ in (5.80) converge to 1 as $j \to \infty$, which is essential for convergence of the product $\prod_{j=1}^{\infty} m_0(\xi/2^j)$.

Observe that m_0 is a 2π-periodic function on $\mathbb{R}$ and

$$m_0(\xi) = \frac{1}{\sqrt{2}} \hat{u}(-\xi), \tag{5.84}$$

where here $\hat{u}$ denotes the Fourier transform in the sense of $\ell^2(\mathbb{Z})$ (see Definition 4.29). For $u \in \ell^1(\mathbb{Z})$, the partial sums of the series in equation (5.75) are continuous and converge uniformly to m_0 (Exercise 4.4.9), hence m_0 is continuous.

Recall (Lemma 4.42 i) that the assumption that $\{R_{2k}u\}_{k\in\mathbb{Z}}$ is orthonormal in $\ell^2(\mathbb{Z})$ is equivalent to the identity

$$|\hat{u}(\theta)|^2 + |\hat{u}(\theta+\pi)|^2 = 2$$

for all θ. In terms of m_0, this is equivalent to the condition

$$|m_0(\xi)|^2 + |m_0(\xi+\pi)|^2 = 1 \quad \text{for all } \xi. \tag{5.85}$$

Obviously, this implies that $|m_0(\xi)| \leq 1$ for all ξ.

Let v be the companion for u defined by $v(k) = (-1)^{k-1}\overline{u(1-k)}$, so that u and v generate a first-stage wavelet system for $\ell^2(\mathbb{Z})$ (Lemma 4.47). Define

$$\psi = \sum_{k\in\mathbb{Z}} v(k)\varphi_{1,k}, \tag{5.86}$$

as in Lemma 5.34. If we take the Fourier transform of both sides of equation (5.86) and use equation (5.73), an argument similar to the derivation of equation (5.74) yields

$$\hat{\psi}(\xi) = \frac{1}{\sqrt{2}} \sum_{k\in\mathbb{Z}} v(k)e^{-ik\xi/2}\hat{\varphi}\left(\frac{\xi}{2}\right). \tag{5.87}$$

By defining

$$m_1(\xi) = \frac{1}{\sqrt{2}} \sum_{k\in\mathbb{Z}} v(k)e^{-ik\xi}, \tag{5.88}$$

We find that equation (5.87) is equivalent to

$$\hat{\psi}(\xi) = m_1\left(\frac{\xi}{2}\right)\hat{\varphi}\left(\frac{\xi}{2}\right). \tag{5.89}$$

If we iterate equation (5.76) as we did in deriving equation (5.78), we obtain

$$\hat{\psi}(\xi) = m_1(\frac{\xi}{2})\prod_{j=2}^{\infty} m_0\left(\frac{\xi}{2^j}\right), \tag{5.90}$$

using the normalization that $\hat{\varphi}(0) = 1$.

Note the similarities between equations (5.80) and (3.81), and between equations (5.90) and (3.80) (compare also with Exercise 4.6.2 (ii)).

Observe that m_1 is 2π-periodic and that $m_1(\xi) = \frac{1}{\sqrt{2}}\hat{v}(-\xi)$, where $\hat{v}$ is as in Definition 4.29. The relation (4.57) between $\hat{u}$ and $\hat{v}$ becomes

$$m_1(\xi) = e^{-i\xi}\overline{m_0(\xi + \pi)}. \tag{5.91}$$

Then equation (5.90) gives an explicit formula for the mother wavelet ψ in terms of m_0.

This result suggests that instead of starting with u, as suggested so far, we could also take m_0 as the starting point for the construction of an MRA. Because both approaches are useful, many of our results are given two formulations, one in terms of u and one in terms of m_0. Of course, $u \in \ell^2(\mathbb{Z})$ determines $m_0 \in L^2([-\pi, \pi))$ by equation (5.75), or, equivalently equation (5.84). Conversely, given $m_0 \in L^2([-\pi, \pi))$, we define

$$u(k) = \frac{\sqrt{2}}{2\pi} \int_{-\pi}^{\pi} m_0(\xi)e^{ik\xi}\, d\xi = \sqrt{2}\check{m}_0(-k), \tag{5.92}$$

for $k \in \mathbb{Z}$ and $u = (u(k))_{k \in \mathbb{Z}}$. Then $u \in \ell^2(\mathbb{Z})$ and

$$\frac{1}{\sqrt{2}} \sum_{k \in \mathbb{Z}} u(k)e^{-ik\xi} = \sum_{k \in \mathbb{Z}} \check{m}_0(-k)e^{-ik\xi} = (\check{m}_0)\hat{\,}(\xi) = m_0(\xi),$$

by Lemma 4.31 or Corollary 4.24 iv (here $\hat{\,}$ and $\check{\,}$ are as defined in chapter 4).

By Theorem 4.46, the assumption that u and v generate a first-stage wavelet basis for $\ell^2(\mathbb{Z})$ is equivalent to the system matrix (4.53) being unitary for all θ. In terms of m_0 and m_1, this is equivalent to the matrix

$$\begin{bmatrix} m_0(\xi) & m_1(\xi) \\ m_0(\xi + \pi) & m_1(\xi + \pi) \end{bmatrix} \tag{5.93}$$

being unitary for all ξ. Because the first row of matrix (5.93) must be a vector of length 1 and $m_0(0) = 1$ by (5.82), we obtain $m_1(0) = 0$. By equation (5.89), this implies that $\hat{\psi}(0) = 0$, or, equivalently,

$$\int_{\mathbb{R}} \psi(x)\, dx = 0. \tag{5.94}$$

Hence, the mother wavelet ψ (and therefore every $\psi_{j,k}$, by a change of variables) has mean 0. This is a standard cancellation property for wavelets. It is counterintuitive in view of the expansion $f =$

$\sum_{j,k\in\mathbb{Z}} \langle f, \psi_{j,k}\rangle \psi_{j,k}$ for an arbitrary $f \in L^2(\mathbb{R})$: equation (5.94) appears to imply that any $f \in L^2(\mathbb{R})$ satisfies $\int_{\mathbb{R}} f(x)dx = 0$. However, this is not the case; see Exercise 5.4.2. Note also that by definition of m_1, the condition $m_1(0) = 0$ implies that

$$\sum_{k\in\mathbb{Z}} v(k) = 0,$$

which is the discrete analog of equation (5.94).

We now start to make all this rigorous. We begin by establishing the convergence of the product in equation (5.80).

Lemma 5.37   *Suppose* $m_0 : \mathbb{R} \to \mathbb{C}$ *satisfies* $m_0(0) = 1$, $|m_0(\xi)| \le 1$ *for all* $\xi \in \mathbb{R}$, *and there exist* $\delta > 0$ *and* $C < +\infty$ *such that*

$$|m_0(\xi) - m_0(0)| \le C|\xi|^\delta \quad \text{for all } \xi \in \mathbb{R}. \tag{5.95}$$

For $n \in \mathbb{N}$, *let*

$$G_n(\xi) = \prod_{j=1}^{n} m_0(\xi/2^j).$$

Then $G_n(\xi)$ *converges, as* $n \to +\infty$, *uniformly on every bounded subset of* $\mathbb{R}$, *hence pointwise at every point* $\xi \in \mathbb{R}$.

Proof
For each n, note that

$$|G_{n+1}(\xi) - G_n(\xi)| = \prod_{j=1}^{n} |m_0(\xi/2^j)||m_0(\xi/2^{n+1}) - 1|$$

$$\le |m_0(\xi/2^{n+1}) - 1| \le C|\xi/2^{n+1}|^\delta,$$

by the assumptions $|m_0(\xi)| \le 1$, $m_0(0) = 1$, and relation (5.95). Therefore, if $m > n$, the triangle inequality gives

$$|G_m(\xi) - G_n(\xi)| \le \sum_{j=n}^{m-1} |G_{j+1}(\xi) - G_j(\xi)| \le C \sum_{j=n}^{m-1} |\xi/2^{j+1}|^\delta$$

$$\le C|\xi|^\delta \sum_{j=n}^{m-1} \frac{1}{2^{(j+1)\delta}} \le C2^{-n\delta}|\xi|^\delta \sum_{j=1}^{+\infty} \frac{1}{2^{j\delta}} \le C'2^{-n\delta}|\xi|^\delta,$$

where C' is another constant, depending on δ, because the last series is a convergent geometric series. This estimate shows that on a bounded set of $\mathbb{R}$, the sequence G_n is uniformly Cauchy, and hence converges uniformly. ∎

An example that can be explicitly computed is given in Exercise 5.4.3 (i). We remark that the assumption $|m_0(\xi)| \leq 1$ in Lemma 5.37 is unnecessary (Exercise 5.4.4), but it holds by equation (5.85) in our case and it simplifies the proof.

When inequality (5.95) holds for some $C > 0$, we say that m_0 satisfies a *Lipschitz condition of order δ at* 0. This implies the continuity of m_0 at 0, but relation (5.95) is a stronger condition because it specifies a rate at which $m_0(\xi)$ approachs $m_0(0)$ as ξ approaches 0. Still, condition (5.95) is mild. For example, suppose m_0 is continuously differentiable on $\mathbb{R}$ with a bounded derivative (i.e., there exists some constant C such that $|m_0'(\xi)| \leq C$ for all ξ). By the mean value theorem, $m_0(\xi) - m_0(0) = m_0'(\eta)(\xi - 0)$ for some η between ξ and 0. Hence,

$$|m_0(\xi) - m_0(0)| \leq C|\xi|,$$

so m_0 satisfies a Lipschitz condition of order 1 at 0. The next lemma gives conditions on the scaling sequence u that imply that m_0 defined by identity (5.75) satisfies a Lipschitz condition of positive order at 0.

Lemma 5.38 *Suppose $u = (u(k))_{k \in \mathbb{Z}}$ satisfies*

$$\sum_{k \in \mathbb{Z}} |k|^\epsilon |u(k)| < +\infty \tag{5.96}$$

for some $\epsilon > 0$. Define m_0 by identitiy (5.75). Then m_0 satisfies a Lipschitz condition of order $\delta = min(1, \epsilon)$.

Proof
We assume $\xi \neq 0$ because there is nothing to prove if $\xi = 0$. We first note the elementary inequality $|e^{i\theta} - 1| \leq |\theta|$ for all $\theta \in \mathbb{R}$. To see this, observe that by the triangle inequality, we always have $|e^{i\theta} - 1| \leq 2$, so the result is trivial if $|\theta| \geq 2$. If $|\theta| < 2$, the arc length of the shortest portion of the unit circle connecting $e^{i\theta}$ to $1 = e^{i0}$ is $|\theta|$, which is greater than the straight line distance $|e^{i\theta} - 1|$.

Hence, if $S = \{k \in \mathbb{Z} : |k| \leq 1/|\xi|\}$ and $T = \{k \in \mathbb{Z} : |k| > 1/|\xi|\}$, then

$$|m_0(\xi) - m_0(0)| = \frac{1}{\sqrt{2}} \left| \sum_{k \in \mathbb{Z}} u(k) e^{-ik\xi} - \sum_{k \in \mathbb{Z}} u(k) \right|$$

$$\leq \frac{1}{\sqrt{2}} \sum_{k \in \mathbb{Z}} |u(k)||e^{-ik\xi} - 1| \leq \frac{1}{\sqrt{2}} \sum_{k \in S} |u(k)||k\xi| + \frac{1}{\sqrt{2}} \sum_{k \in T} |u(k)|2$$

$$\leq \frac{1}{\sqrt{2}} \sum_{k \in S} |u(k)||k\xi|^{\delta} + \sqrt{2} \sum_{k \in T} |u(k)||k\xi|^{\delta} \leq \sqrt{2} \sum_{k \in \mathbb{Z}} |u(k)||k|^{\epsilon}|\xi|^{\delta},$$

using $\delta \leq 1$ in the next to last step and $\delta \leq \epsilon$ in the last. This gives condition (5.96) with $C = \sqrt{2} \sum_{k \in \mathbb{Z}} |u(k)||k|^{\epsilon}$, which is finite by assumption.   ∎

The new condition (5.96) on u is stronger than our usual assumption that $u \in \ell^1(\mathbb{Z})$, but it is still weak enough to cover most interesting examples. Under the conditions of Lemmas 5.37, we have established the existence of the product in equation (5.80) (which is defined to be the pointwise limit of the partial products). Now we consider its properties.

Theorem 5.39   *Suppose $m_0 : \mathbb{R} \to \mathbb{C}$ satisfies a Lipschitz condition of order $\delta > 0$ at 0 (i.e., (5.95) holds), $m_0(0) = 1$, m_0 is 2π-periodic, and $|m_0(\xi)|^2 + |m_0(\xi + \pi)|^2 = 1$ for all ξ. Define $\hat{\varphi}(\xi) = \prod_{j=1}^{\infty} m_0(\xi/2^j)$. Then*
   i. $\hat{\varphi}$ satisfies $\hat{\varphi}(\xi) = m_0(\xi/2)\hat{\varphi}(\xi/2)$ for all $\xi \in \mathbb{R}$.
   ii. $\hat{\varphi} \in L^2(\mathbb{R})$.
         Let $\varphi = (\hat{\varphi})^{\vee}$. Let $u = (u(k))_{k \in \mathbb{Z}}$ be such that equation (5.75) holds (i.e., define $u(k)$ by equation (5.92)).
   iii. If $u \in \ell^1(\mathbb{Z})$, then φ satisfies the scaling relation (5.72).
   iv. The function $\hat{\varphi}$ is continuous at 0.

Proof
For any $\xi \in \mathbb{R}$,

$$m_0(\frac{\xi}{2})\hat{\varphi}\left(\frac{\xi}{2}\right) = m_0\left(\frac{\xi}{2}\right)\prod_{j=1}^{\infty} m_0\left(\frac{\xi}{2 \cdot 2^j}\right) = \prod_{j=1}^{\infty} m_0\left(\frac{\xi}{2^j}\right) = \hat{\varphi}(\xi).$$

So part i is proved.

For part ii, it follows from Lemma 5.37 that the product defining $\hat{\varphi}$ converges uniformly on bounded sets. For $n \in \mathbb{N}$, set $G_n(\xi) =$

$\prod_{j=1}^{n} m_0(\xi/2^j)$ and

$$I_n = \int_{-2^n\pi}^{2^n\pi} \left| G_n(\xi) \right|^2 \, d\xi.$$

Observe that $m_0((\xi - 2^n\pi)/2^j) = m_0(\xi/2^j)$ for $j = 1, 2, \ldots, n-1$, by the 2π-periodicity of m_0, and hence G_{n-1} is $2^n\pi$-periodic. For each $n \geq 2$, $G_n(\xi) = G_{n-1}(\xi)m_0(\xi/2^n)$, so

$$I_n = \int_{-2^n\pi}^{0} \left| G_{n-1}(\xi) \right|^2 |m_0(\xi/2^n)|^2 \, d\xi + \int_{0}^{2^n\pi} |G_{n-1}(\xi)|^2 |m_0(\xi/2^n)|^2 \, d\xi$$

$$= \int_{0}^{2^n\pi} \left| G_{n-1}(y) \right|^2 \left(|m_0(y/2^n - \pi)|^2 + |m_0(y/2^n)|^2 \right) \, dy,$$

by the change of variables $y = \xi + 2^n\pi$ in the first integral and the fact that G_{n-1} is $2^n\pi$-periodic. By assumption, the term inside the parentheses in the last integral is identically 1. Hence,

$$I_n = \int_{0}^{2^{n-1}\pi} \left| G_{n-1}(y) \right|^2 \, dy + \int_{2^{n-1}\pi}^{2^n\pi} \left| G_{n-1}(y) \right|^2 \, dy$$

$$= \int_{-2^{n-1}\pi}^{2^{n-1}\pi} \left| G_{n-1}(\xi) \right|^2 \, d\xi = I_{n-1},$$

using the change of variables $\xi = y - 2^n\pi$ in the second integral and the $2^n\pi$-periodicity of G_{n-1}. Therefore

$$\int_{-2^n\pi}^{2^n\pi} \left| G_n(\xi) \right|^2 \, d\xi = I_n = I_{n-1} = I_{n-2} = \cdots = I_1$$

$$= \int_{-2\pi}^{2\pi} \left| m_0(\xi/2) \right|^2 \, d\xi \leq 4\pi,$$

because $|m_0(\xi)| \leq 1$ for all ξ. Note that

$$|\hat{\varphi}(\xi)| = \left| \prod_{j=1}^{n} m_0(\xi/2^j) \right| \left| \prod_{j=n+1}^{\infty} m_0(\xi/2^j) \right|$$

$$= |G_n(\xi)| \prod_{j=n+1}^{\infty} |m_0(\xi/2^j)| \leq |G_n(\xi)|,$$

because $|m_0(\xi)| \leq 1$. Hence,

$$\int_{-2^n\pi}^{2^n\pi} \left| \hat{\varphi}(\xi) \right|^2 \, d\xi \leq \int_{-2^n\pi}^{2^n\pi} \left| G_n(\xi) \right|^2 \, d\xi \leq 4\pi,$$

by the estimate above. Letting $n \to \infty$, it follows (compare with Exercise 5.1.13) that $\hat{\varphi} \in L^2(\mathbb{R})$.

By part ii and Fourier inversion (Theorem 5.24), $\varphi = (\hat{\varphi})^{\vee}$ is defined and belongs to $L^2(\mathbb{R})$. Using Exercise 5.4.6(i) and equation (5.73),

$$\left(\sum_{k \in \mathbb{Z}} u(k) \varphi_{1,k} \right)^{\wedge}(\xi) = \frac{1}{\sqrt{2}} \sum_{k \in \mathbb{Z}} u(k) e^{-ik\xi/2} \hat{\varphi}\left(\frac{\xi}{2} \right)$$

$$= m_0 \left(\frac{\xi}{2} \right) \hat{\varphi}\left(\frac{\xi}{2} \right) = \hat{\varphi}(\xi),$$

by part i. Taking the inverse Fourier transform of the left and right sides of this last equation shows that φ satisfies the scaling equation, proving part iii.

By equation (5.95), m_0 is continuous at 0, hence so are the partial products G_n. By Lemma 5.37, these functions converge uniformly on bounded sets to $\hat{\varphi}$. By a standard analysis result, this uniform convergences implies that the limit $\hat{\varphi}$ is also continuous at 0. ∎

The corresponding statement in terms of u is as follows.

Corollary 5.40   *Suppose $u = (u(k))_{k \in \mathbb{Z}}$ is a sequence such that the set $\{R_{2k}u\}_{k \in \mathbb{Z}}$ is orthonormal in $\ell^2(\mathbb{Z})$, $\sum_{k \in \mathbb{Z}} u(k) = \sqrt{2}$, and $\sum_{k \in \mathbb{Z}} |k|^{\epsilon} |u(k)| < \infty$ for some $\epsilon > 0$. Define m_0 by equation (5.75) and $\hat{\varphi}$ by equation (5.80). Then $\hat{\varphi} \in L^2(\mathbb{R})$, $\hat{\varphi}$ is continuous at 0, and $\varphi = (\hat{\varphi})^{\vee}$ satisfies the scaling relation (5.72) with scaling sequence u.*

Proof
The assumption $\sum_{k \in \mathbb{Z}} |k|^{\epsilon} |u(k)| < \infty$ implies that $u \in \ell^1(\mathbb{Z})$, which in turn implies that m_0 is continuous (Exercise 4.4.9). By definition, m_0 is 2π-periodic. The assumption that $\{R_{2k}u\}_{k \in \mathbb{Z}}$ is orthonormal in $\ell^2(\mathbb{Z})$ is equivalent to the condition $|m_0(\xi)|^2 + |m_0(\xi + \pi)|^2 = 1$ for all ξ. We have $m_0(0) = 1$ since $\sum_{k \in \mathbb{Z}} u(k) = \sqrt{2}$. By Lemma 5.38, m_0 satisfies a Lipschitz condition of order $\delta = \min(1, \epsilon)$. Thus, Theorem 5.39 applies and yields all conclusions. ∎

Under the conditions on u in Corollary 5.40, we can make our heuristic argument above precise and prove the uniqueness in $L^2(\mathbb{R})$, up to a constant multiple, of the solution to the scaling equation (Exercise 5.4.7).

Recall that our objective is to obtain an MRA from a sequence u such that $\{R_{2k}u\}_{k\in\mathbb{Z}}$ is an orthonormal set in $\ell^2(\mathbb{Z})$. So far we have shown that if we add the conditions that $\sum_{k\in\mathbb{Z}} u(k) = \sqrt{2}$, and $\sum_{k\in\mathbb{Z}} |k|^\epsilon |u(k)| < \infty$ for some $\epsilon > 0$, then we can find a solution $\varphi \in L^2(\mathbb{R})$ to the scaling equation with scaling sequence u. However, another requirement in the definition of an MRA (Definition 5.30) is that the set $\{\varphi_{0,k}\}_{k\in\mathbb{Z}}$ is orthonormal in $L^2(\mathbb{R})$. One might think that this follows from the orthonormality of $\{R_{2k}u\}_{k\in\mathbb{Z}}$ in $\ell^2(\mathbb{Z})$, in the same way that the orthonormality of $\{\psi_{0,k}\}_{k\in\mathbb{Z}}$ in $L^2(\mathbb{R})$ followed from the orthonormality of $\{R_{2k}v\}_{k\in\mathbb{Z}}$ in $\ell^2(\mathbb{Z})$ in the proof of Lemma 5.34. However, a closer look shows that this proof assumed and used the orthonormality of $\{\varphi_{0,k}\}_{k\in\mathbb{Z}}$. In fact, the orthonormality of the set $\{\varphi_{0,k}\}_{k\in\mathbb{Z}}$ does not necessarily follow under these conditions, as Example 5.41 shows.

Example 5.41
Define a sequence $u \in \ell^1(\mathbb{Z})$ by

$$u(k) = \begin{cases} \frac{1}{\sqrt{2}} & \text{if } k = 0 \text{ or } k = 3 \\ 0 & \text{otherwise.} \end{cases}$$

Then it is easy to check that $\{R_{2k}u\}_{k\in\mathbb{Z}}$ is orthonormal in $\ell^2(\mathbb{Z})$ because the even integer translates are nonzero on disjoint sets. Also $\sum_{k\in\mathbb{Z}} u(k) = \sqrt{2}$ and the condition $\sum_{k\in\mathbb{Z}} |k|^\epsilon |u(k)| < \infty$ is trivial for any $\epsilon > 0$ because the sum is finite. So the assumptions of Corollary 5.40 hold. Let

$$\varphi(x) = \begin{cases} 1/3 & \text{if } 0 \le x < 3 \\ 0 & \text{if } x < 0 \text{ or } x \ge 3. \end{cases} \tag{5.97}$$

Then φ is a solution of the scaling equation (5.72) for this u, since $\varphi(x) = \varphi(2x) + \varphi(2x - 3)$, or, equivalently,

$$\varphi = \frac{1}{\sqrt{2}} \varphi_{1,0} + \frac{1}{\sqrt{2}} \varphi_{1,3}.$$

Since $\varphi \in L^2(\mathbb{R})$ and $\hat{\varphi}(0) = \int_\mathbb{R} \varphi(x)\,dx = 1$, by Exercise 5.4.7 φ is the unique L^2 solution to the scaling equation satisfying $\hat{\varphi}(0) = 1$ (also see Exercise 5.4.5). However, the set $\{\varphi_{0,k}\}_{k\in\mathbb{Z}}$ is not orthogonal; for example,

$$\langle \varphi, \varphi_{0,1} \rangle = \int_1^3 \frac{1}{3}\frac{1}{3}\, dx = \frac{2}{9}.$$

Fortunately, if we put an additional restriction (see condition (5.102) below) on the sequence u, we can obtain the orthonormality of $\{\varphi_{0,k}\}_{k\in\mathbb{Z}}$. This is the most delicate part of the theory. We begin by finding criteria for the orthonormality of any set of the form $\{\varphi_{0,k}\}_{k\in\mathbb{Z}}$.

Lemma 5.42 *Suppose* $\varphi \in L^2(\mathbb{R})$. *Then the following conditions are equivalent:*

i. *The set* $\{\varphi_{0,k}\}_{k\in\mathbb{Z}}$ *is orthonormal in* $L^2(\mathbb{R})$.

ii.

$$\int_{\mathbb{R}} |\hat{\varphi}(\xi)|^2 e^{ik\xi}\, d\xi = \begin{cases} 2\pi & \text{if } k = 0 \\ 0 & \text{if } k \neq 0. \end{cases}$$

iii.

$$\sum_{k\in\mathbb{Z}} |\hat{\varphi}(\xi + 2\pi k)|^2 = 1 \text{ a.e.}$$

Proof
By a change of variables, $\langle \varphi_{0,k}, \varphi_{0,\ell} \rangle = \langle \varphi, \varphi_{\ell-k} \rangle$, so the orthonormality of $\{\varphi_{0,k}\}_{k\in\mathbb{Z}}$ is equivalent to the conditions that $\langle \varphi, \varphi_{0,k} \rangle$ is 1 if $k = 0$ and 0 otherwise. By Parseval's relation (Theorem 5.22 (i)),

$$\langle \varphi, \varphi_{0,k} \rangle = (2\pi)^{-1} \langle \hat{\varphi}, \hat{\varphi}_{0,k} \rangle = (2\pi)^{-1} \int_{\mathbb{R}} |\hat{\varphi}(\xi)|^2 e^{ik\xi}\, d\xi,$$

for $k \in \mathbb{Z}$, since $(\hat{\varphi}_{0,k})(\xi) = e^{-ik\xi}\hat{\varphi}(\xi)$. Hence conditions i and ii are equivalent.

We leave the equivalence of conditions ii and iii (which we will not use later) as Exercise 5.4.8 . ∎

It is convenient to introduce the following standard notation.

Definition 5.43 *For any set* $E \subseteq \mathbb{R}$, *define the function* $\chi_E : \mathbb{R} \to \mathbb{R}$ *by*

$$\chi_E(x) = \begin{cases} 1 & \text{if } x \in E \\ 0 & \text{if } x \notin E. \end{cases}$$

We call χ_E *the characteristic function of* E.

Our approach to proving the orthonormality of $\{\varphi_{0,k}\}_{k\in\mathbb{Z}}$ is to obtain φ as a limit of a sequence $\{\varphi_n\}_{n=0}^{+\infty}$ such that for each n, $\{(\varphi_n)_{0,k}\}_{k\in\mathbb{Z}}$ is orthonormal, where $(\varphi_n)_{0,k}(x) = \varphi_n(x - k)$.

Lemma 5.44 *Suppose $m_0 : \mathbb{R} \to \mathbb{C}$ is 2π-periodic and satisfies $|m_0(\xi)|^2 + |m_0(\xi + \pi)|^2 = 1$ for all ξ. Define*

$$\hat{\varphi}_0 = \chi_{[-\pi,\pi)},$$

and, for $n \geq 1$, inductively define

$$\hat{\varphi}_n(\xi) = m_0\left(\frac{\xi}{2}\right)\hat{\varphi}_{n-1}\left(\frac{\xi}{2}\right).$$

Then for each $n \geq 1$,

$$\hat{\varphi}_n(\xi) = \chi_{[-2^n\pi, 2^n\pi)}(\xi)\prod_{j=1}^{n} m_0\left(\frac{\xi}{2^j}\right). \tag{5.98}$$

Define $\varphi_n = (\hat{\varphi}_n)^{\vee}$ for each $n \geq 1$. Then, for each n,

$$\{(\varphi_n)_{0,k}\}_{k\in\mathbb{Z}} \text{ is an orthonormal set in } L^2(\mathbb{R}). \tag{5.99}$$

Proof

We obtain equation (5.98) from the definition of $\hat{\varphi}_n$ and a simple inductive argument (Exercise 5.4.9).

To prove statement (5.99), we also proceed by induction, using an argument similar to the proof of Theorem 5.39 ii (in fact, that argument is the special case $k = 0$ of the following argument). Condition ii in Lemma 5.42 is easy to check for φ_0, so statement (5.99) holds for $n = 0$. Now suppose statement (5.99) holds for $n-1$. For $n \in \mathbb{N}$, set $G_n(\xi) = \prod_{j=1}^{n} m_0(\xi/2^j)$. Since $G_n(\xi) = m_0(\xi/2^n)G_{n-1}(\xi)$,

$$\int_{\mathbb{R}} |\hat{\varphi}_n(\xi)|^2 e^{ik\xi} d\xi = \int_{-2^n\pi}^{2^n\pi} \prod_{j=1}^{n}\left|m_0\left(\frac{\xi}{2^j}\right)\right|^2 e^{ik\xi} d\xi$$

$$= \left(\int_{-2^n\pi}^{0} + \int_{0}^{2^n\pi}\right)\left|m_0\left(\frac{\xi}{2^n}\right)\right|^2 |G_{n-1}(\xi)|^2 e^{ik\xi} d\xi.$$

In the integral over $[-2^n\pi, 0]$, we change variables (let $y = \xi + 2^n\pi$) and use the facts that $G_{n-1}(\xi)$ is $2^n\pi$-periodic (as in the proof of Theorem 5.39 ii) and $e^{ik\xi}$ is 2π-periodic. We then obtain

$$\int_{\mathbb{R}} |\hat{\varphi}_n(\xi)|^2 e^{ik\xi} d\xi$$

$$= \int_0^{2^n \pi} \left[\left| m_0 \left(\frac{y}{2^n} - \pi \right) \right|^2 + \left| m_0 \left(\frac{y}{2^n} \right) \right|^2 \right] |G_{n-1}(y)|^2 e^{iky} \, dy.$$

By assumption, the sum in brackets is identically 1. We break the region of integration into two intervals, $[0, 2^{n-1}\pi]$ and $[2^{n-1}\pi, 2^n\pi]$. In the integral over $[2^{n-1}\pi, 2^n\pi]$, we change variables, setting $\xi = y - 2^n\pi$. Note that the integrand is unchanged, by the $2^n\pi$-periodicity of G_{n-1} and e^{iky}. Hence, we obtain

$$\int_{\mathbb{R}} |\hat{\varphi}_n(\xi)|^2 e^{ik\xi} d\xi = \int_{-2^{n-1}\pi}^{2^{n-1}\pi} |G_{n-1}(\xi)|^2 e^{ik\xi} \, d\xi = \int_{\mathbb{R}} |\hat{\varphi}_{n-1}(\xi)|^2 e^{ik\xi} \, d\xi,$$

by equation (5.98). This last integral is 2π when $k = 0$ and 0 otherwise, by the induction hypothesis and Lemma 5.42. Thus, we have obtained the condition in Lemma 5.42 ii for $\hat{\varphi}_n$. Hence, Lemma 5.42 implies statement (5.99) for n, completing the induction. $\blacksquare$

Observe that by equation (5.98), $\hat{\varphi}_n(\xi)$ converges pointwise as $n \to \infty$ to $\hat{\varphi}(\xi)$ as defined in equation (5.80). If we could show that

$$\int_{\mathbb{R}} |\hat{\varphi}(\xi)|^2 e^{ik\xi} \, d\xi = \lim_{n \to \infty} \int_{\mathbb{R}} |\hat{\varphi}_n(\xi)|^2 e^{ik\xi} \, d\xi, \qquad (5.100)$$

for all $k \in \mathbb{Z}$, then we could conclude by Theorem 5.42 that $\{\varphi_{0,k}\}_{k \in \mathbb{Z}}$ is orthonormal, since

$$\lim_{n \to \infty} \int_{\mathbb{R}} |\hat{\varphi}_n(\xi)|^2 e^{ik\xi} \, d\xi = \lim_{n \to \infty} \begin{cases} 2\pi & \text{if } k = 0 \\ 0 & \text{if } k \neq 0 \end{cases} = \begin{cases} 2\pi & \text{if } k = 0 \\ 0 & \text{if } k \neq 0, \end{cases} \qquad (5.101)$$

using Lemma 5.44 and Lemma 5.42. One might think that equation (5.100) always holds because it is just a matter of letting $n \to \infty$. However, we know from Example 5.41 that this cannot always work. In fact, interchanging a limit with an integral is a delicate matter (see Exercise 5.4.11 for some simple examples for which it cannot be done). Various conditions under which it can be done are usually covered in a beginning graduate course in real analysis. We accept without proof the following fundamental result, the proof of which can be found in any text on Lebesgue integration.

Theorem 5.45 (Lebesgue's dominated convergence theorem, or DCT) *Suppose* $\{f_n\}_{n=1}^\infty$ *is a sequence of functions that converges a.e. to a function* f. *Suppose there exists a function* $g \geq 0$ *such that*

$\int_{\mathbb{R}} g(x)\, dx < +\infty$ *and*

$$|f_n(x)| \le g(x) \text{ for all } n \in \mathbb{N} \text{ and a.e. } x \in \mathbb{R}.$$

Then

$$\lim_{n \to +\infty} \int_{\mathbb{R}} f_n(x)\, dx = \int_{\mathbb{R}} f(x)\, dx.$$

The assumption in Theorem 5.45 is that all of the functions f_n are simultaneously dominated at an a.e. point by a function g that has a finite integral. In some sense, this means that all of the action is under control, which suggests why the conclusion holds.

We know from Example 5.41 that the assumptions of Theorem 5.39 are not enough to guarantee the orthonormality of $\{\varphi_{0,k}\}_{k \in \mathbb{Z}}$. Thus we need some additional assumption, namely condition (5.102). This condition is not so intuitive, but its role is to provide us with an estimate that allows us to apply the DCT to justify equation (5.100). Condition (5.102) is not the sharpest possible, but it is relatively easy to check and is sufficient for the applications we consider here.

Lemma 5.46 *Suppose $m_0 : \mathbb{R} \to \mathbb{C}$ satisfies a Lipschitz condition of order $\delta > 0$ at 0 (i.e., inequality (5.95) holds), $m_0(0) = 1$, m_0 is 2π-periodic, $|m_0(\xi)|^2 + |m_0(\xi + \pi)|^2 = 1$ for all ξ, and*

$$\inf_{|\xi| \le \pi/2} |m_0(\xi)| > 0. \tag{5.102}$$

Define $\hat{\varphi}(\xi) = \prod_{j=1}^{\infty} m_0(\xi/2^j)$. Let $\varphi = (\hat{\varphi})^{\vee}$. Then $\{\varphi_{0,k}\}_{k \in \mathbb{Z}}$ is an orthonormal set in $L^2(\mathbb{R})$.

Proof
The pointwise convergence of $\prod_{j=1}^{+\infty} m_0(\xi/2^j)$ follows from Lemma 5.37, so $\hat{\varphi}$ is defined. Define $\hat{\varphi}_n$ by equation (5.98). This definition shows that $\hat{\varphi}_n(\xi)$ converges pointwise to $\hat{\varphi}(\xi)$. By Theorem 5.39, $\hat{\varphi} \in L^2(\mathbb{R})$.

We show that there exists a constant C_1 independent of ξ such that for all $n \in \mathbb{N}$,

$$|\hat{\varphi}_n(\xi)| \le C_1 |\hat{\varphi}(\xi)|. \tag{5.103}$$

Assuming this inequality momentarily, then

$$\left| |\hat{\varphi}_n(\xi)|^2 e^{ik\xi} \right| \le C_1^2 |\hat{\varphi}(\xi)|^2,$$

which is a function (that will play the role of g in the DCT) with a finite integral since $\hat{\varphi} \in L^2(\mathbb{R})$. Note that $|\hat{\varphi}_n(\xi)|^2 e^{ik\xi}$ converges pointwise as $n \to \infty$ to $|\hat{\varphi}(\xi)|^2 e^{ik\xi}$ (by the pointwise convergence of $\hat{\varphi}_n$ to $\hat{\varphi}$), hence equation (5.100) follows from the DCT. Therefore equation (5.101) implies the orthonormality of $\{\varphi_{0,k}\}_{k\in\mathbb{Z}}$, by Lemma 5.42.

Observe that relation (5.103) is trivial if $|\xi| > 2^n\pi$ because the left side is 0. By definition of $\hat{\varphi}$ and $\hat{\varphi}_n$, for $|\xi| \le 2^n\pi$,

$$|\hat{\varphi}(\xi)| = \prod_{j=1}^{n} |m_0\left(\frac{\xi}{2^j}\right)| \cdot \prod_{j=n+1}^{\infty} |m_0\left(\frac{\xi}{2^j}\right)| = |\hat{\varphi}_n(\xi)| \prod_{j=n+1}^{\infty} |m_0\left(\frac{\xi}{2^j}\right)|. \tag{5.104}$$

Note that

$$\prod_{j=n+1}^{\infty} |m_0\left(\frac{\xi}{2^j}\right)| = \prod_{j=1}^{\infty} |m_0\left(\frac{2^{-n}\xi}{2^j}\right)| = |\hat{\varphi}(2^{-n}\xi)|. \tag{5.105}$$

By equations (5.104) and (5.105), inequality (5.103) is reduced to showing that $|\hat{\varphi}(2^{-n}\xi)| \ge 1/C_1 > 0$ (with C_1 independent of n) for all $\xi \in [-2^n\pi, 2^n\pi]$, or, equivalently, that

$$|\hat{\varphi}(\xi)| \ge \frac{1}{C_1} > 0 \quad \text{for all} \quad \xi \in [-\pi, \pi]. \tag{5.106}$$

For $|\xi| \le \pi$, we have

$$|m_0\left(\frac{\xi}{2^j}\right) - 1| \le C\frac{|\xi|^\delta}{2^{j\delta}} \le C\frac{\pi^\delta}{2^{j\delta}},$$

by the Lipschitz assumption on m_0 and the condition $m_0(0) = 1$. Select N sufficiently large that $C\pi^\delta/2^{N\delta} \le 1/2$. For $j > N$, $C\pi^\delta/2^{j\delta} \le C_2\pi^\delta/2^{N\delta} \le 1/2$, by our choice of N. For $0 \le x \le 1/2$, $1 - x \ge e^{-2x}$ (Exercise 5.4.12) , so by the triangle inequality we obtain

$$|m_0\left(\frac{\xi}{2^j}\right)| \ge 1 - |1 - m_0\left(\frac{\xi}{2^j}\right)| \ge 1 - C\frac{\pi^\delta}{2^{j\delta}} \ge e^{-2C\pi^\delta 2^{-j\delta}}. \tag{5.107}$$

Let

$$C_2 = \inf_{|\xi| \le \pi/2} |m_0(\xi)|,$$

which is positive by assumption. For $\xi \in [-\pi, \pi]$ and $j \in \mathbb{N}$, $|\xi/2^j| \le \pi/2$, so $|m_0(\xi/2^j)| \ge C_2$. Using this result and relation (5.107) yields, for $|\xi| \leq \pi$,

$$|\hat{\varphi}(\xi)| = \prod_{j=1}^{N} \left| m_0\left(\frac{\xi}{2^j}\right) \right| \cdot \prod_{j=N+1}^{\infty} \left| m_0\left(\frac{\xi}{2^j}\right) \right| \geq C_2^N \prod_{j=N+1}^{\infty} e^{-2C\pi^\delta 2^{-j\delta}}$$

$$= C_2^N e^{-2C\pi^\delta \sum_{j=N+1}^{\infty} 2^{-j\delta}} \equiv \frac{1}{C_1} > 0,$$

because the series is convergent. We have proved relation (5.106), and therefore relation (5.103). ∎

This result gives us an explanation of our normalization $|\hat{\varphi}(0)| = 1$. (For another explanation, based only on the orthonormality of $\{\varphi_{0,k}\}_{k\in\mathbb{Z}}$ and the scaling relation (5.72), see Exercise 5.4.14.) Namely, by Lemma 5.46 this normalization yields a function with L^2 norm 1, so any other choice would yield a different L^2 norm, contradicting the orthonormality of $\{\varphi_{0,k}\}_{k\in\mathbb{Z}}$.

For Example 5.41,

$$m_0(\xi) = \frac{1}{2}(1 + e^{-3i\xi}),$$

which is 0 at $\xi = \pi/3$, and hence fails to satisfy condition (5.102). This example shows that we cannot replace $\pi/2$ in relation (5.102) with any number less than $\pi/3$, which indicates that this condition is somewhat delicate. We also remark that m_0 is continuous, under the condition $u \in \ell^1(\mathbb{R})$, in which case condition (5.102) is equivalent to the condition that m_0 has no zeros on $[-\pi/2, \pi/2]$ (recall that a continuous function on a compact set attains its infimum on that set).

Lemma 5.46 represents the most difficult work going into our construction of an MRA. With this lemma we can define $\hat{\varphi}$ by equation (5.80) and obtain the orthonormality of $\{\varphi_{0,k}\}_{k\in\mathbb{Z}}$ and the scaling relation (5.72). By defining V_j by equation (5.50), properties i, ii, and iii in Definition 5.30 of a multiresolution analysis follow easily. However, we still need to consider properties iv and v. Lemma 5.4.7 says that the trivial intersection property iv of Definition 5.30 is redundant; it follows from properties ii and iii.

Lemma 5.47 *Suppose $\varphi \in L^2(\mathbb{R})$, and for each $j \in \mathbb{Z}$, $\{\varphi_{j,k}\}_{k\in\mathbb{Z}}$ is an orthonormal set. Define $\{V_j\}_{j\in\mathbb{Z}}$ by equation (5.50). Then*

$$\bigcap_{j\in\mathbb{Z}} V_j = \{0\}.$$

Proof

Suppose $f \in \cap_{j \in \mathbb{Z}} V_j$. Let $\epsilon > 0$. Since $f \in L^2(\mathbb{R})$, by Exercise 5.1.13 we can select R large enough that

$$\int_{\{x : |x| > R\}} |f(x)|^2 \, dx < \epsilon^2.$$

Let

$$g(x) = \begin{cases} f(x) & \text{if } |x| \leq R \\ 0 & \text{if } |x| > R. \end{cases}$$

By choice of R and g,

$$\|f - g\| < \epsilon.$$

For each $j \in \mathbb{Z}$, define the orthogonal projection operator $P_j : L^2(\mathbb{R}) \to V_j$ by equation (5.66). By Lemmas 4.7 and 4.8,

$$\|P_j(h)\|^2 = \sum_{k \in \mathbb{Z}} |\langle h, \varphi_{j,k} \rangle|^2 \leq \|h\|^2,$$

for any $h \in L^2(\mathbb{R})$. Also, by Lemma 4.14 iii, $P_j(f) = f$ since $f \in V_j$ by assumption. Therefore,

$$\|f - P_j(g)\| = \|P_j(f - g)\| \leq \|f - g\| < \epsilon, \tag{5.108}$$

for any j.

Since $g(x) = 0$ for $|x| > R$, we have

$$|\langle g, \varphi_{j,k} \rangle|^2 = \left| \int_{\mathbb{R}} f(x) \chi_{[-R,R]}(x) \overline{\varphi_{j,k}(x)} \, dx \right|^2 \leq \|f\|^2 \|\chi_{[-R,R]} \overline{\varphi_{j,k}}\|^2,$$

by the Cauchy-Schwarz inequality (5.3). However,

$$\|\chi_{[-R,R]} \overline{\varphi_{j,k}}\|^2 = \int_{-R}^{R} |2^{j/2} \varphi(2^j x - k)|^2 \, dx = \int_{-k-2^j R}^{-k+2^j R} |\varphi(y)|^2 \, dy,$$

by the change of variable $y = 2^j x - k$. Now select $J \in \mathbb{Z}$ sufficiently negative such that $2^J R < 1/2$. Then for $j < J$, we have from the above estimates that

$$\|P_j(g)\|^2 = \sum_{k \in \mathbb{Z}} |\langle g, \varphi_{j,k} \rangle|^2 \leq \|f\|^2 \sum_{k \in \mathbb{Z}} \int_{-k-2^j R}^{-k+2^j R} |\varphi(y)|^2 \, dy$$

$$= \|f\|^2 \int_{\mathbb{R}} \chi_{\cup_{k \in \mathbb{Z}} [-k-2^j R, -k+2^j R]}(y) |\varphi(y)|^2 \, dy,$$

since the intervals $[-k - 2^j R, -k + 2^j R]$ do not overlap because $2^j R < 1/2$ for $j < J$. Let

$$h_j(y) = \chi_{\cup_{k \in \mathbb{Z}}[-k-2^j R, -k+2^j R]}(y)|\varphi(y)|^2.$$

Then $h_j(y) \to 0$ a.e. as $j \to -\infty$. Note that each h_j satisfies $|h_j(y)| \le |\varphi(y)|^2$, and $\int_{\mathbb{R}} |\varphi(y)|^2 \, dy < +\infty$ since $\varphi \in L^2(R)$. Therefore the DCT (Theorem 5.45) applies and shows that

$$\int_{\mathbb{R}} \chi_{\cup_{k \in \mathbb{Z}}[-k-2^j R, -k+2^j R]}(y)|\varphi(y)|^2 \, dy \to 0 \text{ as } j \to -\infty.$$

In particular, by the estimates just discussed, we can select j so that $\|P_j(g)\| < \epsilon$.

Combining this inequality with relation (5.108), the triangle inequality (5.4) gives us

$$\|f\| \le \|f - P_j(g)\| + \|P_j(g)\| < \epsilon + \epsilon = 2\epsilon.$$

Since $\epsilon > 0$ is arbitrary, this shows that $\|f\| = 0$, and hence $f = 0$. ∎

The density property v of Definition 5.30 is not quite automatic, but it holds under mild conditions.

Lemma 5.48 *Suppose $\varphi \in L^2(\mathbb{R})$ is such that $\hat{\varphi}$ is bounded, $\hat{\varphi}$ is continuous at 0, and $\hat{\varphi}(0) = 1$. Also, suppose that for each $j \in \mathbb{Z}$, $\{\varphi_{j,k}\}_{k \in \mathbb{Z}}$ is an orthonormal set. Define $\{V_j\}_{j \in \mathbb{Z}}$ by equation (5.50). Then*

$$\bigcup_{j \in \mathbb{Z}} V_j \text{ is dense in } L^2(\mathbb{R}).$$

Proof
Suppose $f \in L^2(\mathbb{R})$. Let $\epsilon > 0$. By Plancherel's formula (Theorem 5.22 ii), $\hat{f} \in L^2(\mathbb{R})$, so there exists R large enough (see Exercise 5.1.13) so that

$$\int_{\{\xi : |\xi| > R\}} |\hat{f}(\xi)|^2 \, d\xi < \epsilon^2.$$

Define $\hat{g}$ by

$$\hat{g}(\xi) = \begin{cases} \hat{f}(\xi) & \text{if } |\xi| \le R \\ 0 & \text{if } |\xi| > R, \end{cases}$$

and let $g = (\hat{g})^{\vee}$. Then by Plancherel's formula again,

$$\|f - g\| = \frac{1}{\sqrt{2\pi}} \|\hat{f} - \hat{g}\| < \frac{\epsilon}{\sqrt{2\pi}} < \epsilon. \qquad (5.109)$$

Define the orthogonal projection operator $P_j : L^2(\mathbb{R}) \to V_j$ by equation (5.66). Then $P_j(g) \in V_j$ (Lemma 4.14 i), and $g - P_j(g)$ is orthogonal to every element of V_j (Lemma 4.14 iv). Hence (by equation (1.44)),

$$\|g\|^2 = \|g - P_j(g) + P_j(g)\|^2 = \|g - P_j(g)\|^2 + \|P_j(g)\|^2. \qquad (5.110)$$

We claim that $\|P_j(g)\| \to \|g\|$ as $j \to +\infty$. To see this, first note that, by Parseval's relation (Theorem 5.22 i),

$$|\langle g, \varphi_{j,k} \rangle|^2 = \frac{1}{(2\pi)^2} |\langle \hat{g}, \hat{\varphi}_{j,k} \rangle|^2.$$

By changing variables in the integral,

$$\hat{\varphi}_{j,k}(\xi) = \int_{\mathbb{R}} 2^{j/2} \varphi(2^j x - k) e^{-ix\xi} \, d\xi = 2^{-j/2} e^{-ik\xi/2^j} \hat{\varphi}\left(\frac{\xi}{2^j}\right).$$

Hence,

$$\begin{aligned}
|\langle g, \varphi_{j,k} \rangle|^2 &= \frac{1}{(2\pi)^2} 2^{-j} \left| \int_{\mathbb{R}} \hat{g}(\xi) \overline{\hat{\varphi}(\xi/2^j)} e^{ik\xi/2^j} \, d\xi \right|^2 \\
&= \frac{1}{(2\pi)^2} 2^j \left| \int_{\mathbb{R}} \hat{g}(2^j y) \overline{\hat{\varphi}(y)} e^{iky} \, dy \right|^2 \\
&= \frac{1}{(2\pi)^2} 2^j \left| \sum_{\ell \in \mathbb{Z}} \int_{(2\ell-1)\pi}^{(2\ell+1)\pi} \hat{g}(2^j y) \overline{\hat{\varphi}(y)} e^{iky} \, dy \right|^2 \\
&= 2^j \left| \sum_{\ell \in \mathbb{Z}} \frac{1}{2\pi} \int_{-\pi}^{\pi} \hat{g}(2^j(\theta + 2\pi\ell)) \overline{\hat{\varphi}(\theta + 2\pi\ell)} e^{ik\theta} \, d\theta \right|^2,
\end{aligned}$$

which is 2^j times the square of the $-k^{\text{th}}$ Fourier coefficient on $[-\pi, \pi)$ of

$$H(\theta) = \sum_{\ell \in \mathbb{Z}} \hat{g}(2^j(\theta + 2\pi\ell)) \overline{\hat{\varphi}(\theta + 2\pi\ell)}. \qquad (5.111)$$

Hence, by Corollary 4.24 ii (Plancherel's formula for Fourier series), we obtain

$$\|P_j(g)\|^2 = \sum_{k \in \mathbb{Z}} |\langle g, \varphi_{j,k} \rangle|^2 = 2^j \frac{1}{2\pi} \int_{-\pi}^{\pi} |H(\theta)|^2 \, d\theta.$$

We select J such that $2^J > R/\pi$. If $j > J$, we claim that at every point θ, there is at most one term in the sum on the right side of equation (5.111) which is not zero. To see this, suppose at some point θ that $\hat{g}(2^j(\theta+2\pi n)) \neq 0$ and $\hat{g}(2^j(\theta+2\pi m)) \neq 0$. Since $\hat{g}(\xi) = 0$ for $|\xi| > R$, this implies $|\theta + 2\pi n| \leq R/2^j$ and $|\theta + 2\pi m| \leq R/2^j$. Then by the triangle inequality, for $j > J$,

$$|2\pi(m - n)| \leq |2\pi m + \theta| + |-\theta - 2\pi n| \leq \frac{R}{2^j} + \frac{R}{2^j} = \frac{2R}{2^j} < \frac{2R}{2^J} < 2\pi.$$

This is impossible unless $m = n$. Hence as claimed, the sum reduces to a single nonzero term (at most) at each point. Therefore the square of the sum is the sum of the squares, that is,

$$|H(\theta)|^2 = \sum_{\ell \in \mathbb{Z}} |\hat{g}(2^j(\theta + 2\pi\ell))\hat{\varphi}(\theta + 2\pi\ell)|^2.$$

By substituting this equation into the expression for $\|P_j(g)\|^2$, we get

$$\|P_j(g)\|^2 = 2^j \frac{1}{2\pi} \int_{-\pi}^{\pi} \sum_{\ell \in \mathbb{Z}} |\hat{g}(2^j(\theta + 2\pi\ell))\hat{\varphi}(\theta + 2\pi\ell)|^2 \, d\theta$$

$$= 2^j \frac{1}{2\pi} \sum_{\ell \in \mathbb{Z}} \int_{(2\ell-1)\pi}^{(2\ell+1)\pi} |\hat{g}(2^j y)|^2 |\hat{\varphi}(y)|^2 \, dy$$

$$= 2^j \frac{1}{2\pi} \int_{\mathbb{R}} |\hat{g}(2^j y)|^2 |\hat{\varphi}(y)|^2 \, dy$$

$$= \frac{1}{2\pi} \int_{\mathbb{R}} |\hat{g}(\xi)|^2 \left|\hat{\varphi}\left(\frac{\xi}{2^j}\right)\right|^2 \, d\xi.$$

We want to apply the DCT to this last integral. By assumption, $\hat{\varphi}$ is bounded, say $|\hat{\varphi}(\xi)| \leq C$ for all ξ. Thus, for all j,

$$|\hat{g}(\xi)|^2 \left|\hat{\varphi}\left(\frac{\xi}{2^j}\right)\right|^2 \leq C|\hat{g}(\xi)|^2,$$

and $\int_{\mathbb{R}} C|\hat{g}(\xi)|^2 \, d\xi < +\infty$ since $g \in L^2(\mathbb{R})$. Thus we have an appropriate dominating function. We also assumed that $\hat{\varphi}$ is continuous at 0 and $\hat{\varphi}(0) = 1$. Therefore, for each ξ, $|\hat{g}(\xi)|^2 |\hat{\varphi}(\xi/2^j)|^2$ converges to $|\hat{g}(\xi)|^2$ as $j \to +\infty$. Thus by the dominated convergence theorem, $\|P_j(g)\|^2$ converges, as $j \to +\infty$, to

$$\frac{1}{2\pi} \int_{\mathbb{R}} |\hat{g}(\xi)|^2 \, d\xi = \frac{1}{2\pi} \|\hat{g}\|^2 = \|g\|^2,$$

by Plancherel's formula. By equation (5.110), this equation implies that $\|g - P_j(g)\|^2$ converges to 0 as $j \to +\infty$. In particular, there must be some j such that $\|g - P_j(g)\| < \epsilon$. By relation (5.109) and the triangle inequality, we obtain $\|f - P_j(g)\| < 2\epsilon$. Since $P_j(g) \in V_j$ and f and ϵ are arbitrary, this result shows that $\cup_{j \in \mathbb{Z}} V_j$ is dense in $L^2(\mathbb{R})$. ∎

We remark that for the Haar system, the definition of φ (see equation (5.49)) and a calculation show that

$$\hat{\varphi}(\xi) = e^{-i\xi/2} \frac{\sin(\frac{\xi}{2})}{\frac{\xi}{2}},$$

for $\xi \neq 0$, and $\hat{\varphi}(0) = 1$. Thus $|\hat{\varphi}(\xi)|$ is bounded by 1 and is continuous at 0. Hence, Lemma 5.48 applies and completes the proof that $\{V_j\}_{j \in \mathbb{Z}}$ forms a multiresolution analysis, as promised back in Example 5.31.

We can now state and prove the main results of this section.

Theorem 5.49   *Suppose $m_0 : \mathbb{R} \to \mathbb{C}$ is 2π-periodic,*

  i. $|m_0(\xi)|^2 + |m_0(\xi + \pi)|^2 = 1$ for all $\xi \in \mathbb{R}$.

  ii. $m_0(0) = 1$.

  iii. m_0 satisfies a Lipschitz condition of order $\delta > 0$ at 0 (i.e., inquality (5.95) holds for some $\delta, C > 0$).

  iv. $\inf_{|\xi| \leq \pi/2} |m_0(\xi)| > 0$.

Let $u = (u(k))_{k \in \mathbb{Z}}$ be such that $m_0(\xi) = \frac{1}{\sqrt{2}} \sum_{k \in \mathbb{Z}} u(k) e^{-ik\xi}$ (i.e., define $u(k)$ by equation (5.92)), and suppose $u \in \ell^1(\mathbb{Z})$.

Then $\prod_{j=1}^{\infty} m_0(\xi/2^j)$ converges uniformly on bounded sets to a function $\hat{\varphi} \in L^2(\mathbb{R})$. Let $\varphi = (\hat{\varphi})^{\vee}$. Then φ satisfies the scaling equation (5.72) and $\{\varphi_{0,k}\}_{k \in \mathbb{Z}}$ is an orthonormal set in $L^2(\mathbb{R})$. For $j \in \mathbb{Z}$, define

$$V_j = \left\{ \sum_{k \in \mathbb{Z}} z(k) \varphi_{j,k} : z = (z(k))_{k \in \mathbb{Z}} \in \ell^2(\mathbb{Z}) \right\}. \tag{5.112}$$

Then $\{V_j\}_{j \in \mathbb{Z}}$ is a multiresolution analysis with scaling function φ and scaling sequence u.

Proof

By properties i, ii, and iii, Theorem 5.39 implies that $\prod_{j=1}^{\infty} m_0(\xi/2^j)$ converges uniformly on bounded sets to $\hat{\varphi} \in L^2(\mathbb{R})$, and that φ satisfies the scaling equation (5.72). By property iv, Lemma 5.46 implies the orthonormality of $\{\varphi_{0,k}\}_{k \in \mathbb{Z}}$. Thus the definition of V_0

gives property ii in Definition 5.30 (the definition of a MRA). The definition of V_j implies the dilation property (Definition 5.30 iii). By a dilation argument, $\{\varphi_{j,k}\}_{k\in\mathbb{Z}}$ is an orthonormal set in $L^2(\mathbb{R})$ for each $j \in \mathbb{Z}$. The scaling equation (5.72) and another dilation argument show that

$$\varphi_{j,k} = \sum_{m\in\mathbb{Z}} u(m - 2k)\varphi_{j+1,m}$$

for all $j, k \in \mathbb{Z}$. Hence Exercise 4.6.5 implies that $V_j \subseteq V_{j+1}$ for every $j \in \mathbb{Z}$ (Definition 5.30 i). Lemma 5.47 gives Definition 5.30 iv automatically. Note that $|\hat{\varphi}(\xi)| \leq 1$ for all $\xi \in \mathbb{R}$ since $|m_0(\xi)| \leq 1$ by property ii. By property iii, m_0 is continuous at 0, and hence $\hat{\varphi}$ is also by Theorem 5.39 iv. Also, by definition, $\hat{\varphi}(0) = \prod_{j=1}^{\infty} m_0(0) = \prod_{j=1}^{\infty} 1 = 1$. Hence, Lemma 5.48 shows that Definition 5.30 v holds. ∎

Theorem 5.50 *Suppose $u = (u(k))_{k\in\mathbb{Z}}$ is a sequence satisfying*
  i. $\sum_{k\in\mathbb{Z}} |k|^{\epsilon} |u(k)| < +\infty$ *for some $\epsilon > 0$.*
  ii. $\sum_{k\in\mathbb{Z}} u(k) = \sqrt{2}$.
  iii. $\{R_{2k}u\}_{k\in\mathbb{Z}}$ *is an orthonormal set in $\ell^2(\mathbb{Z})$.*
  iv. $\inf_{|\xi|\leq\pi/2} |m_0(\xi)| > 0$, *for $m_0(\xi) = \frac{1}{\sqrt{2}} \sum_{k\in\mathbb{Z}} u(k)e^{-ik\xi}$.*
  Then $\prod_{j=1}^{\infty} m_0(\xi/2^j)$ converges uniformly on bounded subsets of $\mathbb{R}$ to a function $\hat{\varphi} \in L^2(\mathbb{R})$. Let $\varphi = (\hat{\varphi})^{\vee}$. For each $j \in \mathbb{Z}$, define V_j by equation (5.112). Then $\{V_j\}_{j\in\mathbb{Z}}$ is a multiresolution analysis with scaling function φ and scaling sequence u.

Proof
By definition, m_0 is 2π-periodic. By property i, $u \in \ell^1(\mathbb{Z})$, and Theorem 5.49 iii holds by Lemma 5.38. Property ii is equivalent to Theorem 5.49 ii, and property iii is equivalent to Theorem 5.49 i. Property iv is the same as Theorem 5.49 iv. Hence all conclusions follow from Theorem 5.49. ∎

Thus we have reduced the construction of a multiresolution analysis (and hence, by Theorem 5.35, the construction of wavelet systems) to the constuction of a sequence u satisfying conditions i–iv in Theorem 5.50. We will see some examples in Section 5.5.

Exercises

5.4.1. Prove equation (5.73).

5.4.2. i. For $n \in \mathbb{N}$, define a function f_n on $\mathbb{R}$ by

$$f_n(x) = \begin{cases} 0 & \text{if } x < 0 \\ 1 & \text{if } 0 \le x < 1 \\ -\frac{1}{n} & \text{if } 1 \le x < n+1 \\ 0 & \text{if } x \ge n+1. \end{cases}$$

Define $f(x)$ by $f(x) = 1$ for $0 \le x < 1$ and $f(x) = 0$ otherwise. Show that $f \in L^1(\mathbb{R}) \cap L^2(\mathbb{R})$, $f_n \in L^1(\mathbb{R}) \cap L^2(\mathbb{R})$ for all $n \in \mathbb{N}$,

$$\int_{\mathbb{R}} f_n(x)\,dx = 0 \quad \text{for all} \quad n \in \mathbb{N},$$

and

$$\|f_n - f\| \to 0$$

as $n \to \infty$, even though

$$\int_{\mathbb{R}} f(x)\,dx = 1.$$

ii. Let f be any function such that $f \in L^1(\mathbb{R}) \cap L^2(\mathbb{R})$. Generalize part i by constructing a sequence of functions $\{f_n\}_{n=1}^{\infty}$ such that $\|f_n - f\| \to 0$ as $n \to \infty$ and

$$\int_{\mathbb{R}} f_n(x)\,dx = 0$$

for all $n \in \mathbb{N}$.

Remark: This result shows that the condition $\int_{\mathbb{R}} f_n(x)\,dx = 0$ is not maintained in the limit when $\{f_n\}_{n=1}^{\infty}$ converges to f in the L^2 sense. This explains how the wavelet identity (5.43) can hold for all $f \in L^2(\mathbb{R})$, despite $\psi_{j,k}$ satisfying $\int_{\mathbb{R}} \psi_{j,k}(x)\,dx = 0$ for each $j, k \in \mathbb{Z}$ (by equation (5.94) and a change of variable): any finite partial sum of equation (5.43) has integral 0, but these partial sums converge in the L^2 sense, so the limit may not have integral 0.

iii. Suppose $\{f_n\}_{n \in \mathbb{N}}$ is a sequence of functions in $L^1(\mathbb{R})$ and $\int_{\mathbb{R}} f_n(x)\,dx = 0$ for all $n \in \mathbb{N}$. Suppose $\{f_n\}_{n \in \mathbb{N}}$ converges to

f in L^1 norm, which means that $\|f_n - f\|_1 \to 0$ as $n \to \infty$. Prove that $\int_{\mathbb{R}} f(x)\,dx = 0$. This implies that for $f \in L^1(\mathbb{R})$ with $\int_{\mathbb{R}} f(x)\,dx \neq 0$, the wavelet identity (5.43) will not converge in L^1 norm.

5.4.3. Prove the formula

$$\prod_{j=1}^{\infty} \cos\left(\frac{x}{2^j}\right) = \frac{\sin x}{x} \qquad (5.113)$$

two ways:

i. by writing

$$\sin x = \sin 2\left(\frac{x}{2}\right) = 2\cos\left(\frac{x}{2}\right)\sin\left(\frac{x}{2}\right)$$

and iterating, and

ii. by applying equation (5.78) in the case of the Haar system (this uniqueness argument is justified by Exercise 5.4.7 below). Hint: For u as in Example 5.36,

$$m_0(\xi) = \frac{1}{2}(1 + e^{-i\xi}) = e^{-i\xi/2}\cos\left(\frac{\xi}{2}\right).$$

Write $\hat{\varphi}(\xi)$ similarly.

5.4.4. Prove Lemma 5.37 without the assumption $|m_0(\xi)| \leq 1$. Hint: Show that

$$\left|m_0\left(\frac{\xi}{2^j}\right)\right| \leq 1 + C2^{-j\delta}|\xi|^\delta \leq e^{C2^{-j\delta}|\xi|^\delta}.$$

5.4.5. Let u be the sequence in Example 5.41. Compute $m_0(\xi)$. Define $\hat{\varphi}$ by equation (5.80). By computation (i.e., without appealing to the uniqueness of the solution of equation (5.72)), show that φ is as given in equation (5.97). Hint: As in the hint to Exercise 5.4.3 ii, show that $m_0(\xi) = e^{-i3\xi/2}\cos(3\xi/2)$ and apply equation (5.113).

5.4.6. i. Suppose $\{f_k\}_{k\in\mathbb{Z}}$ is a sequence of functions that is bounded in $L^2(\mathbb{R})$; that is, $\sup_{k\in\mathbb{Z}} \|f_k\| < \infty$. Suppose $a = (a(k))_{k\in\mathbb{Z}} \in \ell^1(\mathbb{Z})$. Prove that $\sum_{k\in\mathbb{Z}} a(k)f_k \in L^2(\mathbb{R})$ and

$$\left(\sum_{k\in\mathbb{Z}} a(k)f_k\right)^{\wedge} = \sum_{k\in\mathbb{Z}} a(k)\hat{f}_k.$$

    ii. Show that we cannot replace the assumption $a \in \ell^1(\mathbb{Z})$ in part i with the assumption $a \in \ell^2(\mathbb{Z})$. Hint: For example, let $a(k) = 1/k$ for $k \in \mathbb{N}$. Let $f \in L^2(\mathbb{R})$ be a function with $\|f\| \neq 0$. Let $f_k = f$ for all k. Show that the sequence $\sum_{k=1}^{N} a(k)f_k$ does not converge in $L^2(\mathbb{R})$ as $N \to \infty$.

5.4.7. Suppose $u = (u(k))_{k \in \mathbb{Z}}$ satisfies the assumptions of Corollary 5.40. Suppose $\varphi \in L^2(\mathbb{R})$ is a solution of the scaling equation (5.72) such that $\hat{\varphi}(0) = 1$. Prove that φ is the solution obtained in Corollary 5.40. Hint: Use Exercise 5.4.6 to justify equation (5.76). Iterate to obtain equation (5.77). Then use some of the conclusions of Corollary 5.40 to justify equation (5.78).

5.4.8. Prove the equivalence of conditions ii and iii in Lemma 5.42. Hint: Write

$$\int_{\mathbb{R}} |\hat{\varphi}(\xi)|^2 e^{ik\xi}\, d\xi = \sum_{\ell \in \mathbb{Z}} \int_{(2\ell-1)\pi}^{(2\ell+1)\pi} |\hat{\varphi}(\xi)|^2 e^{ik\xi}\, d\xi$$

$$= \int_{-\pi}^{\pi} \sum_{\ell \in \mathbb{Z}} |\hat{\varphi}(\theta + 2\pi\ell)|^2 e^{ik\theta}\, d\theta.$$

5.4.9. Prove equation (5.98).

5.4.10. Prove the main induction step in the proof of Lemma 5.44 (if statement (5.99) holds for $n-1$, then it works for n) in the following ways.

    i. Define a sequence $u = \{u(k)\}_{k \in \mathbb{Z}}$ by equation (5.92), so that equation (5.75) holds. Fourier inversion shows that

$$\varphi_n = \sum_{k \in \mathbb{Z}} u(k)\, (\varphi_{n-1})_{1,k}.$$

It is a bit tricky to justify the interchange of the sum and the Fourier transform, because we are only assuming $u \in \ell^2(\mathbb{Z})$ (compare to Exercise 5.4.6). However, the facts that $|\hat{\varphi}_{n-1}(\xi)| \leq 1$ for all ξ and supp $\hat{\varphi} \subseteq [-2^{n-1}\pi, 2^{n-1}\pi]$ allow one to apply Plancherel's theorem and show that the partial sums of $\sum_{k \in \mathbb{Z}} u(k)\, (\varphi_{n-1})_{1,k}$ are Cauchy in $L^2(\mathbb{R})$.) Use the formula for φ_n to prove that

$$\langle \varphi_n, (\varphi_n)_{0,k} \rangle = \sum_{\ell \in \mathbb{Z}} u(\ell) \sum_{m \in \mathbb{Z}} \overline{u(m)} \langle \varphi_{n-1}, (\varphi_{n-1})_{0,2k+m-\ell} \rangle$$

$$= \langle u, R_{2k}u \rangle.$$

ii. Use the criterion in Lemma 5.42 iii. Consider $\sum_{k \in \mathbb{Z}} |\hat{\varphi}_n(\xi + 2\pi k)|^2$. Apply the definition of $\hat{\varphi}_n$ and break the sum on k into its even and odd parts, say k of the form 2ℓ and k of the form $2\ell + 1$. Using the 2π-periodicity of m_0, rewrite the sum on k as

$$|m_0(\xi/2)|^2 \sum_{\ell \in \mathbb{Z}} |\hat{\varphi}_{n-1}((\xi/2) + 2\pi\ell)|^2$$

$$+ |m_0((\xi/2) + \pi)|^2 \sum_{\ell \in \mathbb{Z}} |\hat{\varphi}_{n-1}((\xi/2) + 2\pi\ell + \pi)|^2,$$

and apply the induction hypothesis.

5.4.11. i. For each $n \in \mathbb{N}$, define $f_n \in L^1(\mathbb{R})$ by

$$f_n(x) = \begin{cases} 0 & \text{if } x \le 0 \\ n & \text{if } 0 < x < \frac{1}{n} \\ 0 & \text{if } x \ge \frac{1}{n}. \end{cases}$$

Prove that $\lim_{n \to \infty} f_n(x) = f(x)$ exists for all $x \in \mathbb{R}$, and $\lim_{n \to \infty} \int_{\mathbb{R}} f_n(x)\, dx$ exists, but

$$\lim_{n \to \infty} \int_{\mathbb{R}} f_n(x)\, dx \ne \int_{\mathbb{R}} f(x)\, dx.$$

Why does the DCT (Theorem 5.45) not apply here? What is the smallest function g satisfying $f_n(x) \le g(x)$ for all $x \in \mathbb{R}$ and all $n \in \mathbb{N}$?

ii. Answer the same questions as in part i except with

$$f_n(x) = \begin{cases} 0 & \text{if } x \le 0 \\ \frac{1}{n} & \text{if } 0 < x < n \\ 0 & \text{if } x \ge n. \end{cases}$$

5.4.12. For $0 \le x \le 1/2$, prove that $1 - x \ge e^{-2x}$. Hint: Use calculus.

5.4.13. Suppose $\varphi \in L^2(\mathbb{R})$ and $\{\varphi_{0,k}\}_{k \in \mathbb{Z}}$ is an orthonormal set. Suppose $m_0(\xi)$ is 2π-periodic and satisfies equation (5.76). Use Lemma 5.42 iii to prove directly (i.e., without appealing to Lemma 4.42) that equation (5.85) holds. Hint: The proof is similar to the proof in Exercise 5.4.10 ii.

5.4.14. Assume that $\hat{\varphi}$ is continuous and satisfies $|\hat{\varphi}(\xi)| \le C/|\xi|$ for all $\xi \in \mathbb{R}$. Derive the condition $|\hat{\varphi}(0)| = 1$ from the orthonormality of $\{\varphi_{0,k}\}_{k \in \mathbb{Z}}$, equation (5.76), and the 2π-periodicity of m_0, by using the criterion in Lemma 5.42

iii. Hint: By Exercise 5.4.13, equation (5.85) holds. Then $m_0(0) = 1$ follows as in the text. By this result, equation (5.85), and the periodicity of m_0, $m_0(j\pi)$ is 0 if j is odd and 1 if j is even. If k is odd, then equation (5.76) implies that $\hat{\varphi}(2\pi k) = 0$. If k is even and not 0, write $k = 2^\ell j$ for some $\ell \geq 1$ and j odd. Then equation (5.76) gives

$$\hat{\varphi}(2\pi 2^\ell j) = \hat{\varphi}(2\pi 2^{\ell-1} j).$$

Iterate this and obtain $\hat{\varphi}(2\pi k) = 0$ for $k \neq 0$.

5.4.15. Suppose $u = (u(k))_{k\in\mathbb{Z}} \in \ell^1(\mathbb{Z})$ satisfies $\sum_{k\in\mathbb{Z}} u(k) = \sqrt{2}$ and $\{R_{2k}u\}_{k\in\mathbb{Z}}$ is orthonormal in $\ell^2(\mathbb{Z})$. Define $m_0(\xi)$ by equation (5.75).

i. Prove that $\sum_{k\in\mathbb{Z}}(-1)^k u(k) = 0$. Hint: Use equations (5.85) and (5.82).

ii. Prove that $\sum_{k\in\mathbb{Z}} u(2k) = \frac{\sqrt{2}}{2} = \sum_{k\in\mathbb{Z}} u(2k+1)$.

5.4.16. Suppose $u = (u(k))_{k\in\mathbb{Z}}$ satisfies $\sum_{k\in\mathbb{Z}} u(k) = \sqrt{2}$ and $\{R_{2k}u\}_{k\in\mathbb{Z}}$ is orthonormal in $\ell^2(\mathbb{Z})$. Suppose φ satisfies the scaling equation (5.72). Suppose also that φ is continuous and decays rapidly enough ($|\varphi(x)| \leq C(1+|x|)^{-1-\epsilon}$ for some $\epsilon > 0$ is sufficient, but do not bother to prove this) so that

$$g(x) = \sum_{\ell\in\mathbb{Z}} \varphi(x - \ell)$$

converges everywhere and is continuous. Prove that $g(x)$ is constant on $\mathbb{R}$. Hint: Substitute for $\varphi(x - \ell)$ in the definition of g by using equation (5.72). Change the summation index in the interior sum to obtain

$$g(x) = \sum_{\ell\in\mathbb{Z}} \sum_{m\in\mathbb{Z}} u(m - 2\ell)\sqrt{2}\varphi(2x - m).$$

Interchange the order of summation, and apply Exercise 5.4.15 (ii) to deduce that $g(x) = g(2x)$ for all $x \in \mathbb{R}$. Hence $g(x) = g(2^{-n}x)$ for any $n \in \mathbb{N}$.

5.4.17. Let $\{V_j\}_{j\in\mathbb{Z}}$ be the MRA in Exercise 5.3.4. Find the wavelet ψ associated with this MRA. Hint: Apply Theorem 5.49. Recall from Exercise 5.3.4 that $\hat{\varphi}(\xi) = \chi_{[-\pi,\pi]}(\xi)$. Show that equation (5.76) implies that the restriction to $[-\pi, \pi)$ of m_0 is given by $m_0(\xi) = \chi_{[-\pi/2,\pi/2]}(\xi)$. Use equation (5.91) to obtain m_1.

Then obtain $\hat\psi$ from equation (5.89). Answer:

$$\psi(x) = \frac{1}{\pi(x - 1/2)}\left(\sin 2\pi\left(x - \frac{1}{2}\right) - \sin\pi\left(x - \frac{1}{2}\right)\right).$$

Remark: This exercise shows that it may be easier to determine the wavelet corresponding to an MRA by using equations (5.89) and (5.91) if m_0 is known explicitly, rather than by the prescription based on knowing u in Theorem 5.35. In fact, Theorems 5.35 and 5.49, as we have stated them, do not apply here, because the scaling sequence u can not belong to $\ell^1(\mathbb{Z})$, since $m_0(\xi)$ is not continuous (see Exercise 4.4.9). Nevertheless, $\{\psi_{j,k}\}_{j,k\in\mathbb{Z}}$ is a wavelet system in $L^2(\mathbb{R})$. This can be most easily checked by looking at $\hat\psi(\xi) = e^{-i\xi/2}\chi_{\{\xi\,:\,\pi\,\leq\,|\xi|\,\leq\,2\pi\}}(\xi)$. The factor $e^{-i\xi/2}$ can be omitted in this case (in general it is needed to obtain orthogonality between different levels, but here this is immediate because of the supports of the Fourier transforms). This leads to a simpler mother wavelet ψ having x in place of $x - 1/2$ above. These wavelets are known as *Shannon wavelets for $L^2(\mathbb{R})$*, because they can be constructed by methods related to the Shannon sampling theorem (see Exercise 5.3.2). They were known long before modern wavelet theory.

5.4.18. (Meyer's wavelets) Let $h : \mathbb{R} \to \mathbb{R}$ be a C^2 function (recall that this means that h has at least two continuous derivatives at every point) such that $h(x) = 0$ for all $x \leq 0$, $h(x) = \pi/2$ for all $x \geq \pi$, and h is increasing on $[0, \pi]$.

i. Show that h defined by

$$h(x) = \begin{cases} 0 & \text{if } x < 0 \\ \frac{x}{2} - \frac{1}{4}\sin 2x & \text{if } 0 \leq x \leq \pi \\ \frac{\pi}{2} & \text{if } x > \pi \end{cases}$$

is an example of a C^2 function satisfying the conditions stated above.

Let $m_0 : \mathbb{R} \to \mathbb{R}$ be the 2π-periodic function whose restriction to $[-\pi, \pi)$ is defined by

$$m_0(\xi) = \begin{cases} \sin[h(3\xi + 2\pi)] & \text{if } -\pi \leq \xi < 0 \\ \cos[h(3\xi - \pi)] & \text{if } 0 \leq \xi < \pi. \end{cases}$$

ii. Prove that m_0 is a C^2 function on $\mathbb{R}$, $m_0(\xi) = 0$ if $2\pi/3 \le |\xi| \le \pi$ and $m_0(\xi) = 1$ if $|\xi| \le \pi/3$.

iii. Prove that m_0 satisfies conditions i, ii, iii, and iv in Theorem 5.49.

iv. Define $u = (u(k))_{k \in \mathbb{Z}}$ by equation (5.92). Prove that $u \in \ell^1(\mathbb{Z})$. Hint: Integrate by parts twice in equation (5.92), using the fact that m_0 is C^2, to show that $|u(k)| \le C/k^2$ for $k \ne 0$.

v. Define $\hat{\varphi}$ by equation (5.80). Show that in this case, $\hat{\varphi}(\xi) = m_0(\xi/2)$ for $|\xi| \le 2\pi$ and $\hat{\varphi}(\xi) = 0$ for $|\xi| > 2\pi$. Hint: Observe that $m_0(\xi/2^j) = 0$ for $2\pi 2^j/3 \le |\xi| \le 4\pi 2^j/3$ for each $j \in \mathbb{N}$, whereas for $j \ge 2$, $m_0(\xi/2^j) = 1$ for $|\xi| \le 4\pi/3$.

vi. Define m_1 by equation (5.91) and $\hat{\psi}$ by equation (5.90). By Theorem 4.49, $\psi = (\hat{\psi})^\vee$ is a mother wavelet for a wavelet system. Sketch the graph of $|\hat{\psi}|$. In particular, show that the graph consists of two bumps, one supported on $-8\pi/3 \le \xi \le -2\pi/3$, the other on $2\pi/3 \le |\xi| \le 8\pi/3$.

vii. Suppose h satisfies $h(\pi - x) = \pi/2 - h(x)$ for all $x \in \mathbb{R}$ (check that this is true for the h given in part i). Prove that m_0 is an even function (i.e., $m_0(-\xi) = m_0(\xi)$ for all ξ). Also prove that $|m_1|$ is even and hence $|\hat{\psi}|$ is even.

Remarks: It is possible to find an infinitely differentiable function h satisfying the conditions of this exercise (including vii). The corresponding wavelets are known as Meyer's wavelets, after Yves Meyer, who constructed this wavelet system in 1985, before the theory of multiresolution analyses. In fact, MRA theory was developed largely to explain Meyer's construction and put it in a more general framework. The only wavelets known prior to Meyer's wavelets were the Haar wavelets (Example 5.36), the Shannon wavelets (Exercise 5.4.17), and a wavelet system developed by Strömberg in 1980 in the study of Hardy spaces. Note that Meyer's wavelets are obtained in some sense by smoothing out the Shannon wavelets on the Fourier transform side. The result is that Meyer's wavelets decay faster than the reciprocal of any polynomial at ∞, unlike the Shannon wavelets, which decay only like $1/|x|$ at ∞. Unlike the Haar wavelets, however, Meyer's wavelets are infinitely differentiable (because their Fourier transforms have compact support; differentiate under the integral sign in the Fourier inversion formula (5.24)). Strömberg's wavelets could be made C^k for any finite k but not infinitely differentiable. Note that Meyer's wavelets decay rapidly at ∞ and have compactly supported Fourier transforms; hence they are localized in both space and frequency.

5.5 Wavelets with Compact Support and Their Computation

We begin by summarizing what we have obtained from Theorems 5.35 and 5.50. Suppose $u = (u(k))_{k \in \mathbb{Z}}$ satisfies conditions i–iv in Theorem 5.50. Define $m_0(\xi) = \frac{1}{\sqrt{2}} \sum_{k \in \mathbb{Z}} u(k) e^{-ik\xi}$. Then $\prod_{j=1}^{+\infty} m_0(\xi/2^j)$ converges to a function $\hat{\varphi} \in L^2(\mathbb{R})$ such that $\varphi = (\hat{\varphi})^{\vee}$ is the scaling function for a multiresolution analysis $\{V_j\}_{j \in \mathbb{Z}}$, by Theorem 5.50. Define a sequence $v \in \ell^1(\mathbb{Z})$ by $v(k) = (-1)^{k-1}\overline{u(1-k)}$, and a function ψ on $\mathbb{R}$ by $\psi = \sum_{k \in \mathbb{Z}} v(k)\varphi_{1,k}$. Then (by Theorem 5.35) ψ is the mother wavelet for a wavelet system; that is, $\{\psi_{j,k}\}_{j,k \in \mathbb{Z}}$ is a complete orthonormal set in $L^2(\mathbb{R})$.

In Example 4.57, a wavelet system for $\ell^2(\mathbb{Z})$ was constructed with generators u and v that had only six nonzero components (Daubechies's D6 wavelets for $\ell^2(\mathbb{Z})$). In Example 5.52 we see that u satisfies conditions i–iv in Theorem 5.50 and hence that we can use the recipe just described to construct a wavelet system in $L^2(\mathbb{R})$. First we note in the next theorem that the property that u has only finitely many nonzero components implies that the scaling function φ and the mother wavelet ψ have compact support. We do not give the proof of this result in full detail because it requires some aspects of measure theory that are beyond the prerequisites for this text. Nevertheless, it is instructive to give an outline of the proof, assuming certain facts that can be found in standard graduate-level analysis texts.

Theorem 5.51   *Suppose $m_0 : \mathbb{R} \to \mathbb{C}$ is a trigonometric polynomial of the form*

$$m_0(\xi) = \frac{1}{\sqrt{2}} \sum_{k=0}^{N} u(k) e^{-ik\xi}$$

for some positive integer N. Suppose $|m_0(\xi)| \leq 1$ for all $\xi \in \mathbb{R}$ and $m_0(0) = 1$. Let $\hat{\varphi}(\xi) = \prod_{j=1}^{\infty} m_0(\xi/2^j)$. Then $\varphi = (\hat{\varphi})^{\vee}$ has compact support, with

$$supp\, \varphi \subseteq [0, N].$$

We begin with a summary of the facts about measures needed for the proof of Theorem 5.51. A (signed Borel) *finite measure* μ on $\mathbb{R}$ is a map assigning to each set in a certain collection (the *Borel sets*, which we do not define here) of subsets of $\mathbb{R}$ a real number, with the property that μ is *σ-additive*. This means that if $\{E_j\}_{j=1}^{\infty}$ is a sequence of Borel sets that are disjoint (i.e., $E_i \cap E_j = \emptyset$ if $i \neq j$), then $\mu(\cup_{j=1}^{\infty} E_j) = \sum_{j=1}^{\infty} \mu(E_j)$. The assumption that μ is finite means that there is a constant C such that $\mu(E) \leq C$ for every Borel set E. A key example of a measure is the *point mass* δ_{x_0} at a point $x_0 \in \mathbb{R}$, defined by

$$\delta_{x_0}(E) = \begin{cases} 1 & \text{if } x_0 \in E \\ 0 & \text{if } x_0 \notin E. \end{cases}$$

This measure is also known as the *delta function*, even though it is not a function. A finite linear combination $\mu = \sum_{j=1}^{N} c_j \mu_j$ of finite measures $\mu_1, \ldots, \mu_N$, where $c_1, \ldots, c_N$ are real numbers, is also a finite measure defined by $\mu(E) = \sum_{j=1}^{N} c_j \mu_j(E)$.

Given a finite measure μ, one can define the integral of a reasonable function f (one that is *μ-integrable*, which we do not define here) with respect to μ; this is denoted $\int f d\mu$ or $\int f(x) d\mu(x)$. For example, a function $g \in L^1(\mathbb{R})$ yields a finite measure μ by the formula $\mu(E) = \int_E g(x) \, dx$. Then $d\mu$ is denoted $g dx$, and $\int f \, d\mu = \int f(x) g(x) \, dx$ for reasonable functions f. For the point mass δ_{x_0},

$$\int f d\delta_{x_0} = f(x_0).$$

For a finite measure μ, any bounded continuous function f will be μ-integrable, so $\int f d\mu$ will be defined. In particular, this allows us to define the Fourier transform of a finite measure μ by

$$\hat{\mu}(\xi) = \int e^{-ix\xi} \, d\mu(x).$$

If $d\mu = g \, dx$ for $g \in L^1(\mathbb{R})$, as described, then $\hat{\mu} = \hat{g}$. So the definition of the Fourier transform of a finite measure is consistent with the usual definition for integrable functions. It turns out that in some general sense (the sense of distributions), Fourier inversion carries over to finite measures: $\mu = (\hat{\mu})^\vee$. We also have the following version of Parseval's relation: if $g, \hat{g} \in L^1(\mathbb{R})$, g is continuous, and μ is a finite measure, then

$$\int_{\mathbb{R}} \hat{\mu}(\xi) \overline{\hat{g}(\xi)} \, d\xi = 2\pi \int \overline{g(x)} \, d\mu(x). \tag{5.114}$$

To prove this result, note that the left side of equation (5.114) is

$$\int_{\mathbb{R}} \overline{\hat{g}(\xi)} \int_{\mathbb{R}} e^{-ix\xi} \, d\mu(x) \, d\xi = \int_{\mathbb{R}} \int_{\mathbb{R}} \overline{\hat{g}(\xi)} e^{-ix\xi} \, d\xi \, d\mu(x)$$

$$= 2\pi \int \overline{(\hat{g})^\vee (x)} \, d\mu(x) = 2\pi \int \overline{g(x)} \, d\mu(x),$$

by interchanging the order of integration and applying Fourier inversion (Theorem 5.15).

The convolution of two finite measures μ and v is the finite measure defined by

$$\mu * v(E) = \int \int \chi_E(x + y) \, d\mu(x) \, dv(y).$$

(One can check (Exercise 5.5.1) that this result agrees with the usual definition of $\int_E f * g(x) \, dx$ when $d\mu = f \, dx$ and $dv = g \, dx$ for $f, g \in L^1(\mathbb{R})$.) Then

$$(\mu * v)^\wedge(\xi) = \hat{\mu}(\xi)\hat{v}(\xi),$$

just as in the case of integrable functions. This fact and Fourier inversion show that convolution of measures is commutative and associative.

A null set for a measure μ is a μ-measurable set B such that $\mu(E) = 0$ for every μ-measurable subset $E \subseteq B$. The support of a measure μ is

$$\text{supp} \, \mu = \mathbb{R} \setminus \cup\{O \subseteq \mathbb{R} : O \ \text{ is an open null set of } \mu\}.$$

The support of a finite measure μ is contained in a closed set E if and only if $\int f\,d\mu = 0$ for all C^2 functions with compact support whose support is disjoint from E. If μ is a finite measure supported in the interval $[a, b]$, and v is a finite measure supported in $[c, d]$, then $\mu * v$ is supported in $[a+c, b+d]$ (for the corresponding result for integrable functions, see Exercise 5.5.2). With this preparation, we can outline the proof of Theorem 5.51.

Proof Sketch
For $j \in \mathbb{N}$, let μ_j be the finite measure

$$\mu_j = \frac{1}{\sqrt{2}} \sum_{k=0}^{N} u(k)\delta_{2^{-j}k},$$

where $\delta_{2^{-j}k}$ is the point mass at the point $2^{-j}k$. Observe that supp $\mu_j \subseteq [0, N/2^j]$. By linearity of the integral,

$$\hat{\mu}_j(\xi) = \frac{1}{\sqrt{2}} \sum_{k=0}^{N} u(k) \int e^{-ix\xi}d\delta_{2^{-j}k}(x) = \frac{1}{\sqrt{2}} \sum_{k=0}^{N} u(k)e^{-ik\xi/2^j} = m_0(\xi/2^j).$$

The rough idea of the proof is that

$$\hat{\varphi}(\xi) = \prod_{j=1}^{\infty} m_0\left(\frac{\xi}{2^j}\right) = \prod_{j=1}^{\infty} \hat{\mu}_j(\xi) = (\mu_1 * \mu_2 * \mu_3 * \cdots)\hat{}(\xi),$$

so intuitively

$$\varphi = \mu_1 * \mu_2 * \mu_3 * \cdots,$$

and hence

$$\text{supp}\,\varphi \subseteq \text{supp}\,\mu_1 + \text{supp}\,\mu_2 + \text{supp}\,\mu_3 + \cdots$$
$$= [0, N/2] + [0, N/4] + [0, N/8] + \cdots = [0, N].$$

However, we have not defined such infinite convolutions, so to be more precise we proceed via a limiting argument. For $n \in \mathbb{N}$, define

$$\gamma_n = \mu_1 * \mu_2 * \cdots * \mu_n.$$

By the support property of convolutions previously noted and an induction argument,

$$\text{supp}\,\gamma_n \subseteq \left[0, \frac{N}{2} + \frac{N}{4} + \cdots + \frac{N}{2^n}\right] = \left[0, N - \frac{N}{2^n}\right] \subseteq [0, N],$$

for each $n \in \mathbb{N}$. Also

$$\hat{\gamma}_n(\xi) = \prod_{j=1}^{n} \hat{\mu}_j(\xi) = \prod_{j=1}^{n} m_0\left(\frac{\xi}{2^j}\right).$$

By Lemma 5.37, $\hat{\gamma}_n$ converges pointwise to $\hat{\varphi}$ (note that equation (5.95) is satisfied with $\delta = 1$ because m_0 is a trigonometric polynomial, hence it is continuously differentiable). The rest of the proof consists of showing that γ_n converges to φ as $n \to \infty$ in a sense that allows us to conclude that $\operatorname{supp}\varphi \subseteq [0, N]$ from the fact that $\operatorname{supp}\gamma_n \subseteq [0, N]$ for each n.

Note that $|\hat{\gamma}_n(\xi)| \le 1$ for all ξ (because the same is true of m_0). If $G \in L^1(\mathbb{R})$, it follows by the DCT (Theorem 5.45) that

$$\lim_{n \to \infty} \int_{\mathbb{R}} \hat{\gamma}_n(\xi) G(\xi)\, d\xi = \int_{\mathbb{R}} \hat{\varphi}(\xi) G(\xi)\, d\xi, \tag{5.115}$$

since $|\hat{\gamma}_n(\xi) G(\xi)| \le |G(\xi)| \in L^1(\mathbb{R})$.

Let f be a C^2 function with compact support such that $\operatorname{supp} f \cap [0, N] = \emptyset$. Then $f, \hat{f} \in L^1(\mathbb{R})$ by Exercise 5.2.7. Let $g = \hat{f}$. Then by Parseval's relation and equations (5.115) and (5.114),

$$\int_{\mathbb{R}} \varphi(x) f(x)\, dx = \int_{\mathbb{R}} \varphi(x) \overline{g(x)}\, dx = \frac{1}{2\pi} \int_{\mathbb{R}} \hat{\varphi}(\xi) \overline{\hat{g}(\xi)}\, d\xi$$

$$= \lim_{n \to \infty} \frac{1}{2\pi} \int_{\mathbb{R}} \hat{\gamma}_n(\xi) \overline{\hat{g}(\xi)}\, d\xi = \lim_{n \to \infty} \int_{\mathbb{R}} \overline{g(x)}\, d\gamma_n(x)$$

$$= \lim_{n \to \infty} \int_{\mathbb{R}} f(x)\, d\gamma_n(x).$$

But $\int_{\mathbb{R}} f(x)\, d\gamma_n(x) = 0$ for each n because $\operatorname{supp}\gamma_n \subseteq [0, N]$ and $\operatorname{supp} f \cap [0, N] = \emptyset$. Therefore $\int_{\mathbb{R}} \varphi(x) f(x)\, dx = 0$ for all C^2 functions with compact support that is disjoint from $[0, N]$. This implies that $\operatorname{supp}\varphi \subseteq [0, N]$. ∎

The proof of Theorem 5.51 gives us a certain intuition regarding the relation between wavelets on $\mathbb{Z}_N$ or $\mathbb{Z}$ and wavelets on $\mathbb{R}$. The finite measure $\mu_j = \frac{1}{\sqrt{2}} \sum_{k=0}^{N} u(k) \delta_{2^{-j}k}$ in the proof corresponds in some sense to the vector $u \in \ell^1(\mathbb{Z})$ with values $u(k)$ for $k = 0, 1, \ldots, N$ and 0 for other k, only renormalized by a factor of $1/\sqrt{2}$ and rescaled to take values on the grid $2^{-j}\mathbb{Z} = \{2^{-j}k : k \in \mathbb{Z}\}$ instead of $\mathbb{Z}$. We could make similar definitions with v in place of u. Recall that wavelets on $\mathbb{Z}_N$ or $\mathbb{Z}$ were constructed by repeated convolution of such sequences (see equations (3.43)–(3.44) and (4.74)–(4.75)). The proof of Theorem 5.51 shows that φ is a limit of $\gamma_n = \mu_1 * \mu_2 * \cdots * \mu_n$, which corresponds to equation (3.44) or (4.75). From equation (5.90), the formula for ψ corresponds to equation (3.43) or (4.74) in the same way. Thus wavelets on $\mathbb{R}$ are obtained as a limit as $j \to \infty$ of wavelets on $2^{-j}\mathbb{Z}$. We will see this principle again when we discuss how to compute φ and ψ numerically (Lemma 5.56).

Suppose as in Theorem 5.51 that $u(k) = 0$ for $k < 0$ and $k > N$. Since $v(k) = (-1)^{k-1}\overline{u(1-k)}$, it follows that $v(k)$ is nonzero only when $0 \le 1-k \le N$, or $-N+1 \le k \le 1$. Thus $\psi = \sum_{k=-N+1}^{1} v(k)\varphi_{1,k}$. Since $\operatorname{supp}\varphi \subseteq [0,N]$, it follows that $\operatorname{supp}\varphi_{1,k} \subseteq [k/2, (N+k)/2]$ (since $\varphi(2x-k) = 0$ if $2x-k \notin [0,N]$). Because k ranges from $-N+1$ to 1, we obtain

$$\operatorname{supp}\psi \subseteq [-N/2 + 1/2, N/2 + 1/2]. \tag{5.116}$$

Note that this is still an interval of length N.

We remark that if we have a sequence u satisfying conditions i–iv in Theorem 5.50, then $R_\ell u$, the translate of u by ℓ, still satisfies conditions i–iv. The ultimate effect is only that φ is translated by ℓ, and the mother wavelet is multiplied by -1 if ℓ is odd (Exercise 5.5.3). Thus, given $u(k)$, which is nonzero only for $N_1 \le k \le N_2$, we can replace u by $R_{-N_1}u$ to obtain a vector that is nonzero only for $0 \le k \le N_2 - N_1$. By convention, Daubechies's DN wavelets are chosen with $u(k) \neq 0$ only for $0 \le k < N$.

We now show that Daubechies's D6 wavelets on $\mathbb{Z}$ can be used to construct an example of a wavelet system on $\mathbb{R}$ with compactly supported wavelets.

Example 5.52

(Daubechies's wavelets for $L^2(\mathbb{R})$) In Example 4.57 we constructed vectors $u, v \in \ell^1(\mathbb{Z})$ with only six nonzero components, which generate a first-stage wavelet system in $\ell^2(\mathbb{Z})$. Recall that

$$(u(0), u(1), u(2), u(3), u(4), u(5))$$
$$= \frac{\sqrt{2}}{32}(b+c, 2a+3b+3c, 6a+4b+2c,$$
$$6a+4b-2c, 2a+3b-3c, b-c), \tag{5.117}$$

where

$$a = 1 - \sqrt{10}, b = 1 + \sqrt{10}, \quad \text{and} \quad c = \sqrt{5 + 2\sqrt{10}}.$$

In the notation of Example 4.57, we have

$$m_0(\xi) = \frac{1}{\sqrt{2}} \sum_{k=0}^{5} u(k)e^{-ik\xi} = \frac{1}{\sqrt{2}}\hat{u}(-\xi),$$

where

$$\hat{u}(\xi) = \sqrt{2}e^{5i\xi/2} \cos^3\left(\frac{\xi}{2}\right) \left[\cos^2\left(\frac{\xi}{2}\right) - \sqrt{10}\sin^2\left(\frac{\xi}{2}\right)\right.$$
$$\left. -ic\cos\left(\frac{\xi}{2}\right)\sin\left(\frac{\xi}{2}\right)\right]. \quad (5.118)$$

From equation (4.96), we also have

$$|m_0(\xi)|^2 = \frac{1}{2}|\hat{u}(-\xi)|^2 = b(-\xi) = b(\xi),$$

for

$$b(\xi) = \cos^{10}\left(\frac{\xi}{2}\right) + 5\cos^8\left(\frac{\xi}{2}\right)\sin^2\left(\frac{\xi}{2}\right) + 10\cos^6\left(\frac{\xi}{2}\right)\sin^4\left(\frac{\xi}{2}\right).$$

It is easy to check conditions i–iv in Theorem 5.50. Condition i is trivial because there are only finitely many nonzero coefficients $u(k)$. Recall that condition ii is equivalent to $m_0(0) = 1$, or $\hat{u}(0) = \sqrt{2}$, which follows from equation (5.118). Condition iii holds because of the orthonormality of Daubechies's D6 wavelets on $\ell^2(\mathbb{Z})$. By the definition of b above, $|m_0|^2 = b$ is a sum of nonnegative terms, the first of which is bounded away from 0 for $-\pi/2 \leq \xi \leq \pi/2$. Thus condition iv holds.

Now we follow Theorem 5.50 to define φ, v, and ψ. Then $\{\psi_{j,k}\}_{j,k\in\mathbb{Z}}$ is a wavelet system for $L^2(\mathbb{R})$. This is Daubechies's D6 wavelet system for $L^2(\mathbb{R})$. Note that the coefficients $u(k)$ are all real, which is convenient in calculations.

Thus we obtain a wavelet system such that φ is supported in $[0, 5]$ and ψ is supported in $[-2, 3]$ (by relation (5.116)). Also φ and ψ are real-valued. One way to see this is to note that the measures γ_n in the proof sketch of Theorem 5.51 are all real, hence so is their limit φ. Then ψ is real because $v(k)$ is real for each k, and $\psi = \sum_{k\in\mathbb{Z}} v(k)\varphi_{1,k}$.

In a similar way, for each $N \in \mathbb{N}$ one can use Daubechies's DN wavelets for $\ell^2(\mathbb{Z})$ (for which u has N nonzero components) to construct compactly supported wavelets for $L^2(\mathbb{R})$, known as *Daubechies's DN wavelets for $L^2(\mathbb{R})$*.

Figure 44 shows the graphs of the father wavelets φ and the mother wavelets for Daubechies's D4, D6, and D12 wavelet systems. Note that these functions appear to be (and in fact are) continuous, unlike the Haar functions in Examples 5.31 and 5.36.

Ingrid Daubechies made a detailed study of compactly supported wavelets, creating more than one infinite family of them. Her wavelets are the most commonly used in applications. Her book (Daubechies, 1992) contains tables of the nonzero coefficients of the scaling sequences u for several of these wavelets in a couple of these families, and graphs of a several of the resulting scaling functions and wavelets. (We remark that in Figure 44, our scaling functions φ are the same as in Daubechies's book, but our wavelets ψ are the negatives of hers, because she uses a slightly different convention that results in her sequence v being the negative of ours.) As N increases, Daubechies's DN wavelets have larger support intervals. However, this is compensated for in many applications by the facts that the DN wavelets for larger N tend to have more cancellation (e.g., $\int_{\mathbb{R}} x^j \psi(x) \, dx = 0$ for some positive integers, instead of just $j = 0$ as guaranteed by equation (5.94), which is why the graphs look more "wiggly") and more smoothness (as suggested by the graphs). Daubechies and others have made a systematic study of the smoothness of these wavelets. (See Exercises 5.5.8 and 5.5.9 for the first steps in one approach to this study.) It turns out that in a precise sense, the smoothness of Daubechies's wavelets grows linearly with their support, and linear growth is the best possible.

Now we turn to the question of how one actually computes with these wavelets. We see that everything is done using only the scaling sequence u and the associated sequence v.

Lemma 5.53 *Suppose $\{V_j\}_{j\in\mathbb{Z}}$ is an MRA with scaling function φ and scaling sequence $u = (u(k))_{k\in\mathbb{Z}}$. Suppose $v = (v(k))_{k\in\mathbb{Z}}$ is defined by $v(k) = (-1)^{k-1}\overline{u(1-k)}$, and $\psi = \sum_{k\in\mathbb{Z}} v(k)\varphi_{1,k}$. Suppose $f \in L^2(\mathbb{R})$ and, for each $j \in \mathbb{Z}$, define sequences $x_j = (x_j(k))_{k\in\mathbb{Z}}$ and $y_j = (y_j(k))_{k\in\mathbb{Z}}$*

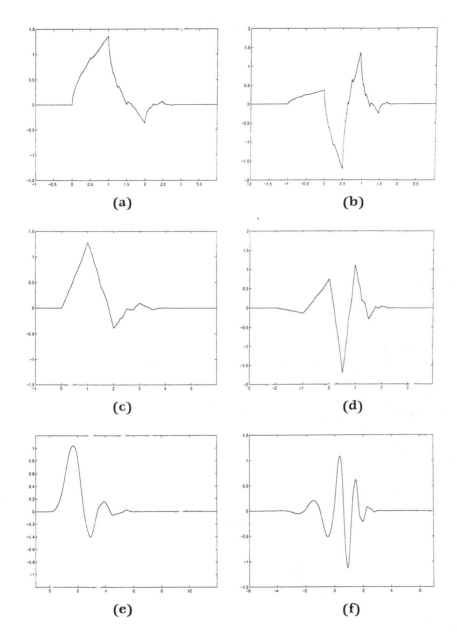

FIGURE 44 **(a)** φ for D4, **(b)** ψ for D4, **(c)** φ for D6, **(d)** ψ for D6, **(e)** φ for D12, **(f)** ψ for D12

by

$$x_j(k) = \langle f, \psi_{j,k} \rangle \tag{5.119}$$

and

$$y_j(k) = \langle f, \varphi_{j,k} \rangle. \tag{5.120}$$

Then

$$x_j = D(y_{j+1} * \tilde{v}) \tag{5.121}$$

and

$$y_j = D(y_{j+1} * \tilde{u}), \tag{5.122}$$

where D is the downsampling operator (Definition 4.43) and the sequences and convolution are in $\ell^2(\mathbb{Z})$, as in chapter 4.

Also

$$y_{j+1} = U(y_j) * u + U(x_j) * v, \tag{5.123}$$

where U is the upsampling operator (Definition 4.43) on $\ell^2(\mathbb{Z})$.

Proof

From equation (5.52) and a rescaling argument (Exercise 5.5.6) we obtain, for all $j, k \in \mathbb{Z}$,

$$\varphi_{j,k} = \sum_{m \in \mathbb{Z}} u(m - 2k)\varphi_{j+1,m}. \tag{5.124}$$

Similarly, from equation (5.59),

$$\psi_{j,k} = \sum_{m \in \mathbb{Z}} v(m - 2k)\varphi_{j+1,m}, \tag{5.125}$$

for all $j, k \in \mathbb{Z}$. By equation (5.124),

$$y_j(k) = \langle f, \varphi_{j,k} \rangle = \left\langle f, \sum_{m \in \mathbb{Z}} u(m - 2k)\varphi_{j+1,m} \right\rangle = \sum_{m \in \mathbb{Z}} \overline{u(m - 2k)}\langle f, \varphi_{j+1,m} \rangle$$

$$= \sum_{m \in \mathbb{Z}} \tilde{u}(2k - m)y_{j+1}(m) = y_{j+1} * \tilde{u}(2k) = D(y_{j+1} * \tilde{u})(k).$$

This proves equation (5.122). In a similar way, using equation (5.125) instead of equation (5.124), we obtain equation (5.121).

To prove equation (5.123), recall the identity (5.62). By a dilation argument (Exercise 5.5.6), we obtain, for any $j, m \in \mathbb{Z}$,

$$\varphi_{j+1,m} = \sum_{k \in \mathbb{Z}} \tilde{u}(2k - m)\varphi_{j,k} + \sum_{k \in \mathbb{Z}} \tilde{v}(2k - m)\psi_{j,k}. \qquad (5.126)$$

Hence,

$$
\begin{aligned}
y_{j+1}(m) = \langle f, \varphi_{j+1,m} \rangle &= \sum_{k \in \mathbb{Z}} \overline{\tilde{u}(2k - m)}\langle f, \varphi_{j,k} \rangle + \sum_{k \in \mathbb{Z}} \overline{\tilde{v}(2k - m)}\langle f, \psi_{j,k} \rangle \\
&= \sum_{k \in \mathbb{Z}} u(m - 2k)y_j(k) + \sum_{k \in \mathbb{Z}} v(m - 2k)x_j(k) \\
&= \sum_{k \in \mathbb{Z}} u(m - 2k)U(y_j)(2k) + \sum_{k \in \mathbb{Z}} v(m - 2k)U(x_j)(2k) \\
&= \sum_{\ell \in \mathbb{Z}} u(m - \ell)U(y_j)(\ell) + \sum_{\ell \in \mathbb{Z}} v(m - \ell)U(x_j)(\ell) \\
&= U(y_j) * u(m) + U(x_j) * v(m),
\end{aligned}
$$

where the next to last equality holds because $U(y_j)(\ell) = U(x_j)(\ell) = 0$ for the odd values of ℓ by the definition of upsampling. This proves equation (5.123). ∎

The reader should recognize these formulas (compare with equations (3.33) and (3.34), for which the indexing is reversed). Equations (5.121) and (5.122) say that to pass from y_{j+1} to x_j and y_j, we apply one segment of the analysis phase of a filter bank, as shown in the left half of Figure 45. Conversely, by equation (5.123), to recover y_{j+1} from x_j and y_j, we apply one segment of the reconstruction phase of a filter bank, exhibited in the right half of Figure 45.

Because the wavelet expansion in general involves infinitely many terms, we can never compute it exactly. So we must

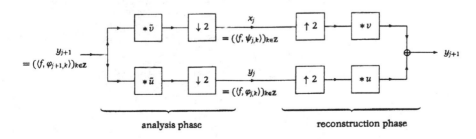

FIGURE 45

approximate at some stage. By the definition of a multiresolution analysis, $\cup_{j \in \mathbb{Z}} V_j$ is dense in $L^2(\mathbb{R})$. So given $f \in L^2(\mathbb{R})$, we first pick m sufficiently large that $\|P_m(f) - f\|$ is negligible, where $P_m(f) = \sum_{k \in \mathbb{Z}} \langle f, \varphi_{m,k} \rangle \varphi_{m,k}$ is the projection of f on V_m. This is possible, since $\|P_m(f) - f\| \to 0$ as $m \to +\infty$, by Exercise 5.3.10. The next lemma tells us how to approximate the sequence $y_m = (\langle f, \varphi_{m,k} \rangle)_{k \in \mathbb{Z}}$ in Lemma 5.53, for sufficiently large m. Its proof is an approximate identity argument (compare with Theorem 5.11).

Lemma 5.54   *Suppose $f \in L^2(\mathbb{R})$ satisfies a Lipschitz condition of order ϵ, for some $\epsilon \in (0, 1]$, which means that there exists a constant $C_1 < \infty$ such that for all $x, y \in \mathbb{R}$,*

$$|f(x) - f(y)| \le C_1 |x - y|^\epsilon. \tag{5.127}$$

Suppose $\varphi \in L^1(\mathbb{R}) \cap L^2(\mathbb{R})$ satisfies

$$\int_{\mathbb{R}} \varphi(x) \, dx = 1, \tag{5.128}$$

and

$$\int_{\mathbb{R}} |x|^\epsilon |\varphi(x)| \, dx = C_2 < \infty. \tag{5.129}$$

Then

$$\left| 2^{m/2} \langle f, \varphi_{m,k} \rangle - f(2^{-m}k) \right| \le C_1 C_2 2^{-m\epsilon}, \tag{5.130}$$

or, equivalently,

$$\left| \langle f, \varphi_{m,k} \rangle - 2^{-m/2} f(2^{-m}k) \right| \le C_1 C_2 2^{-m(\epsilon+1/2)}. \tag{5.131}$$

Proof
Note that

$$\int_{\mathbb{R}} 2^{m/2} \overline{\varphi(2^m x - k)} \, dx = 2^{-m/2} \int_{\mathbb{R}} \overline{\varphi(t)} \, dt = 2^{-m/2} \overline{1} = 2^{-m/2},$$

by the change of variables $t = 2^m x - k$ and by using equation (5.128). Hence,

$$\left| \langle f, \varphi_{m,k} \rangle - 2^{-m/2} f(2^{-m}k) \right|$$

$$= \left| \int_{\mathbb{R}} \left(f(x) - f(2^{-m}k) \right) 2^{m/2} \overline{\varphi(2^m x - k)} \, dx \right|$$

$$\leq \int_{\mathbb{R}} \left| f(x) - f(2^{-m}k) \right| 2^{m/2} \left| \varphi(2^m x - k) \right| dx$$

$$= 2^{-m/2} \int_{\mathbb{R}} \left| f(2^{-m}t + 2^{-m}k) - f(2^{-m}k) \right| \left| \varphi(t) \right| dt$$

$$\leq C_1 2^{-m/2} \int_{\mathbb{R}} \left| 2^{-m}t \right|^\epsilon \left| \varphi(t) \right| dt = C_1 C_2 2^{-m(\epsilon + 1/2)},$$

by relations (5.127) and (5.129). ∎

Lemma 5.54 explains why we normalize the argument of the scaling function φ so that $\hat{\varphi}(0) = \int_{\mathbb{R}} \varphi(x)\,dx = 1$. Any reasonable scaling function will satisfy relation (5.129) for all $0 < \epsilon \leq 1$. For example, if φ is bounded with compact support, relation (5.129) is obvious.

For m sufficiently large, in the notation of Lemma 5.53,

$$y_m(k) = \langle f, \varphi_{m,k} \rangle \approx 2^{-m/2} f(2^{-m}k).$$

Lemma 5.54 gives an estimate for the error in this approximation if f satisfies a Lipshitz condition. This condition is mild: any differentiable function with bounded derivative satisfies relation (5.127) with $\epsilon = 1$, by the mean value theorem.

In some cases, one may not want to go to sufficiently large m so that the approximation in inequality (5.130) is satisfactory. In that case, one can estimate $\langle f, \varphi_{m,k} \rangle$ by some numerical method (for example, some quadrature formula as in Sweldens and Piessens (1994)) to sufficient accuracy.

After obtaining our approximation to y_m, we calculate the wavelet coefficients of a function f as follows. We apply equations (5.121) and (5.122) iteratively (the analysis phase of the filter bank in Figure 46) to obtain, after ℓ steps, the wavelet coefficients $x_j(k) = \langle f, \psi_{j,k} \rangle$ for $j = m - 1, m - 2, \ldots, m - \ell$, and the remaining coefficients $y_{m-\ell}(k) = \langle f, \varphi_{m-\ell,k} \rangle$. At this stage the corresponding formula for $P_m(f)$ (our approximation to f) is (by Exercise 5.5.10)

$$P_m(f) = \sum_{j=m-\ell}^{m-1} \sum_{k \in \mathbb{Z}} \langle f, \psi_{j,k} \rangle \psi_{j,k} + \sum_{k \in \mathbb{Z}} \langle f, \varphi_{m-\ell,k} \rangle \varphi_{m-\ell,k}. \tag{5.132}$$

Conversely, if we have the wavelet coefficients $x_j(k) = \langle f, \psi_{j,k} \rangle$ for $j = m-1, m-2, \ldots, m-\ell$, and the remaining coefficients $y_{m-\ell}(k) =$

$\langle f, \varphi_{m-\ell,k} \rangle$, we can recover y_m (and hence our approximation to the values $f(2^{-m}k) \approx 2^{m/2}\langle f, \varphi_{m,k} \rangle = 2^{m/2}y_m(k)$) by applying equation (5.123) recursively (the reconstruction phase of the filter bank in Figure 46) to obtain $y_{m-\ell+1}, y_{m-\ell+2}, \ldots, y_{m-1}$, and finally y_m.

In Figure 46, the symbol "..." in the left half of the diagram (the analysis phase) represents a convolution (with either $\tilde{v}$ or $\tilde{u}$) followed by a downsampling, whereas in the right half (the reconstruction phase), it represents an upsampling followed by a convolution (with either v or u).

Thus we never need to know the scaling function φ or the mother wavelet ψ. All that is needed in the computation are the values of the scaling sequence u (because v is determined by u via $v(k) = (-1)^{k-1}\overline{u(1-k)}$). We illustrate this with Example 5.55.

Example 5.55

We compute the Daubechies's D6 wavelet transform of the function

$$f(x) = \begin{cases} \sin(4\pi x) & \text{if } 0 \leq x \leq 1 \\ 0 & \text{otherwise.} \end{cases}$$

We cannot compute the entire transform because it is infinite, and the part we do compute will be an approximation. We begin by approximating the scaling function coefficients $\langle f, \varphi_{m,k} \rangle$ by the sample values $2^{-m/2}f(2^{-m}k)$, with an error that decreases as we increase k, as in Lemma 5.54. Suppose we decide that $m = 9$ gives an acceptable error. Thus we take

$$y_9(k) = 2^{-9/2}f(2^{-9}k) = \begin{cases} \frac{1}{\sqrt{512}}\sin(\frac{\pi k}{128}) & \text{if } 0 \leq k \leq 512 \\ 0 & \text{otherwise.} \end{cases}$$

Note that this is $1/\sqrt{512}$ times the vector z whose $\ell^2(\mathbb{Z})$ wavelet coefficients are shown in Figure 42. We now compute the vectors x_8, x_7, x_6, x_5, and y_5 (say we do a 4-level transform), by the filter bank arrangement in the left half of Figure 46. Then $x_j(k)$ is our approximation to the wavelet coefficient $\langle f, \psi_{j,k} \rangle$ for $j = 8, 7, 6$, and 5, and y_5 is our approximation to the scaling function coefficient $\langle f, \varphi_{5,k} \rangle$. Note that the vectors u and v for the D6 wavelets on $L^2(\mathbb{R})$ are the same as those for the D6 wavelet transform on $\ell^2(\mathbb{Z})$ (this is how Daubechies's D6 wavelets for $L^2(\mathbb{R})$ were constructed in Example 5.52). Hence, a comparison with the algorithm for

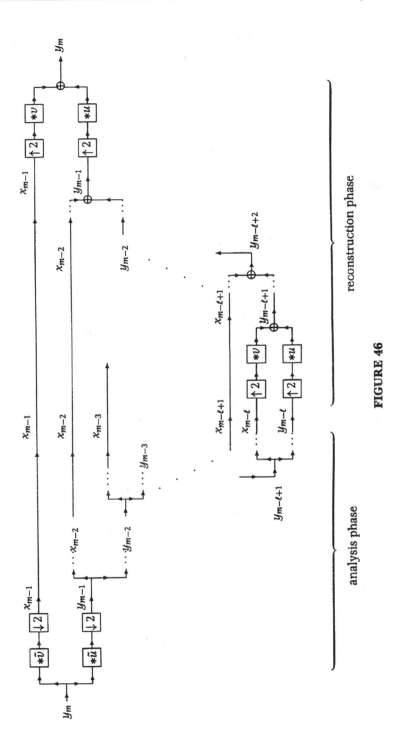

FIGURE 46

computing wavelets in $\ell^2(\mathbb{Z})$ shows that finding x_8, x_7, x_6, x_5, and y_5 is exactly the same computation (except for the factor of $1/\sqrt{512}$) as for the 4th level $\ell^2(\mathbb{Z})$ wavelet transform for the vector $z \in \ell^2(\mathbb{Z})$. The only difference is the shift in indices; here, for example, x_8 corresponds to the -1 level D6 wavelet coefficients of z. Thus, except for the factor of $1/\sqrt{512}$, the wavelet coefficients $\langle f, \psi_{j,k} \rangle$ for $j = 8, 7, 6$, and 5 are plotted in Figures 42b, c, d, and e, respectively, and the scaling function coefficients $\langle f, \varphi_{5,k} \rangle$ are plotted in Figure 42 f, with the understanding that the point $n \in \mathbb{Z}$ in Figure 42 corresponds to the point $n/512 \in \mathbb{R}$.

This gives another interpretation of the computation of wavelets on $L^2(\mathbb{R})$. We can sample the given function f at a sufficiently fine resolution, yielding the vector $z \in \ell^2(\mathbb{Z})$ of sample values. Then we computes the (exact) wavelet transform of z in the sense of $\ell^2(\mathbb{Z})$ (as in chapter 4) to approximate the $L^2(\mathbb{R})$ wavelet transform of f. This is a useful interpretation, especially because we are given only the sample values of f in many experimental circumstances.

On the other hand, some students may feel let down by this. They may think that we have just arrived at what we already had in the context of $\mathbb{Z}$, and in the process we have had to go through some difficult analysis and add some conditions on an $\ell^2(\mathbb{Z})$ wavelet-generating sequence u to guarantee that the associated ψ is the mother wavelet for a wavelet system in $L^2(\mathbb{R})$. However, there are crucial advantages in knowing that our calculations correspond to a wavelet system in the continuous setting of $\mathbb{R}$ instead of just the discrete setting $\mathbb{Z}$. We see this in chapter 6 when we apply wavelets to differential equations, where the nondiscrete nature of the underlying space is essential.

Although we have noted that we do not need to know the function values of φ and ψ to compute the wavelet transform, the reader may still wonder how the graphs of these functions (as in Figure 44) can be obtained. Moreover, it aids our intuition to have these graphs. These graphs are obtained as an application of our reconstruction algorithm (5.123) as follows.

We cannot compute φ and ψ everywhere explicitly, but we can compute them to good approximation at points of the form $2^{-m}k$ for

$k \in \mathbb{Z}$ and m large, and then fill in between these values by continuity (e.g., linear interpolation). We apply Lemma 5.54 with f replaced by φ and ψ to conclude that for m large,

$$\varphi(2^{-m}k) \approx 2^{m/2}\langle \varphi, \varphi_{m,k}\rangle \text{ and } \psi(2^{-m}k) \approx 2^{m/2}\langle \psi, \varphi_{m,k}\rangle. \qquad (5.133)$$

Lemma 5.56 tells us how to compute the inner products in approximation (5.133).

Lemma 5.56 *Suppose φ is the scaling function and ψ is the mother wavelet, obtained as in Theorem 5.35, for some multiresolution analysis, with scaling sequence $u \in \ell^1(\mathbb{Z})$ and associated sequence v defined by $v(k) = (-1)^{k-1}\overline{u(1-k)}$. For $k \geq 1$, define sequences $c_k, d_k \in \ell^1(\mathbb{Z})$ inductively by*

$$c_1 = u, \quad d_1 = v,$$

and, for $m \geq 2$,

$$c_m = U(c_{m-1}) * u, \text{ and } d_m = U(d_{m-1}) * u. \qquad (5.134)$$

Equivalently, define

$$c_m = u * U(u) * U^2(u) * \cdots * U^{m-1}(u) \qquad (5.135)$$

and

$$d_m = u * U(u) * \cdots * U^{m-2}(u) * U^{m-1}(v). \qquad (5.136)$$

Then for all $j \geq 1$,

$$c_j(k) = \langle \varphi, \varphi_{j,k}\rangle \qquad (5.137)$$

and

$$d_j(k) = \langle \psi, \varphi_{j,k}\rangle, \qquad (5.138)$$

for all k.

Proof
To prove equation (5.137), we apply Lemma 5.53. Define x_j and y_j as in equations (5.119) and (5.120), with $f = \varphi$. Thus equation (5.137) is equivalent to the statement that $y_j = c_j$ for $j \geq 1$. Note that

$$y_0(k) = \langle \varphi, \varphi_{0,k}\rangle = \delta(k),$$

or $y_0 = \delta$. Also note that for $j \geq 0$, $\psi_{j,k} \in W_j$, and $\varphi \in V_0 \subseteq V_j$. Because $V_j \perp W_j$, it follows that for $j \geq 0$

$$x_j(k) = \langle \varphi, \psi_{j,k} \rangle = 0,$$

for all $k \in \mathbb{Z}$. Hence $x_j = 0$ for all $j > 1$. Therefore in this case, equation (5.123) reduces to

$$y_{j+1} = U(y_j) * u, \tag{5.139}$$

for $j \geq 0$. When $j = 1$, we obtain

$$y_1 = U(y_0) * u = U(\delta) * u = \delta * u = u.$$

This gives $y_1 = u = c_1$. Then $y_m = c_m$ follows for each $m \geq 2$ by equations (5.134) and (5.139).

Now define x_j and y_j in Lemma 5.54 by equations (5.119) and (5.120) with $f = \psi$. To prove equation (5.138), we must show that $y_j = d_j$ for $j \geq 1$. Note that

$$x_0(k) = \langle \psi, \psi_{0,k} \rangle = \delta(k),$$

and

$$y_0(k) = \langle \psi, \varphi_{0,k} \rangle = 0,$$

since $\psi \in W_0$, $\varphi_{0,k} \in V_0$, and $W_0 \perp V_0$. Hence, equation (5.123) gives

$$y_1 = U(x_0) * v = U(\delta) * v = \delta * v = v = d_1. \tag{5.140}$$

For $j \geq 1$,

$$x_j(k) = \langle \psi, \psi_{j,k} \rangle = 0,$$

since $W_0 \perp W_j$ for $j \neq 0$. Hence, for $j \geq 1$, equation (5.123) reduces to $y_{j+1} = U(y_j) * u$. By induction, equations (5.134) and (5.140) show that $y_j = d_j$ for all j. ∎

Observe that equation (5.135) is an analog of equations (3.44) and (4.75), whereas equation (5.136) is analogous to equations (3.43) and (4.74). This confirms our statement after Theorem 5.51 that wavelets on $\mathbb{R}$ are limits, in some sense, of wavelets on $\mathbb{Z} \approx 2^{-j}\mathbb{Z}$, as $j \to \infty$.

Exercises

5.5.1. Suppose $f, g \in L^1(\mathbb{R})$. Let $d\mu = f\, dx$ and $dv = g\, dx$. Prove that

$$\int \int \chi_E(x+y)\, d\mu(x)\, dv(y) = \int_E f * g(x)\, dx.$$

Hint: Change variables (let $t = x + y$) and interchange the order of integration. Note that for a general h, $\int \chi_E(t) h(t)\, dt = \int_E h(t)\, dt$.

5.5.2. Suppose $f, g \in L^1(\mathbb{R})$ with $\operatorname{supp} f \subseteq [a, b]$ and $\operatorname{supp} g \subseteq [c, d]$. Prove that $\operatorname{supp} f * g \subseteq [a+c, b+d]$. (Compare with Exercise 4.7.6).

5.5.3. Suppose $u = (u(k))_{k \in \mathbb{Z}}$ satisfies conditions i–iv in Theorem 5.50. Define $\hat{\varphi}(\xi) = \prod_{j=1}^{\infty} m_0(\xi/2^j)$, $v(k) = (-1)^{k-1}\overline{u(1-k)}$, and $\psi = \sum_{k \in \mathbb{Z}} v(k)\varphi_{1,k}$. Let $\ell \in \mathbb{Z}$.

  i. Show that $R_\ell u$ (defined as in chapter 4 by $(R_\ell u)(n) = u(n-\ell)$) also satisfies conditions i–iv in Theorem 5.50, for $m_0(\xi)$ replaced by $M_0(\xi) = \frac{1}{\sqrt{2}} \sum_{k \in \mathbb{Z}} (R_\ell u)(k) e^{-ik\xi}$. Hint: Observe that $M_0(\xi) = e^{-i\ell\xi} m_0(\xi)$.

  ii. Define φ_ℓ, v_ℓ, and ψ_ℓ associated with $R_\ell u$ and M_0 in the same way that φ, v, and ψ are associated with u and m_0. Prove that

$$\varphi_\ell(x) = \varphi(x - \ell)$$

and

$$\psi_\ell = (-1)^\ell \psi.$$

Hint: Use equations (5.89) and (5.91), and Lemma 5.26.

5.5.4. Suppose $u = (u(k))_{k \in \mathbb{Z}}$ is a sequence satisfying conditions ii and iii in Theorem 5.50. Suppose also that there are only two values of $k \in \mathbb{Z}$ such that $u(k) \neq 0$.

  i. Prove that there exists $m \in \mathbb{Z}$ and an odd number $j \in \mathbb{Z}$ such that

$$u(k) = \begin{cases} \frac{1}{\sqrt{2}} & \text{if } k = m \text{ or } k = m+j \\ 0 & \text{otherwise.} \end{cases}$$

Hint: The fastest way to see this is to use Exercise 5.4.15 (ii).

ii. By defining m_0 by equation (5.75) and $\hat{\varphi} = \prod_{j=1}^{\infty} m_0(\xi/2^j)$, prove that

$$\varphi = \frac{1}{j} \chi_{[m, m+j]},$$

for m and j as in part i. Hint: Show that $m_0(\xi) = e^{-im\xi} e^{-ij\xi/2} \cos(j\xi/2)$ and apply equation (5.113), as in Exercise 5.4.5.

iii. Show that the set $\{\varphi_{0,k}\}_{k \in \mathbb{Z}}$ is orthogonal only when $j = 1$. Note that this is also the only case in which condition iv in Theorem 5.50 holds. Observe that in this case the resulting wavelets are the Haar wavelets (up to a possible negative sign, by Exercise 5.5.3).

5.5.5. Show that Daubechies's D4 wavelets for $\ell^2(\mathbb{Z})$ (Exercise 4.7.9) lead to wavelets for $L^2(\mathbb{R})$ in the same way as in Example 5.52.

5.5.6. Prove equations (5.124), (5.125), and (5.126).

5.5.7. Suppose $f : \mathbb{R} \to \mathbb{C}$ satisfies a Lipschitz condition of order $\alpha > 1$; that is, there exists a constant C such that, for all $x, y \in \mathbb{R}$,

$$|f(x) - f(y)| \leq C|x - y|^{\alpha}.$$

Prove that f must be a constant function. Hint: Show that f is differentiable everywhere and calculate the derivative.

Remark: This exercise explains the restriction $\epsilon \leq 1$ in Lemma 5.54.

5.5.8. Suppose $0 < \alpha \leq 1$, $f \in L^1(\mathbb{R})$, and there exist constants $\epsilon > 0$ and $C > 0$ such that, for all $\xi \in \mathbb{R}$,

$$|\hat{f}(\xi)| \leq C(1 + |\xi|)^{-1-\alpha-\epsilon}.$$

i. Show that $\hat{f} \in L^1(\mathbb{R})$. By Theorem 5.15, by modifying f on a set of measure 0, we can assume the $f(x) = (\hat{f})^{\vee}(x)$ at every point $x \in \mathbb{R}$. Deduce that f is bounded; that is, there exists a constant C_1 (depending on f, α, ϵ and C, but not on x) such that

$$|f(x)| \leq C_1,$$

for all $x \in \mathbb{R}$.

ii. Prove that f satisfies a Lipschitz condition of order α; that is, for some other constant C_2, and for all $x, y \in \mathbb{R}$,

$$|f(x) - f(y)| \le C_2 |x - y|^\alpha.$$

Hint: By part i, the estimate is easy if $|x-y| > 1$. Suppose $|x - y| \le 1$. By Fourier inversion, obtain

$$|f(x) - f(y)| \le \frac{1}{2\pi} \int_{\mathbb{R}} |\hat{f}(\xi)| |e^{-i\xi x} - e^{-i\xi y}| \, d\xi$$

$$= \frac{1}{2\pi} \int_{\mathbb{R}} |\hat{f}(\xi)| |e^{-i\xi(x-y)} - 1| \, d\xi.$$

On $\{\xi : |\xi(x - y)| \le 1\}$, use the estimate

$$|e^{-i\xi(x-y)} - 1| \le |\xi(x - y)| \le |\xi(x - y)|^\alpha.$$

On the complementary set, make the trivial estimate $|e^{-i\xi(x-y)} - 1| \le 2$.

Remark: This result is natural because it says that the faster $|\hat{f}(\xi)|$ decays as $|\xi| \to \infty$ (i.e., the smaller the weights of the high-frequency terms in equation (5.24)), the smoother f is. In particular, this provides a way of estimating the smoothness of f based on knowledge of the decay of $\hat{f}$. This is one method used to prove smoothness estimates for wavelets. Higher-order smoothness can be obtained from more rapid decay of $\hat{f}(\xi)$, as in Exercise 5.5.9.

5.5.9. Suppose $f \in L^1(\mathbb{R})$, and there exist constants $\epsilon > 0$ and $C > 0$ such that, for all $\xi \in \mathbb{R}$,

$$|\hat{f}(\xi)| \le C(1 + |\xi|)^{-2-\epsilon}.$$

i. Prove that $f'(x)$ exists for all $x \in \mathbb{R}$, and

$$f'(x) = \frac{1}{2\pi} \int_{\mathbb{R}} \hat{f}(\xi)(-i\xi) e^{-ix\xi} \, d\xi.$$

Hint: For any sequence $\{h_n\}_{n=1}^\infty$ such that $\lim_{n \to \infty} h_n = 0$ and $h_n \ne 0$ for all n, write

$$\frac{f(x + h_n) - f(x)}{h_n} = \frac{1}{2\pi} \int_{\mathbb{R}} \hat{f}(\xi) \frac{e^{-i(x+h_n)\xi} - e^{-ix\xi}}{h_n} \, d\xi.$$

Make an estimate for $|(e^{-i(x+h_n)\xi} - e^{-ix\xi})/h_n|$ that is independent of n and apply the DCT (Theorem 5.45) to justify differentiating under the integral.

ii. Let m be a positive integer. Assume that for some constants $\delta > 0$ and $C_1 > 0$,

$$|\hat{f}(\xi)| \leq C_1(1 + |\xi|)^{-m-1-\delta}.$$

Prove that the mth derivative $f^{(m)}(x)$ of f exists at every point $x \in \mathbb{R}$ and

$$f^{(m)}(x) = \frac{1}{2\pi} \int_{\mathbb{R}} \hat{f}(\xi)(-i\xi)^m e^{-ix\xi} \, d\xi.$$

Hint: Use part i and induction.

5.5.10. Let φ be the scaling function for an MRA $\{V_j\}_{j=1}^{\infty}$ with scaling sequence $u \in \ell^1(\mathbb{Z})$. Let ψ be the mother wavelet determined as in Theorem 5.35. For $m \in \mathbb{Z}$, and $f \in L^2(\mathbb{R})$, let

$$P_m(f) = \sum_{k \in \mathbb{Z}} \langle f, \varphi_{m,k} \rangle \varphi_{m,k}$$

and

$$Q_m(f) = \sum_{k \in \mathbb{Z}} \langle f, \psi_{m,k} \rangle \psi_{m,k}$$

be the orthogonal projections of f onto V_m and W_m, respectively, where W_m is defined in equation (5.63).

i. Prove that $P_m(f) = P_{m-1}(f) + Q_{m-1}(f)$. Hint: Use equation (5.65).

ii. Deduce equation (5.132).

6

C H A P T E R

Wavelets and Differential Equations

6.1 The Condition Number of a Matrix

Many applications of mathematics require the numerical approximation of solutions of differential equations. In this chapter we give a brief introduction to this topic. A thorough discussion is beyond the scope of this text. Instead, by simple examples, we give an idea of the contribution wavelet theory can make to this subject.

The methods we discuss for numerically solving a linear ordinary differential equation (which is all we discuss here) come down to solving a linear system of equations, or equivalently, a matrix equation $Ax = y$. Theoretically, such a system is well understood: for a square matrix A, there is a unique solution x for every y if and only if A is an invertible matrix. However, in applications there are further issues that are of crucial importance. One of these has to do with the condition number of the matrix. We begin with an example.

Example 6.1
Consider the linear system $Ax = y$, where $x, y \in \mathbb{C}^2$, and

$$A = \begin{bmatrix} 5.95 & -14.85 \\ 1.98 & -4.94 \end{bmatrix}.$$

The determinant of A is .01, which is not 0, so A is invertible. For

$$y = \begin{bmatrix} 3.05 \\ 1.02 \end{bmatrix},$$

the unique solution to $Ax = y$ is

$$x = \begin{bmatrix} 8 \\ 3 \end{bmatrix},$$

as the reader can check. Now suppose

$$y' = \begin{bmatrix} 3 \\ 1 \end{bmatrix}.$$

Then the solution to $Ax' = y'$ is

$$x' = \begin{bmatrix} 3 \\ 1 \end{bmatrix}.$$

Note that y and y' are close but x and x' are far apart. A linear system for which this happens is called *badly conditioned*. In this situation, small errors in the data y can lead to large errors in the solution x. This is undesirable in applications, because in nearly all computations with real data there is an error either due to rounding off (computers can give only a finite degree of accuracy) or due to imperfect measurements of the data. For a badly conditioned system, the apparent solution can be virtually meaningless physically.

By considering the diagonalization of A, we get an indication of what is going on. We have already seen that $Ax' = x'$, that is, that x' is an eigenvector of A with eigenvalue 1. Subtracting x' from x, we see that

$$x'' = \begin{bmatrix} 5 \\ 2 \end{bmatrix}$$

satisfies $Ax'' = .01x''$, so x'' is an eigenvector of A with eigenvalue .01. Thus A is similar to the diagonal matrix with diagonal entries 1 and .01. In particular, in the x' direction, A behaves as the identity, so perturbing y in the x' direction results in a perturbation of the solution x by the same amount. However, in the x'' direction, A acts by multiplication by .01, so perturbing the component of y in the x''

direction by some amount results in a perturbation of x by 100 times this amount. This difference in behavior of A in different directions is the source of the problem.

From the previous example, one might conclude that the way to determine whether a system is badly conditioned is to look at the determinant or maybe the smallest eigenvalue. However, this is not correct because these quantities do not scale properly. For example, if we multiply matrix A in Example 6.1 by 10, we multiply each eigenvalue by 10 and the determinant by 100, but the basic nature of the matrix has not changed. This suggests that we should look for some quantity that is scale invariant. The next quantity is not scale invariant, but it will be used to define the right quantity, which is.

Definition 6.2 Let A be an $n \times n$ matrix. Define $\|A\|$, called the operator norm, or just the norm, of A by

$$\|A\| = \sup \frac{\|Az\|}{\|z\|}, \tag{6.1}$$

where the supremum is taken over all nonzero vectors in $\mathbb{C}^n$.

Equivalently (Exercise 6.1.2),

$$\|A\| = \sup\{\|Az\| : \|z\| = 1, z \in \mathbb{C}^n\}. \tag{6.2}$$

The norm of a linear transformation $T : H_1 \to H_2$, for Hilbert spaces H_1 and H_2 was given in Definition 4.25. This corresponds to equation (6.1) in the case $H_1 = H_2 = \mathbb{C}^n$ in the sense that if we define $T_A : \mathbb{C}^n \to \mathbb{C}^n$ by $T_A(z) = Az$, then $\|T_A\| = \|A\|$.

Because of the finite-dimensionality of $\mathbb{C}^n$, the supremum in equation (6.1) or (6.2) is always finite (Exercise 6.1.3). Note that from the definition of supremum, $\|A\|$ is an upper bound for $\{\|Az\|/\|z\| : z \neq 0\}$. Hence we obtain the boundedness (Definition 4.25) of $T_A : \mathbb{C}^n \to \mathbb{C}^n$: for all $z \in \mathbb{C}^n$,

$$\|T_A(z)\| = \|Az\| \leq \|A\| \|z\|. \tag{6.3}$$

It is not true that every set of real numbers contains an element attaining its supremum. However, for an $n \times n$ matrix, the finite dimensionality of $\mathbb{C}^n$ guarantees (Exercise 6.1.4) that there is always a nonzero vector z such that

$$\|T_A(z)\| = \|Az\| = \|A\| \|z\|. \tag{6.4}$$

We remark that for a bounded linear transformation $T : H \rightarrow H$ on an infinite dimensional Hilbert space H, there does not have to be a nonzero vector $v \in H$ such that $\|T(v)\| = \|T\|\|v\|$ (Exercise 6.1.5).

Definition 6.3　*Let A be an invertible $n \times n$ matrix. Define $C_{\#}(A)$, the condition number of A, by*

$$C_{\#}(A) = \|A\|\|A^{-1}\|.$$

If A is not invertible, set $C_{\#}(A) = +\infty$.

It is not difficult (Exercise 6.1.6) to show that for $c \neq 0$, $C_{\#}(cA) = C_{\#}(A)$; that is, the condition number is scale invariant. Also (Exercise 6.1.7), for any matrix A,

$$C_{\#}(A) \geq 1. \tag{6.5}$$

Lemma 6.4　*Suppose that A is an $n \times n$ normal (Definition 1.108) invertible matrix. Let*

$$|\lambda|_{max} = max \{|\lambda| : \lambda \text{ is an eigenvalue of } A\} \tag{6.6}$$

and

$$|\lambda|_{min} = min \{|\lambda| : \lambda \text{ is an eigenvalue of } A\}. \tag{6.7}$$

Then

$$C_{\#}(A) = \frac{|\lambda|_{max}}{|\lambda|_{min}}. \tag{6.8}$$

Proof
By the spectral theorem for matrices (Theorem 1.109), A is unitarily similar to a diagonal matrix D. By Exercise 6.1.8 (ii), $C_{\#}(A) = C_{\#}(D)$. The diagonal entries of D are the eigenvalues of A (Lemma 1.74 ii). Hence by Exercise 6.1.9, $\|D\| = |\lambda|_{max}$. The matrix D^{-1} is the diagonal matrix whose diagonal entries are the reciprocals of the corresponding diagonal entries of D (none of which are 0 because A is assumed invertible). Therefore, by Exercise 6.1.9 again, $\|D^{-1}\| = 1/|\lambda|_{min}$. Putting this together,

$$C_{\#}(A) = C_{\#}(D) = \|D\|\|D^{-1}\| = \frac{|\lambda|_{max}}{|\lambda|_{min}},$$

as desired.　∎

The condition number of a matrix A measures the stability of the linear system $Ax = y$ under perturbations of y. Think of perturbing y by δy to obtain $y + \delta y$. Let x be the solution of $Ax = y$ and δx the solution to $A\delta x = \delta y$. Then by linearity, $A(x + \delta x) = y + \delta y$. Thus perturbing y by δy results in perturbing the solution x by δx. The stability of the linear system is most naturally described by comparing the relative size $\|\delta x\|/\|x\|$ of the change in the solution to the relative size $\|\delta y\|/\|y\|$ of the change in the given data. The condition number is the maximum value of this ratio.

Lemma 6.5 *Suppose A is an $n \times n$ invertible matrix, $x, y, \delta x, \delta y \in \mathbb{C}^n$, $x \neq 0$, $Ax = y$, and $A\delta x = \delta y$. Then*

$$\frac{\|\delta x\|}{\|x\|} \leq C_\#(A)\frac{\|\delta y\|}{\|y\|}. \tag{6.9}$$

Moreover, there exist nonzero $x, y, \delta x, \delta y \in \mathbb{C}^n$ such that $Ax = y$, $A\delta x = \delta y$ and equality is attained in relation (6.9). Hence $C_\#(A)$ cannot be replaced in relation (6.9) by any smaller number.

Proof
Since $y = Ax$, by relation (6.3) we have

$$\|y\| \leq \|A\|\,\|x\|. \tag{6.10}$$

Similarly, since $\delta x = A^{-1}\delta y$,

$$\|\delta x\| \leq \|A^{-1}\|\,\|\delta y\|. \tag{6.11}$$

By multiplying inequalities (6.10) and (6.11), and using the definition of the condition number, we get

$$\|y\|\,\|\delta x\| \leq C_\#(A)\|x\|\,\|\delta y\|,$$

which is equivalent to inequality (6.9) (note that $y \neq 0$ because $Ax = y$ and A is invertible).

To prove the optimality of the number $C_\#(A)$ in relation (6.9), note that by Exercise 6.1.4 (ii), there must exist a nonzero vector x such that

$$\|Ax\| = \|A\|\,\|x\|,$$

and a nonzero vector δy such that

$$\|A^{-1}\delta y\| = \|A^{-1}\|\,\|\delta y\|.$$

Let $y = Ax$ and $\delta x = A^{-1}\delta y$. Then $y \neq 0$ and $\delta x \neq 0$ because A and A^{-1} are invertible. Multiplying the two equations above gives

$$\|y\|\|\delta x\| = \|A\|\|A^{-1}\|\|x\|\|\delta y\| = C_\#(A)\|x\|\|\delta y\|.$$

Hence we have equality in relation (6.9). ∎

The condition number of A measures how unstable the linear system $Ax = y$ is under perturbation of the data y. In applications, therefore, a small condition number (i.e., near 1) is desirable. If the condition number of A is high, we would like to replace the linear system $Ax = y$ by an equivalent system whose matrix has a low condition number, for example, by multiplying by a *preconditioning matrix* B to obtain the equivalent system $BAx = By$, where $C_\#(BA)$ is smaller than $C_\#(A)$. For an invertible matrix A, this is always possible in theory, by taking $B = A^{-1}$. But this is cheating in most applications because usually it is impractical to compute A^{-1}. However, in some cases it is possible to find a simple preconditioning matrix.

Exercises

6.1.1. Let

$$A = \begin{bmatrix} 2.6 & .8 \\ -4.8 & -1.4 \end{bmatrix}.$$

    i. Check that

$$A\begin{bmatrix} 1 \\ -2 \end{bmatrix} = \begin{bmatrix} 1 \\ -2 \end{bmatrix} \quad \text{and} \quad A\begin{bmatrix} .5 \\ -.5 \end{bmatrix} = \begin{bmatrix} .9 \\ -1.7 \end{bmatrix}.$$

    ii. Prove that $C_\#(A) \geq 5$. Hint: Let

$$x = \begin{bmatrix} 1 \\ -2 \end{bmatrix}, \quad x + \delta x = \begin{bmatrix} .5 \\ -.5 \end{bmatrix},$$

    and apply relation (6.9).

6.1.2. Let A be an $n \times n$ matrix. Prove equation (6.2).

6.1.3. Let A be an $n \times n$ matrix. Prove that

$$\sup_{\{z : z \neq 0\}} \frac{\|Az\|}{\|z\|} < +\infty.$$

Hint: One approach is to use the fact that $\|Ae_j\| < \infty$ for $j = 1, 2, \ldots, n$.

6.1.4. Let A be an $n \times n$ matrix. Define $T_A : \mathbb{C}^n \to \mathbb{C}^n$ by $T_A(z) = Az$.

   i. Show that T_A is uniformly continuous on $\mathbb{C}^n$: if $\epsilon > 0$ is given, there exists $\delta > 0$ such that $\|T_A(z) - T_A(w)\| < \epsilon$ for all $z, w \in \mathbb{C}^n$ such that $\|z - w\| < \delta$. Hint: Apply relation (6.3).

   ii. Prove that there exists a nonzero vector $z \in \mathbb{C}^n$ such that $\|Az\| = \|A\|\|z\|$. Hint: By equation (6.2) and the definition of the supremum, for each $k \in \mathbb{N}$, there must be $z_k \in \mathbb{C}^n$ such that $\|z_k\| = 1$ and

$$\|Az_k\| \geq \left(1 - \frac{1}{k}\right) \|A\|.$$

   Assume the Bolzano-Weierstrass theorem for $\mathbb{C}^n$ (any bounded sequence of vectors in $\mathbb{C}^n$ has a convergent subsequence) and apply part i.

6.1.5. Define $T : L^2([-\pi, \pi)) \to L^2([-\pi, \pi))$ as follows. For each $f \in L^2([-\pi, \pi))$, represent f by its Fourier series $\sum_{k \in \mathbb{Z}} \langle f, e^{ik\theta} \rangle e^{ik\theta}$, and let

$$(T(f))(\theta) = \sum_{k \in \mathbb{Z}, k \neq 0} \left(1 - \frac{1}{k}\right) \langle f, e^{ik\theta} \rangle e^{ik\theta}.$$

   i. Prove that $T(f) \in L^2([-\pi, \pi))$ for all $f \in L^2([-\pi, \pi))$.

   ii. Prove that $\|T\| = 1$, where

$$\|T\| = \sup \{\|T(f)\|/\|f\| : f \in L^2([-\pi, \pi)) \quad \text{and} \quad f \neq 0\}.$$

   iii. Prove that for all $f \in L^2([-\pi, \pi))$ such that $f \neq 0$,

$$\|T(f)\| \neq \|T\|\|f\|.$$

6.1.6. Let A be an $n \times n$ invertible matrix, and $c \in \mathbb{C}$ with $c \neq 0$.

   i. Prove that $\|cA\| = |c|\|A\|$ and $\|(cA)^{-1}\| = \|A^{-1}\|/|c|$.

   ii. Prove that $C_\#(cA) = C_\#(A)$.

6.1.7. i. Let A and B be $n \times n$ matrices. Prove that $\|AB\| \leq \|A\|\|B\|$.

   ii. Let A be an $n \times n$ invertible matrix. Prove that $C_\#(A) \geq 1$.

6.1.8. Suppose A and B are $n \times n$ matrices that are unitarily similar (Definition 1.107).

    i. Prove that $\|A\| = \|B\|$. Hint: Use Lemma 1.105 v.

    ii. Suppose also that A is invertible. Prove that B is invertible and $C_\#(A) = C_\#(B)$.

6.1.9. Suppose D is an $n \times n$ diagonal matrix with diagonal entries $d_1, d_2, \ldots, d_n$. Prove that

$$\|D\| = \max\{|d_1|, |d_2|, \ldots, |d_n|\}.$$

6.1.10. Let

$$A = \begin{bmatrix} 8.1 & -3.8 \\ -3.8 & 2.4 \end{bmatrix}.$$

    i. Find $C_\#(A)$.

    ii. Find nonzero vectors $x, y, \delta x, \delta y \in \mathbb{C}^2$ such that $Ax = y$, $A\delta x = \delta y$, and equality holds in relation (6.9).

6.1.11. Suppose U is an $n \times n$ unitary matrix. Prove that $C_\#(U) = 1$.

6.1.12. Let A and δA be $n \times n$ matrices, and $x, y, \delta x \in \mathbb{C}^n$. Suppose A is invertible,

$$Ax = y, \quad \text{and} \quad (A + \delta A)(x + \delta x) = y.$$

Prove that

$$\frac{\|\delta x\|}{\|x + \delta x\|} \le C_\#(A)\frac{\|\delta A\|}{\|A\|}.$$

Hint: Show that the two equations imply

$$\delta x = -A^{-1}\delta A(x + \delta x),$$

and apply Exercise 6.1.7 (i).

    Remark: This exercise has the following interpretation. Starting with the system $Ax = y$, perturb the matrix A slightly to obtain $A + \delta A$. Leaving the data y unchanged, consider the solution $x + \delta x$ of the perturbed equation $(A + \delta A)(x + \delta x) = y$. The conclusion of Exercise 6.1.12 is that the relative error (measured here as $\|\delta x\|/\|x + \delta x\|$) in the solution is bounded by $C_\#(A)$ times the relative change in the norm of A. Thus the lower the condition number, the greater the stability of the system $Ax = y$ under rounding off or perturbing the matrix A, as well as under perturbing y.

6.2 Finite Difference Methods for Differential Equations

Suppose $f : [0, 1] \to \mathbb{C}$ is continuous. Our goal is to obtain a C^2 (i.e., twice continuously differentiable) function u that is a solution to the equation

$$-u''(t) = f(t) \quad \text{for} \quad 0 \le t \le 1, \tag{6.12}$$

with *Dirichlet boundary conditions*

$$u(0) = 0 \quad \text{and} \quad u(1) = 0. \tag{6.13}$$

(Nonzero boundary values can be dealt with easily once this case is understood (Exercise 6.2.1).) In equation (6.12), derivatives are interpreted in the one-sided sense at the endpoints 0 and 1.

The theory of this equation is well understood. It is easy to see that a unique solution u of equations (6.12) and (6.13) exists (Exercise 6.2.2). However, if f is a function whose antiderivative cannot be expressed in terms of elementary functions, it may not be possible to explicitly evaluate the formula in Exercise 6.2.2. One approach to approximating the solution u is to numerically estimate the integrals in this formula. Another method, which is more general because it applies to equations whose solutions are not so easy to write explicitly, is the finite difference method. It is based on approximating the derivatives in equation (6.12) by differences evaluated on a finite set of points in the interval $[0, 1]$.

By the definition of the derivative,

$$u'(t) \approx \frac{u(t + \Delta t) - u(t)}{\Delta t}$$

for small Δt. For reasons of symmetry, let $h > 0$, consider both $\Delta t = h/2$ and $\Delta t = -h/2$, and average:

$$u'(t) \approx \frac{1}{2} \left[\frac{u\left(t + \frac{h}{2}\right) - u(t)}{\frac{h}{2}} + \frac{u(t) - u\left(t - \frac{h}{2}\right)}{\frac{h}{2}} \right]$$

$$= \frac{u\left(t + \frac{h}{2}\right) - u\left(t - \frac{h}{2}\right)}{h}.$$

Applying this to u' leads to

$$u''(t) \approx \frac{u'(t + \frac{h}{2}) - u'(t - \frac{h}{2})}{h} \approx \frac{u(t + h) - 2u(t) + u(t - h)}{h^2}. \quad (6.14)$$

We consider the points

$$t_j = \frac{j}{N}, \quad j = 0, 1, \ldots, N.$$

On this partition, the smallest step we can take is $1/N$, so we let $h = 1/N$. We set

$$x(j) = u\left(\frac{j}{N}\right) \text{ and } y(j) = \frac{1}{N^2}f\left(\frac{j}{N}\right), \quad \text{for } j = 0, 1, \ldots, N. \quad (6.15)$$

To solve $-u''(t_j) = f(t_j)$, we approximate $u''(t_j)$ using approximation (6.14) and consider the system of equations

$$-u\left(\frac{j+1}{N}\right) + 2u\left(\frac{j}{N}\right) - u\left(\frac{j-1}{N}\right) = \frac{1}{N^2}f\left(\frac{j}{N}\right),$$

with boundary conditions $u(0) = u(1) = 0$. When $j = 0$ or $j = N$, this equation does not make sense because $u(-1/N)$ and $u(1 + 1/N)$ are undefined, so we restrict ourselves to $1 \le j \le N - 1$. Thus, we consider

$$-x(j+1) + 2x(j) - x(j-1) = y(j), \quad \text{for } j = 1, \ldots, N - 1, \quad (6.16)$$

with the boundary conditions

$$x(0) = 0 \quad \text{and} \quad x(N) = 0. \quad (6.17)$$

Notice that for $j = 1$, equation (6.16) reduces to $-x(2) + 2x(1) = y(1)$ because $x(0) = 0$, and when $j = N - 1$, equation (6.16) reduces to $2x(N - 1) - x(N - 2) = 0$ because $x(N) = 0$. Thus equation (6.16) is a linear system of $N - 1$ equations in the $N - 1$ unknowns $x(1), \ldots, x(N - 1)$ represented by the matrix equation

$$
\begin{bmatrix}
2 & -1 & 0 & \cdot & \cdot & \cdot & 0 & 0 \\
-1 & 2 & -1 & 0 & \cdot & \cdot & \cdot & 0 \\
0 & -1 & 2 & -1 & 0 & \cdot & \cdot & 0 \\
\cdot & \cdot & \cdot & \cdot & \cdot & \cdot & \cdot & \cdot \\
\cdot & \cdot & \cdot & \cdot & \cdot & \cdot & \cdot & \cdot \\
0 & 0 & \cdot & \cdot & 0 & -1 & 2 & -1 \\
0 & 0 & \cdot & \cdot & \cdot & 0 & -1 & 2
\end{bmatrix}
\begin{bmatrix}
x(1) \\
x(2) \\
\cdot \\
\cdot \\
\cdot \\
\cdot \\
x(N-1)
\end{bmatrix}
$$

$$
= \begin{bmatrix} y(1) \\ y(2) \\ \cdot \\ \cdot \\ \cdot \\ y(N-1) \end{bmatrix}, \tag{6.18}
$$

which we denote

$$
A_N x = y.
$$

One can check (Exercise 6.2.3) that $\det A_N = N$. Hence A_N is invertible, and there is a unique solution x to equation (6.18) for each vector y. As we let $h \to 0$, that is, $N \to +\infty$, we expect our solution x to approximate the true values of u in equation (6.15) with greater accuracy. However, we saw in section 6.1 that it is important numerically for a linear system to be well conditioned. So next we consider the condition number of A_N.

Note that A_N is real and symmetric, hence Hermitian (Definition 1.110). Therefore, A_N is normal. By Lemma 6.4, $C_\#(A) = |\lambda|_{\max}/|\lambda|_{\min}$, where $|\lambda|_{\max}$ and $|\lambda|_{\min}$ are defined by equations (6.6) and (6.7). At first it is not clear how to compute the eigenvalues of A_N. However, consider the matrix (which we denote B_{N-1}) that agrees with A_N except that the entries of B_{N-1} in the top right and lower left corners are -1 instead of 0. Then B_{N-1} is circulant (Definition 2.20). Hence, we can diagonalize B_{N-1} and determine its eigenvalues using Theorem 2.19.

Another way to view the relation between A_N and the circulant variant just noted is to observe that A_N is the $N-1 \times N-1$ submatrix obtained by deleting the first row and column of the $N \times N$ matrix B_N in the matrix equation $B_N x' = y'$ as follows:

$$
\left[\begin{array}{c|cccccccc} 2 & -1 & 0 & 0 & \cdot & \cdot & \cdot & 0 & -1 \\ \hline -1 & 2 & -1 & 0 & \cdot & \cdot & \cdot & 0 & 0 \\ 0 & -1 & 2 & -1 & 0 & \cdot & \cdot & & 0 \\ \cdot & \cdot & \cdot & \cdot & \cdot & \cdot & \cdot & \cdot & \cdot \\ \cdot & \cdot & \cdot & \cdot & \cdot & \cdot & \cdot & \cdot & \cdot \\ -1 & 0 & 0 & \cdot & \cdot & \cdot & 0 & -1 & 2 \end{array}\right] \left[\begin{array}{c} x(0) \\ \hline x(1) \\ x(2) \\ \cdot \\ \cdot \\ x(N-1) \end{array}\right]
$$

$$= \begin{bmatrix} \underline{y(0)} \\ y(1) \\ y(2) \\ \cdot \\ \cdot \\ y(N-1) \end{bmatrix}, \tag{6.19}$$

where $x' = (x(0), x(1), \ldots, x(N-1))$ and similarly for y'. The matrix B_N arises naturally in the periodic formulation of equation (6.12) (see Exercise 6.2.5). However, this formulation is less natural than equations (6.12) and (6.13) (Exercise 6.2.5) and B_N is not invertible (Exercise 6.2.6 (i)), so we leave this variation to the exercises. Instead, we see that we can get information regarding the eigenvalues and eigenvectors of A_N from those of B_N.

Suppose $x' = (x(0), x(1), \ldots, x(N-1))$ is an eigenvector of B_N such that $x(0) = 0$. Let λ be the associated eigenvalue. Let $x = (x(1), x(2), \ldots, x(N-1))$. Then $B_N x'(j) = A_N x(j)$ for $j = 1, 2, \ldots, N-1$ because the condition $x(0) = 0$ guarantees that the first column of B_N has no effect on the value of $B_N x'(j)$. Therefore,

$$A_N x(j) = B_N x'(j) = \lambda x'(j) = \lambda x(j),$$

for $j = 1, 2, \ldots, N-1$. In other words, $A_N x = \lambda x$, so x is an eigenvector of A_N with eigenvalue λ.

Because B_N is circulant, its eigenvectors are $F_0, F_1, \ldots, F_{N-1}$, the elements of the Fourier basis (see Definition 2.7 and Theorem 2.18). Because $F_j(0) = 1/N$ for every j, it may seem that no eigenvector x' of B_N satisfies $x'(0) = 0$. However, in Example 2.36 we computed the eigenvalues of B_N (actually we considered $A = -B_N$ in Example 2.36). We found that $B_N F_j = \lambda_j F_j$, where $\lambda_j = 4 \sin^2(\pi j/N)$. Hence, if $1 \leq j < N/2$, we have $\lambda_{N-j} = \lambda_j$, so the eigenspace corresponding to λ_j is two-dimensional, spanned by F_j and F_{N-j}. Therefore a linear combination of F_j and F_{N-j} belongs to this eigenspace. For $1 \leq j < N/2$, define $H_j \in \ell^2(\mathbb{Z}_N)$ by

$$H_j(n) = \frac{N}{2i} \left(F_j(n) - F_{N-j}(n) \right)$$

$$= \frac{1}{2i} \left(e^{2\pi i j n/N} - e^{-2\pi i j n/N} \right) = \sin\left(\frac{2\pi j n}{N} \right).$$

Then $H_j(0) = 0$ and $B_N H_j = \lambda_j H_j$. By the earlier discussion, this implies that the vector of length $N - 1$ obtained by deleting the first component from H_j is an eigenvector of A_N with eigenvalue λ_j. For reasons that will be clear momentarily, we denote this vector G_{2j} and more generally define vectors G_m of length $N - 1$ for $1 \leq m \leq N - 1$ by

$$G_m(n) = \sin\left(\frac{\pi mn}{N}\right) \quad \text{for} \quad n = 1, 2, \ldots, N - 1. \tag{6.20}$$

In this notation, we have seen that $G_2, G_4, G_6, \ldots$ up to either G_{N-1} if N is odd or G_{N-2} if N is even are eigenvectors of A_N.

It is worth checking this explicitly:

$$(A_N G_m)(\ell) = -G_m(\ell - 1) + 2G_m(\ell) - G_m(\ell + 1)$$

$$= -\sin\left(\frac{\pi m(\ell - 1)}{N}\right) + 2\sin\left(\frac{\pi m\ell}{N}\right) - \sin\left(\frac{\pi m(\ell + 1)}{N}\right)$$

$$= -\left[\sin\frac{\pi m\ell}{N}\cos\frac{\pi m}{N} - \cos\frac{\pi m\ell}{N}\sin\frac{\pi m}{N}\right] + 2\sin\frac{\pi m\ell}{N}$$

$$\quad - \left[\sin\frac{\pi m\ell}{N}\cos\frac{\pi m}{N} + \cos\frac{\pi m\ell}{N}\sin\frac{\pi m}{N}\right]$$

$$= \left[2 - 2\cos\left(\frac{\pi m}{N}\right)\right]\sin\left(\frac{\pi m\ell}{N}\right) = 4\sin^2\left(\frac{\pi m}{2N}\right)G_m(\ell),$$

where the first equality is correct even if $\ell = 1$ because $G_m(0) = 0$. When $m = 2k$, we recover the fact noted earlier that $A_N G_{2k} = 4\sin^2(\pi k/N)G_{2k}$. However, note that this computation did not require the condition that m is even. Thus we see that the eigenvectors of A_N are $G_m, m = 1, 2, \ldots, N - 1$, with corresponding eigenvalues $4\sin^2(\pi m/2N)$. (Note that these eigenvalues are distinct, hence so are the eigenvectors, and thus this is a complete set of eigenvectors for the $(N - 1) \times (N - 1)$ matrix A_N.)

A comparison of the problem represented by equations (6.12) and (6.13) with the periodic formulation in Exercises 6.2.5 and 6.2.6 is instructive. The property that

$$-G_m(\ell - 1) + 2G_m(\ell) - G_m(\ell + 1) = 4\sin^2\left(\frac{\pi m}{2N}\right)G_m(\ell)$$

is the discrete analog of the property that the function $g(x) = \sin \pi m x$ satisfies $g'' = \pi^2 m^2 g$. The eigenvectors G_m of A_N for even values of m correspond to an integer number of full periods of the sine function, which arise in the periodic as well as nonperiodic settings. For m odd, G_m corresponds to $m/2$ (which is not an integer) periods of the sine function, which is missed when looking only for periodic solutions, in the same way (Exercise 6.2.5 (ii)) that $u(x) = \pi^{-2} \sin \pi x$ satisfies equations (6.12) and (6.13) for $f(x) = \sin \pi x$ but u is not 1-periodic.

The condition number of A_N is

$$C_{\#}(A_N) = \frac{|\lambda|_{max}}{|\lambda|_{min}} = \frac{4 \sin^2 \left(\frac{\pi(N-1)}{2N} \right)}{4 \sin^2 \left(\frac{\pi}{2N} \right)}.$$

As $N \to \infty$, $\sin^2(\pi(N-1)/2N) \to 1$, whereas $\sin^2(\pi/2N)$ behaves like $\pi^2/4N^2$. Thus

$$C_{\#}(A_N) \approx \frac{4N^2}{\pi^2}.$$

Thus the condition number of A_N goes to ∞ proportionally to N^2. So although increasing N should increase the accuracy of the approximation to the solution u of equations (6.12) and (6.13), the linear system $A_N x = y$ becomes increasing unstable and the solution becomes more and more unreliable.

For the simple equations (6.12) and (6.13), we were able to explicitly diagonalize the matrix A_N arising in the finite difference approximation. Partially this was due to the fact (see the remark at the end of Exercise 6.2.6) that the operator L defined by $Lu = -u''$ is translation invariant (as noted near the end of section 5.2, for example). Consequently the matrix A_N was close to circulant, in the sense that A_N is closely related to the circulant matrix B_N, which arises in the periodic formulation of the problem (Exercise 6.2.6). This and a bit of luck enabled us to obtain the eigenvectors of A_N. More generally, any *linear constant coefficient ordinary differential operator*, that is, an operator L of the form

$$(Lu)(t) = L(u)(t) = \sum_{j=0}^{N} b_j \frac{d^j}{dt^j} u(t), \tag{6.21}$$

where each b_j is a constant, is translation invariant (Exercise 6.2.7).

If the coefficients b_j in equation (6.21) are allowed to vary with t (such an operator is called a *linear variable coefficient ordinary differential operator*), the operator L will not be translation invariant. (For example, if $R = td/dt$, then $R(u(t - s)) = tu'(t - s)$, whereas $(Ru)(t - s) = (t - s)u'(t - s)$. See Exercise 6.2.7 for the general result.) For such an operator, even the periodic problem will not be diagonalized by the Fourier system. When we look at the matrix A arising in the finite difference approximation to the solution of $Lu = f$ on $[0, 1]$ with the boundary conditions $u(0) = u(1) = 0$, A will not be close to circulant (see Exercise 6.2.8 for an example), so we cannot follow the methods applied to A_N above. Even if A is diagonalizable, which is not clear, there is little hope to explicitly diagonalize A. We expect the condition number of A to be large because that is the case even in the much simpler case of equations (6.12) and (6.13).

An alternative approach using wavelets that includes the variable coefficient case is considered in section 6.3. This approach leads to linear systems with bounded condition numbers.

Exercises

6.2.1. Let $f : [0, 1] \to \mathbb{C}$ be a continuous function and let $a, b \in \mathbb{C}$. Suppose $u_0 : [0, 1] \to \mathbb{C}$ is a C^2 function that satisfies equations (6.12) and (6.13). Find a C^2 function u (expressed in terms of u_0, a, and b) satisfying $-u'' = f$ on $[0, 1]$, $u(0) = a$, and $u(1) = b$. Hint: What equation does $u - u_0$ satisfy?

6.2.2. Suppose $f : [0, 1] \to \mathbb{C}$ is a continuous function. Define

$$u(x) = -\int_0^x \int_0^t f(s)\, ds\, dt + x \int_0^1 \int_0^t f(s)\, ds\, dt,$$

for $0 \le x \le 1$.

i. Prove that u is a C^2 function that satisfies equations (6.12) and (6.13).

ii. Prove uniqueness: u is the only C^2 function satisfying equations (6.12) and (6.13).

6.2.3. Let A_N be the $(N-1) \times (N-1)$ matrix in equations (6.18). Prove that $\det A_N = N$ for $N \geq 2$. Hint: Use induction. Show that $\det A_{N+1} = 2 \det A_N - \det A_{N-1}$.

6.2.4. Prove that $\{e^{2\pi i n t}\}_{n \in \mathbb{Z}}$ is a complete orthonormal set in $L^2([0,1))$. Hint: For completeness, extend f to be defined on $\mathbb{R}$ with period 1, and apply Exercise 4.3.5 with $a = 1/2$. The result for $[-1/2, 1/2]$ gives the result for $[0, 1]$ by translation.

6.2.5. Let $f : \mathbb{R} \to \mathbb{C}$ be a continuous function with period 1 : $f(t+1) = f(t)$ for all $t \in \mathbb{R}$. The periodic formulation of the problem represented by equations (6.12) and (6.13) is to find a C^2 function $u : \mathbb{R} \to \mathbb{C}$ that has period 1 and satisfies $-u'' = f$ on $[0, 1]$ and $u(0) = 0$.

  i. Observe that a solution u to the periodic problem satisfies equations (6.12) and (6.13).

  ii. The periodic formulation above is not equivalent to equations (6.12) and (6.13). For example, let $f(x) = \sin \pi x$. Show that $u(x) = \pi^{-2} \sin \pi x$ is C^2 on $[0, 1]$ and satisfies equations (6.12) and (6.13), but does not have a C^2 1-periodic extension because the derivative of u from the right at 0 does not agree with the derivative of u from the left at 1. By part i and uniqueness (Exercise 6.2.2 (ii)), this periodic problem has no solution.

  iii. Prove that if the periodic problem has a solution u, then $\int_0^1 f(t)\, dt = 0$. This is another way to see that the example in part ii has no 1-periodic solution.

  iv. Suppose the 1-periodic continuous function f satisfies $\int_0^1 f(t)\, dt = 0$. Prove that u as defined in Exercise 6.2.2 has a 1-periodic extension that is a solution to the periodic problem. Hint: One needs only to check that the 1-periodic extension is C^2 at the endpoints 0 and 1. Certainly u and $u'' = -f$ match up at 0 and 1. The assumption $\int_0^1 f(t)\, dt = 0$ is required only when checking u'.

  v. Suppose $\int_0^1 f(t)\, dt = 0$. Then the periodic problem can be solved formally by Fourier methods. Assume that f can be represented by its Fourier series on $[0, 1)$ (see

Exercise 6.2.4):

$$f(t) = \sum_{n \in \mathbb{Z}, n \neq 0} c_n e^{2\pi i n t}.$$

The constant term is $c_0 = (2\pi)^{-1} \int_0^1 f(t)\, dt = 0$ by assumption. Show that

$$u(t) = a + \sum_{n \in \mathbb{Z}, n \neq 0} \frac{c_n}{4\pi^2 n^2} e^{2\pi i n t},$$

where the constant a is chosen so that $u(0) = 0$, is 1-periodic and satisfies $-u'' = f$ at least formally (that is, assuming that it is valid to take the second derivative inside the sum on n).

6.2.6. In this problem we consider the finite difference method for solving the periodic problem formulated in Exercise 6.2.5. Because u and f are defined on all of $\mathbb{R}$, we can extend the definitions in equation (6.15) to all $j \in \mathbb{Z}$. The 1-periodicity of f and u implies that x and y are N-periodic, that is, $x, y \in \ell^2(\mathbb{Z}_N)$. Then equation (6.16) makes sense also for $j = 0$, since $x(-1) = x(N-1)$. Similarly, when $j = N-1$ in equation (6.16), we have $x(j+1) = x(N) = x(0)$. If we write the resulting system of equations for $j = 0, 1, \ldots, N-1$, we obtain $B_N x = y$, where B_N is the matrix in equation (6.19); $x = (x(0), x(1), \ldots, x(N-1))$; $y = (y(0), y(1), \ldots, y(N-1))$; and the boundary condition is $x(0) = 0$.

i. Let $w \in \ell^2(\mathbb{Z}_N)$ be defined by $w = (1, 1, \ldots, 1)$. Prove that $B_N w = 0$. This shows that B_N is not invertible.

ii. Prove that $\ker B_N$ is one-dimensional, hence is spanned by w. Hint: If $B_N x = 0$, then for each j, $x(j+1) - x(j) = x(j) - x(j-1)$. Thus the values $x(j)$ lie on a line. Then use periodicity.

iii. Let $w^\perp = \{z \in \mathbb{C}^N : z \perp w\}$. Prove that range $B_N = w^\perp$. Hint: By the rank theorem (Exercise 1.4.10) and part ii, the dimension of the range of B_N must be $N-1$. If $B_N x = y$, for some x, show that

$$\langle y, w \rangle = \sum_{j=0}^{N-1} x(j-1) - 2 \sum_{j=0}^{N-1} x(j) + \sum_{j=0}^{N-1} x(j+1) = 0.$$

Remark: Thus $B_N x = y$ has a solution x if and only if $\sum_{j=0}^{N-1} y(j) = \langle y, w \rangle = 0$, which is the discrete analogue of the compatibility condition $\int_0^1 f(t)\,dt = 0$ in Exercise 6.2.5 (iii).

iv. Let T_{B_N} be the operator associated with the matrix B_N, that is, $T_{B_N}(x) = B_N x$. Restrict T_{B_N} to $w^\perp$. Prove that $T_{B_N}|_{w^\perp} : w^\perp \to w^\perp$ is $1-1$ and onto, hence invertible.

v. Let $f : \mathbb{R} \to \mathbb{C}$ be continuous, 1-periodic, and satisfy $\int_0^1 f(t)\,dt = 0$ (a necessary condition for solvability of the periodic problem, by Exercise 6.2.5). Define $y = (y(0), y(1), \dots, y(N-1))$ for $y(j)$ as in equation (6.15). Prove that

$$\lim_{N \to +\infty} N \sum_{j=0}^{N-1} y(j) = 0.$$

Hint: Recognize $N \sum_{j=0}^{N-1} y(j)$ as a Riemann sum.

Remark: Although it is not necessarily true that $y \in w^\perp$, part v shows that $\sum_{j=0}^{N-1} y(j)$ approaches 0 relatively rapidly. When trying to solve $B_N x = y$, one can first approximate y with

$$y^\# = y - \left\langle y, \frac{w}{\|w\|} \right\rangle \frac{w}{\|w\|},$$

the orthogonal projection of y onto $w^\perp$. By part iv, the system $B_N x = y^\#$ has a unique solution in $w^\perp$. This solution can be modified by adding a vector in $w^\perp = \ker B_N$ to satisfy the boundary condition $x(0) = 0$.

vi. Because B_N is circulant, it is diagonalized (Theorem 2.18) by the Fourier basis $\{F_0, F_1, \dots, F_{N-1}\}$ in Definition 2.7. This leads to a simple formula for the solution to $B_N x = y^\#$. Note that $w = N F_0$, and hence that $w^\perp = \text{span}\{F_1, F_2, \dots, F_{N-1}\}$. Since $y = \sum_{j=0}^{N-1} \hat{y}(j) F_j$ (by equation (2.15)), we obtain $y^\# = \sum_{j=1}^{N-1} \hat{y}(j) F_j$. Let λ_j be the eigenvalue of B_N associated with the eigenvector F_j, so that $B_N F_j = \lambda_j F_j$. Note that for $1 \le j \le N-1$, we have $\lambda_j \ne 0$, since $F_j \notin \ker B_N$ (to be explicit, $\lambda_j = 4\sin^2(\pi j/N)$, by Example 2.36). Show that the general solution to

$B_N x = y^\#$ is

$$x = cF_0 + \sum_{j=1}^{N-1} \left(\frac{\hat{y}(j)}{\lambda_j} \right) F_j.$$

Determine c so that the initial condition $x(0) = 0$ is satisfied.

Remark: Note the close analogy between the solution of the continuous periodic problem in Exercise 6.2.5 and its finite difference discretization in Exercise 6.2.6. The underlying reason for this analogy is that the second-derivative operator L (defined by $Lu = u''$) is translation invariant and hence is diagonalized by the Fourier system. In the discrete setting, the corresponding feature is the fact that the matrix B_N is circulant (i.e., the associated operator is translation invariant) and therefore is diagonalized by the discrete Fourier transform. The imposition of boundary conditions upsets the translation invariance of the problem, but in the periodic formulation the effect is mild and can be dealt with as we have indicated. The boundary conditions (6.13) in the nonperiodic formulation are more difficult to incorporate, as described in the text.

6.2.7. Let L be a (possibly variable coefficient) differential operator

$$L = \sum_{j=0}^{N} b_j(t) \frac{d^j}{dt^j},$$

defined for $t \in \mathbb{R}$. Prove that L is translation invariant if and only if each coefficient function $b_j(t)$ is constant. Hint: Consider Lu, where $u(t) = t^m, 0 \le m \le N$. Write out $(Lu)(t - s)$ and $L(u(t - s))$. Set $s = t$ to deduce that b_0 is constant. Then cancel this term, divide by $t - s$, and iterate.

6.2.8. Consider the equation

$$tu''(t) + u(t) = f(t)$$

on $[0, 1]$, with boundary conditions $u(0) = u(1) = 0$. Discretize this equation on the partition $t_j = j/N, j = 0, 1, \ldots, N - 1$. Set $x(j) = u(t_j)$. Show that $t_j u''(t_j)$ is approximately $jN [x(j + 1) - 2x(j) + x(j - 1)]$. Set $y(j) =$

$f(j/N)$. As in the case of equations (6.12) and (6.13), obtain the system of equations

$$jNx(j+1)+(1-2jN)x(j)+jNx(j-1) = y(j), \quad \text{for} \quad 1 \le j \le N-1.$$

Show that this can be written as $Ax = y$, where $x = (x(1), x(2), \ldots, x(N-1))$; $y = (y(0), y(1), \ldots, y(N-1))$; and

$$A = \begin{bmatrix} 1-2N & N & 0 & 0 & \cdots & 0 & 0 \\ 2N & 1-4N & 2N & 0 & \cdots & 0 & 0 \\ 0 & 3N & 1-6N & 3N & \cdots & 0 & 0 \\ \cdot & \cdot & \cdot & \cdots & \cdot & \cdot \\ \cdot & \cdot & \cdot & \cdots & \cdot & \cdot \end{bmatrix}.$$

6.3 Wavelet-Galerkin Methods for Differential Equations

In this section we present another approach to the numerical solution of ordinary differential equations, known as the *Galerkin* method. For a certain class of equations, using wavelets in conjunction with the Galerkin method gives the two primary desired features for the associated linear system: sparseness and low condition number.

We consider the class of ordinary differential equations (known as *Sturm-Liouville equations*) of the form

$$Lu(t) = -\frac{d}{dt}\left(a(t)\frac{du}{dt}\right) + b(t)u(t) = f(t), \text{ for } 0 \le t \le 1, \quad (6.22)$$

with Dirichlet boundary conditions

$$u(0) = u(1) = 0.$$

Here a, b, and f are given real-valued functions and we wish to solve for u. We assume f and b are continuous and a has a continuous derivative on $[0, 1]$ (this always means a one-sided derivative at the endpoints). Note that L may be a variable coefficient differential operator because $a(t)$ and $b(t)$ are not necessarily constant. We assume the operator is *uniformly elliptic*, which means that there exist finite constants C_1, C_2, and C_3 such that

$$0 < C_1 \leq a(t) \leq C_2 \text{ and } 0 \leq b(t) \leq C_3. \tag{6.23}$$

for all $t \in [0, 1]$. By a result in the theory of ordinary differential equations, there is a unique function u satisfying equation (6.22) and the boundary conditions $u(0) = u(1) = 0$.

The simplest example is $a(t) = 1$ and $b(t) = 0$, which yields equations (6.12) and (6.13). We will see the relevance of the ellipticity assumption later. It may seem more natural to write out

$$-\frac{d}{dt}\left(a(t)\frac{du}{dt}\right) = -a'(t)u'(t) - a(t)u''(t),$$

by the product rule, but the formulation in equation (6.22) is more convenient when we integrate by parts.

Note (compare with Exercises 4.3.5 and 6.2.4) that $L^2([0, 1])$ is a Hilbert space with inner product

$$\langle f, g \rangle = \int_0^1 f(t)\overline{g(t)}\, dt.$$

For the Galerkin method, we suppose that $\{v_j\}_j$ is a complete orthonormal system for $L^2([0, 1])$, and that every v_j is C^2 on $[0, 1]$ and satisfies

$$v_j(0) = v_j(1) = 0. \tag{6.24}$$

We select some finite set Λ of indices j and consider the subspace

$$S = \text{span}\{v_j : j \in \Lambda\}.$$

We look for an approximation to the solution u of equation (6.22) of the form

$$u_S = \sum_{k \in \Lambda} x_k v_k \in S, \tag{6.25}$$

where each x_k is a scalar. Our criterion for determining the coefficients x_k is that u_S should behave like the true solution u on the subspace S, that is, that

$$\langle Lu_S, v_j \rangle = \langle f, v_j \rangle \text{ for all } j \in \Lambda. \tag{6.26}$$

By linearity, it follows that

$$\langle Lu_S, g \rangle = \langle f, g \rangle \text{ for all } g \in S.$$

Notice that the approximate solution u_S automatically satisfies the boundary conditions $u_S(0) = u_S(1) = 0$ because of equation (6.24). It turns out that u_S determined by equation (6.26) is the best approximation in S to u, with respect to a certain natural norm (not the L^2 norm—see Exercise 6.3.1 (iii)).

If we substitute equation (6.25) in equation (6.26), we obtain

$$\left\langle L\left(\sum_{k \in \Lambda} x_k v_k\right), v_j\right\rangle = \langle f, v_j\rangle, \text{ for all } j \in \Lambda,$$

or

$$\sum_{k \in \Lambda} \langle L v_k, v_j\rangle x_k = \langle f, v_j\rangle, \text{ for all } j \in \Lambda. \tag{6.27}$$

Let x denote the vector $(x_k)_{k \in \Lambda}$, and let y be the vector $(y_k)_{k \in \Lambda}$, where $y_k = \langle f, v_k\rangle$. Let A be the matrix with rows and columns indexed by Λ, that is, $A = [a_{j,k}]_{j,k \in \Lambda}$, where

$$a_{j,k} = \langle L v_k, v_j\rangle. \tag{6.28}$$

Thus, equation (6.27) is the linear system of equations

$$\sum_{k \in \Lambda} a_{j,k} x_k = y_j, \quad \text{for all } j \in \Lambda,$$

or

$$Ax = y. \tag{6.29}$$

In the Galerkin method, for each subset Λ we obtain an approximation $u_S \in S$ to u, by solving the linear system (6.29) for x and using these components to determine u_S by equation (6.25). We expect that as we increase our set Λ is some systematic way, our approximations u_S should converge to the actual solution u (see Exercise 6.3.1 (iv)).

Our main concern is the nature of the linear system (6.29) that results from choosing a wavelet basis for the Galerkin method as opposed to some other basis, for example, some Fourier basis. For numerical purposes, there are two properties that we would like the matrix A in the linear system (6.29) to have. First, as discussed in section 6.1, we would like A to have a small condition number, to obtain stability of the solution under small perturbations in the data. Second, for performing calculations with A quickly, we would like

A to be *sparse*, which means that A should have a high proportion of entries that are 0. The best case is when A is diagonal, but the next best case is when A is sparse.

In this text we have not discussed wavelets on the interval $[0, 1]$. It would take us too far afield to do so, so we will assume the following facts. There is a way of modifying the wavelet system for $L^2(\mathbb{R})$ so as to obtain a complete orthonormal system

$$\{\psi_{j,k}\}_{(j,k)\in\Gamma} \tag{6.30}$$

for $L^2([0, 1])$. The set Γ is a certain subset of $\mathbb{Z} \times \mathbb{Z}$ that we do not specify. The functions $\psi_{j,k}$ are not exactly the same functions as in a wavelet basis for $L^2(\mathbb{R})$, but they are similar. In particular, $\psi_{j,k}$ has a scale of about 2^{-j}, $\psi_{j,k}$ is concentrated near the point $2^{-j}k$, and $\psi_{j,k}$ is 0 outside an interval centered at $2^{-j}k$ of length proportional to 2^{-j}. Wavelets concentrated well into the interior of $[0, 1]$ are nearly the same as usual wavelets, but those concentrated near the boundary points are substantially modified. (In particular, they are no longer all translates and dilates of the original mother wavelet ψ.) For each $(j, k) \in \Lambda$, $\psi_{j,k}$ is C^2 and satisfies the boundary conditions

$$\psi_{j,k}(0) = \psi_{j,k}(1) = 0.$$

The wavelet system $\{\psi_{j,k}\}_{(j,k)\in\Gamma}$ also satisfies the following key estimate: There exist constants $C_4, C_5 > 0$ such that for all functions g of the form

$$g = \sum_{j,k} c_{j,k}\psi_{j,k} \tag{6.31}$$

where the sum is finite, we have

$$C_4 \sum_{j,k} 2^{2j}|c_{j,k}|^2 \leq \int_0^1 |g'(t)|^2 \, dt \leq C_5 \sum_{j,k} 2^{2j}|c_{j,k}|^2. \tag{6.32}$$

An estimate of this form is called a *norm equivalence*; it states that up to the two constants, the quantities $\sum_{j,k} 2^{2j}|c_{j,k}|^2$ and $\int_0^1 |g'(t)|^2 \, dt$ are equivalent. Such estimates show up more and more in analysis at an advanced level. Although we do not prove estimate (6.32) here, we can get a general sense of why it might be true. We know that

$$\int_0^1 |g(t)|^2 \, dt = \|g\|^2 = \sum_{j,k} |c_{j,k}|^2, \tag{6.33}$$

since $g = \sum c_{j,k} \psi_{j,k}$ and $\{\psi_{j,k}\}_{(j,k)\in\Lambda}$ is a complete orthonormal system in $L^2([0,1])$. Now consider g' instead of g. Recall that for standard wavelets in $L^2(\mathbb{R})$,

$$\psi_{j,k}(t) = 2^{j/2} \psi(2^j t - k).$$

By the chain rule,

$$(\psi_{j,k})'(t) = 2^j 2^{j/2} \psi'(2^j t - k) = 2^j (\psi')_{j,k}. \tag{6.34}$$

These wavelets on $[0,1]$ are not standard wavelets on $\mathbb{R}$, but their behavior is similar. In equation (6.34), taking the derivative gives us a factor of 2^j and changes ψ to ψ'. With a little leap of faith, we can believe that ψ' behaves like ψ (in particular they have the same scale), so roughly speaking, $\sum c_{j,k}(\psi_{j,k})'$ behaves like $\sum c_{j,k} 2^j \psi_{j,k}$. Therefore, identity (6.33) suggests estimate (6.32).

Estimate (6.32) is a good example of what was meant at the end of chapter 5 when we said that it is important to have wavelets on $\mathbb{R}$, not just on $\mathbb{Z}$. This estimate shows that wavelets are compatible with the continuous structure on $\mathbb{R}$.

The notation used for applying the Galerkin method with these wavelets is somewhat confusing due to the fact that the wavelets are indexed by two integers. Thus for wavelets we write equation (6.25) as

$$u_S = \sum_{(j,k)\in\Lambda} x_{j,k} \psi_{j,k},$$

and equation (6.27) as

$$\sum_{(j,k)\in\Lambda} \langle L\psi_{j,k}, \psi_{\ell,m} \rangle x_{j,k} = \langle f, \psi_{\ell,m} \rangle \quad \text{for all } (\ell, m) \in \Lambda, \tag{6.35}$$

for some finite set of indices Λ. We can still regard this as a matrix equation $Ax = y$, where the vectors $x = (x_{j,k})_{(j,k)\in\Lambda}$ and $y = (y_{j,k})_{(j,k)\in\Lambda}$ are indexed by the pairs $(j, k) \in \Lambda$, and the matrix

$$A = [a_{\ell,m;j,k}]_{(\ell,m),(j,k)\in\Lambda}$$

defined by

$$a_{\ell,m;j,k} = \langle L\psi_{j,k}, \psi_{\ell,m}\rangle \qquad (6.36)$$

has its rows indexed by the pairs $(\ell, m) \in \Lambda$ and its columns indexed by the pairs $(j, k) \in \Lambda$. Because Λ is a finite set, this could all be reindexed to have the usual form, but there is no natural reindexing, and the traditional wavelet indexing is useful.

As suggested, we would like A to be sparse and have a low condition number. Actually A itself does not have a low condition number, but we can replace the system $Ax = y$ by an equivalent system $Mz = v$, for which the new matrix M has the desired properties. To see this, first define the diagonal matrix $D = [d_{\ell,m;j,k}]_{(\ell,m),(j,k)\in\Lambda}$ by

$$d_{\ell,m;j,k} = \begin{cases} 2^j & \text{if } (\ell, m) = (j, k) \\ 0 & \text{if } (\ell, m) \neq (j, k). \end{cases} \qquad (6.37)$$

Define $M = [m_{\ell,m;j,k}]_{(\ell,m),(j,k)\in\Lambda}$ by

$$M = D^{-1}AD^{-1}. \qquad (6.38)$$

By writing this out, we see that

$$m_{\ell,m;j,k} = 2^{-j-\ell}a_{\ell,m;j,k} = 2^{-j-\ell}\langle L\psi_{j,k}, \psi_{\ell,m}\rangle. \qquad (6.39)$$

The system $Ax = y$ is equivalent to

$$D^{-1}AD^{-1}Dx = D^{-1}y,$$

or, setting $z = Dx$ and $v = D^{-1}y$,

$$Mz = v. \qquad (6.40)$$

The norm equivalence (6.32) has the consequence that the system (6.40) is well conditioned, as we see in Theorem 6.7. The process (when possible) of changing an ill-conditioned system into a well-conditioned system is a variation on the preconditioning process described at the end of section 6.1.

Before stating and proving Theorem 6.7, it is useful to observe the following lemma. It explains the need for the uniform ellipticity assumption (6.23).

Lemma 6.6 *Let L be a uniformly elliptic Sturm-Liouville operator (i.e., an operator as defined in equation (6.22) satisfying relation (6.23)).*

Suppose $g \in L^2([0, 1])$ is C^2 on $[0, 1]$ and satisfies $g(0) = g(1) = 0$. Then

$$C_1 \int_0^1 |g'(t)|^2 \, dt \leq \langle Lg, g \rangle \leq (C_2 + C_3) \int_0^1 |g'(t)|^2 \, dt, \qquad (6.41)$$

where $C_1, C_2,$ and C_3 are the constants in relation (6.23).

Proof

Note that

$$\langle -(ag')', g \rangle = \int_0^1 -(ag')'(t)\overline{g}(t) \, dt$$
$$= \int_0^1 a(t)g'(t)\overline{g'(t)} \, dt = \langle ag', g' \rangle,$$

by integration by parts. (The boundary term is 0 because $g(0) = g(1) = 0$.) Therefore,

$$\langle Lg, g \rangle = \langle -(ag')' + bg, g \rangle = \langle ag', g' \rangle + \langle bg, g \rangle.$$

Hence, by relation (6.23),

$$C_1 \int_0^1 |g'(t)|^2 \, dt \leq \int_0^1 a(t)|g'(t)|^2 \, dt = \int_0^1 a(t)g'(t)\overline{g'(t)} \, dt = \langle ag', g' \rangle. \qquad (6.42)$$

Also by relation (6.23),

$$0 \leq \int_0^1 b(t)|g(t)|^2 \, dt = \langle bg, g \rangle.$$

Adding these two inequalities gives

$$C_1 \int_0^1 |g'(t)|^2 \, dt \leq \langle Lg, g \rangle,$$

which is the left half of relation (6.41). For the other half, note that by relation (6.23),

$$\langle ag', g' \rangle = \int_0^1 a(t)|g'(t)|^2 \, dt \leq C_2 \int_0^1 |g'(t)|^2 \, dt. \qquad (6.43)$$

Also note that because $g(0) = 0$,

$$g(t) = \int_0^t g'(s) \, ds,$$

by the fundamental theorem of calculus. Hence by the Cauchy-Schwarz inequality (5.3) (applied to the functions $g'\chi_{[0,t]}$ and $\chi_{[0,t]}$, with χ as in Definition 5.43),

$$|g(t)|^2 \leq \left(\int_0^t |g'(s)|^2 \, ds\right)\left(\int_0^t 1 \, ds\right) \leq \int_0^1 |g'(s)|^2 \, ds$$

for every $t \in [0, 1]$. Therefore

$$\int_0^1 |g(t)|^2 \, dt \leq \int_0^1 |g'(s)|^2 \, ds \int_0^1 dt = \int_0^1 |g'(s)|^2 \, ds. \qquad (6.44)$$

Hence, by (6.23),

$$\langle bg, g \rangle = \int_0^1 b(t)g(t)\overline{g(t)} \, dt \leq C_3 \int_0^1 |g(t)|^2 \, dt \leq C_3 \int_0^1 |g'(t)|^2 \, dt.$$

This result and relation (6.43) give the right side of relation (6.41). ∎

Theorem 6.7 *Let L be a uniformly elliptic Sturm-Liouville operator (an operator as defined in equation (6.22) satisfying relation (6.23)). Let $\{\psi_{j,k}\}_{(j,k)\in\Gamma}$ be a complete orthonormal system for $L^2([0,1])$ such that each $\psi_{j,k}$ is C^2, satisfies $\psi_{j,k}(0) = \psi_{j,k}(1) = 0$, and such that the norm equivalence (6.32) holds. Let Λ be a finite subset of Γ. Let M be the matrix defined in equation (6.38). Then the condition number of M satisfies*

$$C_\#(M) \leq \frac{(C_2 + C_3)C_5}{C_1 C_4}, \qquad (6.45)$$

for any finite set Λ, where C_1, C_2, and C_3 are the constants in relation (6.23), and C_4 and C_5 are the constants in relation (6.32).

Proof
Let $z = (z_{j,k})_{(j,k)\in\Lambda}$ be any vector with $\|z\| = 1$. For D as in equation (6.37), let $w = D^{-1}z$; that is, $w = (w_{j,k})_{(j,k)\in\Lambda}$, where

$$w_{j,k} = 2^{-j}z_{j,k}.$$

Define

$$g = \sum_{(j,k)\in\Lambda} w_{j,k}\psi_{j,k}.$$

Then by equation (6.39),

$$\langle Mz, z \rangle = \sum_{(\ell,m)\in\Lambda} (Mz)_{\ell,m} \overline{z_{\ell,m}}$$

$$= \sum_{(\ell,m)\in\Lambda} \sum_{(j,k)\in\Lambda} \langle L\psi_{j,k}, \psi_{\ell,m}\rangle 2^{-j} z_{j,k} 2^{-\ell} \overline{z_{\ell,m}}$$

$$= \left\langle L\left(\sum_{(j,k)\in\Lambda} w_{j,k}\psi_{j,k}\right), \sum_{(\ell,m)\in\Lambda} w_{\ell,m}\psi_{\ell,m}\right\rangle = \langle Lg, g\rangle,$$

since $2^{-j}z_{j,k} = w_{j,k}$ and $2^{-\ell}z_{\ell,m} = w_{\ell,m}$. Applying Lemma 6.6 and relation (6.32) gives

$$\langle Mz, z\rangle = \langle Lg, g\rangle \le (C_2+C_3) \int_0^1 |g'(t)|^2 \, dt \le (C_2+C_3)C_5 \sum_{(j,k)\in\Lambda} 2^{2j}|w_{j,k}|^2,$$

and

$$\langle Mz, z\rangle = \langle Lg, g\rangle \ge C_1 \int_0^1 |g'(t)|^2 \, dt \ge C_1 C_4 \sum_{(j,k)\in\Lambda} 2^{2j}|w_{j,k}|^2.$$

However,

$$\sum_{(j,k)\in\Lambda} 2^{2j}|w_{j,k}|^2 = \sum_{(j,k)\in\Lambda} |z_{j,k}|^2 = \|z\|^2 = 1.$$

So for any z with $\|z\| = 1$,

$$C_1 C_4 \le \langle Mz, z\rangle \le (C_2 + C_3)C_5.$$

If λ is an eigenvalue of M, we can normalize the associated eigenvector z so that $\|z\| = 1$, to obtain

$$\langle Mz, z\rangle = \langle \lambda z, z\rangle = \lambda\langle z, z\rangle = \lambda\|z\|^2 = \lambda.$$

Therefore, every eigenvalue λ of M satisfies

$$C_1 C_4 \le \lambda \le (C_2 + C_3)C_5. \tag{6.46}$$

Note that H is Hermitian (Exercise 6.3.2), and hence normal, so by Lemma 6.4, $C_\#(M)$ is the ratio of the largest eigenvalue to the smallest (which are all positive, by relation (6.46)). So by relation (6.46), condition (6.45) holds. ∎

Thus the matrix in the preconditioned system (6.40) has a condition number bounded independently of the set Λ. So as we increase Λ to approximate our solution with more accuracy, the condition number stays bounded. This is much better than the finite difference case in section 6.2 in which the condition number grows as N^2. Thus, with the Galerkin method using wavelets on $[0, 1]$, we do not have to worry about measurement or round-off errors invalidating our solution as we aim for higher and higher accuracy.

There are other complete orthonormal systems for which a similar preconditioning can be done to yield a bounded condition number. The Fourier system is an example. We cannot use the functions $e^{2\pi i n t}$ because they don't satisfy the boundary conditions, but we can use a basis of sine functions that vanish at the endpoints (Exercise 6.3.3).

So although we see the advantage of the Galerkin method over finite differences, the advantage of wavelets over the Fourier system is still not clear. To see this, we should consider the other feature of the matrix M that is desirable: we would like M to be sparse. We can see from equation (6.39) that this is the case, because of the localization of the wavelets. Namely, $\psi_{j,k}$ is 0 outside an interval of length $c2^{-j}$ around the point $2^{-j}k$, for some constant c (depending on the choice of wavelet system). Because L involves only differentiation and multiplication by another function, it does not change this localization property. So $L\psi_{j,k}$ is 0 outside this interval also. Similarly, $\psi_{\ell,m}$ is 0 outside an interval of length $c2^{-\ell}$ around the point $2^{-\ell}m$. As j and ℓ get large, fewer and fewer of these intervals intersect, so more and more of the matrix entries

$$m_{\ell,m;j,k} = 2^{-j-\ell}\langle L\psi_{j,k}, \psi_{\ell,m}\rangle = 2^{-j-\ell}\int_0^1 L\psi_{j,k}(t)\overline{\psi_{\ell,m}(t)}\,dt$$

are 0. So M is sparse, which makes computation with it easier. The basic reason for this sparseness is the compact support of the wavelets.

For the Fourier system in Exercise 6.3.3, the associated matrix is not sparse. We have terms of the form

$$\langle L(\sqrt{2}\sin(2\pi n t)), \sqrt{2}\sin(2\pi m t)\rangle.$$

After writing out L and integrating by parts we obtain the expressions

$$\langle a(t)2\sqrt{2}\pi n \cos(2\pi n t), 2\sqrt{2}\pi m \cos(2\pi m t)\rangle$$

$$= 8\pi^2 nm \int_0^1 a(t) \cos(2\pi nt) \cos(2\pi mt)\, dt \qquad (6.47)$$

and

$$\langle b(t)\sqrt{2}\sin(2\pi nt), \sqrt{2}\sin(2\pi mt)\rangle = 2\int_0^1 b(t)\sin(2\pi nt)\sin(2\pi mt)\, dt. \qquad (6.48)$$

When a and b are constant, the orthonormality of the Fourier system makes these terms 0 unless $m = n$, so the matrix is diagonal. This orthogonality is a precise but delicate property. Multiplying by $a(t)$ or $b(t)$ destroys this, and the result is a matrix that is not necessarily close to diagonal, or even sparse. By integrating by parts in equations (6.47) and (6.48), we can show that these terms decay on the order of $|n - m|^{-k}$, where k is determined by the number of derivatives that a and b have. Especially for relatively nonsmooth functions a and b, this is a slow decay rate in comparison to the sparseness of the wavelet matrix M.

The matrices that we obtain using finite differences, as in section 6.2, are sparse. However, they have large condition numbers. Using the Galerkin method with the Fourier system, we can obtain a bounded condition number, but the matrix is no longer sparse. Using the Galerkin method with a wavelet system, we obtain both advantages.

This demonstrates the basic advance represented by wavelet theory. The Fourier system diagonalizes translation-invariant linear operators, but it does not necessarily come close to diagonalizing non–translation-invariant operators such as variable coefficient differentiable operators. The wavelet system is more crude than the Fourier system in the sense that there are few, if any, naturally occuring operators that are diagonalized by a wavelet basis. But for a very large class of operators, for instance the variable coefficient differential operators considered here, the matrices representing these operators in the wavelet system are sparse, which we regard as being nearly diagonal. Thus, although the wavelet system does not exactly diagonalize much of anything, it nearly diagonalizes a very large class of operators, a much larger class than the translation-invariant operators, which are perfectly diagonalized by the Fourier system.

The fact that a wavelet system nearly diagonalizes a very broad class of operators is one of the key properties of wavelets. We have seen that this property is important in applications to numerical differential equations. Another key property of wavelets is their combination of spatial and frequency localization, which is used in signal analysis applications such as image compression, as we saw in chapter 3. A third key property of wavelets is that norm equivalences for wavelets such as relation (6.32) hold for a much larger class of function spaces than for the Fourier system. This topic, which is beyond the scope of this text (see, e.g., Hernández and Weiss (1996)) is important in many applications of wavelets in pure mathematics.

Exercises

6.3.1. Let $C_0^2([0, 1])$ denote the set of all complex-valued, continuous functions f on $[0, 1]$ such that $f(0) = f(1) = 0$, and f has two continuous derivatives on $[0, 1]$.

   i. Prove that $C_0^2([0, 1])$ is a vector space under the usual addition and scalar multiplication of functions.

   ii. For $f, g \in C_0^2([0, 1])$, let

$$\langle f, g \rangle_0 = \langle L(f), g \rangle = \int_0^1 \left[(-af')'(t) + bf(t) \right] \overline{g(t)} \, dt,$$

for a and b as in relations (6.22) and (6.23). Prove that $\langle \cdot, \cdot \rangle_0$ is an inner product on $C_0^2([0, 1])$. Hint: All properties are clear except I4 in Definition 1.86. To prove this, use relation (6.23) and integration by parts, as in the proof of Lemma 6.6.

Remark: Unfortunately, $C_0^2([0, 1])$ is not complete under the inner product $\langle \cdot, \cdot \rangle_0$, because a sequence of functions in $C_0^2([0, 1])$ can be Cauchy but the apparent limit function may not belong to $C_0^2([0, 1])$. However, there is a space H_0^1 (the *completion* of $C_0^2([0, 1])$ in this norm) containing $C_0^2([0, 1])$ and a way of extending the inner product $\langle \cdot, \cdot \rangle_0$ to H_0^1 so that H_0^1

is complete with this inner product. Also, H_0^1 is a subspace of $L^2([0, 1])$.

iii. Assume the previous remark. Suppose $\{v_j\}_{j=1}^\infty$ is a complete orthonormal set in $L^2([0, 1])$ such that each v_j belongs to $C_0^2([0, 1])$. For some positive integer N, let $\Lambda = \{1, 2, \ldots, N\}$ and set $S = \text{span}\{v_1, v_2, \ldots, v_N\}$. Suppose $Lu = f$ (where L is as in equation (6.22)) with $u \in H_0^1$. Suppose $u_S \in S$ and u_S satisfies equation (6.26). Prove that u_S is the orthogonal projection *in the space* H_0^1 of u onto S. Hint: The orthogonal projection $P_S u$ is characterized by the properties that $P_S u \in S$ and $u - P_S u$ is orthogonal to every vector in S (Exercise 1.6.8). Remark: By the best approximation property (Lemma 1.98 v), this means that the Galerkin approximation u_S is the element of S that is closest to u in the H_0^1 norm.

iv. Let $\| \cdot \|_0$ be the norm induced by the inner product $\langle \cdot, \cdot \rangle_0$ (as in Definition 1.90). (By relation (6.41), $\|g\|^2$ is equivalent to $\int_0^1 |g'(t)|^2\, dt$.) Suppose there exist positive constants C_1 and C_2 and a scalar sequence $\{\lambda_j\}_{j=1}^\infty$ such that for all functions $g = \sum_{j=1}^\infty \alpha_j v_j \in H_0^1$ (here each α_j is a scalar),

$$C_1 \sum_{j=1}^\infty |\lambda_j \alpha_j|^2 \le \|g\|_0^2 \le C_2 \sum_{j=1}^\infty |\lambda_j \alpha_j|^2$$

(for example, relation (6.32)). For $N \in \mathbb{N}$, let u_N be the Galerkin approximation to u for $\Lambda = \{1, 2, \ldots, N\}$, as in part iii. Prove that

$$\|u - u_N\|_0^2 \le C_2 \sum_{j=N+1}^\infty |\lambda_j \beta_j|^2,$$

where $u = \sum_{j=1}^\infty \beta_j v_j$. Deduce that the sequence $\{u_N\}_{N=1}^\infty$ converges to u in the space H_0^1 as $N \to \infty$.

6.3.2. Define $M = [m_{\ell, m; j, k}]_{(\ell, m), (j, k) \in \Lambda}$ by equation (6.39). Prove that M is Hermitian. Hint: Use the form (6.22) of L, and integrate by parts twice in equation (6.39). Recall that a and b are real-valued.

6.3.3. For $n \in \mathbb{N}$, define $s_n \in L^2([0, 1))$ by

$$s_n(t) = \sqrt{2} \sin(\pi n t).$$

i. Prove that $\{s_n\}_{n \in \mathbb{N}}$ is a complete orthonormal set in $L^2([0, 1))$. Note that $s_n(0) = s_n(1) = 0$ for every n. Hint: Apply a rescaling argument to the result in Exercise 4.3.7 (i).

ii. For any g of the form $\sum c_n s_n$, where the sum is finite, prove that

$$\|g'\|^2 = \pi^2 \sum n^2 |c_n|^2.$$

Hint: Prove that the set $\{\sqrt{2} \cos(\pi n t)\}_{n=1}^{\infty}$ is orthonormal in $L^2([0, 1))$. This can be done directly, or by rescaling Exercise 4.3.7 (ii).

iii. By part i, we can apply the Galerkin method with the complete orthonormal system $\{s_n\}_{n \in \mathbb{N}}$ and the finite set $\Lambda_N = \{1, 2, \ldots, N\} \subseteq \mathbb{N}$. Define a matrix $A = [a_{j,k}]_{j,k=1}^{N}$ by setting

$$a_{j,k} = \langle L s_k, s_j \rangle,$$

as in equation (6.28). The preconditioning matrix will be the diagonal matrix $D = [d_{j,k}]_{j,k=1}^{N}$, where $d_{jj} = \pi j$ if $1 \leq j \leq N$, and $d_{jk} = 0$ if $j \neq k$. Define $M = D^{-1} A D^{-1}$. The Galerkin approximation to the the solution of equation (6.22) is obtained by solving $Mz = v$, with z and v as in the text. Prove that

$$C_\#(M) \leq \frac{C_2 + C_3}{C_1},$$

for C_1, C_2, C_3 as in relation (6.23).

Bibliography

There are already hundreds, perhaps thousands, of papers relating to wavelets, so this bibliography is far from complete. It is also not necessarily up-to-date because papers in the field are still being published at a rapid pace. This bibliography is intended only as a suggestion of some possibilities for further investigation. Vast amounts of more current information can be found by entering the query "wavelets" into your favorite Internet search engine.

The two classic texts in wavelet theory are written by two of the founders of the subject, Yves Meyer (1990) and Ingrid Daubechies (1992). Meyer's book requires a research-level background in mathematics, whereas Daubechies's text is accessible to a wider audience. Many of the seminal results in wavelet theory were first presented in papers by Meyer (1985–86), Mallat (1988), and Daubechies (1988). Several subsequent general texts on wavelets (Chui, 1992a; Kaiser, 1995; Koornwinder, 1993) have appeared. A more advanced and comprehensive text focusing on the mathematical theory of wavelets is Hernández and Weiss (1996). Wojtaszczyk (1997) presents a nice treatment from a somewhat advanced mathematical viewpoint. Burke-Hubbard (1996) offers a delightful nontechnical account of wavelets and their history. There are several fine expository articles on wavelets, such as Strichartz (1993), which we have borrowed from in our treatment, and Jawerth and Sweldens (1994), which surveys various developments in wavelet theory arising from the multiresolution analysis point of view. The discrete approach presented in chapters 3 and 4 of this text was described earlier by Frazier and Kumar (1993). Auscher's paper (1995) plays an important role in the theory of multiresolution analyses.

A number of books contain collections of articles related to wavelets (Benedetto and Frazier, 1993; Chui, 1992b; Meyer and Roques, 1993; Ruskai et al., 1992; Schumaker and Webb, 1993). These books contain a wealth of papers covering everything from basic theory to scientific and engineering applications.

There are several important variations on basic wavelet theory. These include wavelet packets (introduced in Coifman et al. (1989), applied to compression of audio signals in Wickerhauser (1992) and covered in detail with accompanying software in Wickerhauser (1994)); biorthogonal wavelets (Cohen, Daubechies, and Feauveau, 1992; Cohen, 1992); wavelets on intervals (Meyer, 1992; Cohen, Daubechies, and Vial, 1993); Wilson bases (Daubechies, Jaffard, and Journé, 1991); local sine and cosine bases (Coifman and Meyer, 1991; Auscher, Weiss, and Wickerhauser, 1992); multiwavelets (Geronimo, Hardin, and Massopust, 1994); and interpolating wavelets (Donoho, 1992).

Proakis and Manolakis (1996) present a standard graduate text on signal processing. The relations between wavelets and signal processing are discussed in the text by Teolis (1998) as well as in several papers. Wavelet theory from the perspective of multirate signal analysis is described in Vetterli and Herley (1992). Applications of wavelets to signal processing are discussed in Rioul and Vetterli (1991). Image compression with wavelets is the topic of a paper by Hilton, Jawerth, and Sengupta (1994). In other papers, wavelets are applied to contrast enhancement in image processing (Lu, Realy, and Weaver, 1994); to mammography (Richardson, Longbotham, and Gokhman, 1993); and to modeling human hearing and acoustic signal compression (Benedetto and Teolis, 1993). Fingerprint image compression, as discussed in the Prologue of this text, is described in Brislawn (1995). Stollnitz, De Rose, and Salesin (1996) discuss applications of wavelets to computer graphics.

Numerical analysis using wavelets was initiated by Beylkin, Coifman, and Rokhlin (1991). Further references include Beylkin, Coifman, and Rokhlin (1992) and Alpert (1992). Numerical aspects of the computation of wavelets are described in Sweldens and Piessens (1994). One of the most fruitful applications of wavelets has been in numerical differential equations (see chapter 6 for an introduction to this topic). A few of the many references include Jaffard (1992), Beylkin (1993), Jawerth and Sweldens (1993), Qian and Weiss (1993), Dahlke and Weinreich (1993), Amaratunga and Williams (1994), and Xu and Shann (1992). A classic background reference on the Galerkin approach (as in section 6.3) is Strang and Fix (1973).

Wavelets in relation to quantum mechanics are discussed in Paul and Seip (1992). Because of their natural self-similarity, wavelets are useful in studying fractals; Massopust (1994) and Meyer (1997) stress this connection in their introductory texts. Important work applying wavelets to signal denoising, also known as curve estimation, has been done by Donoho and Johnstone (1994, among others). A text on the applications of wavelets in statistics is provided by Ogden (1996).

For those interested in learning more about Fourier analysis, Folland (1992) provided a relatively elementary text with a focus on applications. Royden (1988) and Rudin (1987) are standard real analysis graduate mathematics texts; among other things, they cover Lebesque integration theory in detail. Of these two, the presentation in Rudin (1987) is at a more advanced level.

Alpert, B., Wavelets and other bases for fast numerical linear algebra, in C. Chui, ed., *Wavelets: A Tutorial in Theory and Applications*, Academic Press, New York, 1992, 181–216.

Amaratunga, K., and Williams, J., Wavelet-Galerkin solutions for one-dimensional partial differential equations, *International J. Num. Methods in Eng.* 37 (1994), 2703–2716.

Auscher, P., Solution of two problems on wavelets, *J. Geometric Analysis* 5 (1995), 181–236.

Auscher, P., Weiss, G., and Wickerhauser, M., Local sine and cosine bases of Coifman and Meyer and the construction of smooth wavelets, in C. Chui, ed,. *Wavelets: A Tutorial in Theory and Applications*, Academic Press, New York, 1992, 237–256.

Benedetto, J., and Frazier, M., eds., *Wavelets: Mathematics and Applications*, CRC Press, Boca Raton, Fla., 1993.

Benedetto, J., and Teolis, A., A wavelet auditory model and data compression, *Appl. Comp. Harm. Anal.* 1 (1993), 3–28.

Beylkin, G., On wavelet-based algorithms for solving differential equations, in J. Benedetto and M. Frazier, eds., *Wavelets: Mathematics and Applications*, CRC Press, Boca Raton, Fla., 1993, 449–466.

Beylkin, G., Coifman, R., and Rokhlin, V., Fast wavelet transforms and numerical algorithms, *Comm. Pure Appl. Math.* 44 (1991), 141–183.

Beylkin, G., Coifman, R., and Rokhlin, V., Wavelets in numerical analysis, in M. Ruskai at al., eds., *Wavelets and Their Applications*, Jones and Bartlett, Boston, 1992, 181–210.

Brislawn, C., Fingerprints go digital, *Notices of the Amer. Math Soc.* 42 (1995), 1278–1283.

Burke-Hubbard, B., *The World According to Wavelets*, A. K. Peters, Wellesley, Mass., 1996.

Chui, C., *An Introduction to Wavelets*, Academic Press, Boston, 1992a.

Chui, C., ed., *Wavelets: A Tutorial in Theory and Applications*, Academic Press, New York, 1992.

Cohen, A., Biorthogonal wavelets, in C. Chui, ed., *Wavelets: A Tutorial in Theory and Applications*, Academic Press, New Yord, 1992, 123–152.

Cohen, A., Daubechies, I., and Feauveau, J.-C., Biorthogonal bases of compactly supported wavelets, *Comm. Pure Appl. Math.* 45 (1992), 485–500.

Cohen, A., Daubechies, I., and Vial, P., Multiresolution analysis, wavelets and fast algorithms on an interval, *Appl. Comp. Harm. Anal.* 1 (1993), 54–81.

Coifman, R., and Meyer, Y., Remarques sur l'analyse de Fourier á fenêtre, *C. R. Acad. Sci.*, sér 1 4312 (1991), 259–261.

Coifman, R., Meyer, Y., Quake, S., and Wickerhauser, M., Signal processing and compression with wavelet packets, in Y. Meyer and S. Roques, eds., *Proceedings of the International Conference on Wavelets*, (Marseilles), Masson, Paris, 1989.

Dahlke, S. and Weinreich, I., Wavelet-Galerkin methods: An adapted biorthogonal wavelet basis, Constr. Approx. 9 (1993), 237–262.

Daubechies, I., Orthonormal bases of compactly supported wavelets, *Comm. Pure Appl. Math.* 41 (1988), 909–996.

Daubechies, I., *Ten Lectures on Wavelets*, CBMS-NSF Reg. Conf. Series in Appl. Math. 61, Soc. Ind. Appl. Math., Philadelphia, 1992.

Daubechies, I., Jaffard, S., and Journé, J.-L., A simple Wilson orthonormal basis with exponential decay, SIAM J. *Math. Anal.* 22 (1991), 554–572.

Donoho, D., Interpolating wavelet transforms, Department of Statistics, Stanford University, 1992, preprint.

Donoho, D., and Johnstone, I., Ideal spatial adaptation via wavelet shrinkage, *Biometrika* 81 (1994), 425–455.

Folland, G., *Fourier Analysis and Its Applications*, Wadsworth and Brooks/Cole, Belmont, CA, 1992.

Frazier, M., and Kumar, A., An introduction to the orthonormal wavelet transform on discrete sets, in J.Benedetto and M. Frazier, eds., *Wavelets: Mathematics and Applications*, CRC Press, Boca Raton, FL, 1993, 51–95.

Geronimo, J., Hardin, D., and Massopust, P., Fractal functions and wavelet expansions based on several scaling functions, *J. Approx. Theory* 78 (1994), 373–401.

Hernández. E., and Weiss, G., *A First Course on Wavelets*, CRC Press, Boca Raton, FL, 1996.

Hilton, M., Jawerth, B., and Sengupta, A., Compressing still and moving images with wavelets, *Multimedia Systems* 2 (1994), 218–227.

Jaffard, S., Wavelet methods for the fast resolution of elliptic problems, *SIAM J. Numer. Anal.* 29 (1992), 965–986.

Jawerth, B., and Sweldens, W., An overview of wavelet based multiresolution analyses, *SIAM Review* 36 (1994), 377–412.

Jawerth, B., and Sweldens, W., Wavelet multiresolution analyses adapted for the fast solution of boundary value ordinary differential equations, in N. Melson, T. Manteuffel, and S. McCormick, eds., *Sixth Copper Mountain Conference on Multigrid Methods*, NASA Conference Publication 3224 (1993), 259–273.

Kaiser, G., *A Friendly Guide to Wavelets*, Birkhäuser, Boston, 1995.

Koornwinder, T., ed., *Wavelets: an elementary treatment of theory and applications*, Series in Approximations and Decompositions 1, World Scientific, Singapore, 1993.

Lu, J., Healy, D., and Weaver, J., Contrast enhancement of medical images using multiscale edge representation, *Optical Engineering* 33 (1994), 2151–2161.

Mallat, S., A theory for multiresolution signal decomposition: the wavelet representation, *Comm. Pure Appl. Math.* 41 (1988), 674–693.

Massopust, P., *Fractal Functions, Fractal Surfaces, and Wavelets*, Academic Press, San Diego, CA, 1994.

Meyer, Y., Principe d'incertitude, bases Hilbertiennes et algèbres d'opérateurs, *Séminaire Bourbaki* 662 (1985-86), 1–15.

Meyer, Y., *Wavelets and Operators*, Cambridge University Press, Cambridge, 1993, English translation of *Ondelettes et Opérateurs*, Vol. I, Hermann, Paris, 1990.

Meyer, Y., Ondelettes sur l'intervalle, *Rev. Mat. Iberoamericana* 7 (1992), 115–133.

Meyer, Y., *Wavelets, Vibrations and Scalings*, Amer. Math. Soc., Providence, RI, 1997.

Meyer, Y., and Roques, S., eds., *Progress in Wavelet Analysis and Applications: Proceedings of the International Conference "Wavelets and Applications,"* (Toulouse, France - June 1992), Editions Frontières, Gif-sur-Yvette, France, 1993.

Ogden, R. T., *Essential Wavelets for Statistical Applications and Data Analysis*, Birkhäuser, Boston, 1996.

Paul, T., and Seip, K., Wavelets and quantum mechanics, in M. Ruskai et al., *Wavelets and Their Applications*, Jones and Bartlett, Boston, 1992, 302–322.

Proakis, J., and Manolakis, D., *Digital Signal Processing*, 3rd ed., Prentice Hall, Upper Saddle River, NJ, 1996.

Qian, S., and Weiss, J., Wavelets and the numerical solution of partial differential equations, *J. Comp. Phy.* 106 (1993), 155–175.

Richardson, W. Jr., Longbotham, H., and Gokhman, D., Multiscale wavelet analysis of mammograms, in Y. Meyer and S. Roques, eds., *Progress in Wavelet Analysis and Applications: Proceedings of the International Conference "Wavelets and Applications"* (Toulouse, France, June 1992), Editions Frontiéres, Gif-sur-Yvette, France, 1993, 599–608.

Rioul, O., and Vetterli, M., Wavelets and signal processing, *IEEE Signal Proc. Mag.* (1991), 14–38.

Royden, H., *Real Analysis*, 3rd ed., Macmillan, New York, 1988.

Rudin, W., *Real and Complex Analysis*, 3rd ed., McGraw-Hill, New York, 1987.

Ruskai, M., Beylkin, G., Coifman, R., Daubechies, I., Mallat, S., Meyer, Y., and Raphael, L., eds., *Wavelets and Their Applications*, Jones and Bartlett, Boston, 1992.

Schumaker, L., and Webb, G., eds., *Recent Advances in Wavelet Analysis*, Academic Press, New York, 1993.

Stollnitz, E., De Rose, A., and Salesin, D., *Wavelets For Computer Graphics; Theory and Applications*, Morgan-Kaufmann, San Francisco, 1996.

Strang, G., and Fix, G., *An Analysis of the Finite Element Method*, Prentice Hall, Upper Saddle River, NJ, 1973.

Strichartz, R., How to make wavelets, *Amer. Math. Monthly* 100 (1993), 539–556.

Sweldens, W., and Piessens, R., Quadrature formulae and aysmptotic error expansions for wavelet approximations of smooth functions, *SIAM J. Numer. Anal.* 31 (1994), 1240–1264.

Teolis, A., *Computational Signal Processing with Wavelets*, Birkhäuser, Boston, 1998.

Vetterli, M., and Herley, C., Wavelets and filter banks: theory and design, *IEEE Trans. Acoust. Speech Signal Process.* 40 (1992), 2207–2232.

Wickerhauser, M., Acoustic signal compression with wavelet packets, in C. Chui, ed., *Wavelets: A Tutorial in Theory and Applications*, Academic Press, New York, 1992, 679–700.

Wickerhauser, M., *Adapted Wavelet Analysis from Theory to Software*, A. K. Peters, Wellesley, MA, 1994.

Wojtaszczyk, P., *A Mathematical Introduction to Wavelets*, Cambridge University Press, Cambridge, 1997.

Xu, J.-C., and Shann, W.-C., Galerkin-wavelet methods for two-point boundary value problems, *Numer. Math.* 63 (1992), 123–142.

Index

491

Undergraduate Texts in Mathematics

(continued from page ii)

Undergraduate Texts in Mathematics